Current Topics in Microbiology 214 and Immunology

AF182533

Editors

Springer
Berlin
Heidelberg
New York
Barcelona
Budapest
Hong Kong
London
Milan
Paris
Santa Clara
Singapore
Tokyo

Morphogenesis and Maturation of Retroviruses

Edited by H.-G. Kräusslich

With 34 Figures

Springer

formation emphasizing the similarities and differences of specific members of the virus family may thus be more appropriate to provide a generalized picture of the underlying molecular principles. Following the discussion of the retroviral assembly pathway, the last two chapters of this book address the retrotransposons of yeast and drosophila and the pararetrovirus hepatitis B virus. While these viruses are rather distant from bona fide retroviruses in many aspects, it is believed that their discussion within the context of retrovirus morphogenesis will serve to illustrate that similar as well as divergent principles are applied by related viruses and much is still to be learned from the amazing array of viral strategies.

The genomes of all replication competent retroviruses contain the genes *gag*, *pol*, and *env*, which encode the internal structural proteins of the virus particle, the viral replication proteins, and the glycoproteins inserted into the virus membrane, respectively. The products of these three genes are incorporated into the virus particle and, together with the genomic RNA and the envelope derived from the plasma membrane of the infected cell, may in fact be sufficient to form an infectious virion. Identification of more complex retroviruses and in particular detailed analysis of the human immunodeficiency virus revealed additional proteins which are presumed to serve regulatory functions. Some of these proteins are involved in the intricate regulation of viral gene expression while others function at a post-translational level and enhance the production of extracellular particles or the specific infectivity of the virus. Some of these accessory viral proteins are incorporated into the virus while others appear to exert their effect only in the virus-producing cell. Moreover, there are also some cellular proteins incorporated at least into some retroviruses and these host proteins may have profound effects on virus infectivity and pathogenesis. Recent research has identified and characterized many accessory players in retrovirus morphogenesis, but much has yet to be learned regarding their function and future experiments may unravel additional layers of complexity.

In principle, assembly of an infectious retrovirus can be divided into a number of individual steps: Firstly, following synthesis of the individual components of the virus, they need to be localized to a specific site or compartment within the cell where they interact with each other to assemble a virus or core structure. Much has been learned regarding the transport of glycoproteins of the viral envelope, which essentially follows the route of cellular glycoproteins. The Gag- and Gag-Pol polyproteins, on the other hand, are transported via a cytoplasmic route and despite the recent identification of targeting domains on the viral proteins much less is known about their transport pathway and potential interacting cellular partners. Even less information is available regarding transport of the genomic RNA to the assembly site, which reflects the relative lack of knowledge regarding intracellular RNA transport in general. Once the individual players have come together in one site, intermolecular interactions of the components will lead to formation of an early assembly structure, which is the first step observed by electron microscopy. A great deal of information has been obtained in recent years concerning specific interaction domains in various retroviral proteins and nucleic acids and a generalized view of the assembly process is beginning to emerge. Additional insight can be derived from the recent determination of the three-dimensional structures of several retroviral structural proteins. These results provide the basis for higher resolution structure determination of complete viruses or virus-like particles and for attempts to reconstitute retrovirus particle formation in vitro from purified individual components. The last two steps in formation of the infectious virus are release of the immature virion and extracellular maturation: Retroviruses are released from the plasma membrane in a budding process which involves envelopment of the viral core, either during or following assembly of the core. Once the particle is released, major internal rearrangements lead to morphological conversion to the mature infectious virion. This process is dependent on and may be triggered by proteolysis of the Gag- and Gag-Pol polyproteins by the viral proteinase. Proteolysis of viral polyproteins once all required elements are confined into an assembly or budding structure may therefore serve to commit the virus to the disassembly route and to make the morphogenesis pathway functionally irreversible.

This book is an up to date review of the complex and fascinating topic of retrovirus assembly, which has progressed rapidly in recent years. All chapters have been written by leading experts in the field, who have incorporated many new and as yet unpublished results. Consequently, this book does not primarily highlight historical developments but attempts to provide an overview of the current status and likely future developments. Given the importance of retroviruses as pathogens on the one hand and their potential as gene therapy vectors on the other hand, it appears safe to predict that much of the basic research in this area may directly translate into medical benefit. This expectation is illustrated by the potential to disturb virus assembly as a target for antiviral therapy and to produce rationally designed alterations of the host range, stability, and composition of gene therapy vectors.

Hamburg H.-G. KRÄUSSLICH

List of Contents

List of Contributors

(Their addresses can be found at the beginning of their respective chapters.)

Berkowitz, R. 177
Boulanger, P. 237
Cohen, E.A. 219
Craven, R.C. 65
Einfeld, D. 133
Fisher, J. 177
Goff, S.P. 177
Göttlinger, H.G. 219
Hockley, D.J. 1
Jones, I. 237
Kräusslich, H.-G. 25
Menees, T.M. 261
Nassal, M. 297
Nermut, M.V. 1
Parent, L.J. 65
Sandmeyer, S.B. 261
Subbramanian, R.A. 219
Vogt, V.M. 95
Welker, R. 25

Comparative Morphology and Structural Classification of Retroviruses

M.V. Nermut and D.J. Hockley

1 Introduction

Retroviruses are characterised by two exceptional features: reverse transcription and a process of maturation outside the host cell which transforms the primary non-infectious product of assembly into a mature, infectious virus particle – the

National Institute for Biological Standards and Control, Blanche Lane, South Mimms, Potters Bar, Herts, EN6 3QG, United Kingdom virion. Electron microscopy played an important role in morphological characterisation of individual retrovirus groups and the results became a basis for classification of retroviruses. In addition, electron microscopy has become an indispensable companion in biological and molecular biological research. Several excellent reviews have been published recently which provide the reader with the necessary information on retrovirus morphology (GELDERBLOM 1991; GELDERBLOM et al.1991; COFFIN 1992). The aim of our contribution will be to present recent data in relation to morphogenesis and its mechanism. We shall concentrate on structural aspects of the assembly process, considering the three-dimensional structure of the *gag*-encoded polyprotein and the organisation of the gag protein shell which determines the overall shape of the virus.

2 Biochemical and Physical Characteristics

Retroviral structural proteins are encoded by two genes: *gag* and *env*. In both cases a precursor polyprotein is produced in the host cell and assembled into the immature virus particle. During the maturation process the gag polyprotein is cleaved into three or four internal proteins: MA protein forms a shell that lines the inner leaflet of the viral membrane, CA protein forms the core shell and NC protein forms a complex with RNA. The *env* products usually consist of an oligomeric complex of a transmembrane glycoprotein and a highly glycosylated surface glycoprotein. Proteins constitute about 60% of the weight of the virus particle, with about 3.5% carbohydrates and about 30% lipids which are derived from the host cell plasma membrane.

Enzymatic proteins that are incorporated into virus particles are encoded by the *pol* gene (reverse transcriptase, RT, and integrase, IN) or the *pro* gene (protease, PR). Molecular weights of retroviral proteins are summarised in Table 1.

Retroviruses contain two identical copies of single-stranded RNA (positive sense) 7–10 kbp in length. In mature particles these two copies exist as a non-covalently linked dimer.

3 Morphology and Architecture of the Virion

Viruses are organised assemblies of macromolecules which can be considered as basic 'structural elements'; these form 'structural complexes' such as envelopes or cores by association with other macromolecules. It will be useful to name and define the structural complexes of viruses before comparing individual members of the retrovirus family. This terminology is based on early definitions

Table 1. Molecular weights of retroviral proteins (Adapted from NERMUT and STEVEN (1987), COFFIN (1992), MONTELARO et al. (1993))

	env-encoded	*gag*-encoded proteins				*pol*-encoded non-structural proteins			
			MA	CA	NC	PR	IN	RT	?
MMTV	gp52 gp36	pr73	p10	p27	p14	p13			p21
MPMV	gp70 gp22	pr78	p10	p27	p14	p17		p70	p18
MuLV	gp70 p15E	pr65	p15	p30	p10	p14	p46	p80	
ALSV	gp85 gp37	pr76	p19	p27	p12	p15	p34	p68	p10
HTLV	gp46 gp21	pr55	p19	p24	p15	p14			p12
BLV	gp60 gp30	pr53	p15	p24	p12	p14			p13
EIAV	gp90 gp45	pr55	p15	p26	p11	p11	p32	p66	p9
HIV	gp120 gp41	pr55	p17	p24	p7	p14	p34	p66	p6
SIV	gp110 gp32	pr60	p17	p27	p8	p11		p68	p6

MMTV, mouse mammary tumour virus; MPMV, Mason-Pfizer monkey virus; MuLV, murine leukaemia virus; ALSV, avian leukaemia and sarcoma virus; HTLV, human T-cell leukaemia virus; BLV, bovine leukaemia virus; EIAV, equine infections anaemia virus; HIV, human immunodeficiency virus; SIV, simian immunodeficiency virus

(see NERMUT 1987 for references), but has been slightly updated. It differs to some extent from the names of proteins as published by LEIS et al. (1989).

3.1 Naming of the Parts

The virion: intact, fully assembled, infectious virus. Immature forms will therefore be called virus particles not virions.

The envelope: virus membrane with glycoprotein projections sometimes called peplomers.

The capsid: closed protein shell enclosing either nucleic acid only or a nucleoprotein complex or a core.

The nucleocapsid: nucleic acid within an isometric closed protein shell (= capsid) or within a helical protein complex. The latter name obviously contradicts the above definition of the capsid and one can think of other names such as nucleoprotein complex or nucleohelix.

The core: internal body containing nucleoprotein complex (or nucleocapsid) within a protein shell (core shell).

A specific feature of retroviruses is the existence of immature and mature forms. There is at present no unity in the terminology of the submembrane protein shells in the two forms. The names 'capsid' and 'membrane-associated shell' have been used but we have decided to use the following terminology throughout this chapter in order to prevent further confusion.

Immature gag shell: submembrane shell formed by a gag polyprotein in immature virus particles.

Mature gag shell: submembrane shell formed by membrane-associated cleavage product of the gag protein.

Intermediate layer: material intercalated between the gag shell and viral membrane in some retroviruses. This layer is external to the gag shell in A-type particles.

Gag spherical shell: gag precursor shell in the cytoplasm not associated with a membrane and with a possible additional surface layer in the case of A-,B- and D-type particles and also spumavirus.

Lateral bodies: proteinaceous material between the core and the envelope which has no apparent organisation and may represent excess protein.

3.2 Structural Characterisation of Retroviruses

Retroviruses are isometric enveloped viruses typically 90–140 nm in diameter. There is firm evidence of icosahedral symmetry in retroviruses (NERMUT et al. 1972, 1993; OZEL et al. 1988; MARX et al. 1988; TAKAHASHI et al. 1989; GELDERBLOM 1991) and there are two basic types of virus architecture which are

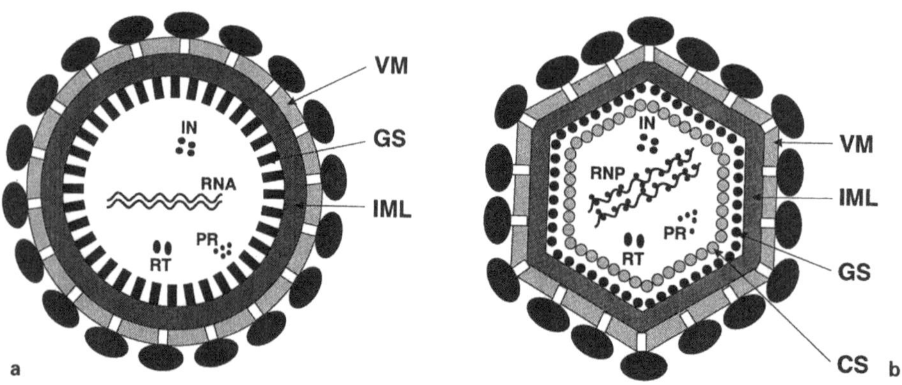

Fig. 1a,b. Diagrammatic representation of type 1 architecture of retroviruses as present in B,C,D types and spumavirus. **a** Immature form, **b** mature form. *VM*, viral membrane; *IML*, intermediate layer; *GS*, gag shell (membrane-associated shell); *RT*, reverse transcriptase; *IN*, integrase; *PR*, protease; *RNP*, ribonucleoprotein complex (nucleocapsid). Virus core in mature form can be icosahedral, cylindrical or conical. *CS*, coreshell. Approximately in scale

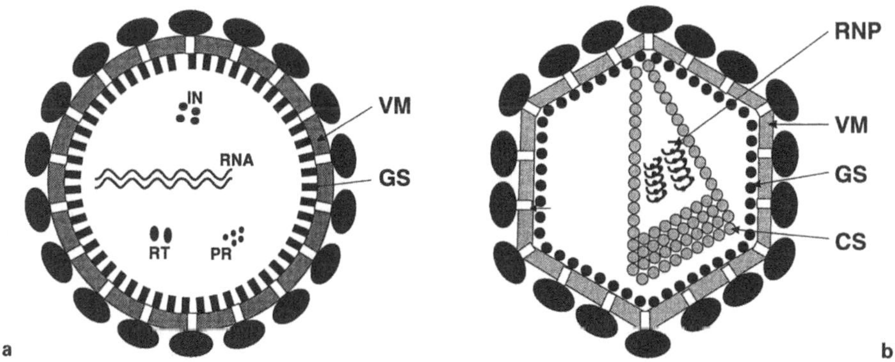

Fig. 2a,b. Diagrammatic representation of type 2 architecture as observed with lentivirus and the HTLV/BLV Group. No intermediate layer is present in this type. **a** Immature form, **b** mature form. Virus core can be conical, cylindrical or icosahedral. Approximately in scale

found in both the immature and mature virus particles. Retroviruses of type B,C,D and spumaviruses belong to type 1, which is illustrated in Fig. 1, while lentiviruses and the human T-cell leukaemia virus/bovine leukaemia virus (HTLV/BLV) group of viruses have type 2 architecture (Fig. 2). The major difference between the two architectures is seen most clearly in electron micrographs of immature particles. Type 1 virus particles have an additional protein layer between the gag shell and viral membrane (Fig. 3a,c); this additional layer is not present in type 2 architecture (Fig. 3d–f). The different staining density of the second layer indicates a different type of protein to the underlying gag protein. A thin inner region of increased electron density has also been recognised in the submembrane layer of HIV gag protein in immature viruses and gag particles but lentiviruses show no evidence of two distinct layers in gag particles or viruses examined by negative staining or sectioning. The second layer has been seen in many published electron micrographs of B-type and C-type viruses (see DALTON and HAGUENAU 1973; NERMUT and STEVEN 1987; GELDERBLOM 1991 for illustrations) but its presence has been largely ignored and no attempt has been made to understand its function in the assembly process. Speculations on the composition of this intermediate layer include the actin 'membrane skeleton' since actin has been found in purified virus preparations of several enveloped viruses including retroviruses (WANG et al. 1976; DAMSKY et al. 1977). Recently actin-binding proteins have been detected in rabies virus (SAGARA et al. 1995). The intermediate layer was also observed in sections (Fig. 3b) and by negative staining (as a distinct second layer) in Moloney murine leukaemia virus (MoMuLV) gag-particles produced by recombinant Semliki Forest virus. A small amount of actin has been found in such virus-like particles (VLPs) by immunoelectron microscopy and immunoblotting of sodium dodecyl sulphate (SDS) gels (NERMUT, WALLENGREN and GAROFF, in preparation). Although the intermediate layer is less obvious in mature type 1 particles, actin has also been localised by gold immunolabelling in wild-type MoMuLV particles (NERMUT and PAGER, in

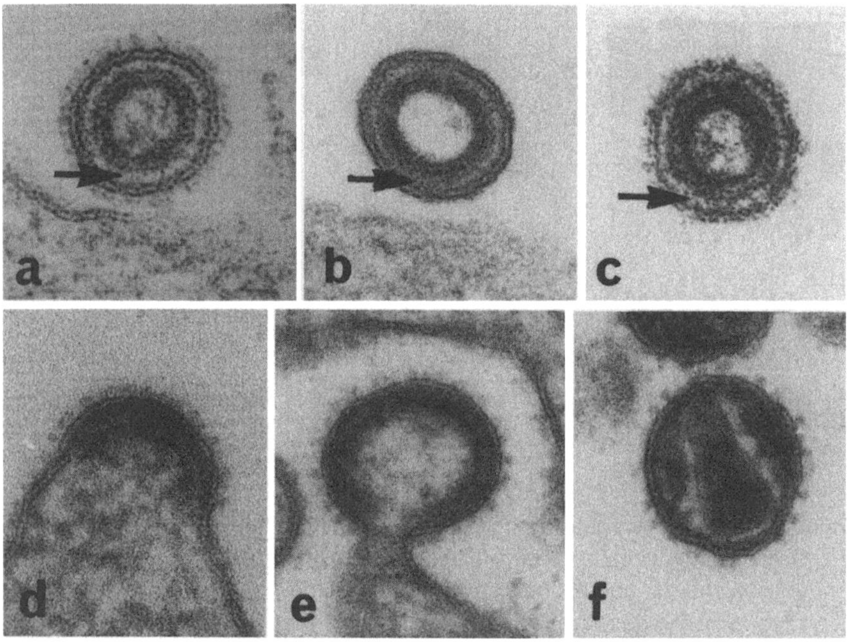

Fig. 3. a Thin section of immature B-type virus particle (MMTV). Note the presence of a less electron-opaque layer between the dark gag-shell and viral membrane. (Micrograph by courtesy of H. Frank). **b** Thin section of VLP produced in BHK cells by recombinant Semliki Forest virus expressing MoMuLV gag protein. (Specimen by K. Wallengren). **c** Immature MuLV particle with densely stained gag layer and less contrasted intermediate layer. (Micrograph by courtesy of H.Frank). **a–c** *Arrows* point to 'intermediate' layer. **d–f** Budding, immature and mature HIV. Note close contact of dark gag layer with plasma membrane. (Micrographs by courtesy of H.R.Gelderblom). All micrographs are x150 000

preparation). The presence of a non-gag layer in most retroviruses could have consequences for our views on retrovirus assembly and specifically on glycoprotein: gag protein interactions.

Although there are only two principal architectures of the virion there are morphological differences among retroviral groups which help discriminate and identify them by electron microscopy. Useful markers are the surface projections and the cores. The size of surface glycoprotein projections suggests that they are oligomers. However, no consensus has been reached whether they are dimers, trimers or tetramers. Surface projections differ also in shape and dimensions. Mouse mammary tumour virus (MMTV) projections are globular, about 6 nm in diameter and 8 nm long. Three subunits have been described (Sarkar 1987). Globular surface projections about 10 nm in diameter have also been observed on MuLV by negative staining and in shadowed replicas (Nermut et al. 1972). The glycoprotein projections on Mason-Pfizer monkey virus (M-PMV) are less conspicuous and they are about 5 nm long (Gelderblom 1987; Coffin 1992). Surface projections on human immunodeficiency virus (HIV) appear globular in shadowed replicas and in scanning electron microscopy

(SEM), and measure about 12 nm in diameter and 10 nm in length. In simian immunodeficiency virus (SIV) a trimeric shape of 'knobs' has been observed by negative staining (Grief et al. 1989). Spumaviruses possess rather slim, needle-like and long projections.

The core of retroviruses is either cylindrical (in D-type viruses), conical (in lentiviruses, occasionally also in D-type) or icosahedral (in other retroviruses). There is some evidence that the nucleoprotein complex is helically organised (Frank 1987a) but no recent data are available.

4 Structural Aspects of Retrovirus Assembly and Maturation

Understanding the molecular organisation of the virion is closely related to knowledge of the assembly process which includes the following steps (Fig. 4).

1. Synthesis of gag precursor protein on free ribosomes
2. Transport of the precursor protein to the site of assembly (in the cytoplasm or at the plasma membrane)
3. Assembly of precursor molecules into a spherical shell either in the cytoplasm or at a cellular membrane concomitant with the budding process
4. Acquisition of lipid membrane with glycoprotein projections (synthesised on rough endoplasmic reticulum (RER) and inserted into plasma membrane) and release of immature virus particle (budding)
5. Formation of mature virion through cleavage of the precursor proteins, which is followed by structural reorganisation of the virus interior

This sequence of events in retrovirus morphogenesis is reflected in the individual chapters of the book.

Over the past few years alterations in the primary structure of gag proteins and their effects on transport, targeting and assembly of the protein have been extensively studied. Here we shall concentrate on structural aspects of gag precursor assembly both in the cytoplasm and during budding and on the structural effects of maturation.

4.1 Assembly and Release of Retrovirus Particles

The formation of a three-dimensional structure (spherical shell) from gag precursor molecules is determined by the self-assembly property of the gag protein. There are, however, two forms of self-assembly: in the B-type of assembly 'gag spherical shells' are formed in the cytoplasm while in the C type the spherical shells are formed at the plasma membrane during budding. All retrovirus genera except lentiviruses and the HTLV/BLV group also acquire an

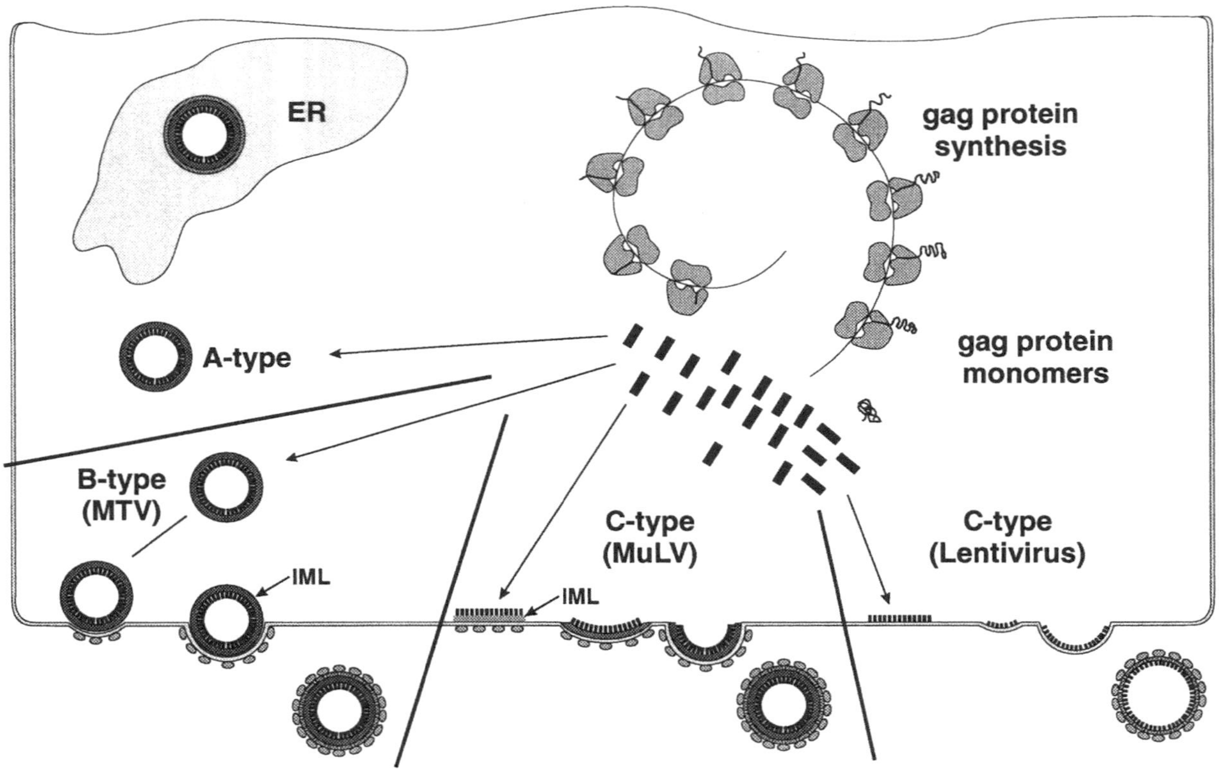

Fig. 4. Retrovirus assembly. Gag polyprotein molecules are synthesised on free ribosomes and assemble into 'spherical shells' to form A-type particles which either bud into endoplasmic reticulum or move towards the plasma membrane, where they acquire viral glycoproteins and form immature B-type virus particles. A thin layer of unidentified protein is associated with 'spherical shells' before budding. C-type viruses assemble at the plasma membrane simultaneously with budding and association with viral glycoproteins. Some C-type retroviruses acquire an 'intermediate layer' similarly to B-type viruses. No such layer is present in lentiviruses and the HTLV/BLV group

additional protein layer, exterior to the gag protein, either before association with the plasma membrane (in B-type assembly) or during the process of budding (C-type assembly).

4.1.1 Assembly of Gag Spherical Shells in the Cytoplasm (B-Type Assembly)

B-type morphogenesis has been observed in the mammary-type B-oncovirus group, D-type retroviruses and spumavirus. Gag spherical shells or A-type particles (BERNHARD 1958) are formed in the cytoplasm; they move towards the plasma membrane, where they become enveloped and released by budding from the plasma membrane, thus forming B-type particles. Particles of A-type morphology are also found in the cisternae of endoplasmic reticulum (ER). These "intracisternal A-type particles" are most probably aberrant forms of other retroviruses such as animal leukaemia viruses or lentiviruses and may assemble at the ER membrane (FÄCKE et al. 1993) or in the cytoplasm (TANAKA et al. 1972). Such particles maintain the immature morphology and are not infectious. A-type particles were studied in cells from lymphoid leukaemias and mammary tumours by TANAKA et al. (1972) and two dense concentric layers were observed in thin sections, indicating that an additional protein layer had become associated with the gag spherical shell. Similar pictures were published by BERNHARD (1973) and SARKAR et al. (1973). It is not known whether the site of assembly in the cytoplasm is random or specific. It was suggested that actin cytoskeleton is involved in budding and release of virus particles (DAMSKY et al. 1977; PEARCE-PRATT et al. 1994).

4.1.2 Assembly at the Plasma Membrane (C-Type Assembly)

This type of assembly is typical for animal and avian C-type viruses, the HTLV/BLV group and lentiviruses. The formation of the spherical gag shell takes place during the process of budding, which starts with association of gag precursor with the plasma membrane. How the 'intermediate layer' is formed during this process is unclear. Most of the recent studies of C-type assembly have utilised lentiviruses (HIV) in which the intermediate layer is absent, which limits comparison with other retroviruses.

The assembly starts with association of gag monomers with the plasma membrane and new data on this interaction have recently been reported. Nuclear magnetic resonance (NMR) studies of the MA protein (p17) of HIV (MATTHEWS et al. 1994; MASSIAH et al. 1994) localised a cluster of basic amino acids near the N-terminus tail, thus supporting the suggestion of ZHOU et al. (1994) for the electrostatic interaction of the gag protein with acidic phospholipids. The insertion of myristic acid in the lipid bilayer does not seem to guarantee stable anchoring since its binding energy is only 8 Kcal/mol (PEITZSCH and MCLAUGHLIN 1993). It thus appears that both myristic acid and a cluster of basic amino acids near the N-terminus of the molecule are needed for correct transport of the gag protein to the plasma membrane and for stable association. It may be that myristic acid provides the first contact with the membrane by inserting itself in the outer membrane leaflet, while a stable association is effected by accumulation of acidic phospholipids in the area and their interaction with positively charged residues.

It is not known whether the self-assembly of the gag polyprotein starts in the cytoplasm or only after association of the molecules with the lipid bilayer. Observations on HIV gag expressed in insect cells indicate that the molecules assemble into small patches of 'rings' at the plasma membrane before budding starts (Fig. 5). Image processing revealed that the rings are organised on a hexagonal network (NERMUT et al. 1994). While this observation supports the idea of an icosahedral symmetry in HIV, it should be noted that very recent cryo-electron microscopic study of virus-like particles containing HIV gag shell indicates a more complex organisation (FULLER, GOWEN, KRÄUSSLICH and VOGT, in preparation).

4.1.3 Mechanism of Gag Protein Assembly into a Spherical Shell

A fundamental aspect of retrovirus assembly is the formation of a spherical protein shell which maintains the three-dimensional structure of the virion and determines its shape and size. Little is known about how the gag molecules assemble and about the molecular organisation of the spherical shell. An early model supported by studies of other enveloped viruses (see reviews by SIMONS

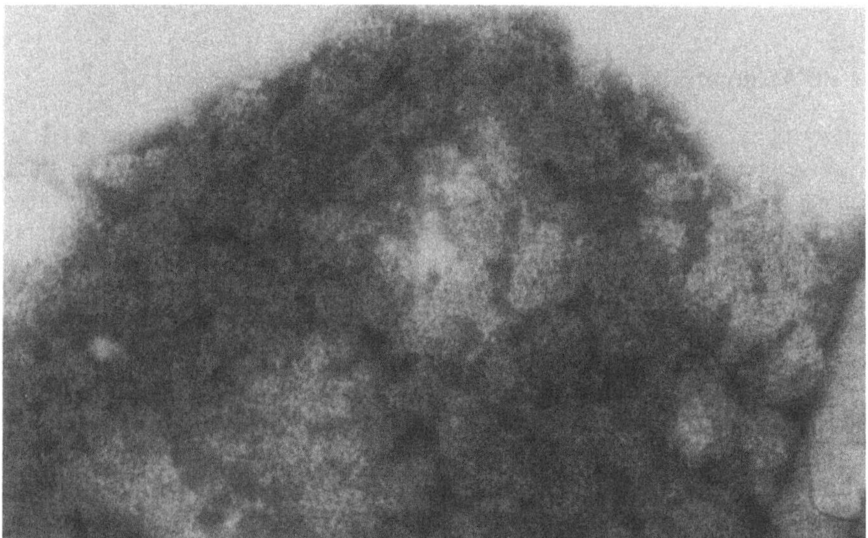

Fig. 5. Patches of ring-like structures observed at cytoplasmic surface of plasma membrane in insect cells producing HIV 'gag particles'. Cells were opened by wet cleaving and obtained membranes were negatively stained with silicotungstate, x200 000

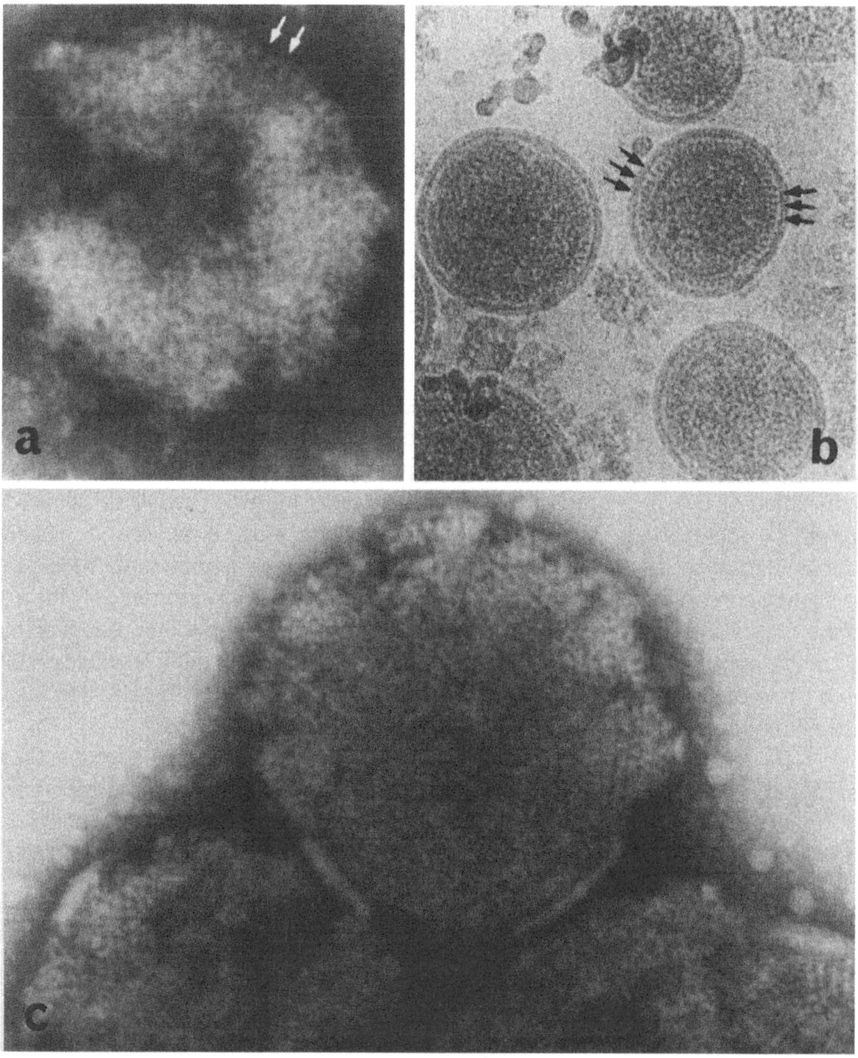

Fig. 6a–c. Delipidised HIV 'gag-particle' produced by insect cells infected with a recombinant baculovirus **a**. Note rod-shaped molecules at periphery (*arrows*). Negative staining electron microscopy, x240 000. **b** Peripheral rod-like molecules observed in frozen hydrated preparation of 'gag particles', x150 000. (Micrograph by courtesy of S. Fuller). **c** Gag particle displaying an end-on view of 'rings'. Negative staining, x300 000

and GAROFF 1980; DUBOIS-DALCQ et al. 1984) suggested that the curvature of the budding virus is a result of links between the transmembrane glycoproteins and the underlying gag sheet. The precondition for this model was the self-assembly property of the membrane-associated protein (M-protein in myxo- or paramyxoviruses). Such capacity was demonstrated for Sendai virus (HEWITT and NERMUT 1977; BUECHI and BACHI 1982) but no such knowledge about retrovirus gag proteins was available at that time. In addition, arguments against an active role of viral glycoproteins in the assembly process have since emerged (reviewed by STEPHENS and COMPANS 1988, HUNTER 1994), the strongest being provided by the discovery that virus-like particles are produced by recombinant viruses expressing gag protein only (see WILLS and CRAVEN 1991, LUCIW and LEUNG 1992 for refs.). The cytoplasmic formation of gag spherical shells and the presence of an additional protein layer (about 8 nm thick) in B-type viruses is also incompatible with the involvement of envelope glycoproteins.

Over the past few years new information about the structure and assembly of the gag protein in lentiviruses has been obtained. Ultrastructural studies have revealed the presence of peripheral striations in immature HIV (HOCKLEY et al. 1988) and also in recombinant adenovirus produced particles (VERNON et al. 1991). Further work on HIV virus-like particles showed that the gag precursor is a rod-like molecule (Fig. 6a,b) and some evidence was presented in favour of rings consisting of five or six rods (NERMUT et al. 1994, and Fig. 5). Cylindrical structures (= rod-shaped proteins), however, cannot form a sphere if closely apposed to each other and it is therefore most likely that the precursor molecule is slightly conical. In this case a side-to-side assembly would automatically form a sphere (HEWITT 1977). In a simple model it is possible to show that reduction by 2° of the right angle at one end of a cylinder 8.5 nm in length (Fig. 7) would result in a circle of 100 nm in diameter which is close to the size of the HIV gag shell (NERMUT et al. 1994).

No such information is available on the gag protein of other retroviruses although all gag proteins have a similar molecular weight. A recent report by KLIKOVA et al. (1995) describes successful assembly of M-PMV gag shells in bacteria and also in vitro. Negative staining and thin sections of obtained VLP

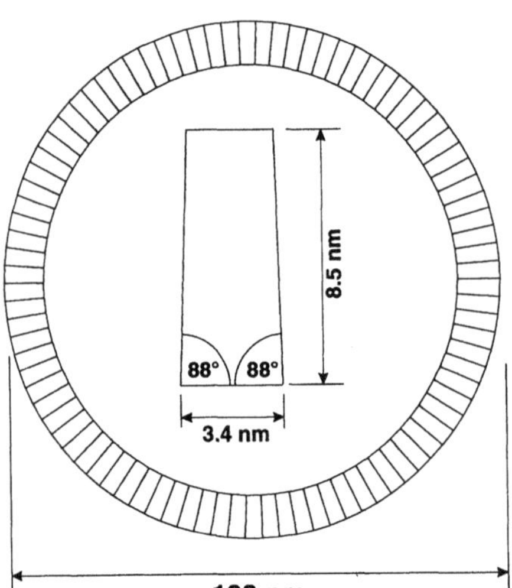

Fig. 7. Diagram showing how conical gag-molecules would form a circle 100 nm in diameter when the right angle was reduced to 88°

showed hexagonal profiles, suggesting a regular icosahedral organisation. No high-resolution electron microscopy has been reported to show the overall shape of the gag precursor in M-PMV but since the obtained gag shells were "indistinguishable from the particles assembled in HeLa cells" the presence of an additional protein layer is likely.

The fullerene-like model as proposed for HIV (NERMUT et al. 1994) suggests that the gag molecules form hexagonal rings with pentagonal rings at vertices of the icosahedron. Such a network requires strong intermolecular interactions. Several studies have recently been directed towards the search for the specific contact sites on the gag protein molecules responsible for association into a 'fullerene-like' network. In the proposed model (NERMUT et al. 1994) one rod-like molecule is shared by two adjacent rings formed by six molecules and as a result there are necessarily four contact sites on each molecule. If two rods are shared by two 'rings' only three binding sites are required on each gag molecule for formation of a hexagonal net (Fig. 8). Attempts to dissociate the gag-shells in VLP using high salt or high pH (unpublished results) have been largely unsuccessful, indicating multiple protein: protein interactions. Mapping of amino acid sequences essential for regular virus assembly has proven only partly effective. The major reason has been the lack of information on conformational changes brought about by alterations in the primary structure. Nevertheless a consensus has been reached that the most probable contact points on the gag molecule are in the MA and CA domains. Using quantitative immuno-electron microscopy and a ligand affinity blotting assay, CARRIERE et al. (1995) pointed to four possible contact sites in HIV gag protein present in the MA and CA domains and also one overlapping the spacer peptide 2 and the NC domain. Using cysteine cross-linking reagents, HANSEN and BARKLIS (1995) localised a large contact site near the C-terminus of CA protein in MuLV overlapping into the NC domain. Multimers of gag protein (but no trimers) have been observed. The presence of multiple binding sites in the MA domain is necessary for the stability of the membrane-associated shell in mature virus particles. It has recently been

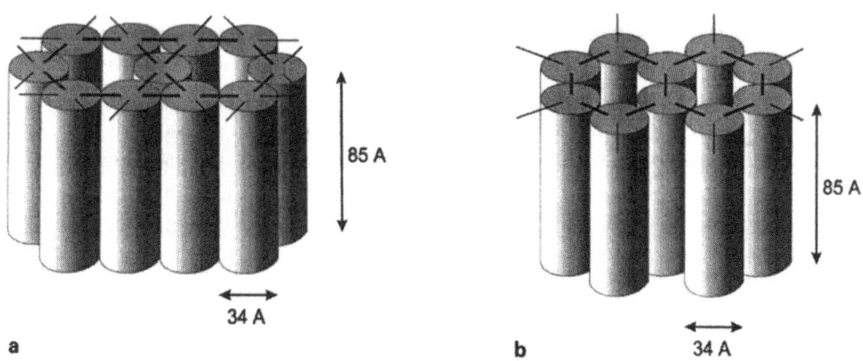

Fig. 8a,b. Diagram of rings of six rods sharing one (**a**) or two (**b**) rods. In **a** every rod has four contact points, in **b** only three contact points

shown that mutations of Cys57 in p17 of HIV that abolished VLP assembly also prevented dimerisation of p17 (Morikawa et al. 1995). The presence of disulphide bonds in HIV gag protein is supported by the observation of Grigoriev et al. (1992) that mercaptoethanol dissociated the immature gag shell in HIV. This, however, does not mean that disulphide bridges are the only types of gag intermolecular bonds. It seems obvious that the correct understanding of the gag protein interactions will be achieved only after the tertiary structure of the protein has been determined.

The formation of an immature virus particle is completed by association of viral glycoproteins (GPs) with the budding gag shell. This takes place in the early stages of budding (Fig. 3d). Although the glycoproteins are dispensable for assembly and budding of virus particles they are indispensable for infectivity. It will be useful to consider the mechanism of transport and anchoring of glycoprotein molecules in the membrane area with underlying gag protein. One of the crucial questions is the specificity of the interaction with gag (or other) protein layer. It is generally accepted that most cellular proteins are excluded from membrane areas where budding is taking place. Transmembrane proteins can be immobilised by interaction with membrane-associated proteins such as the gag protein layer. The exclusion of cellular proteins from the bud indicates that viral GPs possess higher specific affinity for gag protein than cellular proteins. If, however, there are no specific viral proteins in the membrane or if there is space there, some cellular proteins may become incorporated into the virus particle. This would explain the presence of major histocompatibility (MHC) proteins in HIV (Gelderblom et al. 1987).

4.2 Morphological Aberrations of Assembly

Conformational changes brought about by mutations in the *gag* gene can alter the normal assembly process or abolish the capacity of the gag protein to form virus particles. A comparative study of such effects in HIV has recently been published by Hockley et al. (1994). The following questions have been the subject of particular interest: which parts of the gag molecule are dispensable for assembly into spherical particles and can morphological aberrations be accounted for by the fullerene-like model?

It has been shown by several groups that truncations of the HIV gag polyprotein down to AA 390 (mol.wt. 42 K) does not abolish the capacity to form VLP. However, in addition to the formation of spherical VLP some of the truncated gag proteins (44 K and 45 K, AA 410–427) produced tubular forms smaller in diameter than the spherical forms (Fig. 9a,b and Gheysen et al. 1989; Hockley et al. 1994). Moreover, many intracytoplasmic spherical shells budded into cytoplasmic vacuoles which may simply reflect overproduction of gag protein. Another aberrant morphology was spiral gag shells (Fig. 9c). Further truncation of the gag protein down to 41 K resulted in the formation of membrane-associated sheets with no budding, while deletions within the p24

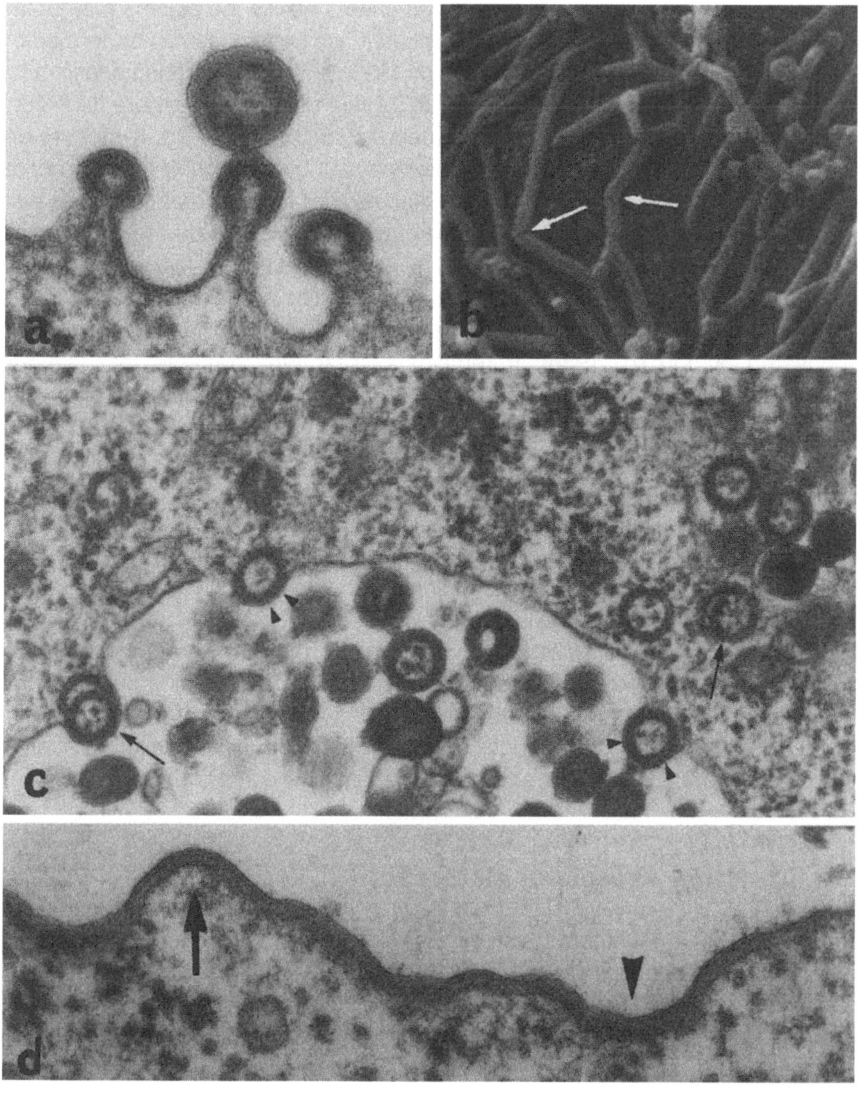

Fig. 9. a Thin section of tubular forms of VLP produced by gag protein truncated to a mol. wt of p44/45. Note smaller diameter of tubular forms as compared with spherical particles. The presence of gag protein in these structures was demonstrated by gold immunolabelling (GHEYSEN et al. 1989; HOCKLEY et al. 1994), x135 000. **b** Scanning electron micrograph showing gag-tubular forms on the cell surface (same specimen as above). *Arrows* point to sharp angles in the filaments, x20 000. **c** p45 mutant showing gag-spherical shells or spirals (*arrows*) in the cytoplasm, some in process of budding into a vacuole (*arrow* heads), x68 000. **d** Convoluted cell surface in cells infected with a mutant with deleted AA 249–261. The plasma membrane is thickened by a layer of gag protein. *Arrow* indicates positive, *arrowhead* negative curvature, x105 000

domain brought about curved, convoluted sheets below the plasma membrane (Fig. 9d). The filaments produced by truncated gag molecules can be formed by a tubular hexagonal network (see CASPAR 1993 for refs.); the round end is most probably closed up by an 'icosahedral cap' of five triangles. It is more difficult to explain the convoluted layer of gag protein as shown in Fig. 9d. It must result from a mismatch in the assembly process caused by the loss of specific bonds due to conformational changes in the molecule. The existence of convex as well as concave forms (positive and negative curvature) could be due to introduction of heptagonal units into a hexagonal network as suggested by IIJIMA et al. (1992) on the basis of studies of carbon fullerenes. Spiral shells (Fig. 9c) can be accounted for by displacement of pentagons in the hexagonal network as suggested by KROTO (1988). It remains unclear why mutants that lost the capacity to form spherical shells (p41 or with p24 deletions) assembled only at cellular membranes and were not detected in the cytoplasm as for example with the p45 mutant.

Induced aberrations in retrovirus assembly and final morphology help in understanding the variety of forms and sizes of virus particles observed even in normal preparations processed using low-temperature technology (NERMUT et al. 1993; GRIEF et al. 1994). Such morphological aberrations including the tear-drop HIV particles described by MORITA et al. (1990) do not contradict the view of the presence of icosahedral symmetry in retroviruses. They can be accounted for by errors in assembly including insertion of heptagons or displacing pentagons in the hexagonal network or by mutations affecting the surface of the gag protein. Large or small virus particles might be formed by a larger or smaller number of gag protein copies, i.e. larger or smaller triangulation number.

4.3 Structural Aspects of Maturation

The process of maturation is initiated by cleavage of the precursor molecule, which starts probably during the budding process or very soon after the release of virus particles (KAPLAN et al. 1994). Little is known about the shape and size of the *gag*-encoded polyprotein in retroviruses with the exception of HIV. It has been reported that the HIV gag precursor is rod shaped, 8.5 nm in length and 3.4 nm in diameter (NERMUT et al. 1994 and Figs. 6,7). Based on the fact that the molecule is approximately a cylinder it is possible to calculate the dimensions of the smaller protein molecules produced by cleavage using the formula published by GREEN (1969). In this particular case the fragments were considered to be 3.2 nm in diameter to correct for a slightly conical shape of the molecule. The following dimensions were obtained: p17 = 2.6 nm, p24 = 4 nm, p7 = 1.1 nm, p6 = 0.9 nm. The sum equals 8.6 nm. These data correlate well with recent NMR studies of p17 of HIV and with the length of p55 as measured from electron micrographs. The three-dimensional structure of p17 published by MATTHEWS et al. (1994) showed a range of diameters from 3.0 nm to 3.5 nm with an average of 3.1 nm. The length of the main body of the molecule was

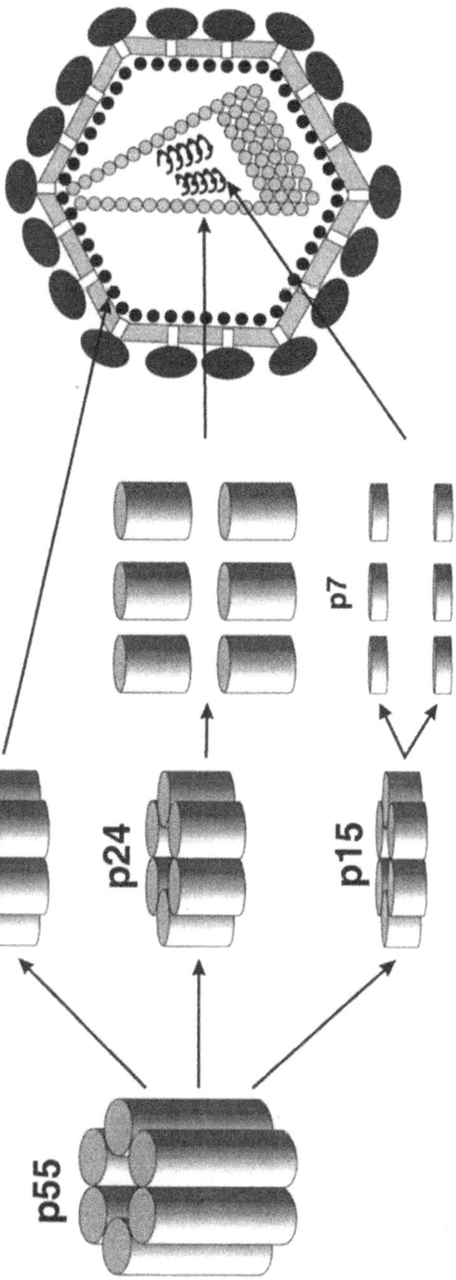

Fig. 10. Diagrammatic representation of the maturation process in HIV. Note that hexagons of p24 and p15 are probably dissociated into monomers before re-assembly in the mature virus particle. Only three (instead of six) p7 and p6 fragments are shown

2.9–3.5 nm, which is longer than calculated and is probably caused by relaxation of the molecule after cleavage.

After cleavage p17 (MA) remains associated with the membrane but this thin protein layer is difficult to visualize by electron microscopy (Fig. 3f). In some C-type viruses a thin submembrane layer (inner coat) was visualized by prolonged negative staining and in sections as a thickened inner membrane leaflet (FRANK 1987a,b). The largest cleavage product (CA) forms the surface of the central body called the core. The different forms of cores are obviously influenced by the shape and surface bonds of the protein. Hexagonal rings of p24 in HIV will probably be dissociated into monomers before re-assembly into the conical core shell (Fig. 10). The mechanism of this event is unknown but it will probably depend on the type of intermolecular bonds in the gag shell and their susceptibility to the intraviral environment.

In an average-sized HIV particle the number of gag precursor molecules was determined to be 1890 using the fullerene-like model (NERMUT et al. 1994). Very little is known about the number of p24 copies in the core shell or the other proteins except that the molar ratios of individual cleavage products in SIV are close to 1 (HENDERSON et al. 1988). However, the presence of 'lateral' bodies observed in conventional sections (GELDERBLOM et al. 1987, 1991) and the uniform high electron density of freeze-substituted virus particles (GRIEF et al. 1994) indicate that 'unused' protein material is present within the central cavity. This could mean that the number of protein molecules in the core shell and possibly in the ribonucleoprotein (RNP) complex is less than the number of the precursor copies. The exact location of p6 in HIV is not known but gold immunolabelling (GRIEF et al. 1994) has shown that the protein is not concentrated exclusively in the 'core/envelope' link (HOGLUND et al.1992).

The virus capsids or envelopes provide efficient protection for the virus genome during transfer from cell to cell. It is known that immature retroviruses are not infectious and one reason for this may be that in immature virus particles the RNA is not in a complex with protein and therefore might be easily destroyed by cellular RNAses after uptake. The molecular organisation of retrovirus RNP is not known although some observations indicate that it is a helical filament (FRANK 1987a). Helical organisation of nucleoproteins is frequent in viruses and it can assume two forms: a typical 'continuous' helix with the RNA on top of, or within, a protein core, or it can be a discontinuous helix in the form of nucleosomes. A model of a continuous version applied to HIV shows that such a filament would be 0.6 μm long and 7.5 nm in diameter. It would fill the wider end of the conical core at about one-third of the length of the core. This is in good correlation with the image of the core in thin sections of HIV showing higher stain density in the wide part of the cone.

5 Structural Classification of Retroviruses

Morphological features and type of assembly as described by BERNHARD (1958) have become basic criteria for classification of retroviruses (COFFIN 1992) and also have a diagnostic value in clinical virology.

5.1 Morphology of the Virion

The major morphological criteria include the overall size of the virion (not always useful because of a rather big variability within individual genera); shape, size and frequency of glycoprotein surface projections; shape of cores in mature virus forms; presence or absence of an intermediate layer between the gag precursor shell and viral membrane in immature forms.

5.2 Type of Virus Assembly

There are two basic types of assembly in retroviruses: the B type and the C type as described on the previous pages. It is, however, important to stress that there are two 'subgroups' in C-type retroviruses. In mammalian and avian C-type virus groups an intermediate layer is acquired at the plasma membrane; this layer is not seen in lentivirus and the HTLV/BLV groups and the gag protein layer appears to be in direct contact with the lipid bilayer (Figs. 3, 4).

Based on the above description of individual retrovirus genera we can now propose updated morphological characteristics of retroviruses.

5.3 General Characteristics

Retroviruses are enveloped isometric, approximately spherical particles with a diameter ranging from 90 to 140 nm. Their surface is studded with glycoprotein projections varying in shape and dimensions. Immature virus particles possess one or two membrane-associated shells, and mature particles are characterised by a prominent central core containing the nucleoprotein complex. Cores are icosahedral, cylindrical or conical.

5.4 Specific Characteristics of Individual Genera

5.4.1 Mammary Type-B Oncovirus

Type species: mouse mammary tumour virus. This virus is characterised by a typical B-type assembly producing immature virus particles with prominent surface spikes and a protein layer between the gag-shell and viral membrane.

Mature virions are characterised by a dense spherical-appearance core, often eccentrically located. Intracytoplasmic 'double shell' particles are about 75 nm in diameter, while enveloped particles measure 90–120 nm in diameter.

5.4.2 Mammalian Type-C Retrovirus

Type species: murine leukaemia virus. Here a type-C assembly is seen at the plasma membrane with an intermediate protein layer between the gag-shell and viral membrane. Surface projections are often shed and thus not easily seen in sections. The virus core is probably icosahedral. There is some evidence that the nucleocapsid is helical. The diameter of the virus particles is 100–120 nm.

5.4.3 Avian Type-C Retrovirus

Type species: avian leukosis virus. This virus is characterised by a type-C assembly and morphology as for mammalian type-C retrovirus.

5.4.4 Type-D Retrovirus

Type species: Mason-Pfizer monkey virus. This virus has a type-B assembly and morphology except that the surface spikes are less prominent and the core is cylindrical or conical. Particle diameter ranges from 110 to 130 nm. Preformed gag spherical shells are about 76 nm in diameter.

5.4.5 Spumavirus

Type species: human foamy virus. In principle this virus has a B-type assembly and morphology, but budding of gag spherical shells takes place more frequently at the endoplasmic reticulum (or Golgi?) cisternae than at the plasma membrane. The fringe of needle-like spikes is usually prominent, and cores appear spherical in sections. Virus particles measure 100–130 nm in diameter.

5.4.6 HTLV/BLV Group

Type species: human T-cell lymphotropic virus type 1. The HTLV/BLV group of viruses is characterised by a type-C assembly at the plasma membrane with no intermediate layer between the gag-shell and viral membrane. Cores are icosahedral. Surface projections are less conspicuous and about 5 nm high. Virus particles measure about 120 nm.

5.4.7 Lentivirus

Type species: human immunodeficiency virus. The lentivirus has a type-C assembly at the plasma membrane, with no intermediate protein layer observed. Surface projections are globular, and frequently lost from mature particles in some strains (HIV-1). Cores are conical, only occasionally cylindrical. Virus particle diameter ranges from 100 to 130 nm.

6 Conclusions and Future Tasks

Molecular biology and genetics have recently provided much new information about retroviruses. We have attempted to relate structural studies to the new data and thus extend the molecular information towards consideration of the three-dimensional organisation of viral proteins and of the virus particle. Our underlying philosophy is based on the recognition that retroviruses are regular structures built on the principles of icosahedral symmetry. The assembly process should be looked upon as a dynamic probability event based on interactions between participating proteins and determined by their self-assembly capacity. We conclude that aberrant forms are consequences of errors in assembly or products of mutations. Based on updated structural analysis we propose a simplified classification of retrovirus groups based on the fact that there are two types of virus assembly (B and C) and two types of virus architecture: type 1 common to most retroviruses and characterised by an additional protein layer and type 2 present in lentiviruses and the HTLV/BLV group, where only the gag layer is present.

Structural and morphological studies of retroviruses have been revitalised by recent progress in molecular biology and the consequent availability of recombinant virus-like particles. Some of the important questions that may be answered in this way concern the three-dimensional structure of the gag precursor and effects of mutations on the process of assembly. Other interesting problems are: (a) the composition, organisation and function of the intermediate layer and its role in virus assembly; (b) the formation and molecular organisation of mature viral cores by rearrangement of cleaved gag proteins; and (c) the mechanism of transport and targeting of gag molecules in B-type and C-type assembly.

Acknowledgements. We acknowledge support from the UK MRC AIDS Programme and the collaboration of Ian Jones (NERC, Oxford), Henrik Garoff and Kristina Wallengren (Stockholm). We also thank colleagues who provided us with illustrations: Hans Gelderblom (Berlin), Stephen Fuller (EMBL, Heidelberg) and Hermann Frank (Tübingen). Steve Matthews (Oxford) kindly provided coordinates of the NMR structure of HIV p17. Critical comments on the manuscript and discussions with E. Hunter are greatly appreciated. Computer graphics were produced by Andrew Davies, photography by Robert Longuehaye.

Note Added in Proof. The recent X-ray crystallography data of SIV p17 protein [Rao et al. (1995) Nature 378: 743–747] have shown that "the molecule forms a trimer consistent with oligomerization in vitro, the observed virion architecture..." and that three trimers may assemble into a ring.

References

Bernhard W (1958) Electron microscopy of tumor cells and tumor viruses. A review. Cancer Res 18: 491–509

Bernhard W (1973) Oncorna viruses. 2. Type A and C virus particles in murine and other mammalian leukemias and sarcomas. In: Dalton AF, Hagenau F (eds) Ultrastructure of animal viruses and bacteriophages. Academic, New York, pp 293–305

Buechi M, Bachi T (1982) Microscopy of internal structures of Sendai virus associated with the cytoplasmic surface of host membranes. Virology 120: 349–359

Carriere C, Gay B, Chazal N, Morin N, Boulanger P (1995) Sequence requirements for encapsidation of deletion mutants and chimeras of human immunodeficiency type 1 gag precursor into retrovirus-like particles. J Virol 69: 2366–2377

Caspar DLD (1993) Deltahedral views of fullerene polymorphism. In: Kroto HW, Walton DRM (eds) The fullerenes. Cambridge University Press, Cambridge, pp 133–144

Coffin JM (1992) Structure and classification of retroviruses.In: Levy J A (ed) The Retroviridae. Plenum, New York, pp 19–50

Dalton AD, Hagenau F (1973) Ultrastructure of animal viruses and bacteriophages. Academic, New York

Damsky CH, Sheffield JB, Tuszynski GP, Warren L (1977) Is there a role for actin in virus budding? J Cell Biol 75: 593–605

Dubois-Dalcq M, Holmes K, Rentier B (1984) Assembly of enveloped RNA viruses. Springer, Vienna, New York

Fäcke M, Janetzko A, Shoeman RL, Kräusslich H-G (1993) A large deletion in the matrix domain of the human immunodeficiency virus gag gene redirects virus particle assembly from the plasma membrane to the endoplasmic reticulum. J Virol 67: 4972–4980

Frank H (1987a) Oncovirinae:type C oncovirus. In: Nermut MV, Steven AC (eds) Animal virus structure. Elsevier, Amsterdam, pp 273–288

Frank H (1987b) Lentivirinae. In: Nermut MV, Steven AC (eds) Animal virus structure. Elsevier, Amsterdam, pp 295–304

Gelderblom HR (1987) Oncovirinae: type D oncovirus. In: Nermut MV, Steven AC (eds) Animal virus structure. Elsevier, Amsterdam, pp 289–294

Gelderblom HR (1991) Assembly and morphology of HIV: potential effect of structure on function. AIDS 5: 617–638

Gelderblom HR, Hausmann EHS, Ozel M, Pauli G, Koch MA (1987) Fine structure of human immunodeficiency virus (HIV) and immunolocalization of structural proteins. Virology 156: 171–172

Gelderblom HR, Ozel M, Winkel T, Morath B, Grund C, Pauli G (1991) Ultrastructural studies on lentiviruses. In: Racz P, Dijkstra CD, Gluckman JC (eds) Accessory cells in HIV and other retroviral infections. Karger, Basel, pp 50–68

Gheysen D, Jacobs E, de Foresta F, Thiriart C, Francotte M, Thines D, de Wilde M (1989) Assembly and release of HIV-1 precursor Pr55gag virus-like particles from recombinant baculovirus-infected insect cells. Cell 59: 102–112

Green NM (1969) Electron microscopy of immunoglobulins. Adv Immunol 11: 1–30

Grief C, Hockley DJ, Fromholc CE, Kitchin PA (1989) The morphology of simian immunodeficiency virus as shown by negative staining electron microscopy. J Gen Virol 70: 2215–2219

Grief C, Nermut MV, Hockley DJ (1994) A morphological and immunolabelling study of freeze-substituted human and simian immunodeficiency viruses. Micron 25: 119–128

Grigoriev V, Vzorov A, Escaig-Haye F, Fournier JG, Bukrinskaya A, Kushch A, Klimenko S (1992) Immunomorphological studies of proteolytic cleavage of "GAG" polyprotein of HIV. Proceedings of the 10th European congress on electron microscopy, vol III, pp 293–294

Hansen MST, Barklis E (1995) Structural interactions between retroviral gag proteins examined by cysteine cross-linking. J Virol 69: 1150–1159

Henderson LE, Benveniste RE, Sowder R, Copeland TD, Schultz AM, Oroszlan S (1988) Molecular characterization of gag proteins from simian immunodeficiency virus (SIV mne). J Virol 62: 2587–2595

Hewitt JA (1977) On the influence of polyvalent ligands on membrane curvature. J Theor Biol 69: 2455–2469

Hewitt JA, Nermut MV (1977) A morphological study of the M-protein of Sendai virus. J Gen Virol 34: 127–136

Hockley DJ, Wood RD, Jacobs JP, Garrett AJ (1988) Electron microscopy of human immunodeficiency virus. J Gen Virol 69: 2455–2469
Hockley DJ, Nermut MV, Grief C, Jowett JBM, Jones IM (1994) Comparative morphology of gag protein structures produced by mutants of the gag gene of human immunodeficiency virus type 1. J Gen Virol 75: 2985–2997
Hoglund S, Ofverstedt LG, Nilson A, Lundquist P, Gelderblom HR, Ozel M, Skoglund U (1992) Spatial visualization of the maturing HIV-1 core and its linkage to the envelope. AIDS Res Hum Retroviruses 8: 1–7
Hunter E (1994) Macromolecular interactions in the assembly of HIV and other retroviruses. Semin Virol 5: 71–83
Iijima S, Toshinari I, Ando Y (1992) Pentagons, heptagons and negative curvature in graphite microtubule growth. Nature 356: 776–778
Kaplan AH, Manchester M, Swanstrom R (1994) The activity of the protease of human immunodeficiency virus type 1 is initiated at the membrane of infected cells before the release of viral proteins and is required for release to occur with maximum efficiency. J Virol 68: 6782–6786
Klikova M, Rhee SS, Hunter E, Ruml T (1995) Efficient in vivo assembly of retroviral capsids from gag precursor expressed in bacteria. J Virol 69: 1093–1098
Kroto H (1988) Space,stars,C 60 and soot. Science 242: 1139–1145
Leis J, Baltimore D, Bishop JM, Coffin J, Fleissner E, Goff SP, Oroszlan S, Robinson H, Skalka AM, Temin HM, Vogt V (1988) Standardized and simplified nomenclature for proteins common to all retroviruses. J Virol 62: 1808–1809
Luciw PA, Leung NJ (1992) Mechanisms of retrovirus replication. In: Levy JA (ed) The Retroviridae, vol 1. Plenum, New York, pp 159–298
Marx PA, Munn RJ, Joy KI (1988) Computer emulation of thin section electron microscopy predicts an envelope associated icosadeltahedral capsid for human immunodeficiency virus. Lab Invest 58: 112–118
Massiah MA, Starich MR, Paschal C, Summers MF, Christensen AM, Sundquist WI (1994) Three-dimensional structure of the human immunodeficiency virus type 1 matrix protein. J Mol Biol 244: 198–223
Matthews S, Barlow P, Boyd J, Barton G, Russell R, Mills H,Cunningham M, Meyers N, Burns N, Clark N, Kingsman S, Kingsman A, Campbell I (1994) Structural similarity between the p17 matrix protein of HIV and interferon-gamma. Nature 370: 666–668
Montelaro RC, Ball JM, Rushlow KE (1993) Equine retroviruses. In: Levy JA (ed) The Retroviridae. Plenum, New York, pp 257–270
Morikawa Y, Kishi T, Zhang WH, Nermut MV, Hockley DJ, Jones IM (1995) A molecular determinant of HIV particle assembly located in the matrix antigen p17. J Virol 69: 4519–4523
Morita Ch, Ikuta K, Goto T, Sano K, Nakai M, Hirai K, Kato S (1990) Isolation and characterisation of cell clones persistently producing teardrop-shaped particles of human immunodeficiency virus type 1. J AIDS 3: 231–237
Nermut MV (1987) General principles of virus architecture. In: Nermut MV, Steven AC (eds) Animal virus structure. Elsevier, Amsterdam, pp 3–20
Nermut MV, Steven AC (eds) (1987) Animal virus structure. Elsevier, Amsterdam
Nermut MV, Frank H, Schafer W (1972) Properties of mouse leukemia virus. III.Electron microscopic appearance as revealed after conventional preparation techniques as well as freeze-drying and freeze-etching. Virology 49: 345–358
Nermut MV, Grief C, Hashmi S, Hockley DJ (1993) Further evidence of icosahedral symmetry in human and simian immunodeficiency virus. AIDS Res Hum Retroviruses 9: 929–938
Nermut MV, Hockley DJ, Jowett JB, Jones IM, Garreau M, Thomas D (1994) Fullerene-like organization of HIV gag-protein shell in virus-like particles produced by recombinant baculovirus. Virology 198: 288–296
Ozel M, Pauli G, Gelderblom HR (1988) The organization of the envelope projections on the surface of HIV. Arch Virol 100: 255–266
Pearce-Pratt R, Malamud D, Phillips DM (1994) Role of the cytoskeleton in cell-to-cell transmission of human immunodeficiency virus. J Virol 68: 2898–2905
Peitzsch RM, McLaughlin S (1993) Binding of acylated peptides and fatty acids to phospholipid vesicles: pertinence to myristoylated proteins. Biochemistry 32: 10436–10443
Sagara J, Tsukita S, Yonemura S, Tsukita Sh, Kawai A (1995) Cellular actin-binding ezrin-radixin-moesin (ERM) family proteins are incorporated into the rabies virion and closely associated with viral envelope proteins in the cell. Virology 206: 485–494

Sarkar NH (1987) Oncovirinae: type B oncovirus. In: Nermut MV, Steven AC (eds) Animal virus structure. Elsevier, Amsterdam, pp 257–272
Sarkar NH, Moore DH, Kramarsky B, Chopra HC (1973) Oncornaviruses 3. The mammary tumor virus. In: Dalton AJ, Haguenau F (eds) Ultrastructure of animal viruses and bacteriophages. Academic, New York, pp 307–321
Simons K, Garoff H (1980) The budding mechanisms of enveloped animal viruses. J Gen Virol 50: 1–21
Stephens EB, Compans RW (1988) Assembly of animal viruses at cellular membranes. Annu Rev Microbiol 42: 489–516
Takahashi I, Takama M, Ladhoff A-M, Scholz D (1989) Envelope structure model of human immunodeficiency virus type 1. J Acquir Immune Defici Syndr 2: 136–140
Tanaka H, Tamura A, Tsujimura D (1972) Properties of intracytoplasmic A particles purified from mouse tumors. Virology 49: 61–78
Vernon SK, Murthy S, Wilhelm J, Chanda PK, Kalyan N, Lee S-G, Hung PP (1991) Ultrastructural characterization of human immunodeficiency virus type 1 Gag-containing particles assembled in a recombinant adenovirus vector system. J Gen Virol 72: 1243–1251
Wang E, Wolf BA, Lamb RA, Choppin PW, Goldberg AR (1976) The presence of actin in enveloped viruses. In: Goldbaum R, Pollard T, Rosenbaum J (eds) Cell motility. Cold Spring Harbor Press, Cold Spring Harbor, pp 589–599
Wills JW, Craven RC (1991) Form, function and use of retroviral Gag proteins. AIDS 5: 639–654
Zhou W, Parent LJ, Wills JW, Resh MD (1994) Identification of a membrane-binding domain within the amino-terminal region of human immunodeficiency virus type 1 gag protein which interacts with acidic phospholipids. J Virol 68: 2556–2569

Intracellular Transport of Retroviral Capsid Components

H.-G. KRÄUSSLICH and R. WELKER

1 Introduction

Formation of a virus particle protects the viral genome from exogenous agents and permits release of its content upon delivery to the target cell. Retroviruses generally cause persistent infections and do not usually lyse their host cells. In this case virus morphogenesis requires transport of individual virion components to a specific site within the producer cell, with subsequent assembly and release by budding from the plasma membrane. Extracellular infectious retroviruses are composed of an inner core containing the RNA genome and the viral replication enzymes (products of the *pol* gene) enclosed in a host-derived lipid membrane that contains the viral glycoproteins (products of the *env* gene; Fig. 1). Morphogenesis of any retrovirus therefore requires the morphopoietic function of

Department of Cell Biology and Virology, Heinrich Pette Institute of Experimental Virology and Immunology, Martinistr. 52, 20251 Hamburg, Germany

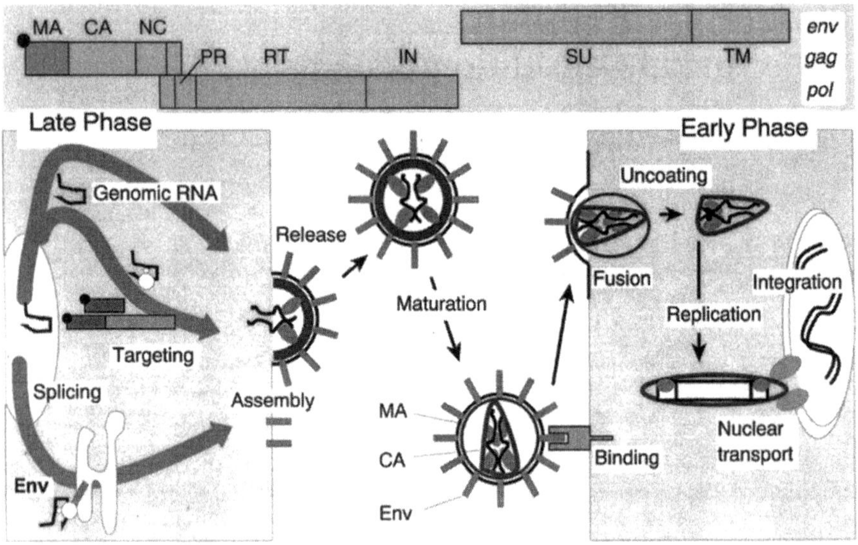

Fig. 1. Assembly and disassembly (late and early) phases of retroviral replication. In the *upper* part, the three essential genes for particle formation (*gag, pol, env*) are depicted. The domain organization is shown for the polyproteins of human immunodeficiency virus (HIV)-1, but is generally similar for all retroviruses. *MA*, matrix; *CA*, capsid; *NC*, nucleocapsid; *PR*, proteinase; *RT*, reverse transcriptase; *IN*, integrase; *SU*, surface glycoprotein; *TM*, transmembrane glycoprotein. The N-terminal MA domain is *highlighted*; the *circle* indicates modification by myristic acid. In the *lower* part, the late events leading to assembly, budding, and release of immature viruses are shown on the *left*, extracellular maturation to the infectious virion in the *middle*, and the early phase of infection (including virus uptake, uncoating, genome replication, and integration) on the *right*

the core proteins encoded by the *gag* gene at the 5′ end of the genome. Incorporation of the other components (Pol and Env proteins and genomic RNA) is not necessary for particle formation, but is necessary for infectivity and may also be important in targeting and assembly. Additional viral and cellular proteins have been identified in infectious retrovirus preparations, most notably in the case of human immunodeficiency virus type 1 (HIV-1). Some of these accessory proteins serve important functions (see Chap. 7), while others are fortuitous contaminants, due to the fact that retroviruses cannot be purified to homogeneity. Moreover, particle assembly in the infected cell can lead to nonspecific incorporation of cytoplasmic proteins or nucleic acids if these are available at the assembly site and are not actively excluded. Similarly, retroviral envelopes are generally thought to contain exclusively viral glycoproteins, but integral proteins of the plasma membrane are also incorporated in at least some cases (ARTHUR et al. 1992).

General aspects of retrovirus assembly and disassembly are outlined in Fig. 1. The structural components of the core and the replication enzymes are synthesized on cytosolic polysomes and are transported and assembled as two polyprotein precursors (Gag and Gag-Pol). Proteolytic cleavage of polyproteins into individual domains occurs predominantly in the nascent particle and is achieved by the viral proteinase (PR; see Chap. 4). Retroviruses are initially formed as immature, noninfectious particles containing a spherical electron-dense core underneath the viral envelope with an electron-lucent interior. Maturation of the virion depends on polyprotein cleavage and leads to conversion to the condensed core, which exhibits different shapes in individual retroviruses (see Chaps. 1 and 4). In mature virus, the N-terminal matrix (MA) domain of Gag is directly apposed to the lipid bilayer, as shown by lipid–protein cross-linking (PEPINSKY and VOGT 1979, 1984) and immunoelectron microscopy (GELDERBOLM et al. 1987). The core shell is formed by the capsid (CA) protein, while the nucleocapsid (NC) protein remains associated with the genomic RNA as ribonucleoprotein complex. The localization of other components within the mature virion is currently not known.

Synthesis of stable polyproteins enables the virus to target several particle constituents to the assembly site using a single targeting signal, but requires proteolytic separation of functional domains once confined within the virion. Since particles are assembled from polyproteins and not from mature core proteins, assembly domains do not necessarily adhere to the boundaries of final cleaved products, but may be present only in the polyprotein(s) (WILLS and CRAVEN 1991; KRÄUSSLICH et al. 1995). Proteolytic maturation should inactivate such assembly domains and prepare the virus for disassembly. Thus polyprotein targeting, together with regulated proteolysis, is an essential feature governing the functionally unidirectional flow from assembly to disassembly (Fig. 1). Individual virion components associate in the producer cell to form a stable, immature core which buds from the cell. Subsequent to proteolytic maturation, it develops into an extracellular metastable core which is protected and directed by the lipid envelope containing the glycoproteins. Upon fusion with the target cell membrane, this core structure is ready for replication of the viral genome, thereby initiating a new infectious cycle.

1.1 Budding Type Depends on Intracellular Targeting

Originally, retroviruses were classified according to their morphological appearance and budding phenotype as observed by electron microscopy (see Chap. 1). In the case of B-type and D-type oncoviruses, spherical cores are assembled in the cytoplasm and subsequently transported to the plasma membrane, where budding and maturation take place. The intracytoplasmic immature cores represent intermediates in the assembly pathway, termed A-type particles. A peculiar class of retroviruses are the intracisternal A-type particles (IAP) of rodent cells, a class of endogenous retroviruses which assemble and bud into cisternae of the endoplasmic reticulum (ER), where they remain as immature particles (KUFF and LUEDERS 1988). C-type oncoviruses and lentiviruses do not appear to assemble larger structures in the cytoplasm, and formation of their immature core occurs concomitantly with budding at the plasma membrane.

Differences in the morphogenesis pathway may be caused by different requirements for assembly or may reflect targeting of the individual polyprotein. The latter hypothesis is supported by the observation that a point mutation in the MA domain of Mason Pfizer monkey virus (MPMV) converted its budding phenotype from D type to C type without detrimental effect on particle yield (RHEE and HUNTER 1990b). Thus, except for polyprotein targeting, there may be no fundamental difference in the various pathways of morphogenesis. More recent experiments have shown that alteration of the MA domain of an IAP polyprotein can lead to particle assembly and release at the ER membrane, at the plasma membrane, or exclusively at the nuclear envelope (WELKER et al. 1996). Moreover, deletion of the presumed ER targeting region led to assembly of intranuclear particles which did not acquire a lipid envelope (WELKER et al. 1996). Similarly, overexpression of nonmyristoylated Gag proteins of the simian immunodeficiency virus (SIV) or HIV led to intracytoplasmic or intranuclear assembly of unenveloped immature shells, depending on the specific construct made (DELCHAMBRE et al. 1989; ROYER et al. 1991; SPEARMAN et al. 1994). These results indicate that immature cores can assemble in many regions within the cell and that the site of assembly is determined primarily by polyprotein targeting, leading to accumulation of assembly substrates in one specific area.

The assembly pathway can be viewed as a stepwise mechanism. Initially, polyproteins are transported to and concentrated at specific sites within the cell. Once they have reached the membrane, C-type and lentivirus Gag and Gag-Pol proteins are retained most likely by a combination of hydrophobic and electrostatic interactions, restricting their movement within the plane of the membrane. This restriction should cause an alignment of polyproteins and increases the probability of Gag–Gag interactions leading to assembly and release. In the case of B- and D-type viruses, assembly occurs in the cytoplasm without the alignment function of the membrane and is therefore predicted to require stronger interacting and aligning forces within the respective polyproteins or participation of cellular factors. In principle, however, any retrovirus core can assemble independent of membrane association, and even in a test tube (KLIKOVA et al. 1995; CAMPBELL and VOGT 1995), provided its local concentration is sufficiently high.

1.2 The Gag Polyprotein Determines the Assembly Site

As indicated above, the retroviral Gag protein is both necessary and sufficient for formation of a noninfectious particle. Gag proteins of various retroviruses have been expressed from viral and plasmid vectors in a large variety of host cells, leading to assembly and release of immature particles (reviewed in Chap. 8). No other virion component was required, and assembly of Gag particles in most instances faithfully reproduced morphogenesis of the respective virus. Thus the targeting signal on Gag appears to be dominant in determining the site and process of virion assembly. Moreover, there is no fundamental difference in various host cells, suggesting that cellular requirements for retrovirus assembly are generally conserved, although subtle differences depending on the host cell have been reported (see Sect. 3.3).

Targeting of Gag proteins is primarily determined by the N-terminal MA domain (see Sect. 2), which is also present on Gag-Pol. Thus Gag and Gag-Pol should be concentrated at the same site. Recent experiments have indicated that Gag-Pol proteins lacking their own targeting signal can be rescued into virus particles by interaction with Gag (PARK and MORROW 1992; SMITH et al. 1993). A similar observation has been made for targeting-defective Gag proteins or chimeric Gag proteins, which can be rescued by coexpression of wild-type Gag proteins (WELDON et al. 1990, JONES et al. 1990; WANG et al. 1994). This phenotype is dependent on the presence of intact interaction (assembly) domains in the CA-NC region of both proteins (JONES et al. 1990; SRINIVASAKUMAR et al. 1995; WANG et al. 1994). Thus viral polyproteins may form multimers even in those cases in which no detectable structure is observed prior to budding (see Chap. 3). Rescue of targeting-defective polyproteins has not been successful in all cases; examples are the exclusion of myristoylation-defective mutants of murine leukemia virus (MLV) and spleen necrosis virus (SNV) Gag proteins from particle incorporation (SCHULTZ and REIN 1989; WEAVER and PANGANIBAN 1990). Formation of mixed multimers may rescue targeting-defective proteins in some cases but is unlikely to be efficient, and incorporation of Gag-Pol by rescue only may not be sufficient to sustain productive retrovirus infection in vivo. Interestingly, recent experiments indicated that no Gag-Pol polyprotein is made in the case of foamy viruses (KONVALINKA et al. 1995b; LÖCHELT and FLÜGEL 1996; YU et al. 1996). In this case, incorporation of Pol has to be achieved by a different mechanism, since neither the presumed targeting domain in MA nor the interaction domain in the CA-NC region is present on the Pol polyprotein. This Pol protein may therefore contain its own targeting signal or a separate domain for interaction with Gag. Alternatively, foamy virus Pol may specifically bind to virion RNA to be rescued into the virus by RNA incorporation. In some retroviruses, additional proteins are incorporated into virus particles. In the case of HIV-1, they include the viral protein Vpr (LU et al. 1993) and the cellular protein cyclophilin A (FRANKE et al. 1994; THALI et al. 1994). These proteins are incorporated by interaction with Gag as well; Vpr binds to the C-terminal p6 domain and cyclophilin to the CA domain (LUBAN et al. 1993).

Specific incorporation of the two copies of genomic RNA is dependent on the RNA packaging signal and on the NC domain of Gag (see Chap. 6). RNA incorporation should therefore follow the targeting and accumulation of Gag. Besides genomic RNA, a specific tRNA is needed as a primer for replication and is specifically incorporated into the virus. The RT domain of Gag-Pol is required for specific tRNA incorporation, while little specificity is detected in the tRNA binding of purified RT (MAK et al. 1994). This may be yet another example of an essential assembly function present only in the viral polyprotein and destroyed once virus constituents are locked in the nascent virion. In summary, although quite different domains and mechanisms are involved in the incorporation of various particle components, they all interact directly or indirectly with the Gag polyprotein, and interaction with Gag determines their incorporation.

In principle, several possible routes can be envisioned for the transport of Gag. Following translation, the polyprotein may migrate through the cytoplasm by diffusion and be concentrated at membranes due to its inherent membrane-binding capacity. However, this model cannot explain why budding occurs almost exclusively at the plasma membrane and not at intracellular membranes, despite the fact that the ER membrane is most abundant inside the cell. In addition, it cannot explain the phenotype of B-type or D-type viruses which assemble in the cytoplasm and subsequently migrate to the plasma membrane. Since conversion of the budding type is easily achieved (see above), there should be no fundamental difference between various pathways. Alternatively, Gag may associate with a cellular transport receptor or pathway which directs it to a specific intracellular site. Transport may occur by diffusion or – more likely – may be associated with elements of the cytoskeleton or the vesicular transport route. In addition, Gag may be trapped in specific locations by interaction with cellular or viral partners (e.g., the viral glycoproteins) that restrict further migration. In addition to proteins, lipid transport and lipid composition of individual membranes or membrane segments may also play a role in Gag polyprotein sorting. Finally, post-translational modification of Gag appears to be involved in regulation of targeting. Although the mechanism of polyprotein targeting has not been worked out in detail, currently available evidence suggests that several of the described features play a role. In the following sections, we shall discuss specific aspects regarding targeting signals on viral polyproteins, the cellular pathways involved, and potential roles of glycoproteins and lipids.

2 Targeting Signals on Retroviral Polyproteins

Most available evidence points towards an essential role of the N-terminal MA region of Gag in polyprotein targeting. MA is localized at the inner layer of the viral envelope in close proximity to the lipid membrane. Various MA mutations in several retroviruses impaired or abolished targeting or led to an altered budding type. Moreover, fusion of MA or N-terminal segments thereof with targeting-defective Gag proteins or heterologous proteins localized chimeric proteins to the plasma membrane, at least in some cases (Bennett et al. 1993; Weldon et al. 1990; Jones et al. 1990; Wang et al. 1994; Zhou et al. 1994). These results suggest that dominant targeting signals reside in retroviral MA proteins which facilitate accumulation of Gag and Gag-Pol at specific intracellular locations. SIV MA by itself has been shown to be sufficient for assembly and release of particles (González et al. 1993), but it is not known whether MA particles were formed by the same mechanism as wild-type virus. Furthermore, particle formation by MA alone was not observed for other retroviral MA proteins (e.g., HOSHIKAWA et al. 1991).

The features of retroviral MA proteins that promote membrane association are only partly understood. Most, but not all, Gag proteins are cotranslationally linked to the 14-carbon fatty acid myristic acid, and several MA domains contain clusters of basic amino acids which may form ionic interactions with negatively charged phospholipid head groups in the membrane. Results from mutagenesis studies suggest, however, that additional elements of Gag are involved in polyprotein targeting, and these elements may reside within MA or in downstream domains of Gag. Release of extracellular particles has been reported for a deletion mutant of HIV-1 Gag lacking the entire MA domain but containing a myristoylation signal at its N terminus (LEE and LINIAL 1994). In this case, membrane association might be due exclusively to the hydrophobic nature of the N-terminal fatty acid. Alternatively, the truncated Gag polyprotein may provide sufficient targeting information to reach the plasma membrane; myristoylation would then be mainly required for retaining the protein at the membrane.

2.1 Matrix Domain

Matrix proteins are found in enveloped viruses almost invariably in the same location of the particle, closely apposed to the lipid membrane. In the extracellular virus, matrix proteins provide structural support to the virus membrane. During morphogenesis, they should link the envelope to the internal core. In addition, they are likely to play a role in acquisition of envelope proteins and in maintenance of their structure and conformation within the virus membrane, which is important for binding and fusion of the target cell membrane. Finally, as discussed above, matrix proteins can have essential functions in directing the transport of viral structural components.

MA domains of different retroviruses have been subjected to mutational analysis. In the absence of structural information, these mutants served primarily to underscore the important contribution of MA towards polyprotein targeting, virus assembly, and Env incorporation. Deletion analysis of MA proteins of MPMV and MLV mostly yielded unstable polyproteins, presumably because of protein misfolding and mislocalization (JORGENSEN et al. 1992; RHEE and HUNTER 1990a; GRANOWITZ and GOFF 1994). In the case of HIV-1, many deletion and substitution mutants resulted in stable polyproteins with various defects in morphogenesis (Table 1). These defects included complete loss of particle assembly, redirected particle assembly to intracellular compartiments, and loss of Env incorporation, respectively. Additional defects of HIV MA mutations were reported in the early phase of virus infection (YU et al. 1992a) and in nuclear transport of the preintegration complex (BUKRINSKY et al. 1993a; GALLAY et al. 1995a). More recently, the three-dimensional structure of HIV MA has been determined, and mutant phenotypes can now be viewed in the context of structural information.

Table 1. Phenotypic effects of human immunodeficiency virus (HIV)-1 matrix protein mutations (adapted from Massiah et al. 1994)

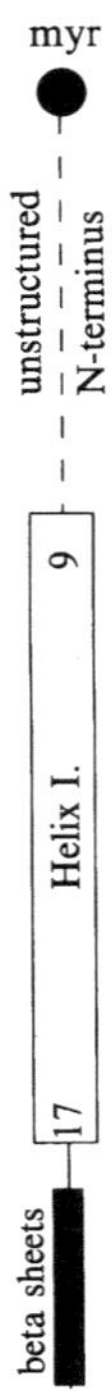

Mutation[a]	Infectivity[b]	Particle release[b]	ENV inc.[b]	Gag[b] stability	Localization by EM or IF[c]	Comments, structure[c]
M						
G2A[d–f]	–	–		Stable	No particle assembly; IF: diffuse cytoplasm staining	myr(–)
G2A (bac)[g]				Stable	Particles in nucleus and cytoplasm	myr(–)
A5D[e]	D	+	+	Stable	No particle assembly	myr(–)
S6I[e]	–	–		Stable	No particle assembly	myr(–)
V7R[e]	D	+	+	Stable	No particle assembly	myr(–)
Δ7–142[h]	–	+	–	Stable	Budding of immature particles at PM	
L8S, S9R[i]	–	+++	–			S9OH...E12NH
G10E[e]	+	+++	+			
Δ11–20[j]	–	++++	–	Reduced		
Δ11–20 (bac)[g]		++		Stable	PM, irregular morphology	
L13E[e]	–	+++	–			
R15, 20, 22, K18, 26–28, 30N[k]		+	+	Stable		
15 Ins CRHR[l]	+[m]	+++		Stable	IF: wild-type staining	
Δ16–18[i]	–	++++	–		Early budding at PM, few intracellular particles	
Δ16–99[n]	–	+	–	Stable	ER localization of immature particles	
Δ16–120[l]	+[m]	++	–	Reduced	IF for β-Gal: "Golgi", punctuate staining	
Δ16–132 (vacc)[o]		–		Stable	myr(+)/(–): immature particles in cytoplasm	
K18T,R20L[e]	+	+++	+			
R20L[e]	+	+++	+			Exposed, basic
Δ20–39[k]		+	+	Stable		
L21E[e]	+	++	+			HC: K27, Y29, H33

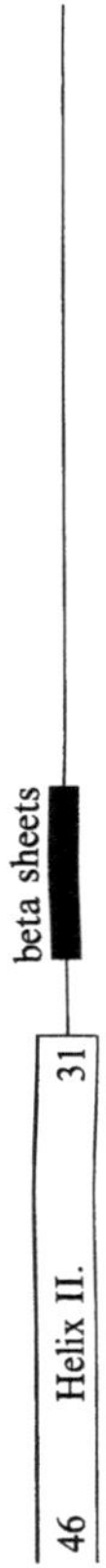

Δ21–31[j]	–	++++	–	Reduced		
Δ21–31(bac)[g]		+		Stable	Smooth-surfaced vesicles in cytoplasm	
Δ22–32[j]		+	+	Stable	IF: punctuate, "Golgi" staining	
K18, 26–28, 30, 32N, R20, 22G[j]		–		Stable		
R22T, K26–28, 32E[j]		–			IF: punctuate, "Golgi" staining	Exposed, basic
R22L[e]	+	++++	+			Exposed, basic
22 Ins LELE[p]	–	–	–	Reduced		
P23E[e]	+	+++	+			
G25D[e]	+	+++	+			
K26, 27T[e, f, q]	N	++	+			Exposed, basic/NLS
K26I, K27N[n]	D	++	+		Budding at PM only	Exposed, basic
K26D, K27E[n]	D	++	+		Budding at PM and immature particles in ER	Exposed, basic
K26–28E[j]		+++		Reduced		Exposed, basic
K27T[q]	N	+++	+			Exposed, basic
Δ27–30[i]	–	+++	–			
Y29F[f]	N	+++	+			Y29...T97
K30A[f]	N	+++	+			Exposed, basic
K30, 32T[e]	+	+	(+)			Exposed, basic
L31E[e]	–	+++	–			
K32A[e]	+	++	+			Exposed, basic
Δ32–40, L41I[j]	–	++++	–	Reduced		
Δ32–41 (bac)[g]		+		Stable	Smooth-surfaced vesicles in cytoplasm	
W36S, A37R[i]	–	+++	–			HC: A37, E40, E74, S77, L78, T81
A37E[e]	D	+	+	Unstable		HC: W36, L78, T81
40 Ins LEFQ (bac)[g]		–		Reduced	myr(+): cytoplasmic particles without MA, myr(–): intranuclear inclusions, no particles	

Table 1. (*Contd.*)

	Mutation[a]	Infectivity[b]	Particle release[b]	ENV inc.[b]	Gag[b] stability	Localization by EM or IF[c]	Comments, structure[c]
31	40 Ins LELE[p]	–	++	+			
	Δ41–43[i]	–	+++	–		Early budding at PM, few intracellular particles	
Helix II.	Δ42–56[j]	–	+		Reduced		
	Δ42–56 (bac)[g]		–		Reduced	Dense material at PM, amorphous intra-nuclear inclusions, transdominant: irregular particles	
	Δ42–99 (bac)[g]		–		Reduced		
46	R43A[e]	D	+++	+	Reduced		Exposed, basic
	L50D[e]	D	++	+	Stable		
	E52G[e]	+	+++	+			
	Δ55–57[i]	–	+				
53	G56E[e]	–	–		Unstable	No particle assembly	
	C57D[e]	–	–		Unstable	No particle assembly	HC: I82, L85, H89, Y86
	C57S[e]	–	–		Stable	No particle assembly	
	C57S (bac)[r]		–			No particles, dense material at PM	
	Δ57–67[j]	–	+		Reduced		
Helix III.	Δ57–67 (bac)[g]		–		Stable	Dense material at PM, amorphous intranuclear inclusions	
	57 Ins SGIP[l]	–	+		Stable	IF: "Golgi" staining with large granules	
	I60E[e]	–	–		Reduced	No particle assembly	HC: L41, I82, L64, L51
	Δ63–65[i]	–	+++	–			
65	63 Ins LELE[p]	D	++	+			
	L64A (bac)[r]		–			Dense material at PM, amorphous intranuclear inclusions	

	Δ68–77, L78I[j]	–	+		Reduced		
	Δ68–77 (bac)[g]		–		Stable	Dense material at PM, amorphous intranuclear inclusions	
Helix IV. 72	S72I[e]	+	++++	+			S72O...75NH MC
	Δ77–80[i]	–	+++	–			
	L78A (bac)[r]		+			Wild-type particle and dense material at PM	
	Δ79–90[j]	–	++++	–	Reduced		
	Δ79–90 (bac)[g]		+		Stable	PM: irregular particle morphology	
	N80G[e]	+	+++	+			
	L85R[e]	–	+		Unstable	Particles in intracellular vesicles	HC: V7, L8, W16, I34
	Y86G[e]	–	+		Stable	Mature + immature particles in intracellular vesicles	HC: C57, R58, L61, Q90, I104
	Y86S, C87R[i]	–	+++	–			
	C87D[e]	–	+		Unstable	Particles in intracellular vesicles	HC: I92, V94, I104
	C87S[e]	D	++	+	Stable	Particles in intracellular vesicles	
	C87S (bac)[r]		+			Wild-type particles and dense material at PM	
	V88E[e]	–	+		Reduced	Mature + immature particles in intracellular vesicles	HC: E12, W16, L8
	H89G[e]	–	+		Stable	Particles in intracellular vesicles	HC: C57; H89...O51MC, E12
92	D91–103[j]	–	++++	–	Reduced		
	D91–103 (bac)[g]		++		Stable	PM: irregular particle morphology	

Table 1. (*Contd.*)

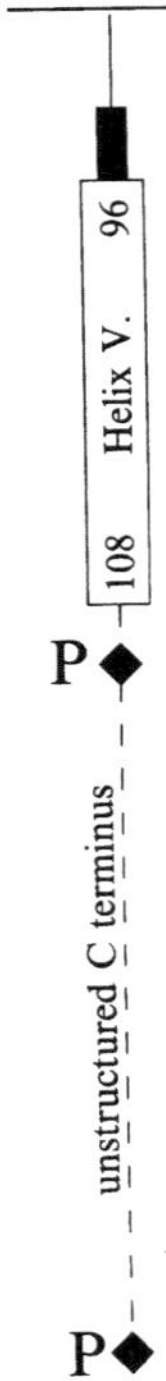

Mutation[a]	Infec-tivity[b]	Particle release[b]	ENV inc.[b]	Gag[b] stability	Localization by EM or IF[c]	Comments, structure[c]
D96L[e]	–	+		Unstable		D96OH...99NMC, 98NMC
K98G[e]	+	++++	+			Exposed, basic
D98–100[i]	–	+++	–			
D99–154 (vacc)[s]	–	–			Large cytoplasmic vacuoles	
A100E[e]	–	+		Unstable		HC: V84, C87, V94
100 Ins GIPA (bac)[g]		+		Stable	myr(+): wild type, myr(–): no particles	
100 Ins RIRA[l]	+[m]	++		Stable	IF: wild-type staining	
Δ105–114[j]	D	+++	+			
Δ105–114 (bac)[g]		++			Smooth-surfaced vesicles in cytoplasm	
E106V[e]	+	++++	+			
S111A[t]						Phosphorylation site
Δ112–114[i]	+	+++	+			
K113T, K114T[e]	+	+++	+			
Δ116–128[j]	D	+++	+	Stable		
Δ116–128 (bac)[g]		++		Stable	Wild type	
119 Ins ARAR[p]	D	+++	+			
A120E[e]	+	+++	+			
120 Ins DRRRS[l]	+[m]	++		Stable		
120 Ins GIPA (bac)[g]		+		Stable	myr(+): wild type myr(–): amorphous intranuclear inclusions	
D121L[e]	+	++	+			
N126R[i]	+	+++	+			
V128E[e]	+	++	+			
Q130G[e]	+	+++	+			
Y132F[f]	D	+++	+	Stable		Phosphorylation site

[a] Substitution mutations are described as follows: wild-type amino acid; amino acid number; mutant amino acid. Residues listed for deletions (Δ) are absent from mutant proteins. Insertions are described: amino acid number, "Ins," inserted amino acids. *Note*: different human immunodeficiency virus (HIV) strains were used

by various groups which may differ in amino acid sequence and/or function of accessory genes. "(bac)" and "(vacc)" denote baculovirus and vaccinia virus expression, respectively.

[b] Virus infectivity is denoted qualitatively as follows: +, comparable to wild-type; D, significantly delayed relative to wild type; –, significantly reduced or undetectable; N, normal in dividing cells, but significantly impaired in quiescent cells. Particle release: ++++, above wild-type level; +++, wild-type level; ++, 31%–75% of wild-type level; +, 5%–30% of wild-type level, –, no particle release; glycoprotein (Env) incorporation into particles is indicated qualitatively with + or –, (+) denotes reduced Env incorporation; Gag stability refers to the stability of the polyprotein and not to individual cleavage products.

[c] Intracellular localization of particles analyzed by electron microscopy (EM) or of Gag proteins analyzed by immunofluorescence (IF). PM, Plasma membrane; NLS, nuclear localization sequence. Structural details related to specific positions refer to the published nuclear magnetic resonance (NMR) structure model (Massiah et al. 1994) and to the model derived from X-ray structure analysis (W. Sundquist, personal communication). Potential hydrogen bonds are depicted as dashed lines, main chain positions are denoted MC, and hydrophobic contacts HC.

[d] Bryant and Ratner 1990; Göttlinger et al. 1989.

[e] Freed and Martin 1994; Freed et al. 1994, 1995.

[f] von Schwedler et al. 1994; Gallay et al. 1995a.

[g] Royer et al. 1991, 1992; Chazal et al. 1994, 1995.

[h] Lee and Linial 1994.

[i] Dorfman et al. 1994.

[j] Yu et al. 1992a, b; Yuan et al. 1993.

[k] Zhou et al. 1994.

[l] Wang and Barklis 1993; Wang et al. 1993, 1994.

[m] Mutant HIV particles pseudotyped with amphotropic murine leukemia virus (MLV) Env proteins.

[n] Fäcke et al. 1993; H.G. Kräusslich, unpublished observation.

[o] Spearman et al. 1994.

[p] Reicin et al. 1995.

[q] Bukrinsky et al. 1993; Heinzinger et al. 1994.

[r] Morikawa et al. 1995.

[s] Wagner et al. 1994.

[t] Burnette et al. 1993; Yu et al. 1995.

2.1.1 Structure–Function Relationship of Human Immunodeficiency Virus-1 Matrix Protein

The three-dimensional structure of the folded, mature HIV-1 MA protein obtained by recombinant expression in *Escherichia coli* has been determined in two laboratories by nuclear magnetic resonance (NMR) methods (MATTHEWS et al. 1994; MASSIAH et al. 1994). Except for minor differences, the two NMR structure models correlate well with each other and are also in good agreement with the recently determined structure model from X-ray crystallography of HIV and SIV MA proteins (HILL et al. 1996, RAO et al. 1995). The HIV-1 MA protein structure is composed of five α-helices, a short 3_{10} helical stretch, and a three-stranded mixed β-sheet (Fig. 2; MASSIAH et al. 1994). Helices I–III and the 3_{10} helix cluster around the central helix IV to form the compact globular core of the protein. This core is capped by the three-stranded β-sheet. The C-terminal helix V

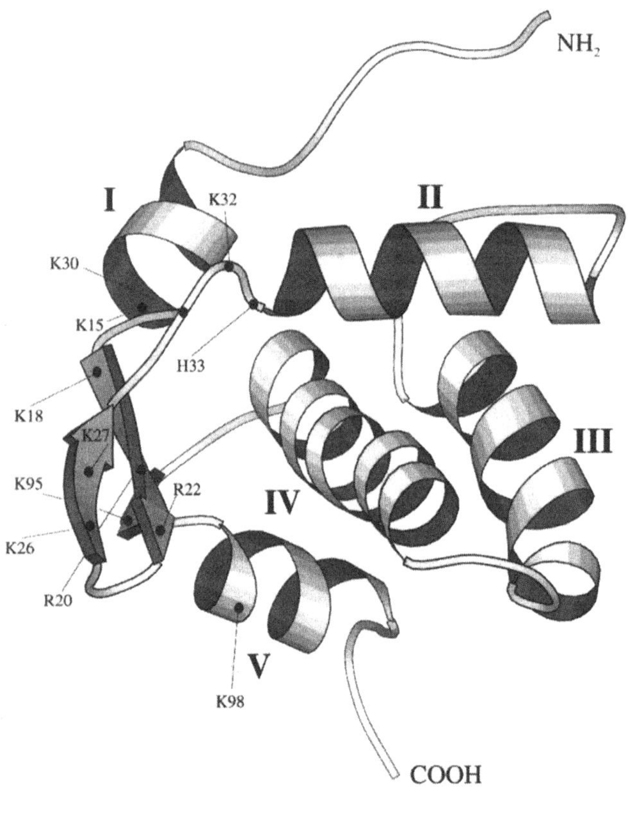

Fig. 2. Three-dimensional structure model of human immunodeficiency virus (HIV)-1 matrix (MA) protein based on the coordinates determined by nuclear magnetic resonance (NMR) methods (MASSIAH et al. 1994). The five helices in MA are indicated by *I-V*. Basic residues that are clustered around the cationic (*left*) surface of the protein are labeled. The C-terminal region of MA, which was disordered in NMR analysis, is not shown

projects away from the β-sheet to expose carboxyl-terminal residues, which have been suggested to be important in early infection (Yu et al. 1992a). Both the N-terminal segment of MA (which was not myristoylated in the proteins analyzed) and its C terminus appeared unstructured (Fig. 2). It should be pointed out, however, that these two regions may fold differently during morphogenesis when the C-terminal part of MA is still linked to the remainder of the polyprotein and its N terminus is myristoylated. Moreover, although the compact fold of mature MA suggests that it adopts an equivalent conformation when part of the Gag polyprotein, conformational differences even in the central folded part cannot be ruled out at present.

Genetic analysis of HIV-1 MA has been performed by many laboratories. Table 1 shows locations of MA mutations with regard to the three-dimensional structure and their phenotypes regarding virus infectivity, particle release, Env incorporation, and intracellular localization of Gag. In summary, several phenotypes can be attributed to specific regions or structural elements of MA, although no unequivocal picture has emerged to date. Mutations in the N-terminal codons of MA prevented myristoylation of the polyprotein and abolished particle formation (see Sect. 2.1.2). Mutations in helix V and in the C-terminal segment generally did not significantly affect intracellular localization of the polyprotein, particle release, or Env incorporation. Several of these mutations did, however, affect steps in early virus infection, suggesting that the C terminus of MA is important at a different stage in the viral life cycle (see Sect. 2.2). Deletion and insertion mutations throughout the globular domain of MA (helices I–IV) almost invariably reduced or abolished incorporation of Env proteins. The dispersion of Env(–) mutations throughout this region suggests that they act indirectly and alter local folding, thereby interfering with close packing of MA and Env proteins. Thus loss of Env incorporation in these mutants may not suffice to define an Env-interacting pocket in MA.

The most dramatic effects on virus assembly and release were reported for mutations affecting either the myristoylation signal or the central part of helix III (Table 1). Complete loss or severe reduction of particle assembly was found for substitution or deletion mutants in the region of amino acids 55–60. In several instances, these mutations led to unstable polyproteins. However, particle assembly was also lost in those cases in which mutant polyproteins were stable. No intracellular particles were observed by electron microscopy analysis (Freed et al. 1994), but dense material accumulated at the plasma membrane upon expression of several mutant polyproteins via recombinant baculovirus vectors (Chazal et al. 1995; Morikawa et al. 1995). This phenotype suggests that mutations in helix III do not primarily affect targeting but cause misfolding of MA, which interferes with ordered assembly. These mutations may either directly address sites of protein–protein interaction or have a more general effect on MA or Gag protein folding. Interestingly, peptides corresponding to helix III of HIV-1 MA (amino acids 47–59) have been shown to inhibit particle assembly (Niedrig et al. 1994), and peptides corresponding to C-terminally adjacent regions inhibited MA protein interaction in vitro (Morikawa et al. 1995).

Another interesting phenotype was observed for mutations in the central part of helix IV. Several mutations affecting amino acids between positions 85 and 96 of MA caused a significant reduction of extracellular release. Virus particles accumulated in intracellular membrane compartments, most likely derived from the Golgi apparatus. Since budding did occur and intravesicular particles appeared morphologically similar to wild-type virus, it is unlikely that these mutations led to misfolding with a subsequent assembly defect. Conceivably, helix IV may be part of a signal governing Gag protein transport specifically to the plasma membrane. Alternatively, this helix, which is part of the globular core of MA, may provide important structural support for a surface-exposed targeting signal. Interestingly, a large deletion within MA which removed helices II–IV plus the mixed β-sheet and linked helix I to helix V led to intracellular budding and accumulation of immature viral particles within the cisternae of the ER with little extracellular virus (FÄCKE et al. 1993; GALLINA et al. 1994). These particles appeared morphologically identical to immature HIV-1, indicating no major alterations in virus assembly. The observed phenotype suggests that deleterious effects of point mutations and small deletions in helix III were mainly caused by local constraints and were overcome by removal of additional parts of MA. An even larger deletion, constructed by WANG et al. (1993), removed helix V and part of the C-terminal segment of MA in addition. Interestingly, this mutation caused only a moderate reduction in particle release, but also redirected the polyprotein to intracellular membranes. Taken together, these results suggest that internal regions in MA play a role in directing Gag to the plasma membrane as opposed to intracellular membranes, while the N-terminal myristoylation signal, probably in conjunction with the cationic face of the three-stranded β-sheet, anchors it to the membrane (see Sect. 2.1.2). Additional regions of Gag may also be involved, modulating or enhancing this process. While projecting individual mutations onto the three-dimensional structure model of HIV-1 MA has helped to rationalize the effects of certain mutations, the precise contribution of individual structure and sequence elements to polyprotein targeting, membrane attachment, and particle assembly is still poorly understood.

2.1.2 Role of Post-translational Modifications and Basic Residues in Membrane Targeting

Most retroviral MA proteins are modified by addition of a myristic acid moiety (HENDERSON et al. 1983), which is thought to function as a membrane attachment site for MA and the MA domains of Gag and Gag-Pol, respectively. Myristoylation occurs at an invariable Gly residue which is N-terminal after removal of the initiator methionine. Substitutions of this Gly residue have been analyzed in a number of retroviruses, including HIV, MPMV, MLV, and SNV. Invariably, lack of myristoylation resulted in a defect in particle release. In the case of mutant MPMV, intracytoplasmic cores were assembled but not transported to the membrane and not released (RHEE and HUNTER 1987). In the case of C-type viruses and lentiviruses, particle assembly was blocked completely

(Rein et al. 1986; Weaver and Panganiban 1990; Bryant and Ratner 1990; Göttlinger et al. 1989; Pal et al. 1990). Subcellular fractionation studies of mutant HIV-transfected cells indicated that nonmyristoylated Gag proteins were more easily extracted from the insoluble fraction, which may indicate a weaker interaction with the cell membrane (Bryant and Ratner 1990; Spearman et al. 1994).

Several lines of evidence suggest that determinants other than myristic acid participate in plasma membrane attachment of retrovirus polyproteins. First, many viral and cellular proteins are modified by myristic acid but do not associate with the plasma membrane (Schultz et al. 1988). Second, Gag proteins of several retroviruses, including avian leukosis virus, foamy virus, and the lentiviruses equine infectious anemia virus and visna virus, are not myristoylated. However, except in the case of foamy viruses, these polyproteins are efficiently targeted to the plasma membrane, and there is no obvious difference regarding their morphogenesis pathway. Most importantly, however, studies with peptides have shown that the binding energy of a myristoyl group in a membrane bilayer is only 8 kcal/mol (Peitzsch and McLaughlin 1993). This corresponds to a K_d of 10^{-4} M, which appears hardly sufficient to anchor MA stably to the membrane (Zhou et al. 1994).

MA of HIV-1 contains a conspicuous stretch of basic amino acids in its N-terminal part (^{15}KWEKIRLRPGGKKQYK^{30}) which is largely conserved in HIV and SIV isolates. It has been suggested that this basic sequence, in conjunction with the myristoylated N terminus, forms a bipartite membrane-binding signal (Zhou et al. 1994). In this model, basic residues interact with anionic phosphatidylserine and phosphatidylinositol head groups concentrated on the inner leaflet of the lipid bilayer. The additional energy provided by the ionic interactions serves to stably attach the protein to the membrane. In support of this model, the N-terminal 31 amino acids of HIV-1 MA have been shown to correct the membrane-targeting defect of a cytoplasmically localized Src protein and to target the cytoplasmic protein dihydrofolate reductase to the membrane when fused to its N terminus (Zhou et al. 1994). Membrane transport of the chimeric proteins was dependent on myristoylation and was considerably reduced when eight basic residues were substituted by uncharged asparagine residues. Since interaction of the basic side chain of lysine with an acidic phospholipid contributes a binding energy of 1.4 kcal/mol (Kim et al. 1991), the total binding energy of eight basic residues – if all of them are participating in ionic interactions – would be 11 kcal/mol. This combined energy is slightly higher than that of the N-terminal myristic acid. Based on these results, it has been suggested that the two signals together are needed to anchor Gag to the membrane (Zhou et al. 1994)

An important role for surface-exposed cationic charges is supported by the three-dimensional structure model of HIV-1 MA (Massiah et al. 1994; Matthews et al. 1994). The basic stretch of amino acids forms part of a three-stranded β-sheet which protrudes from the protein and exposes basic residues Arg20, Arg22, Lys26, and Lys27 to the environment (Fig. 2). This cationic projection is surrounded by a collar of basic residues, including Lys18, Lys30, Lys32, His33, Lys95, and Lys98 (Fig. 2). The position of basic residues in HIV-1 MA is compatible with their contribution to the putative bipartite membrane-binding signal. However, the results of mutagenesis experiments targeting basic amino acids are still controversial. Transfection of an HIV-1 proviral construct with eight basic amino acids replaced by asparagine residues yielded 5% particle release compared with wild-type (ZHOU et al. 1994). A similar decrease in particle formation was observed for mutants replacing eight or five basic residues in MA, while replacement of three basic by acidic residues yielded wild-type particle release (YUAN et al. 1993). An internal deletion in MA removing amino acids 20–39 also yielded a considerable decrease in particle release (ZHOU et al. 1994). In contrast, YU et al (1992b) reported a two-to fivefold increase in particle-associated RT activity for a MA deletion mutant lacking amino acids 21–31 compared with wild-type virus. No obvious decrease in particle-associated antigen was observed for this mutant, which lacked the entire basic stretch. This result is of particular importance, since the same group subsequently reported a severe reduction in virus production for a mutant lacking amino acids 22–32 of MA with only one additional basic residue removed (YUAN et al. 1993). It appears questionable whether the dramatic difference is entirely due to a single basic charge, as these authors suggested. More likely, different effects on protein structure may occur in the two mutants, since residue 32 maps to the beginning of helix II (Fig. 2) and the deletion 22-32 probably affects folding of this helix. No electron microscopy analysis was reported for any of the described mutants, but immunofluorescence and subcellular fractionation experiments suggested association of Gag with intracellular membranes (YUAN et al. 1993). Other investigators analyzed more subtle mutations, deleting or substituting only one or several basic residues, and generally observed a mild decrease in particle release or no phenotype at all (Table 1).

Despite controversial results of mutational analyses, it appears likely that the lipid-binding capacity of MA ties the polyprotein to a membrane once it has reached it. It is more difficult to imagine that nonspecific membrane binding should be the only determinant specifying polyprotein transport to and virus release from the plasma membrane. Additional factors may be required to facilitate the first step of transport. The observed phenotypes for MA mutants might therefore represent defects in two separate, but consecutive steps in membrane targeting (see Sect. 2.1.3). Interestingly, replacing only two basic residues in HIV-1 MA (Lys26, Lys27) by acidic amino acids yielded infectious virus which was released at the plasma membrane to some extent, but showed significant budding at intracellular membranes with accumulation of immature particles within the ER (H.-G. Kräusslich, unpublished observation; Table 1). Accumulation of viral proteins at or particle formation within intracellular membrane compartments was also observed for other mutants in the mixed β-sheet (Table 1). Thus the cationic protrusion of MA may determine the site of virus assembly and release and may therefore be part of a true targeting signal and not only a lipid-binding site.

It should also be considered that both the hydrophobic myristate moiety and the basic surface charges are neutralized by their respective partners in the lipid bilayer once the protein has reached the membrane. However, retroviral Gag and Gag-Pol polyproteins are synthesized on cytosolic polysomes which are not associated with intracellular membranes. Thus hydrophobic areas and ionic charges, subsequently needed to attach the protein to the membrane, should initially be exposed at the surface of the protein. They may be initially obscured by conformational alterations of Gag itself or by interaction with viral or cellular proteins serving a chaperone function and may be revealed on arrival at the membrane. If conformational changes of Gag or release of interacting partners are needed when reaching the budding site, there should be a triggering event. Interestingly, phosphorylation of HIV-1 MA at serine residues and, to a small extent, at tyrosine residues has been observed, and there is some evidence that phosphorylation plays a role in transport. Phosphorylation of MA was also observed for Rous sarcoma virus (RSV; Leis et al. 1989), but no further analysis has been reported. The primary site of phosphorylation of HIV-1 MA in vitro and in cells expressing MA or complete Gag is Ser-111 (Burnette et al. 1993). Phosphorylation is catalyzed by protein kinase C (PKC) or a member of the PKC family (Burnette et al. 1993). Gag phosphorylation can be induced by stimulation with phorbol ester (Burnette et al. 1993), and phorbol ester treatment of cells expressing wild-type HIV-1 MA produced a rapid and significant shift of MA from the cytosol to the plasma membrane followed by its slow, quantitative dissociation back into the cytosol (Yu et al. 1995). This effect correlated with phosphorylation of Ser-111 and was dependent on myristoylation (Yu et al. 1995). These results indicate that a reversible switch involving myristoylation and phosphorylation of MA might be involved in membrane targeting of Gag, as described for other proteins (Blenis and Resh 1993; McLaughlin and Aderem 1995).

The regulation of MA membrane association is reminiscent of the situation for the N-myristoylated alanine-rich PKC (MARCKS). MARCKS is another protein containing the bipartite membrane-targeting signal with a myristoylated N terminus and a stretch of basic amino acids (Taniguchi and Manenti 1993). However, phosphorylation of MARCKS has the opposite effect and dissociates the protein from the membrane (Thelen et al. 1991). This phenotype would be predicted, assuming that surface-exposed cationic charges are important for membrane binding and phosphorylation reduces the net positive charge. In the case of HIV-1 MA, however, addition of an anionic phosphate group promotes membrane association. The structure model of HIV-1 MA cannot aid in providing rational explanations for this puzzling observation, since Ser-111 is part of the unstructured C terminus of MA, whose position in the intact Gag polyprotein is unknown. Considering the effects of MA phosphorylation, it should be pointed out that the viral phenotype of a substitution mutant replacing only Ser-111 has not been reported. However, deletion of residues 105–114 of MA yielded only a mild reduction in virus release and delayed kinetics of infectivity (Yu et al. 1992a), and deletion of three C-terminally adjacent residues which most likely destroys the consensus PKC recognition site had no significant phenotype (DORFMAN et al. 1994). While these results do not rule out a role played by the "myristoyl–phosphate switch", they strongly argue against it being an essential feature in targeting Gag to the plasma membrane.

2.1.3 Phenotypic Effects of Matrix Protein Mutations in Other Retroviruses

Extensive mutagenesis was initially reported for MA of the D-type retrovirus MPMV. This MA domain also belongs to the class of myristoylated proteins with a cluster of basic amino acids close to the N terminus. A systematic mutational analysis of N-terminal basic amino acids has not been reported in the case of MPMV, but results obtained for random point mutations introduced into MPMV MA yielded a number of interesting phenotypes. Three sequential events were distinguished in this analysis, and mutations affecting each step were described: (1) folding of Gag into a stable conformation for assembly ; (2) transport of the immature core to the plasma membrane and (3) association of the core with the membrane and extrusion of the membrane during budding (RHEE and HUNTER 1991). Point mutations affecting each step were scattered throughout MA, with no obvious clusters in any specific region. Since the three-dimensional structure of MPMV MA is not known, effects of specific mutants on protein folding cannot be predicted. It appears likely, however, that mutants of the first type, including deletion mutants (RHEE and HUNTER 1990a) and substitution mutants involving proline residues (RHEE and HUNTER 1991), caused misfolding of the protein and targeted it for degradation. Mutants of the second class led to assembly of intracytoplasmic cores, but abolished or slowed their transport to the plasma membrane (RHEE and HUNTER 1991). Interestingly, a myristoylation-defective mutant showed a similar phenotype (RHEE and HUNTER 1987), indicating that its defect was not only in membrane association, but also in transport. The third class of MA mutants showed normal assembly and transport, but core particles remained at the interior face of the membrane and did not initiate budding (RHEE and HUNTER 1991). These results support the concept of sequential targeting, at least in the case of D-type viruses.

In the analysis of MPMV MA mutants, a point mutation was identified which altered the normal D-type phenotype to that of a C-type retrovirus (RHEE and HUNTER 1990b). No intracytoplasmic immature cores were observed, and particle assembly occurred concomitantly with budding at the plasma membrane. This mutation (Arg-57Trp) did involve a basic residue and reduced the net positive charge of the MA domain. It appears unlikely, however, that the phenotype is mainly due to an alteration in charge, since replacement of the nearby residue Arg-55 by cysteine resulted in wild-type morphogenesis and release (RHEE and HUNTER 1991). Based on their results, RHEE and HUNTER (1990b) suggested that a dominant targeting signal within MA specifies intracytoplasmic assembly of the wild-type polyprotein. Deletion of this signal, as in the case of the Arg-57Trp mutant, reveals the plasma membrane-targeting information, which is also contained within the wild-type protein but is normally overridden by the dominant signal. In line with this hypothesis, it might be suggested that the same plasma membrane-targeting signal is also revealed in the case of the wild-type polyprotein, but only after assembly of the immature core is completed. However, formal proof for this attractive hypothesis requires that individual signals and interacting partners are experimentally defined.

Analysis of the membrane-targeting signal has also been attempted for RSV Gag. In this case, MA has neither a myristoylated N terminus nor a basic stretch of amino acids close to the N terminus. Addition of a myristoylation signal to RSV Gag resulted in enhanced virus release from mammalian cells, but did not appear important in avian cells (Wills et al. 1989). Deletions in the N-terminal part of RSV MA were shown to abolish particle formation. These mutants could be rescued by addition of a heterologous plasma membrane-targeting signal, indicating that their effect was indeed on protein transport (Wills et al. 1991). However, these results cannot distinguish between direct effects on the targeting signal and effects on MA folding, which may destroy a remote targeting signal. The N-terminal segment of RSV MA was reported to rescue a targeting-defective Src protein when grafted onto its N terminus (Zhou et al. 1994), but a considerably larger part of MA (approximately 90 amino acids) was needed than in the case of HIV-1. In summary, the results of mutagenesis studies on a number of retroviral MA domains suggest that targeting signals specifying accumulation of polyproteins at the assembly site reside in MA. Clearly, these targeting signals differ in some aspects, but some of the underlying features and mechanisms may also be conserved throughout the family of retroviruses.

2.2 Alternative Targeting of Gag Polyproteins or Products

Besides localization to the assembly site, several retroviral Gag proteins are also transported to alternative sites in the virus-producing or in the newly infected cell. Alternative targeting signals specifying these transport routes have been detected within MA and in downstream regions of Gag. It is likely that the alternative routes are normally overruled by the dominant signal during virion morphogenesis. One example for dominant and recessive targeting is the stepwise transport of D-type polyproteins to an intracytoplasmic assembly site and subsequently to the cell membrane (see Sect. 2.1.3). Another example is the nuclear localization of several retroviral Gag proteins, including HIV. In the case of foamy virus, almost the entire pool of cell-associated Gag is found in the nucleus at certain times after infection (Schliephake and Rethwilm 1994). Nuclear localization in this case appears to be mediated by basic amino acids in the NC domain. Earlier or later in the infection process, foamy virus Gag is observed predominantly in the cytoplasm, and neither the importance and functional role of nuclear transport nor the mechanism of targeting regulation is currently known. Nuclear localization of Gag and CA was also observed for MLV (Nash et al. 1993). In this case, nuclear transport occurs rapidly after Gag synthesis and involves a subset of Gag. The MLV nuclear targeting signal has been suggested to reside in the CA domain (NASH et al. 1993). A secondary signal targeting Gag to the nucleus was found for the endogenous retrovirus IAP. In this case, nuclear targeting and nuclear assembly of unenveloped immature cores was revealed when the dominant ER targeting signal was removed (WELKER et al. 1996). Thus nuclear targeting of Gag appears to be a common feature that occurs in a variety of different retroviruses and is likely to be functionally important.

Many studies have addressed the nuclear transport of Gag and in particular of MA in the case of HIV. Nuclear HIV MA is believed to play a role in early infection of nondividing cells. Most retroviruses require breakdown of the nuclear envelope to transport their preintegration complex to the nucleus following replication of the genome. In contrast, HIV and other lentiviruses can infect nondividing cells. This propensity has been suggested to depend on karyophilic properties of MA which is associated with the preintegration complex (BUKRINSKY et al. 1992, 1993a;b). A nuclear localization signal in MA has been mapped to the same basic stretch of amino acids considered to be important in membrane attachment (BUKRINSKY et al. 1993a). More recently, GALLAY et al. (1995a) reported that the karyophilic potential of MA is regulated by phosphorylation of the C-terminal tyrosine of MA. MA containing phosphotyrosine localized to the nucleus of the newly infected cells, while nonphosphorylated MA remained attached to the membrane. The same authors also suggested that MA binds to integrase via its free phorphorylated C terminus, thereby mediating MA association with the HIV nucleoprotein complex (GALLAY et al. 1995b). These results indicate that the karyophilic potential of MA in the context of the preintegration complex is only revealed upon phosphorylation of Tyr-132.

Several studies have analyzed mutations within the basic region of MA and showed reduced but detectable viral infectivity, even in nondividing cells (FREED and MARTIN 1994; see Table 1). An additional contribution by the viral Vpr protein was suggested in later studies (HEINZINGER et al. 1994). However, even a mutated HIV-1, lacking both the putative nuclear targeting signal and *vpr*, infected terminally differentiated macrophages, albeit at a reduced level (FREED et al. 1995). Furthermore, substitution of the MLV MA domain by that of HIV-1 did not confer the potential to infect nondividing cells to the resulting virus (DEMINIE and EMERMAN 1994), although both the basic domain and the C-terminal tyrosine were transferred onto the chimeric Gag polyprotein. It should be kept in mind that in the case of MLV the CA domain has been suggested to contain a nuclear targeting signal (NASH et al. 1993), and indeed CA and not MA was found to be associated with the high molecular weight MLV preintegration complex (BOWERMAN et al. 1989). Taken together, neither the putative karyophilic potential of MA nor the phosphoacceptor tyrosine at its C terminus alone may be sufficient to facilitate nuclear transport of the preintegration complex in the absence of mitosis. Moreover, myristoylated MA proteins with similar basic stretches are found in many other Gag polyproteins, and the C-terminal amino acid of MA is tyrosine in most retroviruses. HIV may conceivably be distinguished by the incorporation of a kinase activity capable of phosphorylating the cleaved MA protein (GALLAY et al. 1995a). In this case, it would be predicted that packaging of this as yet unknown kinase into another retrovirus may reveal similar properties upon infection of nondividing cells.

3 Intracellular Transport Pathway

Given the complexity of the intracellular environment, it appears unlikely that Gag proteins find their way to the plasma membrane by diffusion only. More likely, viral polyproteins make use of cellular transport routes delivering cargo to the cytoplasmic face of the plasma membrane. In contrast to vesicular transport of viral and cellular glycoproteins, little information is available regarding the routes for cytoplasmic proteins. Obvious candidate structures to mediate polyprotein transport are the cytoskeleton and the vesicular transport pathway. Evidence has been presented supporting involvement of either one or both of these subcellular structures in Gag transport, but detailed pathways have not been elucidated.

3.1 Vesicular Transport Pathways and Potential Contribution of Glycoprotein and Lipid Sorting

Expression of Gag alone leads to release of virus-like particles. Although additional viral proteins are nonessential for particle formation, they may contribute to it. Retroviruses are exceptional in that they do not require viral glycoproteins for particle formation. In most other enveloped viruses, budding is dependent on envelope proteins, and it is likely that interactions between matrix and envelope proteins serve to organize the internal and external structure of the virus (e.g., GAROFF et al. 1994). Envelope glycoproteins also play a role in retrovirus assembly. Many mutations in HIV-1 MA abolish glycoprotein incorporation (see Table 1). This phenotype can be reverted, at least in some cases, if the long intracellular tail of the transmembrane glycoprotein is truncated (FREED and MARTIN 1995; MAMMANO et al. 1995). Because of the close spatial proximity of the MA layer and the Env-containing virus membrane (GEBHARDT et al. 1984), small structural alterations may obstruct the dense packing of MA and glycoprotein tails. Further support for this hypothesis comes from the observation that MA mutations can suppress PR-mediated cleavage of the cytoplasmic tail of the MPMV transmembrane glycoprotein (BRODY et al. 1992). This cleavage is important for virus infectivity, because it removes an Env incorporation signal and activates fusion activity (BRODY et al. 1994).

Interestingly, HIV-1 envelope proteins can alter the site of virus budding if they are present. In the absence of viral glycoproteins, Gag particles are released both from the apical and basolateral sides of polarized cells (OWENS et al. 1991).

HIV glycoproteins are sorted only to the basolateral membrane of these cells (OWENS and COMPANS 1989), and in their presence particle budding is restricted to the basolateral side as well (OWENS et al. 1991). MA mutants that fail to incorporate envelope glycoproteins also fail to restrict budding to only one side of the cell (LODGE et al. 1994), which indicates that the restriction is specific. Determination of the budding site by Env could be caused by an interaction of Gag and Env before they reach the assembly site. Since protein sorting occurs in the trans-Golgi network (TGN), such interaction should occur before or during transit through the TGN. In the absence of Env, Gag proceeds to the plasma membrane without any sorting. Alternatively, Gag polyproteins are transported to the plasma membrane independent of Env, but are retained and concentrated only at the basolateral side. In this model, glycoproteins are restricted to the basolateral surface because of their transmembrane domains, while membrane-associated Gag, which does not contain a transmembrane domain, can traverse the tight junction and is retained by Env interaction. Similar to the first hypothesis, an interaction between Gag and Env is required, but not necessarily before reaching the membrane. In non-polarized cells, accumulation of Env has been observed in regions of the plasma membrane where HIV particle budding occurs (BUGELSKI et al. 1995). This observation indicates that Gag and Env interact directly or are efficiently sorted to the same region of the membrane in non-polarized cells as well. Recently, it has been suggested that Gag–Env interaction may also play a role in foamy virus morphogenesis. Foamy virus shows little budding at the plasma membrane and accumulates in intracellular vesicles. Intracellular budding may be determined by a putative ER retrieval signal in the transmembrane part of the foamy virus–Env protein complex (GOEPFERT et al. 1995). Mutational analysis of the presumed retrieval signal will be necessary to determine whether this attractive hypothesis is correct.

HIV-1 Gag release is not blocked by treatment with brefeldin A (PAL et al. 1991), which inhibits vesicular transport through the Golgi compartment. However, no quantitative analysis of a potential brefeldin A effect and no analysis in polarized cells has been reported. Brefeldin-insensitive budding suggests that Gag and Env may only interact after glycoproteins have traveled through brefeldin A-sensitive compartments. Alternatively, weak interaction between Gag and Env may allow sufficient Gag proteins to escape from the brefeldin block. Electron microscopy of brefeldin A-treated MLV infected cells indicated that virus budding was partially redirected from the plasma membrane to an intracellular budding compartment which corresponds to the collapsed ER–Golgi complex after drug treatment (ULMER and PALADE 1991). Redistribution and "vesicular staining" of Gag proteins was also reported for MLV-infected cells treated with monensin, which also inhibits vesicular transport (HANSEN et al. 1990). Vesicle-associated transport of Gag could explain the contribution of Env proteins and the observation that many MA mutations caused budding into intracellular membrane compartments, a phenotype also observed for wild-type virus in certain host cells (e.g., HANSEN et al. 1993). However, no increase in intracellular virus was observed in monensin-treated cells in comparison to untreated cells (Hansen et al. 1993). Moreover, no vesicle-associated transport was detected in B- and D-type virus-infected cells, where assembly of immature capsids occurs in the cytoplasm. No effect of monensin on particle release, but loss of Env transport to the plasma membrane was found in the case of MPMV (Chatterjee et al. 1982). Taken together, there is evidence suggesting direct interaction of Gag and Env proteins, at least in the case of HIV, and some indications point towards a role for transport vesicles in Gag transport. The limited effects of monensin and brefeldin A, on the other hand, argue against an essential role of vesicular transport. Moreover, particle release occurs with equal efficiency in the absence of Env, and it is unclear whether there is any fundamental difference in transport and particle formation depending on the presence of Env.

Besides viral glycoproteins, other viral proteins also play a role in HIV morphogenesis (see Chap. 7). Most notably, expression of the Vpu protein significantly enhances release of extracellular particles. Vpu-deletion mutants show reduced particle production with accumulation of particles in intracellular membrane compartments (Klimkait et al. 1990; Terwilliger et al. 1989). Interestingly, Vpu also enhances the release of other retroviruses (Göttlinger et al. 1993). HIV-2, which does not have a Vpu, appears to provide a Vpu-like function by its envelope protein (Bour et al. 1996). Recently, it has been found that Vpu can function as an ion channel, similar to influenza protein M2, which shares some homology with it (Schubert et al. 1996b). The release function of Vpu appears to be associated with its ion channel activity (Schubert et al. 1996a), but the underlying mechanism is as yet unknown.

In contrast to analysis of viral proteins and nucleic acids, little attention has been given to the lipid composition of retrovirus membranes, which represent the third major constituent of the infectious virion. Detailed analysis of lipid composition and fluidity is only available for the HIV membrane. Initially, it was shown to be a rigid and highly ordered membrane with a lipid profile resembling that of other RNA viruses and erythrocytes (Aloia et al. 1988). Comparison of the lipid envelope of the virus with that of the host cell plasma membrane indicated, however, that significant differences exist in the cholesterol to phospholipid ratio and in the sphingomyelin content (Aloia et al. 1993). The virus membrane contains considerably more cholesterol and sphingomyelin and less phosphatidylcholine and phosphatidylinositol compared to the plasma membrane of the infected cell it emerges from. Differences in membrane composition have also been observed for other retrovirus membranes (Pessin and Glaser 1983). The concentrations of cholestrol and sphingomyelin play important roles in fusion of enveloped viruses (Nieva et al. 1994), and the relative enrichment of these lipids in the HIV membrane may serve a similar function. A difference in lipid composition of viral and plasma membranes can be explained by viral proteins selectively binding and recruiting lipids to the envelope, thereby creating a distinct lipid environment. Alternatively, Gag and Env may associate with preexisting lipid patches of defined composition and fluidity. These patches may reside at the plasma membrane or be transported through the vesicular pathway. If binding to preexisting patches occurs prior to transport to the plasma membrane, it is conceivable that lipids are important in intracellular sorting of retroviral proteins. Specific association of Env with lipid patches and concomitant or subsequent binding of Gag would ensure that all necessary components are incorporated into the virion. This might be particularly important if the lipid composition of the virion does indeed have a function in the early entry process.

Considerable evidence has been obtained suggesting that cellular membranes do not constitute homogeneous lipid bilayers, but contain subdomains of defined lipid composition which restrict diffusion of lipids and of certain membrane proteins (SIMONS and VAN MEER 1988). In polarized cells, lipid sorting, similar to protein sorting, occurs in the TGN, yielding a different lipid composition in the apical and basolateral plasma membranes (SIMONS and VAN MEER 1988). It has been hypothesized that microdomains exist in the membrane of transport vesicles which are sorted with their associated cargo rather than as individual proteins and lipids. Functional barriers separate these microdomains and restrict lateral movement and mixing (reviewed in SIMONS and VAN MEER 1988). It would be of interest to determine whether retroviral proteins can indeed associate with such microdomains, leading to the distinct lipid composition of the viral envelope.

3.2 Evidence for Involvement of the Cytoskeleton

Several lines of evidence support a functional role of the cytoskeleton in retrovirus morphogenesis. MLV Gag polyproteins were found to be associated with a detergent-insoluble fraction of the cell and were extracted by ionic detergents or high salt (EDBAUER and NASO 1983). Colocalization of actin and MLV structural proteins was observed following microfilament disruption by cytochalasin D, which reduced particle release by 70%–80% (LUFTIG and LUPO 1994). MLV-infected cells also exhibited considerable changes in microfilament distribution and loss of stress fibers (LUFTIG and LUPO 1994). Depolymerization of microtubules by colchicine or nocodazole, on the other hand, reduced MLV release by 30%–40%, with accumulation of intracellular virus proteins and particle formation in cytoplasmic vesicles (SATAKE and LUFTIG 1982). Moreover, intracytoplasmic A-type particles, the precursors of B- and D-type viruses, were found in close association with microtubule-organizing centers (HEINE et al. 1985). These particles were redistributed into arrays of tubulin upon microtubule depolymerization, which also reduced the yield of extracellular particles (HEINE et al. 1985). Since microtubules are involved in intracellular movement of organelles, these authors suggested that retroviral polyprotein transport involves microtubules as well. However, the described experiments cannot address the question of whether drug treatment influenced polyprotein transport directly. Indirect effects could be due to association of retroviral proteins with a cellular substructure or to defects in transport of cellular factors important for virus morphogenesis.

The cytoskeleton has also been suggested to be involved in HIV morphogenesis. HIV release can be strikingly polar under certain conditions, which may be due to changes in the cytoskeleton. Adding HIV-infected T cell lines to epithelial cells causes rapid T cell polarization with altered cell morphology. Although normally released from many areas of the plasma membrane, HIV was subsequently secreted in a unidirectional manner at the site of cellular adherence (Bourinbaiar and Phillips 1991; Phillips and Bourinbaiar 1992; Phillips and Tan 1992). Colchicine treatment of infected T cells led to the same alteration in morphology and also induced polarized unidirectional release of HIV (Pearce-Pratt et al. 1994). No effect on the total amount of extracellular HIV was observed. Initial effects of microtubule depolymerization included redistribution of F actin from the cytoplasm to focal localizations within newly formed pseudopods and microvilli at the plasma membrane (Pearce-Pratt et al. 1994). Subsequently, virus release occurred from these pseudopods. Morphological changes were similar to those normally observed upon lymphocyte activation or on interaction of cytotoxic T cells with their target. The finding that F actin distribution was also affected by colchicine treatment may be explained by a role of microtubules in maintaining the cytoplasmic microfilament network. The phenotypic alterations in virus release could therefore be due to disruption of either one or both of these cytoskeletal filaments.

The mechanism restricting virus release is currently not understood. It may involve transport of assembly substrates to certain membrane areas or local restrictions at the budding site. Membrane-proximal elements of the cytoskeleton are likely to be involved in defining the assembly site and in the budding process. The plasma membrane is supported by and anchored to a tightly structured membrane skeleton (Pumpin and Bloch 1993). This structure, which has been analyzed primarily in erythrocytes but is also found in other cells, presents a barrier to virus budding. The membrane skeleton and its association with the plasma membrane may also play a role in inhibition of retrovirus formation by interferon-α. In addition to effects on various other steps in retroviral replication, interferon-α specifically blocks HIV-1 production at a post-translational stage (Poli et al. 1989; Fernie et al. 1991; Gendelman et al. 1990; Smith et al. 1991). Inhibition of particle formation was also observed in cells expressing only Gag (Sen and Pinter 1983; K.Mergener and H.G.Kräusslich, unpublished observation) and in the case of several other retroviruses (Bilello et al. 1982; Oka et al. 1990), sometimes leading to accumulation of intracellular virions (Aboud et al. 1982; Canivet et al. 1983). Interestingly, interferon treatment of uninfected cells increases the abundance of submembraneous microfilaments (Wang et al. 1981) and the rigidity of the plasma membrane bilayer (Pfeffer et al. 1981), presumably because of a change in cell surface lipid–protein interactions. The decreased membrane fluidity causes inhibition of the redistribution of surface immunoglobulins (capping) after treatment with anti-immunoglobulin (Pfeffer and Landsberger 1990). Cap formation by other cell surface molecules was also inhibited by interferon-α (Pfeffer et al. 1980). In addition, interferon treatment induced focal redistribution of the membrane skeleton protein spectrin in lymphoid cells (Evans et al. 1993). Increased membrane rigidity can obstruct retroviral budding, and decreased lateral movement of membrane proteins or membrane-associated proteins can reduce the probability of Gag interactions. Effects on plasma membrane–cytoskeleton interactions may therefore be the cause of the interferon-α-mediated block of retrovirus assembly (Pitha et al. 1981). A similar interferon block may also occur as an early event preventing virus uptake by altering the target cell membrane fluidity (Pitha et al. 1981; Vieillard et al. 1994).

One way to overcome the barrier of the membrane skeleton is the proteolytic cleavage of components of this structure. At least in the case of HIV-1, there is evidence for PR-mediated cleavage of cytoskeletal proteins. Substrates of HIV PR in vitro include vimentin, desmin, and other cellular proteins (Shoeman et al. 1990, 1991). Microinjection of HIV-1 PR into cultured fibroblasts led to a profound alteration of the cytoskeleton and caused cytotoxic effects (Höner et al. 1991). Cleavage of vimentin has been confirmed in HIV-1-infected cells (Konvalinka et al. 1995a), and it is likely that other cytoskeletal proteins are also cleaved. PR-dependent alterations of the cytoskeleton are not restricted to HIV: redistribution of the microfilament network is only observed in wild-type MLV-infected cells, but not in the case of an inactive PR mutant (Luftig and Lupo 1994). Mutational inactivation of HIV-1 PR (Peng et al. 1989) or treatment with specific PR inhibitors (Schätzl et al. 1991; Kaplan et al. 1993) leads to release of aberrantly shaped particles. These particles do not exhibit the spherical shell of immature virus, but incomplete closure with "tear-drop" morphology (Peng et al. 1989). Moreover, inactivation of HIV-1 PR reduced the efficiency of particle release (Kaplan et al. 1994). Aberrantly shaped particles were also observed upon mutational inactivation of foamy virus PR (Konvalinka et al. 1995b). Based on these results, it is suggested that local cleavage of membrane skeleton components by PR relaxes the rigidity of submembrane structures, allowing rapid and complete formation of the virus bud. Budding through the membrane skeleton is not unique to retroviruses, and incorporation of actin-binding proteins of the ezrin-radixin-moesin family, which are part of the submembrane structure, has been found in rhabdovirus and other enveloped viruses (Sagara et al. 1995). Actin itself was detected in retrovirus particles (Wang et al. 1976; Damsky et al. 1977), and proteolytic fragments of actin-binding proteins were reported to be observed in purified HIV preparations (Luftig and Lupo 1994). It is not clear, however, whether incorporation and proteolytic cleavage of these proteins is an important factor in retrovirus release.

Currently available evidence does not provide a clear picture regarding interactions of viral proteins with the cytoskeleton and the role of the cytoskeleton in retrovirus morphogenesis. Several results suggest its involvement at various steps from protein and RNA targeting to assembly and site-specific release of the virus from the cell surface. More detailed determination of the interactions of retroviral components with this important cellular substructure will certainly improve our understanding of retrovirus morphogenesis and contribute to the analysis of intracellular transport pathways.

3.3 Role of Proteolysis

Retroviral PR are made as monomeric subunits on the Gag-Pol polyprotein and require dimerization to generate an active enzyme. Concentration-dependent dimerization is one way to maintain stable polyproteins in the cytoplasm and achieve proteolysis by increasing PR concentration on virus assembly. Accordingly, expression of a PR dimer as part of retroviral polyproteins abolished particle formation because of premature cleavage and subsequent separation of individual domains from the targeting signal (KRÄUSSLICH 1991; BURSTEIN et al. 1991). Addition of a specific inhibitor of HIV PR reverted this phenotype and restored particle production (KRÄUSSLICH 1992). A similar loss of particle production was observed when HIV Gag and Pol proteins were artificially placed in the same reading frame, causing functional overexpression of PR (MERGENER et al. 1992; KARACOSTAS et al. 1993). Moreover, significant polyprotein processing occurred in the cytoplasm of acutely HIV-infected cells, and resulting cleavage products failed to be incorporated into virus particles (KAPLAN and SWANSTROM 1991).

Several lines of evidence suggest that activation of proteolysis is not only dependent on polyprotein concentration, but also linked to intracellular transport. First, intracytoplasmic A-type particles consist of uncleaved polyproteins, while budding of the resulting B- or D-type viruses triggers proteolysis. No change in local concentration of viral proteins occurs in this process. Second, IAP budded into the ER remain immature and their Gag polyproteins are not cleaved by PR. If IAP polyproteins are targeted to the plasma membrane by addition of a heterologous signal, specific PR activity of the viral polyprotein is triggered and extracellular particles contain cleaved products (WELKER et al. 1996). Whether transport-dependent activation of PR is caused by specific post-translational modifications of the enzyme or the substrate, by the local environment at the budding site, or by specific conditions in the extracellular environment is not yet known.

3.4 Host Range Effects in Intracellular Targeting

In general, retrovirus morphogenesis has been shown to be fairly independent of the host cell, and release of virus-like particles is observed following expression of various retroviral Gag proteins in many eukaryotic cells (see Chap. 8). Even baculovirus-mediated expression of Gag in insect cells yielded efficient release of particles indistinguishable from immature virus (GHEYSEN et al. 1989). In contrast, expression in yeast cells leads to myristoylation of Gag without assembly or release of virus particles (BATHURST et al. 1989; HAYAKAWA et al. 1992; JACOBS et al. 1989; VLASUK et al. 1989). Bacterial expression of viral major capsid proteins has been shown to produce virus-like capsids, e.g., in the case of hepatitis B virus (see Chap. 10). Although formation of some viral cores has been observed upon bacterial expression of MPMV Gag (Klikova et al. 1995), retroviral cores in general do not appear to assemble efficiently in bacterial cells.

Although particle formation is observed in most cells, the specific host cell may modulate morphogenesis and play a role in determining the site of assembly and release. A well-documented example is infection of T cells and macrophages by HIV-1: HIV buds from the plasma membrane of infected T cells with little, if any, intracellular accumulation of virus particles. A similar phenotype was observed following proviral transfection or recombinant virus-mediated infection of various cell lines. HIV infection of monocytes or macrophages, on the other hand, leads to significant budding of virus at intracellular membranes and accumulation of mature infectious virus within intracellular compartments, most likely derived from the Golgi apparatus (Orenstein 1992; Orenstein et al. 1988). A similar phenotype was observed for the monocyte-derived cell line U937 (Pautrat et al. 1990) and the T lymphocytic line JM (Grief et al. 1991). Most likely, virus is released from these cells by fusion of virus-containing vesicles with the plasma membrane. Intracellular budding of HIV-1 particles in a promonocytic line was enhanced by treatment with interferon-γ (Biswas et al. 1992), but it is not known whether interferon-γ also plays a role in primary macrophages or monocytes.

As already discussed, the budding phenotype of MPMV is altered from D type to C type by a point mutation in MA (Rhee and Hunter 1990b). A similar switch in budding type was also observed when the D-type virus squirrel monkey retrovirus (SMRV) was analyzed in different host cells. In canine thymocytic cell lines, which are used for propagation of SMRV, it behaved as a typical D-type retrovirus with intracellular assembly of capsids. Infection of a human B cell line with SMRV, on the other hand, led to predominantly C-type budding with few intracellular capsids (Kräusslich et al. 1996). MPMV infection of human B cell lines did not show altered morphogenesis, indicating that there is a subtle balance in the pathway, which is dependent on both the virus and the host cell.

4 RNA Transport and Targeting

Besides viral polyproteins, genomic RNA is also needed at the site of assembly. It is generally believed that RNA is transported in association with viral proteins, but there is little evidence for or against this concept at present. Moreover, the question of how genomic RNA is selected from apparently identical Gag and Gag-Pol mRNA remains unanswered. Viral or cellular proteins may conceivably bind to RNA already in the nucleus, and this could be another reason for nuclear localization of Gag proteins. Intranuclear transport of RNA occurs along localized tracks which have been observed for several RNAs, including HIV-1 genomic RNA (Lawrence et al. 1989, 1990). Moreover, various mRNA have been shown to be directed to specific subcellular regions due to mRNA-targeting signals in their 3′ untranslated regions (reviewed in JOHNSTON 1995). Directed RNA transport appears to also involve the cytoskeleton, and mRNA has been found associated with microfilaments and microtubuli. Moreover, there is evidence for regulated transport of mRNA inside the cell (reviewed in JOHNSTON 1995).

RNA transport is also regulated for retroviruses. HIV and other complex retroviruses contain a viral Rev protein which interacts with a Rev response element on the genomic RNA and facilitates export to the cytoplasm. As an intron containing RNA, the viral genome would otherwise be retained in the nucleus. The activation domain of Rev functions as a nuclear export signal in this process which serves to direct genomic RNA into the transport pathway for cellular 5S rRNA (FISCHER et al. 1995; WEN et al. 1995). Other retroviruses are faced with the same problem and have solved it by evolving a constitutive transport element on the genomic RNA which is presumed to function by a similar mechanism (BRAY et al. 1994; ZOLOTHUKIN et al. 1994). Thus considerable evidence has accumulated suggesting that RNA transport within the cell and in particular transport of retroviral RNA do not occur at random, but are regulated and directed. Given the intimate interaction of retroviral elements with cellular RNA transport pathways, it appears likely that targeting and transport of retrovirus RNA may also play a role in virion morphogenesis.

5 Future Prospects

In their landmark paper, BOLOGNESI et al. (1978) presented a model for the assembly of C-type retroviruses. The salient features of this model were that uncleaved polyproteins should possess specific recognition sites for viral glycoproteins inserted in the host cell membrane and for genomic RNA. After orderly alignment at the budding site, virus maturation proceeds via PR-mediated cleavage of the precursor and leads to association of cleavage products, forming concentric protein shells which constitute the mature virion. In principle, this model still holds true today, and many of its predictions have been verified experimentally. Moreover, although it was originally proposed for C-type viruses, other retroviruses appear to follow similar pathways, with different types of morphogenesis merely being variations of a theme. The major aspects of the model were derived from ultrastructural analyses, which implies that this model could not address aspects of protein or RNA transport not visible to the electron microscope. Accordingly, the pathways leading to accumulation of virion components at the assembly site and their interactions with cellular partners and substructures are still only partly understood.

Advances in this area of research have come primarily from genetic analyses and, to a lesser extent, from biochemical studies. Considerable evidence has accumulated indicating that transport, assembly, release, and maturation are multistep processes which are closely interconnected and may be regulated by viral and cellular factors. Moreover, transport of several individual virion components appears to be intimately linked. Initial glimpses regarding structure and function of targeting signals and interacting partners have emerged, but much has yet to be learned about the molecular mechanism of transport and targeting. Additional insights are likely to come from structural analysis of additional MA proteins and mutants thereof and in particular of complete Gag polyproteins. Important contributions are also expected from ultrastructural analyses using novel techniques which provide higher resolution and are more likely to preserve the native architecture of the cell. Another exciting area of research involves the biochemical analysis of viral lipids and the role of specific lipids in retrovirus formation. Recent advances in mass spectroscopy should permit detailed analysis of lipid composition using small quantities of virus or virus-like particles and may greatly aid in this process. It is likely that we will also see a number of additional Gag-binding proteins and a detailed description of the intracellular pathway of Gag and RNA transport and of transport regulation. Elucidation of these transport pathways will constitute yet another example of how viruses exploit existing cellular pathways and of how we, in turn, can exploit viral systems to delineate these pathways.

Acknowledgements. We are grateful to W. Sundquist for providing the coordinates of the HIV-1 MA structure model and for help in preparing Table 1. We are also grateful to U. Schubert and K. Strebel for communicating results prior to publication. We thank C. Klebe for help in preparing Fig. 2 and K. Wiegers for critically reading the manuscript. Work in the authors' laboratory was supported by a grant from the German Ministry of Research and Technology.

References

Aboud M, Shoor R, Bari S, Hasan Y, Shurtz R, Malik Z, Salzberg S (1982) Biochemical analysis and electron microscopic study on intracellular virions in NIH/3T3 mouse cells chronically infected with Moloney murine leukemia virus: effect of interferon. J Gen Virol 62: 219–225

Aloia RC, Jensen FC, Curtain CC, Mobley PW, Gordon LM (1988) Lipid composition and fluidity of the human immunodeficiency virus. Proc Natl Acad Sci USA 85: 900–904

Aloia RC, Tian H, Jensen FC (1993) Lipid composition and fluidity of the human immunodeficiency virus envelope and host cell plasma membranes. Proc Natl Acad Sci USA 90: 5181–5185

Arthur LO, Bess JW Jr, Sowder RC II, Benveniste RE, Mann DL, Chermann JC, Henderson LE (1992) Cellular proteins bound to immunodeficiency viruses: implications for pathogenesis and vaccines. Science 258: 1935–1938

Bathurst IC, Chester N, Gibson HL, Dennis AF, Steimer KS, Barr PJ (1989) N myristylation of the human immunodeficiency virus type 1 gag polyprotein precursor in Saccharomyces cerevisiae. J Virol 63: 3176–3179

Bennett RP, Nelle TD, Wills JW (1993) Functional chimeras of the Rous sarcoma virus and human immunodeficiency virus gag proteins. J Virol 67: 6487–6498

Bilello JA, Wivel NA, Pitha PM (1982) Effect of interferon on the replication of mink cell focus inducing virus in murine cells: synthesis, processing, assembly, and release of viral proteins. J Virol 43: 213–222

Biswas P, Poli G, Kinter AL, Justement JS, Stanley SK, Maury WJ, Bressler P, Orenstein JM, Fauci AS (1992) Interferon gamma induces the expression of human immunodeficiency virus in persistently

infected promonocytic cells (U1) and redirects the production of virions of intracytoplasmic vacuoles in phorbol myristate acetate differentiated U1 cells. J Exp Med 176: 739–750
Blenis J, Resh MD (1993) Subcellular localization specified by protein acylation and phosphorylation. Curr Opin Cell Biol 5: 984–989
Bolognesi DP, Montelaro RC, Frank H, Schafer W (1978) Assembly of type C oncornaviruses: a model. Science 199: 183–186
Bour S, Schubert U, Peden K, Strebel K (1996) The envelope glycoprotein of human immunodeficiency virus type 2 has a Vpu-like activity that enhances viral particle release. J Virol 70: 820–829
Bourinbaiar AS, Phillips DM (1991) Transmission of human immunodeficiency virus from monocytes to epithelia. J Acquir Immune Defic Syndr 4: 56–63
Bowerman B, Brown PO, Bishop JM, Varmus HE (1989) A nucleoprotein complex mediates the integration of retroviral DNA. Genes Dev 3: 469–478
Bray M, Prasad S, Dubay WJ, Hunter E, Jeang K-T, Rekosh D, Hammarskjöld M-L (1994) A small element from the Mason-Pfizer monkey virus genome makes human immunodeficiency virus type 1 expression and replication Rev-independent. J Virol 91: 1256–1260
Brody BA, Rhee SS, Sommerfelt MA, Hunter E (1992) A viral protease mediated cleavage of the transmembrane glycoprotein of Mason Pfizer monkey virus can be suppressed by mutations within the matrix protein. Proc Natl Acad Sci USA 89: 3443–3447
Brody BA, Rhee SS, Hunter E (1994) Postassembly cleavage of a retroviral glycoprotein cytoplasmic domain removes a necessary incorporation signal and activates fusion activity. J Virol 68: 4620–4627
Bryant M, Ratner L (1990) Myristoylation dependent replication and assembly of human immunodeficiency virus 1. Proc Natl Acad Sci USA 87: 523–527
Bugelski PJ, Maleeff BE, Klinkner AM, Ventre J, Hart TK (1995) Ultrastructural evidence of an interaction between Env and Gag proteins during assembly of HIV type 1. AIDS Res Hum Retroviruses 11: 55–64
Bukrinsky MI, Sharova N, Dempsey MP, Stanwick TL, Bukrinskaya AG, Haggerty S, Stevenson M (1992) Active nuclear import of human immunodeficiency virus type 1 preintegration complexes. Proc Natl Acad Sci USA 89: 6580–6584
Bukrinsky MI, Haggerty S, Dempsey MP, Sharova N, Adzhubel A, Spitz L, Lewis P, Goldfarb D, Emerman M, Stevenson M (1993a) A nuclear localization signal within HIV 1 matrix protein that governs infection of non dividing cells. Nature 365: 666–669
Bukrinsky MI, Sharova N, McDonald TL, Pushkarskaya T, Tarpley WG, Stevenson M (1993b) Association of integrase, matrix, and reverse transcriptase antigens of human immunodeficiency virus type 1 with viral nucleic acids following acute infection. Proc Natl Acad Sci USA 90: 6125–6129
Burnette B, Yu G, Felsted RL (1993) Phosphorylation of HIV 1 gag proteins by protein kinase C J Biol Chem 268: 8698–8703
Burstein H, Bizub D, Skalka AM (1991) Assembly and processing of avian retroviral gag polyproteins containing linked protease dimers. J Virol 65: 6165–6172
Campbell S, Vogt VM (1995) Self-assembly in vitro of purified CA-NC proteins from Rous sarcoma virus and human immunodeficiency virus type 1. J Virol 69: 6487–6497
Canivet M, Jouanny C, Fourcade A, Lasneret J, Rhodes-Feuillette A, Peries J (1983) Effect of human interferon on type D retroviruses multiplication in chronically infected cell lines. J Interferon Res 3: 53–64
Chatterjee S, Bradac JA, Hunter E (1982) Effect of monensin or Mason Pfizer monkey virus glycoprotein synthesis. J Virol 44: 1003–1012
Chazal N, Carriére C, Gay B, Boulanger P (1994) Phenotypic characterization of insertion mutants of the human immunodeficiency virus type 1 Gag precursor expressed in recombinant baculovirus-infected cells. J Virol 68: 111–122
Chazal N, Gay B, Carriére C, Tournier J, Boulanger P (1995) Human immunodeficiency virus type 1 MA deletion mutants expressed in baculovirus-infected cells: *cis* and *trans* effects on the Gag precursor assembly pathway. J Virol 69: 365–375
Damsky CH, Sheffield JB, Tuszynski GP, Warren L (1977) Is there a role for actin in virus budding? J Cell Biol 75: 593–605
Delchambre M, Gheysen D, Thines D, Thiriart C, Jacobs E, Verdin E, Horth M, Burny A, Bex F (1989) The Gag precursor of simian immunodeficiency virus assembles into virus-like particles. EMBO J 8: 2653–2660
Deminie CA, Emerman M (1994) Functional exchange of an oncoretrovirus and a lentivirus matrix protein. J Virol 68: 4442–4449

Dorfman T, Mammano F, Haseltine WA, Göttlinger HG (1994) Role of the matrix protein in the virion association of the human immunodeficiency virus type 1 envelope glycoprotein. J Virol 68: 1689–1696
Edbauer CA, Naso RB (1983) Cytoskeleton associated Pr65gag and retrovirus assembly. Virology 130: 415–426
Edbauer CA, Naso RB (1984) Cytoskeleton associated Pr65gag and assembly of retrovirus temperature sensitive mutants in chronically infected cells. Virology 134: 389–397
Evans SS, Wang WC, Gregorio CC, Han T, Repasky EA (1993) Interferon-α alters spectrin organization in normal and leukemic human B lymphocytes. Blood 81: 759–766
Fäcke M, Janetzko A, Shoeman RL, Kräusslich HG (1993) A large deletion in the matrix domain of the human immunodeficiency virus gag gene redirects virus particle assembly from the plasma membrane to the endoplasmic reticulum. J Virol 67: 4972–4980
Fernie BF, Poli G, Fauci AS (1991) Alpha interferon suppresses virion but not soluble human immunodeficiency virus antigen production in chronically infected T lymphocytic cells. J Virol 65: 3968–3971
Fischer U, Huber J, Boelens WC, Mattaj IW, Lührmann R (1995) The HIV-1 Rev activation domain is a nuclear export signal that accesses an export pathway used by specific cellular RNAs. Cell 82: 475–483
Franke EK, Yuan HEH, Luban J (1994) Specific incorporation of cyclophilin A into HIV-1 virions. Nature 372: 4972–4980
Freed EO, Martin MA (1994) HIV-1 infection of non-dividing cells. Nature 369: 107–108
Freed EO, Martin MA (1995) Virion incorporation of envelope glycoproteins with long but not short cytoplasmic tails is blocked by specific single amino acid substitutions in the human immunodeficiency virus type 1 matrix. J Virol 69: 1984–1989
Freed EO, Orenstein JM, Buckler White AJ, Martin MA (1994) Single amino acid changes in the human immunodeficiency virus type 1 matrix protein block virus particle production. J Virol 68: 5311–5320
Freed EO, Englund G, Martin MA (1995) Role of the basic domain of human immunodeficiency virus type 1 matrix in macrophage infection. J Virol 69: 3949–3954
Gallay P, Swingler S, Aiken C, Trono D (1995a) HIV-1 infection of nondividing cells: C-terminal phosphorylation of the viral matrix protein is a key regulator. Cell 80: 379–388
Gallay P, Swingler S, Song J, Bushman F, Trono D (1995b) HIV nuclear import is governed by the phosphotyrosine-mediated binding of matrix to the core domain of integrase. Cell 83: 569–576
Gallina A, Mantoan G, Rindi G, Milanesi G (1994) Influence of MA internal sequences, but not of the myristylated N-terminus sequence, on the binding site of HIV-1 Gag protein. Biochem. Biophys Res Commun 204: 1031–1038
Garoff H, Wilschut J, Liljeström P, Wahlberg JM, Bron R, Suomalainen M, Smyth J, Salminen A, Barth BU, Zhao H et al (1994) Assembly and entry mechanisms of Semliki Forest virus. Arch Virol [Suppl] 9: 329–338
Gebhardt A, Bosch JV, Ziemiecki A, Friis RR (1984) Rous sarcoma virus p19 and gp35 can be chemically crosslinked to high molecular weight complexes: an insight into virus assembly. J Mol Biol 174: 297–317
Gelderblom HR, Hausmann EH, Özel M, Pauli G, Koch MA (1987) Fine structure of human immunodeficiency virus (HIV) and immunolocalization of structural proteins. Virology 156: 171–176
Gendelman HE, Baca L, Turpin JA, Kalter DC, Hansen BD, Orenstein JM, Friedman RM, Meltzer MS (1990) Restriction of HIV replication in infected T cells and monocytes by interferon alpha. AIDS Res Hum Retroviruses 6: 1045–1049
Gheysen D, Jacobs E, de Foresta F, Thiriart C, Francotte M, Thines D, De Wilde M (1989) Assembly and release of HIV 1 precursor Pr55gag virus like particles from recombinant baculovirus infected insect cells. Cell 59: 103–112
Goepfert PA, Wang G, Mulligan MJ (1995) Identification of an ER retrieval signal in a retroviral glycoprotein. Cell 82: 543–544.
González SA, Affranchino JL, Gelderblom HR, Burny A (1993) Assembly of the matrix protein of simian immunodeficiency virus into virus like particles. Virology 194: 548–556
Göttlinger HG, Sodroski JG, Haseltine WA (1989) Role of capsid precursor processing and myristoylation in morphogenesis and infectivity of human immunodeficiency virus type 1. Proc Natl Acad Sci USA 86: 5781–5785
Göttlinger HG, Dorfman, T, Cohen EA, Haseltine WA (1993) Vpu protein of human immunodeficiency virus type 1 enhances the release of capsids produced by gag gene constructs of widely divergent retroviruses. Proc Natl Acad Sci USA 90: 7381–7385

Granowitz C, Goff SP (1994) Substitution mutations affecting a small region of the moloney murine leukemia virus MA Gag protein block assembly and release of virion particles. Virology 205: 336–344

Grief C, Farrar GH, Kent KA, Berger EG (1991) The assembly of HIV within the Golgi apparatus and Golgi derived vesicles of JM cell syncytia. AIDS 5: 1433–1439

Hansen M, Jelinek L, Whithing S, Barklis E (1990) Transport and assembly of Gag proteins into Moloney murine leukemia virus. J Virol 64: 5306–5316

Hansen M, Jelinek L, Jones RS, Stegeman Olsen J, Barklis E (1993) Assembly and composition of intracellular particles formed by Moloney murine leukemia virus J Virol 67: 5163–5174

Hayakawa T, Miyazaki T, Misumi Y, Kobayashi M, Fujisawa Y (1992) Myristoylation-dependent membrane targeting and release of the HTLV-I Gag precursor, Pr53gag, in yeast. Gene 119: 273–277

Heine UL, Demsey AE, Tucker RW, Bykovsky AF (1985) Intracellular type A retrovirus movement associated with an intact microtubule system. J Gen Virol 66: 275–282

Heinzinger NK, Bukrinsky MI, Haggerty SA, Ragland AM, Kewalramani V, Lee MA, Gendelman HE, Ratner L, Stevenson M, Emerman M (1994) The Vpr protein of human immunodeficiency virus type 1 influences nuclear localization of viral nucleic acids in nondividing host cells. Proc Natl Acad Sci USA 91: 7311–7315

Henderson LE, Krutzsch HC, Oroszlan (1983) Myristyl amino-terminal acylation of murine retrovirus proteins: an unusual post-translational protein modification. Proc Natl Acad Sci USA 80: 339–343

Hill CP, Worthylake D, Bancroft DP, Christensen AM, Sundquist WI (1996) Crystal structures of the trimeric HIV-1 matrix protein: implications for membrane association and assembly. Proc Natl Acad Sci USA 93 (in press)

Höner B, Shoeman RL, Traub P (1991) Human immunodeficiency virus type 1 protease microinjected into cultured human skin fibroblasts cleaves vimentin and affects cytoskeletal and nuclear architecture. J Cell Sci 100: 799–807

Hoshikawa N, Kojima A, Yasuda A, Takayashiki E, Masuko S, Chiba J, Sata T, Kurata T (1991) Role of the gag and pol genes of human immunodeficiency virus in the morphogenesis and maturation of retrovirus like particles expressed by recombinant vaccinia virus: an ultrastructural study. J Gen Virol 72: 2509–2517

Jacobs E, Gheysen D, Thines D, Francotte M, de Wilde M (1989) The HIV 1 Gag precursor Pr55gag synthesized in yeast is myristoylated and targeted to the plasma membrane. Gene 79: 71–81

Jones TA, Blaug G, Hansen M, Barklis E (1990) Assembly of gag β-galactosidase proteins into retrovirus particles. J Virol 64: 2265–2279

Johnston DS (1995) The intracellular localization of messenger RNAs. Cell 81: 161–170

Jorgensen EC, Pedersen FS, Jorgensen P (1992) Matrix protein of Akv murine leukemia virus: genetic mapping of regions essential for particle formation. J Virol 66: 4479–4487

Kaplan AH, Swanstrom R (1991) Human immunodeficiency virus type 1 Gag proteins are processed in two cellular compartments. Proc Natl Acad Sci USA 88: 4528–4532

Kaplan AH, Zack JA, Knigge M, Paul DA, Kempf DJ, Norbeck DW, Swanstrom R (1993) Partial inhibition of the human immunodeficiency virus type 1 protease results in aberrant virus assembly and the formation of noninfectious particles J Virol 67: 4050–4055

Kaplan AH, Manchester M, Swanstrom R (1994) The activity of the protease of human immunodeficiency virus type 1 is initiated at the membrane of infected cells before the release of viral proteins and is required for release of occur with maximum efficiency. J Virol 68: 6782–6786

Karacostas V, Wolffe EJ, Nagashima K, Gonda MA, Moss B (1993) Overexpression of the HIV 1 gag pol polyprotein results in intracellular activation of HIV 1 protease and inhibition of assembly and budding of virus like particles. Virology 193: 661–671

Kim J, Mosior M, Chung LA, Wu H, McLaughlin S (1991) Binding of peptides with basic residues to membranes containing acidic phospholipids. Biophys J 60: 135–148

Klikova M, Rhee SS, Hunter E, Ruml T (1995) Efficient in vivo assembly of retroviral capsids from gag precursor expressed in bacteria J Virol 69: 1093–1098

Klimkait T, Strebel K, Hoggan MD, Martin MA, Orenstein JM (1990) The human immunodeficiency virus type 1 specific protein vpu is required for efficient virus maturation and release J Virol 64: 621–629

Konvalinka J, Litterst MA, Welker R, Kottler H, Rippmann F, Heuser AM, Kräusslich HG (1995a) An active-site mutation in the human immunodeficiency virus type 1 proteinase (PR) causes reduced PR activity and loss of PR-mediated cytotoxicity without apparent effect on virus maturation and infectivity J Virol 69: 7180–7186

Konvalinka J, Löchelt M, Zentgraf H, Flügel RM, Kräusslich HG (1995b) Active foamy virus proteinase is essential for virus infectivity but not for formation of a Pol polyprotein. J Virol 69: 7264–7268
Kräusslich HG (1991) Human immunodeficiency virus proteinase dimer as component of the viral polyprotein prevents particle assembly and viral infectivity. Proc Natl Acad Sci USA 88: 3213–3217
Kräusslich HG (1992) Specific inhibitor of human immunodeficiency virus proteinase prevents the cytotoxic effects of a single chain proteinase dimer and restores particle formation. J Virol 66: 567–572
Kräusslich HG, Fäcke M, Heuser AM, Konvalinka J, Zentgraf H (1995) The spacer peptide between human immunodeficiency virus capsid and nucleocapsid proteins is essential for ordered assembly and viral infectivity. J Virol 69: 3407–3419
Kräusslich HG, Pawlita M, Hunter E, Zentgraf H (1996) The budding type of squirrel monkey retrovirus is influenced by the host cell (submitted for publication)
Kuff EL, Lueders KK (1988) The intracisternal A particle gene family: structure and functional aspects. Adv Cancer Res 51: 183–276
Lawrence JB, Singer RH, Marselle LM (1989) Highly localized tracks of specific transcripts within interphase nuclei visualized by in situ hybridization. Cell 57: 493–502
Lawrence JB, Marselle LM, Byron KS, Johnson CV, Sullivan JL, Singer RH (1990) Subcellular localization of low-abundance human immunodeficiency virus nucleic acid sequences visualized by fluorescence in situ hybridization. Proc Natl Acad Sci USA 87: 5420–5424
Lee PP, Linial ML (1994) Efficient particle formation can occur if the matrix domain of human immunodeficiency virus type 1 Gag is substituted by a myristylation signal J Virol 68: 6644–6654
Leis J, Phillips N, Fu X, Tuazon PT, Traugh JA (1989) Phosphorylation of avian retrovirus matrix protein by Ca2+/phospholipid dependent protein kinase. Eur J Biochem 179: 415–422
Löchelt M, Flügel RM (1996) The human foamy virus pol gene is expressed as a Pro-Pol polyprotein and not as a Gag-Pol fusion protein. J Virol 70: 1033–1040
Lodge R, Göttlinger H, Gabuzda D, Cohen EA, Lemay G (1994) The intracytoplasmic domain of gp41 mediates polarized budding of human immunodeficiency virus type 1 in MDCK cells J Virol 68: 4857–4861
Lu YL, Spearman P, Ratner L (1993) Human immunodeficiency virus type 1 viral protein R localization in infected cells and virions. J Virol 67: 6542–6550
Luban J, Bossolt KL, Franke EK, Kalpana GV, Goff SP (1993) Human immunodeficiency virus type 1 Gag protein binds to cyclophilins A and B. Cell 73: 1067–1078
Luftig RB, Lupo LD (1994) Viral interactions with the host cell cytoskeleton: the role of retroviral proteases. Trends Microbiol 2: 178–182
Mak J, Jiang M, Wainberg MA, Hammarskjold ML, Rekosh D, Kleiman L (1994) Role of Pr160gag pol in mediating the selective incorporation of tRNA(Lys) into human immunodeficiency virus type 1 particles. J Virol 68: 2065–2072
Mammano F, Kondo E, Sodroski J, Bukovsky A, Göttlinger HG (1995) Rescue of human immunodeficiency virus type 1 matrix protein mutants by envelope glycoproteins with short cytoplasmic domains. J Virol 69: 3824–3830
Massiah MA, Starich MR, Paschall C, Summers MF, Christensen AM, Sundquist WI (1994) Three-dimensional structure of the human immunodeficiency virus type 1 matrix protein J Mol Biol 244: 198–223
Matthews S, Barlow P, Boyd J, Barton G, Russell R, Mills H, Cunningham M, Meyers N, Burns N, Clark N et al (1994) Structural similarity between the p17 matrix protein of HIV 1 and interferon gamma. Nature 370: 666–668
McLauglin S, Aderem A (1995) The myristoyl-electrostatic switch: a modulator of reversible protein-membrane interactions. TIBS 20: 272–276
Mergener K, Fäcke M, Welker R, Brinkmann V, Gelderblom HR, Kräusslich HG (1992) Analysis of HIV particle formation using transient expression of subviral constructs in mammalian cells. Virology 186: 25–39
Morikawa Y, Kishi T, Zhang WH, Nermut M, Hockley DJ, Jones IM (1995) A molecular determinant of human immunodeficiency virus particle assembly located in matrix antigen p17 J Virol 69: 4519–4523
Nash MA, Meyer MK, Decker GL, Arlinghaus RB (1993) A subset of Pr65gag is nucleus associated in murine leukemia virus infected cells. J Virol 67: 1350–1356
Niedrig M, Gelderblom HR, Pauli G, Marz J, Bickhard H, Wolf H, Modrow S (1994) Inhibition of infectious human immunodeficiency virus type 1 particle formation by Gag protein derived peptides. J Gen Virol 75: 1469–1474

Nieva JL, Bron R, Corver J, Wilschut (1994) Membrane fusion of Semliki Forest virus requires sphingolipids in the target membrane. EMBO J 13: 2797–2804
Oka T, Ohtsuki Y, Sonobe H, Furihata M, Miyoshi (1990) Suppressive effects of interferons on the production and release of human T-lymphotropic virus-I (HTLV-I). Arch Virol 115: 63–73
Orenstein JM (1992) Immunodeficiency virus infection. Ultrastruct Pathol 16: 179–210
Orenstein JM, Meltzer MS, Phipps T, Gendelman HE (1988) Cytoplasmic assembly and accumulation of human immunodeficiency virus types 1 and 2 in recombinant human colony stimulating factor 1 treated human monocytes: an ultrastructural study. J Virol 62: 2578–2586
Owens RJ, Compans RW (1989) Expression of the human immunodeficiency virus envelope glycoprotein is restricted to basolateral surfaces of polarized epithelial cells. J Virol 63: 978–982
Owens RJ, Dubay JW, Hunter E, Compans RW (1991) Human immunodeficiency virus envelope protein determines the site of virus release in polarized epithelial cells. Proc Natl Acad Sci USA 88: 3987–3991
Pal R, Reitz MS Jr, Tschachler E, Gallo RC, Sarngadharan MG, Veronese FD (1990) Myristoylation of gag proteins of HIV 1 plays an important role in virus assembly. AIDS Res Hum Retroviruses 6: 721–730
Pal R, Mumbauer S, Hoke GM, Takatsuki A, Sarngadharan MG (1991) Brefeldin A inhibits the processing and secretion of envelope glycoproteins of human immunodeficiency virus type 1. AIDS Res Hum Retroviruses 7: 707–712
Park J, Morrow CD (1992) The nonmyristoylated $Pr160^{gag\text{-}pol}$ polyprotein of human immunodeficiency virus type 1 interacts with $Pr55^{gag}$ and is incorporated into viruslike particles J Virol 66: 6304–6313
Pautrat G, Suzan M, Salaun D, Corbeau P, Allasia C, Morel G, Filippi P (1990) Human immunodeficiency virus type 1 infection of U937 cells promotes cell differentiation and a new pathway of viral assembly. Virology 179: 749–758
Pearce-Pratt R, Malamud D, Phillips DM (1994) Role of the cytoskeleton in cell-to-cell transmission of human immunodeficiency virus J Virol 68: 2898–2905
Peitzsch RM, McLaughlin S (1993) Binding of acylated peptides and fatty acids to phospholipid vesicles: relevance to myristoylated proteins. Biochemistry 32: 10436–10443
Peng C, Ho BK, Chang TW, Chang NT (1989) Role of human immunodeficiency virus type 1 specific protease in core protein maturation and viral infectivity. J Virol 63: 2550–2556
Pepinsky RB, Vogt VM (1979) Identification of retrovirus matrix proteins by lipid protein cross-linking. J Mol Biol 131: 819–837
Pepinsky RB, Vogt VM (1984) Fine-structure analyses of lipid-protein and protein–protein interactions of gag protein p19 of the avian sarcoma and leukemia viruses by cyanogen bromide mapping. J Virol 52: 145–153
Pessin JE, Glaser M (1983) Budding of Rous sarcoma virus and vesicular stomatitis virus from localized lipid regions in the plasma membrane of chicken embryo fibroblasts. J Biol Chem 255: 9044–9050
Pfeffer LM, Landsberger FR (1990) Interferon-α modulates the plasma membrane-cytoskeletal complex of human lymphoblastoid cells sensitive to the antiproliferative action of interferon-α. J Interferon Res 10: 91–97
Pfeffer LM, Wang E, Tamm I (1980) Interferon inhibits the redistribution of cell surface components. J Exp Med 152: 469–474
Pfeffer LM, Landsberger FR, Tamm I (1981) β-Interferon-induced time-dependent changes in the plasma membrane lipid bilayer of cultured cells. J Interferon Res 1: 613–620
Phillips DM, Bourinbaiar AS (1992) Mechanism of HIV spread from lymphocytes to epithelia. Virology 186: 261–273
Phillips DM, Tan X (1992) HIV-1 infection of the trophoblast cell line BeWo: a study of virus uptake. AIDS Res Hum Retroviruses 8: 1683–1691
Pitha PM, Bilello JA, Riggin CH (1981) Effect of interferon on retrovirus replication. Tex Rep Biol Med 41: 603–609
Poli G, Orenstein JM, Kinter A, Folks TM, Fauci AS (1989) Interferon alpha but not AZT suppresses HIV expression in chronically infected cell lines. Science 244: 575–577
Pumplin DW, Bloch RJ (1993) The membrane skeleton. Trends Cell Biol 3: 113–117
Rao Z, Belyaev AS, Fry E, Roy P, Jones IM, Stuart DI (1995) Crystal structure of SIV matrix antigen and implications for virus assembly. Nature 238: 743–747
Reicin AS, Paik S, Berkowitz RD, Luban J, Lowy I, Goff SP (1995) Linker insertion mutations in the human immunodeficiency virus type 1 gag gene: effects on virion particle assembly, release, and infectivity. J Virol 69: 642–650

Rein A, McClure MR, Rice NR, Luftig RB, Schultz AM (1986) Myristylation site in Pr65gag is essential for virus particle formation by Moloney murine leukemia virus. Proc Natl Acad Sci USA 83: 7246–7250
Rhee SS, Hunter E (1987) Myristylation is required for intracellular transport but not for assembly of D type retrovirus capsids. J Virol 61: 1045–1053
Rhee SS, Hunter E (1990a) Structural role of the matrix protein of type D retroviruses in gag polyprotein stability and capsid assembly. J Virol 64: 4383–4389
Rhee SS, Hunter E (1990b) A single amino acid substitution within the matrix protein of a type D retrovirus converts its morphogenesis to that of a type C retrovirus. Cell 63: 77–86
Rhee SS, Hunter E (1991) Amino acid substitutions within the matrix protein of type D retroviruses affect assembly, transport and membrane association of a capsid. EMBO J 10: 535–546
Royer M, Cerutti M, Gay B, Hong SS, Devauchelle G, Boulanger P (1991) Functional domains of HIV 1 gag polyprotein expressed in baculovirus infected cells. Virology 184: 417–422
Royer M, Hong SS, Gay B, Cerutti M, Boulanger P (1992) Expression and extracellular release of human immunodeficiency virus type 1 Gag precursors by recombinant baculovirus infected cells. J Virol 66: 3230–3235
Sagara J, Tsukita S, Yonemura S, Tsukita S, Kawai A (1995) Cellular actin-binding ezrin-radixin-moesin (ERM) family proteins are incorporated into the rabies virion and closely associated with envelope proteins in the cell. Virology 206: 485–494
Satake M, Luftig RB (1982) Microtubule-depolymerizing agents inhibit Moloney murine leukemia virus production J Gen Virol 58: 339–349
Schätzl H, Gelderblom HR, Nitschko H, von der Helm K (1991) Analysis of non-infectious HIV particles produced in presence of HIV proteinase inhibitor. Arch Virol 120: 71–81
Schliephake AW, Rethwilm A (1994) Nuclear localization of foamy virus Gag precursor protein. J Virol 68: 4946–4954
Schubert U, Bour S, Ferrer-Montiel A, Montal M, Maldarelli F, Strebel K (1996a) The two biological activities of the human immunodeficiency virus type-1 Vpu protein involve two separate structural domains J Virol 70: 809–819
Schubert U, Henklein P, Ferrer-Montiel AV, Oblatt-Montal M, Strebel K, Montal M (1996b) Identification of an ion channel activity of the Vpu transmembrane domain and its function in the regulation of virus release from HIV-1 infected cells (submitted for publication)
Schultz AM, Henderson LE, Oroszlan S (1988) Fatty acylation of proteins. Annu Rev Cell Biol 4: 611–647
Schultz AM, Rein A (1989) Unmyristylated Moloney murine leukemia virus Pr65gag is excluded from virus assembly and maturation events. J Virol 63: 2370–2373
Sen GC, Pinter A (1983) Interferon-mediated inhibition of production of Gazdar murine sarcoma virus, a retrovirus lacking env proteins and containing an uncleaved gag precursor. Virology 126: 403–407
Shoeman RL, Höner B, Stoller TJ, Kesselmeier C, Miedel MC, Traub P, Graves MC (1990) Human immunodeficiency virus type 1 protease cleaves the intermediate filament proteins vimentin, desmin, and glial fibrillary acidic protein. Proc Natl Acad Sci USA 87: 6336–6340
Shoeman RL, Kesselmeier C, Mothes E, Höner B, Traub P (1991) Non viral cellular substrates for human immunodeficiency virus type 1 protease. FEBS Lett 278: 199–203
Simons K, van Meer G (1988) Lipid sorting in epithelial cells. Biochemistry 27: 6197–6202
Smith AJ, Srinivasakumar N, Hammarskjold ML, Rekosh D (1993) Requirements for incorporation of Pr160gag pol from human immunodeficiency virus type 1 into virus like particles. J Virol 67: 2266–2275
Smith MS, Thresher RJ, Pagano JS (1991) Inhibition of human immunodeficiency virus type 1 morphogenesis in T cells by alpha interferon. Antimicrob Agents Chemother. 35: 62–67
Spearman P, Wang JJ, Vander Heyden N, Ratner L (1994) Identification of human immunodeficiency virus type 1 Gag protein domains essential to membrane binding and particle assembly. J Virol 68: 3232–3242
Srinivasakumar N, Hammarskjöld ML, Rekosh D (1995) Characterization of deletion mutations in the capsid region of human immunodeficiency virus type 1 that affect particle formation and Gag-Pol precursor incorporation. J Virol 69: 6106–6114
Taniguchi H, Manenti S (1993) Interaction of myristoylated alanina-rich protein kinase C substrate (MARCKS) with membrane phospholipids. J Biol Chem 268: 9960–9963
Terwilliger EF, Cohen EA, Lu YC, Sodroski JG, Haseltine WA (1989) Functional role of human immunodeficiency virus type 1 vpu. Proc Natl Acad Sci USA 86: 5163–5167
Thali M, Bukovsky A, Kondo E, Rosenwirth B, Walsh C, Sodroski J, Göttlinger HG (1994) Functional association of cyclophilin A with HIV-1 virions. Nature 372: 363–365

Thelen M, Rosen A, Nairn AC, Aderem A (1991) Regulation by phosphorylation of reversible association of myristoylated protein kinase C substrate with the plasma membrane. Nature 351: 320–322

Ulmer JB, Palade GE (1991) Effects of Brefeldin A on the Golgi complex, endoplasmic reticulum and viral envelope glycoproteins in murine erythroleukemia cells. Eur J Cell Biol 54: 38–54

Vieillard V, Lauret E, Rousseau V, de Maeyer E (1994) Blocking of retroviral infection at a step prior to reverse transcription in cells transformed to constitutively express interferon beta. Proc Natl Acad Sci USA 91: 2689–2693

Vlasuk GP, Waxman L, Davis LJ, Dixon RA, Schultz LD, Hofmann KJ, Tung JS, Schulman CA, Ellis RW, Bencen GH, et al (1989) Purification and characterization of human immunodeficiency virus (HIV) core precursor (p55) expressed in Saccharomyces cerevisiae. J Biol Chem 264: 12106–12112

von Schwedler U, Kornbluth RS, Trono D (1994) The nuclear localization signal of the matrix protein of human immunodeficiency virus type 1 allows the establishment of infection in macrophages and quiescent T lymphocytes. Proc Natl Acad Sci USA 91: 6992–6996

Wagner R, Deml L, Fliessbach H, Wanner G, Wolf H (1994) Assembly and extracellular release of chimeric HIV-1 Pr55gag retrovirus-like particles. Virology 200: 162–175

Wang CT, Barklis E (1993) Assembly, processing, and infectivity of human immunodeficiency virus type 1 gag mutants. J Virol 67: 4264–4273

Wang CT, Zhang Y, McDermott J, Barklis E (1993) Conditional infectivity of human immunodeficiency virus matrix domain deletion mutant. J Virol 67: 7067–7076

Wang CT, Stegeman Olsen J, Zhang Y, Barklis E (1994) Assembly of HIV GAG β galactosidase fusion proteins into virus particles. Virology 200: 524–534

Wang E, Wolf BA, Lamb RA, Choppin PW, Goldberg AR (1976) The presence of actin in enveloped viruses. In: Goldman R, Pollard T, Rosenbaums J (eds) Cell motility. Cold Spring Harbor Laboratory, Cold Spring Harbor, pp 589–599

Wang E, Pfeffer LM, Tamm I (1981) Interferon increases the abundance of submembraneous microfilaments in HeLa-S_3 cells in suspension culture. Proc. Natl. Acad. Sc. USA 78: 6281–6285

Weaver TA, Panganiban AT (1990) N myristoylation of the spleen necrosis virus matrix protein is required for correct association of the Gag polyprotein with intracellular membranes and for particle formation. J Virol 64: 3995–4001

Weldon RA Jr, Erdie CR, Oliver MG, Wills JW (1990) Incorporation of chimeric gag protein into retroviral particles. J Virol 64: 4169–4179

Welker R, Janetzko A, Kräusslich HG (1996) Plasma membrane targeting of intracisternal A type particle polyproteins leads to particle release and specifically activates the viral proteinase (submitted for publication)

Wen W, Meinkoth JL, Tsien RY, Taylor SS (1995) Identification of a signal for rapid export of proteins from the nucleus. Cell 82: 463–473

Wills JW, Craven RC (1991) Form, function, and use of retroviral Gag proteins. AIDS 5: 639–654

Wills JW, Craven RC, Achacoso JA (1989) Creation and expression of myristylated forms of Rous sarcoma virus gag protein in mammalian cells. J Virol 63: 4331–4343

Wills JW, Craven RC, Weldon RA Jr, Nelle TD, Erdie CR (1991) Suppression of retroviral MA deletions by the amino terminal membrane binding domain of p60src. J Virol 65: 3804–3812

Yu G, Shen FS, Sturch S, Aquino A, Glazer RI, Felsted RL (1995) Regulation of HIV-1 gag protein subcellular targeting by protein kinase C. J Biol Chem 270: 4792–4796

Yu SF, Baldwin DN, Gwynn SR, Yandapalli S, Linial M (1996) Human foamy virus replication: a pathway distinct from that of retroviruses and hepadnaviruses. Science 271: 1579–1582

Yu X, Yu QC, Lee TH, Essex M (1992a) The C terminus of human immunodeficiency virus type 1 matrix protein is involved in early steps of the virus life cycle. J Virol 66: 5667–5670

Yu X, Yuan X, Matsuda Z, Lee TH, Essex M (1992b) The matrix protein of human immunodeficiency virus type 1 is required for incorporation of viral envelope protein into mature virions. J Virol 66: 4966–4971

Yuan X, Yu X, Lee TH, Essex M (1993) Mutations in the N terminal region of human immunodeficiency virus type 1 matrix protein block intracellular transport of the Gag precursor. J Virol 67: 6387–6394

Zhou W, Parent LJ, Wills JW, Resh MD (1994) Identification of a membrane binding domain within the amino terminal region of human immunodeficiency virus type 1 Gag protein which interacts with acidic phospholipids. J Virol 68: 2556–2569

Zolotukhin AS, Valentin A, Pavlakis GN, Felber BK (1994) Continuous propagation of RRE(–) and Rev(–) RRE(–) human immunodeficiency type 1 molecular clones containing a *cis*-acting element of simian retrovirus type 1 in human peripheral blood lymphocytes. J Virol 68: 7944–7952

Dynamic Interactions of the Gag Polyprotein

R.C. Craven[1] and L.J. Parent[1,2]

1 Introduction

It has been known for 20 years that all retroviruses possess a similar genetic organization and that a single gene product, the Gag protein, is responsible for directing particle assembly (Dickson et al. 1984; Coffin 1984). However it is only very recently that an understanding of how the Gag protein functions has begun to take shape. This chapter reviews recent efforts to understand the assembly and budding functions of the Gag protein with a primary focus on Rous sarcoma virus (RSV, a member of the avian sarcoma-leukosis virus group of oncoviruses) and the human immunodeficiency virus type 1 (HIV-1). These studies have shown that in spite of a striking divergence in amino acid sequ-

[1]Department of Microbiology and Immunology, The Pennsylvania State University School of Medicine, Hershey, PA 17033, USA
[2]Department of Medicine, The Pennsylvania State University School of Medicine, Hershey, PA 17033, USA

ences, the two viruses utilize fundamentally similar mechanisms to direct assembly and budding.

1.1 The Gag Protein as the "Assembly Machine"

The *gag* gene lies at the 5' end of all retroviral genomes and is followed by the pol gene, which encodes the enzymes of replication, and the *env* gene, which directs synthesis of the viral glycoproteins (COFFIN 1984). The product of the *gag* gene, the Gag protein (Fig. 1), is responsible for orchestrating virus assembly and is the precursor to the major structural proteins within the virion (ARCEMENT et al. 1976; VOGT et al. 1975; VOGT and EISENMAN 1973; DICKSON et al. 1984). Gag proteins are synthesized on cytosolic ribosomes and then follow one of two seemingly different morphogenetic pathways to the plasma membrane (TEICH 1984; Chaps. 1, 2, this volume). The C-type oncoviruses, of which RSV is an example, and the lentiviruses such as HIV-1 appear in electron micrographs to

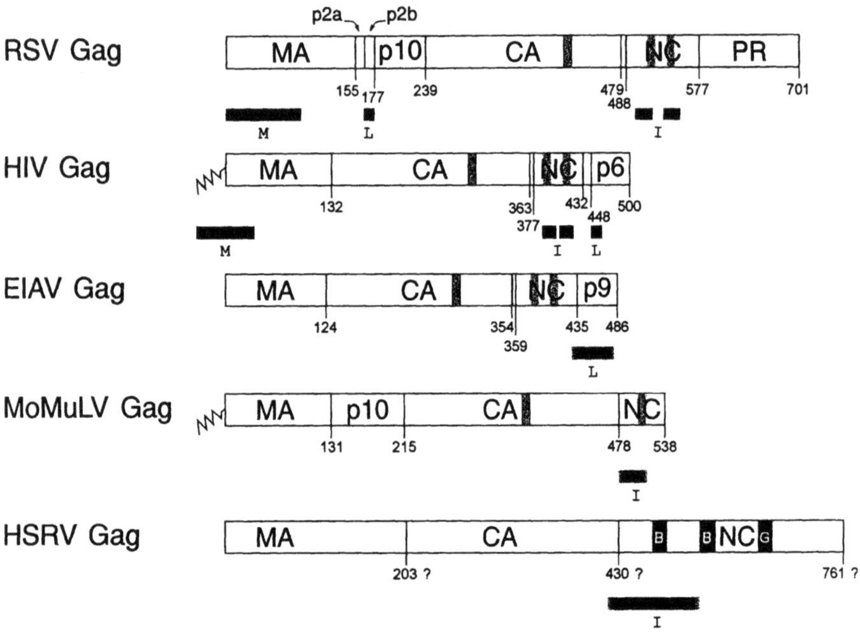

Fig. 1. Comparative organization of retroviral Gag proteins. The proteins are aligned at their amino termini with the zigzag extension representing the myristate modification on HIV Gag and MuLV Gag. The MHR region and zinc finger motifs of RSV, HIV, EIAV and MuLV are indicated by the *shaded regions* within the CA and NC domains, respectively. The HSRV Gag protein possesses neither of these motifs, but does contain within NC two basic regions (*B*) and a glycine-rich sequence (*G*). The precise termini of the mature HSRV proteins are not known. The location of the M, L and I domains of RSV and HIV-1 (*black bars*) have been mapped in detailed fashion, as described in the text. In contrast, the boundaries of the I domains of MuLV and HSRV and L domain of EIAV have not been delineated with the same precision

assemble at the plasma membrane. No electron-dense viral structures are seen in the cytoplasm of infected cells. The Gag proteins first appear to aggregate on the membrane in small electron-dense patches. As assembly of the nascent particles proceeds, the membrane bulges outward until a completely formed spherical protein shell surrounded by a membranous envelope is released from the cell surface. In contrast, the Gag proteins of the B- and D-type oncoviruses (e.g., mouse mammary tumor virus and Mason-Pfizer monkey virus, respectively) appear by electron microscopy (EM) to assemble into intracytoplasmic particles, which then are transported to the plasma membrane where envelopment and release occur. The process of envelopment and release of the B-, C- or D-type particles is known as budding. A third, nonproductive pathway is taken by certain endogenous intracisternal A-type particles, which bud into the endoplasmic reticulum, where they remain as immature, noninfectious particles.

Regardless of which morphogenetic pathway is utilized, the product of assembly and budding is a lipid-enclosed protein shell, the immature virus. Although these particles contain the viral RNA genome and the products of the *pol* and *env* genes, they remain noninfectious until the viral protease (PR) initiates proteolytic processing (Chap. 4, this volume; Dickson et al. 1984; Wills and Craven 1991). Action of PR on the Gag precursor late in the assembly process yields three mature proteins that are common to all retroviruses: MA (matrix), CA (capsid) and NC (nucleocapsid), which are linked in the precursor in that order. Similarly, reverse transcriptase (RT) and integrase, the enzymes of replication in the virion core, are formed by PR-mediated cleavage of the Gag-Pol precursor (the product of translational read-through from the *gag* to the *pol* gene).

Proteolytic maturation of the virion results in a dramatic reorganization of its internal structure, visualized microscopically as the change from an electron-lucent sphere to a particle with an electron-dense core (Dickson et al. 1984; Witte and Baltimore 1978; Luftig and Yoshinaka 1978; Stewart et al. 1990; Chap 4, this volume). In the mature virion, most of the MA protein is believed to remain associated with the inner surface of the virion envelope (Gelderblom 1990; Pepinsky and Vogt 1984). The NC protein and the RNA genome condense into a ribonucleoprotein complex (RNP), which also contains the enzymes of replication, reverse transcriptase and integrase (Bolognesi et al. 1973; Davis and Rueckert 1972; Stromberg et al. 1974). The CA protein, in turn, forms a capsid-like shell around this RNP structure. Together the capsid shell and the RNP compose the core of the virus. The precise organization of the proteins and RNA within the core, as well as the assembly processes that form the core after PR activation, remain largely unknown.

Although essential for viral infectivity, the mature virion proteins are not directly relevant to the processes of Gag assembly and budding. It is their precursor, the Gag polyprotein, that composes the "particle-making machine" (Dickson et al. 1984). Synthesis of the Gag protein alone has been shown to recapitulate the normal assembly pathway and to result in the formation and release of immature particles (Wills et al. 1989; Delchambre et al. 1989; Royer et al. 1991; GHEYSEN et al. 1989; SHIODA and SHIBUTA 1990). Co-expression of PR along with the Gag protein allows these virus-like structures to mature into ones that have both the protein composition and the morphological properties of the virus from which the Gag protein was derived (WELDON et al. 1990; SHIODA and SHIBUTA 1990; MERGENER et al. 1992; STEWART et al. 1990; SMITH et al. 1990).

The Gag proteins of different viruses do not show a strong cell-type specificity but can direct particle assembly and budding in a variety of cell types (ADACHI et al. 1986). It was not evident from the start that this would be true, since as early as 1975 it had been shown that RSV-infected mammalian cells do not produce virus particles (VOGT et al. 1982; EISENMAN et al. 1975). At the time it was not known whether this was due to the low levels of expression or whether the RSV Gag protein was truly nonfunctional in these cells. However, by using heterologous promoters to express the RSV *gag* gene, it was shown that there is no inherent block to RSV Gag assembly in simian or murine cells (WILLS et al. 1989; ERDIE and WILLS 1990). The particles released from mammalian cells are membrane enclosed and are identical to those made in avian cells with respect to their rate of release, particle size and morphology, and kinetics of maturation. Other laboratories have devised successful expression schemes for HIV, SIV, MLV and other retroviral Gag proteins using proviral plasmids, SV40-based vectors, vaccinia virus and baculovirus vectors in a variety of cell types (SHIODA and SHIBUTA 1990; SMITH et al. 1990; GHEYSEN et al. 1989; OVERTON et al. 1989; MERGENER et al. 1992; HOSHIKAWA et al. 1991; Chap. 8, this volume). The lack of an obvious dependence on the host species suggests that cellular factors that are involved in virus assembly, if there are any, must be widely conserved across species boundaries.

As the "assembly machine," the Gag protein itself appears to contain within its structure all the functional elements required for: (a) targeting and binding of the protein to the plasma membrane, (b) establishment of intermolecular interactions between Gag proteins in the immature particles and (c) envelopment and release of the particle. More than this, the assembly of infectious virions requires that the Gag protein must also direct the incorporation of the viral genome (Chap. 6, this volume). The proper organization of this RNA with the Gag-derived proteins in the core of the mature virion is essential if that core is to be competent to catalyze the reverse transcription and integration of the viral genome upon entry into a new host cell. As will be outlined in the reminder of this chapter, considerable progress has been made toward understanding how Gag proteins accomplish these essential functions.

1.2 Comparative Organization of Gag Proteins

Although similar budding functions must be contained within various retroviral Gag proteins, very few clues about where these functional domains lie can be gleaned by a comparison of their amino acid sequences or from the properties of their mature cleavage products. The Gag proteins of related viruses [for example, the members of the avian sarcoma and leukosis viruses (ASLV)] show obvious sequence and antigenic similarities; hence the derivation of the name "Gag," which stands for group-specific antigen. However, sequences of Gag proteins from viruses that are evolutionarily divergent are very different. They are so different, in fact, that computer-aided searches using a given Gag sequence as a probe do not identify all known Gag molecules in protein sequence databases (CRAVEN and WILLS, unpublished observations). A comparison of the CA proteins of RSV and MuLV shows identity at only 10% of the residues, and the other Gag-derived proteins show even less homology.

Nevertheless, there are certain features that are shared by all retroviruses (reviewed by DICKSON et al. 1984; WILLS and CRAVEN 1991) and many of these have now been linked to functional roles as discussed in this chapter. The linear order of the MA, CA and NC proteins within the precursor is invariant among the different retroviruses (Fig. 1). Furthermore, this order is reflected in the position of the mature proteins in the virion where MA forms a layer of protein under the membrane, CA makes an inner shell (the capsid) and NC is located in the center where it is complexed with RNA. It seems likely, although still unproven, that the Gag protein inside the immature particle is also oriented with the MA domain on the outer surface, CA in the middle and NC innermost and that this orientation is essential for the normal budding function of Gag.

When the MA, CA and NC domains are compared from virus to virus, certain motifs are found to be commonly, but not universally, present. The N-terminus of most MA proteins (and therefore their Gag precursors) is modified with the 14 carbon fatty acid myristate, which is added cotranslationally to the second residue, a glycine, after removal of the initiator methionine. However, a number of replication-competent viruses, including the avian retroviruses, lack this feature (KAWAKAMI et al. 1987; SCHWARTZ et al. 1983; SONIGO et al. 1985; MAURER et al. 1988). The majority of Gag proteins also have regions near their N-termini that are enriched in basic amino acids (ZHOU et al. 1994). The myristate group and the positive charges are both important for the membrane-binding properties of the Gag protein (Sect. 2.1). The CA sequence of most retroviruses (and of the yeast transposable element TY3) contains a stretch of conserved sequence, about 20 residues in length, that has been dubbed the major homology region (MHR) since it is the most conserved segment among Gag molecules (PATARCA and HASELTINE 1985; WILLS and CRAVEN 1991; HANSEN et al. 1988). However, the spumaretroviruses do not have this feature, nor do the yeast Ty1 and Ty2 elements. The MHR domain appears to be involved in intermolecular interactions within the virus interior that are important for core maturation and function (Sect. 4). NC, the nucleic acid binding domain of Gag, is distinctive due to the presence of cysteine-histidine arrays, also called zinc fingers because they chelate the metal ion (BESS Jr. et al. 1992; COVEY 1986). These motifs are present in one or two copies in all retroviruses, except the spumaretroviruses. As with the MHR domain, the zinc finger motif is found in the TY3 element (HANSEN et al. 1988), but not in Ty1 or Ty2. The zinc finger structures have been implicated in viral genome selection and packaging (MERIC and SPAHR 1986;

GORELICK et al. 1988; MERIC and GOFF 1989). The regions flanking the zinc fingers in NC contain many basic residues, which are likely to be involved in RNA-binding functions of NC and in directing the dense packing of Gag proteins during assembly (Sect. 2.2).

Aside from the presence of the three invariant cleavage products, there are numerous additional peptides that are unique to each virus and are positioned at different sites with Gag (Fig. 1). Some of these proteins are large (e.g., p16 of MPMV). Others are quite small (e.g., p1 and p2 in HIV-1), requiring sophisticated methods for their detection (HENDERSON et al. 1988b; HENDERSON et al. 1992; TOBIN et al. 1994). The functions of most of these peptides, as well as their positions in the virion, are unknown. Recently, however, a function required for the completion of budding has been attributed to the proteins p2b of RSV Gag, p6 of HIV Gag, and p9 of EIAV Gag (PARENT et al. 1995). In addition, the spacer peptide between CA and NC (a common feature of the lentiviruses and the ASLV group) appears to have a role in establishing protein-protein interactions in the virion leading to proper core maturation (CRAVEN et al. 1993; PEPINSKY et al. 1995).

The diversity in amino acid sequences, as well as in the organization of cleavage products, suggests that the different Gag proteins may have evolved to perform the same functions using a variety of mechanisms, making it risky to try to apply what is known about one Gag protein directly to others without empirically testing each observation. The availability of Gag expression systems has made the genetic dissection of HIV-1 and RSV Gag proteins and their assembly functions both practical and fruitful. Functional elements of both Gag proteins that are involved in membrane association and budding have been identified (the minimal budding machinery, Sect. 2), as have other involved in determining the internal organization of the virion (Sect. 4). These comparative genetic studies, as well as the in vitro assembly studies that are just beginning to come to fruition, have demonstrated that there are more similarities than differences in the assembly functions of the two viruses.

2 The Minimal Budding Machinery

The Gag protein of RSV has proven to be highly amenable to mutational analysis. Comprehensive deletion studies gave the first indications of where within the Gag protein elements essential for assembly and budding might lie. Remarkably large portions of the protein are dispensable for budding, as defined by the ability of the protein to form into particles of normal density and be released at the normal rate (WILLS and CRAVEN 1991; WELDON Jr. et al. 1990; WELDON Jr. and WILLS 1993; BENNETT et al. 1993; WILLS et al. 1994; CRAVEN et al. 1993; CRAVEN et al. 1995). Three regions of the RSV protein (which together comprise not more than 25% of the entire protein) are highly sensitive to mutation, suggesting that functions critical to budding lie within these domains (Fig. 1). This has been confirmed by using a variety of techniques including genetic complementation analyses (Fig. 2A), the construction of chimeric Gag molecules (Fig. 2B), the application of in vitro assays for protein functions and the correlation of genetic analyses with three-dimensional structural information. These three elements, which together comprise the minimal budding machinery, were originally named assembly domains AD1, AD2 and AD3 in reference to their linear order in the RSV Gag protein. However, to make the terminology more useful when comparing homologous elements of different viruses, a new designation for the components of the minimal budding machinery that better reflects their functions has been suggested – the M or membrane-binding domain (AD1), the I or interaction domain (AD3) and the L or late domain (AD2) (PARENT et al. 1995).

Mapping studies of the HIV-1 Gag protein in numerous laboratories have yielded results that are in general agreement with the RSV deletion data. As is true of RSV, large portions of CA and MA can be deleted from the HIV-1 Gag protein without serious consequences for budding efficiencies (SRINIVASAKUMAR et al. 1995; WAGNER et al. 1994; WANG and BARKLIS 1993; YU et al. 1992). Furthermore, regions of HIV-1 Gag have been identified that can substitute for domains of the RSV budding machinery and which presumably serve analogous functions in HIV-1. As would be expected, these regions of the HIV-1 protein are highly sensitive to mutation (see below). The modular elements that make up the budding machinery of RSV and HIV are described in the following sections.

2.1 The Membrane-Binding (M) Domain

The presence of a discrete membrane targeting/binding domain at the amino-terminus of the Gag protein is a common feature of Rous sarcoma virus, HIV-1 and presumably all budding-competent Gag proteins. In RSV, the M domain consists of the first 86 amino acids of Gag (NELLE and WILLS 1996; VERDERAME et al. 1996) (Fig. 1). Deletions anywhere within this region destroy assembly (WILLS et al. 1991; BENNETT et al. 1993) (Fig. 2A). However, all of the RSV M domain deletion mutants are able to be incorporated into particles with a wild-type protein in complementation-rescue experiments (Fig. 2A). Thus, deletions in this N-terminal region do not disrupt the structure of the remainder of the protein, suggesting that it is an independently folded domain. Moreover, the ability of these mutant proteins to be rescued by complementation implies that the major regions involved in Gag-Gag interactions lie downstream.

Evidence that the N-terminal 86 residues of RSV MA do indeed function as a membrane targeting and binding element comes from experiments in which membrane targeting signals derived from the amino-termini of the Src oncoprotein, the Src-related protein Fyn, and HIV-1 Gag were fused to the RSV Gag protein in place of the natural M domain (Fig. 2B) (PARENT et al. 1996; NELLE and WILLS 1996; L.J. PARENT and J.W. WILLS, unpublished data). Each of these was able to supply the missing function and support budding. In the reciprocal

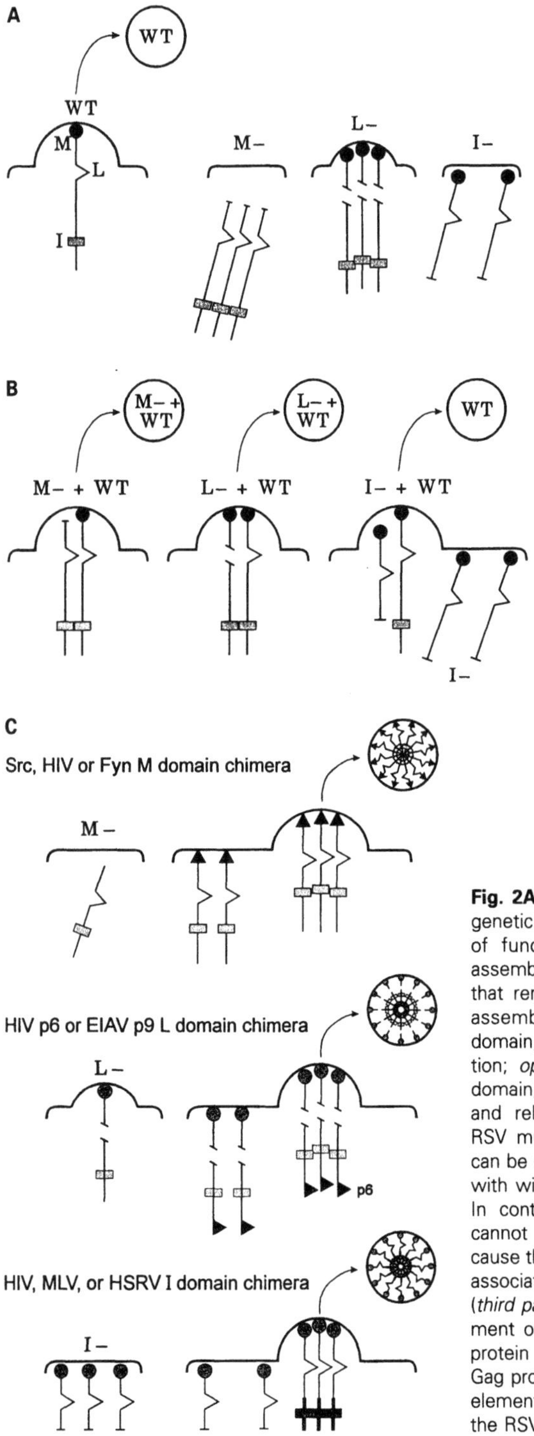

Fig. 2A–C. RSV deletion mutants studied by genetic complementation and by construction of functional chimeras. **A** Phenotypes of assembly domain deletion mutants. Mutations that remove any one of the three essential assembly domains, M (membrane-binding domain; *shaded circle*), L (late budding function; *open triangle*) and I (major interaction domain; *shaded box*) block particle assembly and release. **B** Complementation analysis. RSV mutants lacking the M and L domains can be rescued into particles by coexpression with wild-type Gag protein (*first two panels*). In contrast, mutants lacking the I domain cannot be rescued by complementation because they are missing the region required for association with the full-length Gag protein (*third panel*). **C** Functional chimeras. Replacement of assembly domains of the RSV Gag protein with segments from other retroviral Gag proteins has allowed the identification of elements that are functionally homologous to the RSV M, L or I domains

experiment, replacement of the N-terminus of Src with the RSV M domain results in a chimeric Scr protein that is fully functional and able to transform cells (VERDERAME et al. 1996).

Viruses whose Gag proteins are myristylated also depend upon an N-terminal domain for membrane targeting and binding. Mutations which prevent myristylation of murine leukemia virus (MuLV) (REIN et al. 1986) and HIV-1 (BRYANT and RATNER 1990; GÖTTLINGER et al. 1989) prevent the stable association of Gag molecules with the plasma membrane and interfere with particle formation. In the case of the Mason-Pfizer monkey virus (M-PMV), inactivation of the myristylation signal does not interfere with the formation of intracellular particles but inhibits the migration of these structures to the membrane so they are not released (RHEE and HUNTER 1987). As with the RSV M domain mutants, proteolytic processing of the myristate-minus Gag proteins of MuLV and M-PMV is also blocked (SCHULTZ and REIN 1989; RHEE and HUNTER 1987). Fine mapping of the membrane targeting/binding domain of HIV-1 Gag has shown that, in addition to the myristate moiety, a basic sequence lying between residues 15 and 31 is critical for the assembly and budding function of Gag (ZHOU et al. 1994). The first 31 residues of HIV-1 Gag can function independently as a membrane-targeting domain when fused to heterologous proteins, and both the myristate and the downstream basic residues are required for this activity. The fatty acid provides a hydrophobic anchor in the membrane while the basic charges mediate an electrostatic interaction between the protein and acidic phospholipids on the inner leaflet of the plasma membrane (ZHOU et al. 1994).

In spite of their differences in amino acid sequence and fatty acid modification, the N-terminal 31 residues of HIV-1 (including myristate) and the 86 N-terminal residues of RSV perform the same roles in their respective proteins. This was confirmed by replacing the entire 86-residue M domain of RSV with the N-terminal 31 residues of HIV Gag. Furthermore, the polybasic sequence ^{25}KKKYKLK^{31} from HIV-1 Gag when introduced into an M domain mutant of RSV Gag was able to repair the budding defect (PARENT et al. 1996). The identification of these domains and their roles in membrane targeting and binding, however, cannot explain why Gag proteins are targeted to the plasma membrane rather than to other intracytoplasmic membrane sites (Chap. 2, this volume); thus the specificity of targeting remains unexplained.

The conclusions drawn from the genetic dissection of the M domain are consistent with the three-dimensional structures of an unmyristylated version of the HIV-1 MA protein (MATTHEWS et al. 1994; MASSIAH et al. 1994). Seven basic residues of the HIV-1 M domain are located on the surface of the protein as is the N-terminus, which is normally the site of myristate modification. These features are surrounded by a collar of four additional positive charges contributed by the H^{33}, R^{76}, K^{95} and K^{98} residues, presumably forming the face of the protein that lies against the membrane (MASSIAH et al. 1994). At this writing, no structural information is available for the membrane-binding domain or the entire MA of RSV. Since function of the RSV M domain requires the entire 86 amino acids and no prominent cluster of basic charges can be found in the linear sequence, it is possible that upon tertiary folding of the protein monomer, or in oligomeric units, basic residues from distant regions are brought together to form a positively charged face on the protein surface which could interact with the membrane surface. Furthermore, since it lacks the fatty acid modification, it may require a relatively large charged surface to interact with the membrane.

Biophysical studies of lipid-peptide interactions and of the membrane-binding properties of Src, HIV-1 Gag and other myristylated proteins provide a thermodynamic basis for understanding the membrane-binding activity of the M domain (RESH 1994; McLAUGHLIN and ADERAM 1995) and also its role in particle assembly and budding. In vitro binding studies reveal that myristate alone provides insufficient binding energy to stably partition a peptide into the lipid bilayer (PEITZSCH and McLAUGHLIN 1993). In the case of the HIV-1 Gag and numerous other myristylated membrane proteins, electrostatic interactions between basic amino acids in the protein with acidic phospholipids in the inner leaflet of the plasma membrane provide the required additional affinity. In fact, clusters of basic residues in a peptide or protein apparently work with the myristate group in a cooperative fashion since the interaction of one component (myristate) with the membrane greatly increases the probability that the second component (the basic residues in an adjoining region of the same protein) will also associate with the lipid (reviewed by McLAUGHLIN and ADERAM 1995). In higher arrays of Gag proteins (Fig. 3), the individual M domains form an extended two-dimensional interface with the membrane. The individual molecules in this array presumably act cooperatively with one another to provide a very tight association of the complex with the membrane that may actually be the driving force for envelopment.

2.2 The Interaction (I) Domain

2.2.1 Evidence for I Domain Function

For Gag proteins to form into structures, they must be capable of associating with one another in a multimolecular complex. Although many sites of contact between neighboring Gag molecules may be established, several lines of evidence point to interactions involving the NC domains (acting either through protein-RNA or protein-protein interactions) as being especially critical for the successful assembly of a budding-competent structure.

Analysis of the budding machinery of RSV has shown that the functional elements required for membrane-binding, envelopment and release of particles (the M and L domains) lie near the N-terminus of the Gag protein. This is perhaps most convincingly demonstrated by a mini-Gag molecule (p25) which contains only 180 amino acids from the N-terminus of RSV Gag including the MA, p2 and three residues of p10 (WELDON Jr. and WILLS 1993). The "particles" produced by RSV p25 are released into the medium at an easily detectable rate but do not

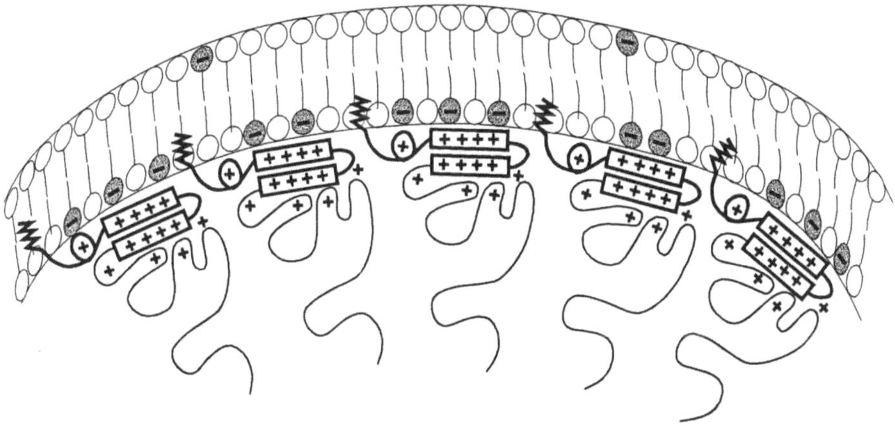

Fig. 3. Hydrophobic and elctrostatic interactions between the HIV Gag M domain and the plasma membrane. The plasma membrane is *at the top.* Acidic phospholipids (*shaded circles containing minus signs*) are enriched on the inner leaflet of the membrane. The M domains of the Gag proteins (*thick black lines*) lie just below the membrane. Myristate, shown as zigzag is buried in the membrane and provides membrane-binding energy through its hydrophobic association with the lipid bilayer. Basic residues (*plus signs*), clustered in two antiparallel beta sheets (*open boxes*) and in a surrounding collar, are on an exposed surface of the protein, presumably making them available to bind to the membrane cooperatively with each binding event, making subsequent ones more energetically favorable. As more Gag proteins bind, a large positively charged protein layer created at the membrane interface. The remainder of the Gag protein is shown as *a thin black line dangling into the cell*

resemble normal virus particles in either their size, which is extremely heterogeneous, or their density, which is considerably lower than normal (WELDON Jr. and WILLS 1993). Thus, the organization of Gag proteins into packages that resemble authentic viruses in size, shape and density appears to be dependent upon elements that lie downstream of p25 (i.e., in the CA and NC regions of Gag). The report of a budding competent mini-Gag consisting of the MA domain of SIV (GONZALEZ et al. 1993) is consistent with the RSV findings, although in that case the particle size was not examined. Fusion of the small RSV p25 protein to a downstream segment of RSV Gag containing the last quarter of CA and all of NC restored the efficiency of particle assembly and density to normal, but the particles remained heterogeneous in size (WELDON Jr. and WILLS 1993). Thus, size and density are physical properties of virus particles that are functionally separable, and it should be possible to define the sequences required for each.

These findings are further supported by the extensive deletion mapping of RSV Gag. The PR domain and carboxy-terminal half of NC can be deleted without detrimental effects on particle release (VOYNOW and COFFIN 1985; WILLS et al. 1994; MERIC and SPAHR 1986) or density (WILLS et al. 1994). However, particle release is destroyed or drastically reduced by deletions that remove all of NC or which extend into the CA region (BENNETT et al. 1993; CRAVEN et al. 1995; DUPRAZ and SPAHR 1992), and the few particles that can be recovered from the medium are of low density, similar to the p25 "particles" (BENNETT et al. 1993).

Removal of all the CA domain significantly improved the efficiency of release (WELDON Jr. and WILLS 1993). The budding-deficient C-terminal truncation mutants, unlike the M and L deletion mutants, cannot be rescued into particles by a budding-competent Gag molecule, suggesting that regions of the Gag protein that are critical for intermolecular interactions might lie in this region. Paradoxically, however, the budding defects of these large deletions could not be mimicked by any of a collection of smaller deletions spanning the CA, SP and NC domains (CRAVEN et al. 1995; MERIC et al. 1988). These results suggest that there may be multiple sites of interaction within the NC region, i.e., there is redundancy of function. However, as with all mutational studies, the possibility must be considered that the folding of the Gag protein is very sensitive to alterations in this region and that the loss of assembly in at least some of these mutants is an indirect result of disturbances in protein structure.

The critical role of NC in assembly has also been indicated by numerous deletion studies of HIV Gag. Progressive deletions of the C-terminus of HIV-1 Gag showed that mutations which removed one zinc finger or both still allowed particle assembly and release (JOWETT et al. 1992). However, consistent with the RSV findings, when both were deleted the particles released were of low density. Alterations to the zinc finger motifs themselves, however, do not interfere with assembly (GORELICK et al. 1990; ALDOVINI and YOUNG 1990). C-terminal truncations that removed all of NC have been found by several laboratories to prevent particle release in HIV-1 (GHEYSEN et al. 1989; JOWETT et al. 1992; CARRIÈRE et al. 1995) and in HIV-2 (LUO et al. 1994). However, the interpretation of such loss-of-function mutations is difficult, especially since the C-terminal truncations are now known to delete multiple elements that are important for assembly and budding – e.g., the regions of protein-RNA interaction, the late or L domain (p6), as well as regions that may be involved in protein-protein interactions (see below).

Confirmation that the NC domains of RSV and HIV are indeed contributing in similar ways to assembly has come from studies of functional chimeric proteins (BENNETT et al. 1993; R.P. BENNETT, N.K. KRISHNA, R.B. BOWZARD, J.W. WILLS, unpublished data) (Fig. 2B). Two separate portions of the HIV NC protein, each bearing one of two zinc finger motifs, are capable of restoring dense particle formation to a budding-defective I domain mutant of RSV. This confirms that there is indeed redundancy of I domain function within certain NC proteins. Most strikingly, a seven-amino-acid peptide bearing a repeat of the RKK sequence from the central basic region (lying between the two zinc fingers) of HIV NC was able to restore dense particle production to the RSV mini-Gag protein p25. Sequences from the RSV or HIV-1 CA or SP are not required for this activity. Furthermore, the NC proteins of two evolutionarily distant viruses, MuLV and the human spumaretrovirus (HSRV), are functionally equivalent to those of RSV and HIV-1 in their ability to promote particle assembly (R.P. BENNETT, N.K. KRISHNA, J.B. BROWZARD, J.W. WILLS, unpublished data; DUPRAZ and SPAHR 1992). In contrast to the HIV-1 chimeras, however, deletion analyses have not yet revealed redundant I domains within the NC sequences of MuLV and HSRV.

Taken together, these studies implicate the NC region of the Gag protein in normal budding efficiency, in the ability of the proteins to interact with one another and in particle density. Are these activities all due to the same function of the same element or are they separable functions? Are any or all of these activities mediated through protein-RNA interactions or only through protein-protein contacts?

2.2.2 Role of NC-RNA Interactions in Particle Density

No models are available to help explain what determines the density of retroviral particles or how differences in density arise. The density of PR-defective particles is the same as their wild-type, mature parents (WILLS et al. 1994; JOWETT et al. 1992), indicating that this property is determined during the assembly process rather than during proteolytic maturation. Particles produced in the absence of viral genome expression still possess densities similar to authentic virions. Thus, the incorporation of viral genomic RNA into the virion is not required for normal density. Furthermore, the characterization of numerous RSV deletion mutants has shown that density is independent of particle size and of the molecular mass of the Gag protein itself (WELDON Jr and WILLS 1993).

The genetic studies indicate that the NC domain of Gag is required for normal particle density, but what are the structural features of NC that are important for this assembly function? The recent determination of the HIV-1 NC structure by nuclear magnetic resonance imaging (MORELLET et al. 1992) has shown that the zinc finger regions form independently folded, globular domains with a chelated metal ion in the center of each. The basic linker between the two fingers, as well as the flexible N- and C-terminal sequences, is solvent exposed and free to interact with RNA. However, the exact structure of the RNA-protein complex, in particular which amino acids of the protein actually contact the nucleic acid in the RNP complex, has not been determined. The zinc finger domains are, within the context of the Gag precursor, involved in the selection of viral genomic RNA from the unspliced mRNA pool and its packaging into virions (GORELICK et al. 1988, 1990; JOWETT et al. 1992; MERIC et al. 1988; DUPRAZ et al. 1990). Furthermore, NC protein has RNA-RNA annealing activity that can be demonstrated in vitro and is believed to catalyze the annealing of the two genomic RNA strands at the dimer linkage site and the tRNA replication primer to the primer binding site on the genome during virus assembly and maturation (DE ROCQUIGNY et al. 1992; PRATS et al. 1988). Finally, the NC protein binds to the RNA via sequence-independent interactions to form the nucleocapsid or ribonucleoprotein complex (RNP) in the virion core (KARPEL et al. 1987). Neither the RNA annealing activity of NC nor the sequence-independent interactions of NC with RNA appear to require functional zinc finger structure (DUPRAZ and SPAHR 1992; DE ROCQUIGNY et al. 1992).

Although the density-conferring sequences appear to flank or overlap the zinc finger motifs, mutation of the cysteine residues or precise deletion of the motifs does not destroy dense particle formation (R.P. BENNETT, S. ERNST,

A. REIN, J.W. WILLS, unpublished data; OTTOMAN et al. 1995). Further mapping is needed to more accurately define the location of the density-conferring elements as well as their number and exact sequences. Nevertheless, the proximity of these elements to the zinc fingers explains why they appear to be duplicated in RSV and HIV (which both possess two fingers) but not in MuLV (which has only one). No sequence consensus is discernible among the NC regions that are capable of promoting dense particle formation. All are highly basic, however, suggesting that these elements coincide with the regions of NC that compose the RNA interface in the mature virion.

These studies suggest a model of assembly in which the RNA acts as the scaffold for condensation of a dense particle (Fig. 4) (BENNETT et al. 1993;

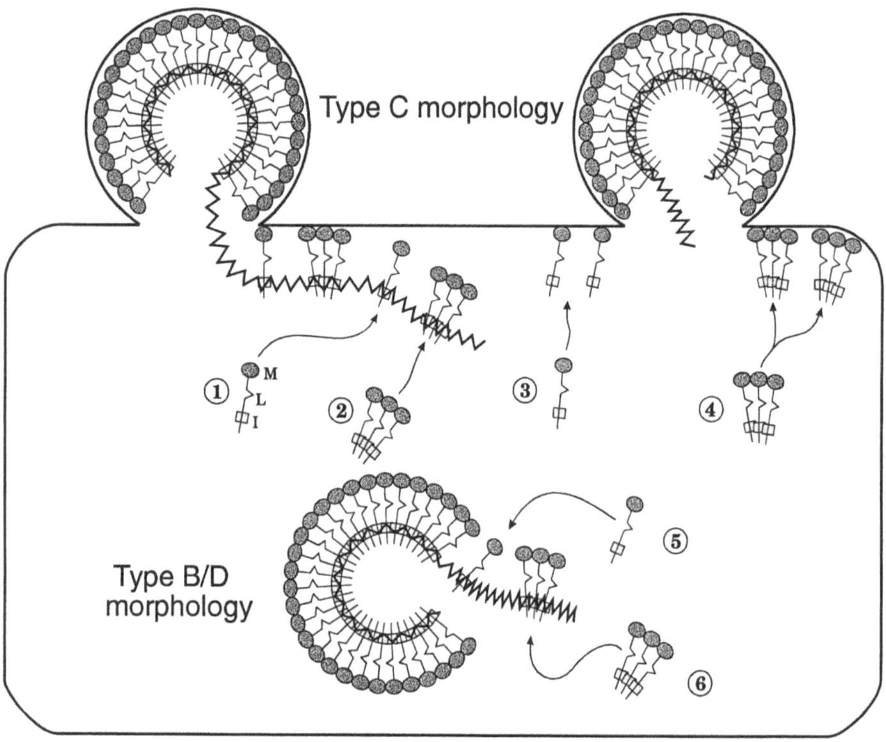

Fig. 4. Cross-sectional view of assembling particles of type C and type B/D retroviruses. Evidence summarized in the text supports a model of retrovirus assembly in which the formation of a dense particle is organized by the viral RNA (*zigzag line*). Protein-protein interactions between Gag molecules determine the size and shape of the nascent particles. It is not known whether Gag molecules are active in assembly as monomers (1, 3 and 5) or whether the protein forms multimers (2, 4 and 6) before binding to the viral RNA or membrane. Clearly, particles of the B- and D-type viruses assemble RNA and Gag proteins into particles before contacting the membrane (5 or 6). In the case of RSV, HIV and MuLV, it is not clear whether Gag monomers or oligomers bind first to RNA (1 and 2) before contacting the membrane or vice versa (3 and 4). However, the rescue of RSV and HIV Gag M domain mutants in complementation experiments implies that intermolecular interactions can occur in the absence of membrane binding (via pathway 1, 2 or 4). MuLV assembly, in contrast, seems to require membrane binding prior to establishment of intermolecular interactions (3)

JOWETT et al. 1992). Electrostatic interactions between the NC domains of Gag proteins and the RNA presumably direct the alignment of the proteins and allow them to pack tightly. Such tight packing of NC domains was predicted by in vitro NC-RNA binding experiments (KARPEL et al. 1987) and has actually been detected by cross-linking experiments in immature particles of MuLV (PEPINSKY 1983; HANSEN and BARKLIS 1995) as well as in mature RSV particles (PEPINSKY et al. 1980). This scheme is supported by in vitro assembly studies using CA-NC proteins derived from RSV and HIV (CAMPBELL and VOGT 1995). In this system, the presence of RNA during assembly is needed for the formation of highly regular protein arrays.

Clearly, the presence of virus RNA is not absolutely required for assembly. Nonspecific RNA can substitute for the viral genome during in vitro particle assembly (CAMPBELL and VOGT 1995). Although it has not been rigorously proven, virus particles that are assembled in cells lacking genomic RNA expression are believed to incorporate miscellaneous cellular RNAs instead (SAKALIAN et al. 1994). In the I domain studies, it is not known to what extent the various NC mutations affect the ability of the proteins to package cellular RNAs. It is possible that the low-density particles contain less than the normal complement of RNA. In an authentic virion, however, the RNA represents only about 1% of the total mass of the particle. Whether the presence or absence of RNA in the particle is alone sufficient to cause the density changes seen in the I domain mutants is not known. An alternative model has been suggested – that the abnormal density of NC mutants results from disordered packing of Gag molecules in the nascent particle, due to the altered structure of the NC domain, causing a reduction in the number of Gag proteins per unit of membrane surface (BENNETT et al. 1993). Both interpretations provide testable predictions, e.g., that light particles would possess either a higher protein:RNA ratio or a lower protein:lipid ratio than normal, dense particles. Also, if the packing of Gag precursors is disordered, this may be detected experimentally as a reduced ability of the MA or CA domains to be chemically cross-linked in light particles.

2.2.3 Protein-Protein Interactions Between Neighboring NC Domains

The lack of obvious sequence dependence suggests that protein-protein interactions between NC domains on neighbouring Gag precursors are not required for normal density. However, it is very probable that such interactions do form during normal assembly and may be important for stabilizing the protein-RNA aggregates during budding or for the interactions detected during complementation-rescue experiments. It is also likely that NC-NC interactions have important functions after budding – in organizing the mature cleavage products within the virion core, in the RNA annealing activity of the NC protein, for protecting the RNA from degradation or for other roles that NC may play during reverse transcription.

At least in the case of RSV, it is becoming clear that budding efficiency and complementation are genetically separable from the density-conferring function in the I domain. In chimeric protein studies, certain NC-derived sequences have proven capable of conferring high density but are not sufficient to restore full efficiency of assembly (R.P BENNETT and J.W. WILLS, unpublished data). Furthermore, certain RSV NC deletions have been found to produce particles efficiently but those particles remain of low density (CRAVEN et al. 1995; N.K. KRISHNA and J.W. WILLS, unpublished data). At least one mutant of this class has proven capable of interaction in complementation-rescue experiments. It seems most likely that protein-protein interactions mapping to this region are important for some of these activities.

Several studies indicate that the NC domains of HIV Gag precursors are capable of forming protein-protein interactions (CARRIÈRE et al. 1995; FRANKE et al. 1994; J.L. DARLIX, 1995 personal communication). A segment of HIV Gag protein bearing the NC protein, as well as the major homology region of CA, is sufficient for multimerization in the yeast two-hybrid system (FRANKE et al. 1994). Also, Gag-Gag interactions that are dependent upon the presence of the SP-NC sequences have been detected by ligand affinity blotting assays (CARRIÈRE et al. 1995). This region of the HIV Gag protein has been shown by many laboratories to be highly sensitive to mutations, consistent with the possible involvement of protein-protein interactions near the SP-NC junction in HIV budding (KRÄUSSLICH et al. 1995; GHEYSEN et al. 1989; JOWETT et al. 1992; PETTIT et al. 1994; SRINIVASAKUMAR et al. 1995; Sect. 4.3). Further studies to map the interactions more precisely and to correlate the protein structure with the in vivo genetic results will ultimately yield a more complete picture.

2.3 The Late (L) Domain

Regardless of the assembly pathways (C vs. B/D) that the Gag proteins follow, budding of virus particles requires the release of the particles from the cell surface. Recent evidence indicates that Gag molecules contain an element named the L (or late) domain that is required for efficient "pinching off." The existence of the L domain in RSV Gag was initially defined by the creation of a large, internal deletion which blocks budding at a late step in the assembly pathway (WILLS et al. 1994). Mutants with this phenotype are targeted to and bind the plasma membrane; however, particles are not released. In contrast to the M and I deletion mutants, these exhibit normal levels of proteolytic processing, which suggests that the block to budding is at a very late step in the process. L domain mutants can be rescued by complementation (Fig. 2A), indicating that regions required for Gag-Gag interaction have not been affected. Further deletion analysis showed that the late function maps to the 11 amino acid p2b region of RSV Gag. Within this region is a proline-rich motif PPP(W/Y), which is found near the MA cleavage site of many retroviruses with the notable exception of the lentiviruses, where MA is directly linked to CA (WILLS et al. 1994).

Mutants of HIV-1 Gag have been described which lack the C-terminal p6 sequence and have a phenotype very similar to the L domain mutants of RSV (GÖTTLINGER et al. 1991). They too are blocked at a late step of budding, and completely formed particles can be seen by EM to accumulate on the plasma membrane. The similarity of phenotypes in the HIV-1 and RSV mutants suggests that the p2 sequence of RSV and p6 of HIV-1 might be performing the same roles in the two viruses. Indeed, in chimeric molecules the C-terminal p6 sequence of HIV-1 is able to suppress internal deletions of the RSV L domain (Fig. 2B). The p9 sequence of equine infectious anemia virus (EIAV) Gag, which is located in a C-terminal position, is similar to the HIV-1 p6, and likewise is able to stimulate release of RSV L domain mutant particles (PARENT et al. 1995). Thus, the oncoviruses and lentiviruses appear to have functionally homologous late budding domains.

The ability of p6 and p9 to replace RSV p2 function is notable for two reasons. For one, in contrast to the major cleavage products MA, CA and NC, whose linear order is preserved in all known retroviruses, the internal position of the p2b sequence of RSV is dramatically different from the C-terminal location of p6 and p9. However, this arrangement is not essential for activity since the p2b sequence is also functional when positioned at the C-terminus in place of p6 in the chimeric protein (PARENT et al. 1995). Likewise, the N-terminal 12 residues of the HIV-1 p6 protein are also able to function when transplanted from the C-terminus of the chimeric protein to the internal position normally occupied by the RSV L domain. Of course, since the tertiary structures of the Gag proteins have not been determined, it remains a possibility that the location of the p6 and p9 domains in the particles may actually be similar to that of p2 in three-dimensional space (PARENT et al. 1995).

In addition to the differences in the position within the Gag protein, there is little or no similarity in amino acid composition or sequence among the three L domains. The active region of the HIV-1 p6 protein resembles the RSV sequence in containing a high frequency of proline residues. A similar proline-rich sequence is present in a number of other retroviruses, including visna virus and in the gibbon ape leukemia virus. However, this correlation does not extend to the EIAV p9 protein, which is not proline-rich and has no apparent sequence similarly to any of the other viruses.

There are at present no clues to suggest how the L domains function, and not enough is known about the actual budding mechanism to make many predictions. It is not known whether these L elements actually work at the membrane to literally stimulate particle release or whether they have a more indirect influence, perhaps affecting the conformation of the Gag protein in a way that makes it able to complete the budding process. It is difficult to propose a strict requirement for a structural motif, since the sequences of p2b, p6 and p9 are very different. Furthermore, p6 is known by CD and NMR analysis to lack a rigid structure (STYS et al. 1993). It may be that a lack of secondary structure is important for function of the L domain. We have speculated that a host factor may be involved in enhancing particle release through an interaction with the L domain (PARENT et al. 1995). This might explain why some studies have failed to show a defect in assembly for HIV-1 p6 mutants (HOSHIKAWA et al. 1991; JOWETT et al. 1992; ROYER et al. 1991; PAXTON et al. 1993), since the cell lines, methods of transfection or expression vectors used could all influence the expression of host factors, and thus reduce the need for p6. Finally, although the late functions of unrelated retroviruses can complement one another, it remains possible that they work via different mechanisms or in association with different host factors.

3 Interactions Between Unrelated Retroviral Gag proteins

The comparative analysis of the RSV and HIV Gag proteins has revealed that the budding machinery of each is composed of similar functional modules. Most strikingly, the ability to exchange individual assembly domains (Sect. 2) or entire MA, CA or NC regions (DEMINIE and EMERMAN 1993; CARRIÈRE et al. 1995) between different Gag molecules without destroying assembly function has made a strong case of the conservation of the basic budding mechanisms across species lines. These similarities of organization and function among diverse retroviruses suggest that the proteins from two very different viruses might be compatible enough to be co-packaged into the same particle. However, numerous attempts to co-package the wild-type MuLV Gag with wild-type RSV Gag protein have been unsuccessful. Cells expressing both molecules appear to produce two discrete and homogeneous populations of particles (BENNETT 1994). Similarly, efforts have also failed to co-package wild-type MuLV and HIV (HANSEN and BARKLIS 1995).

The inability to detect co-assembly of the MuLV protein with other retroviral Gag molecules does not indicate an inherent incompatibility of the proteins themselves, however. When the N-termini of both the RSV and MuLV Gags were replaced by the identical membrane-binding signal (the first ten residues of the Src oncoprotein) and the two proteins expressed in the same cell, mixed aggregates formed (BENNETT 1994). The presence of particles containing both proteins could be detected in the culture medium by co-immunoprecipitation of the two molecules with either an anti-RSV or an anti-MLV antiserum. Genetic evidence provides further support for the ability of the two heterologous proteins to interact. The MuLV (Src) hybrid protein in a complementation experiment is able to rescue an RSV (Src) protein that is budding-defective due to a large L domain deletion.

The fact that the presence of the Src membrane targeting and binding signal at the amino-terminus of each protein could enable co-packaging suggests that it functions by directing the proteins to a common trafficking pathway or a common destination on the plasma membrane, thereby allowing the heterologous proteins a chance to form mixed aggregates via protein-RNA and protein-protein interactions. In the case of MuLV, it has been suggested that membrane-binding is an obligate first step in the association of two MuLV Gag precursors, since their ability to be co-packaged is dependent upon the presence of an intact N-terminus on both molecules (SCHULTZ and REIN 1989). If this interpretation is correct, then the formation of mixed RSV-MuLV multimers during co-expression could presumably only occur once each protein is bound to the membrane.

This stands in contrast to the RSV studies, where it has been shown that M domain mutants (which by themselves presumably do not interact with the membrane) can be rescued by a wild-type Gag protein (WILLS et al. 1991). This complementation could conceivably occur if the two molecules interact first in the cytosolic compartment prior to binding to the membrane or, alternatively, the M domain mutant may be capable of binding to a wild-type protein that is itself already associated with the membrane. In either case, the failure of the wild-type RSV Gag protein to associate into particles with wild-type MuLV Gag proteins appears to be an indirect indicator that the assembly pathways for the two are physically separate and perhaps utilize discretely different portions of the plasma membrane.

4 Role of Gag Elements in the Internal Organization of Particles

The characterization of assembly domains (i.e., the minimal budding machinery) tells only part of the assembly story. Clearly other regions of Gag are essential for the proper organization of the proteins within the particle. The manner in which Gag proteins are organized in the nascent particle, e.g., their packing relative to one another and the establishment of intermolecular bonds in the right places, will determine whether or not the maturation of the core can proceed properly during and after PR processing. This in turn will affect whether or not the resulting core will be functional upon entry into a new host cell. In contrast to the rather small domains required for budding, the portions of the Gag protein that are involved in the internal organization are likely to be very large. The NC domain has been shown to influence the physical structure of particles (Sect. 2.2). Sequences in the first half of Gag have also been implicated.

4.1 The MA-p2-p10 Region of RSV Gag

The release of membrane-enclosed particles formed by the p25 mini-Gag protein of RSV (WELDON Jr. and WILLS 1993) and a similar result with SIV MA (GONZALEZ et al. 1993) indicate that the N-terminal portion of Gag is capable of some degree of self-association on the membrane. However, the heterogeneity in size and density of the particles in the case of the RSV mutant indicates that the protein arrays do not have the regularity of structure that is characteristic of authentic virus particles.

In vitro assembly studies, provide evidence for a contribution of the N-terminal portions of RSV Gag (MA-p2-p10) to the three-dimensional arrangement of Gag molecule within immature particles. A protein consisting of the CA-NC portion of RSV Gag can assemble in vitro with RNA to form tubular ribonucleoprotein complexes, the length of which depends upon the RNA size (Campbell and Vogt 1995). Additional protein sequences upstream of CA, however, are required to form spherical particles that resemble the protein shell of immature viruses (S. Campbell and V.M. Vogt, personal communication). These findings imply that the protein sequences in the MA-p2-p10 region of RSV Gag have a crucial role in determining the symmetry of packing of proteins during the assembly of the immature particle.

4.2 The CA Domain

Purified CA proteins from MuLV and HIV-1 possess the ability to self-associate in vitro into multimers as would be expected for a capsid protein (Ehrlich et al. 1990; Burnette et al. 1976). However, from such studies alone, it cannot be discerned whether the intermolecular forces that allow multimerization in vitro more closely resemble those that occur between the CA domains of the neighboring Gag proteins during assembly or those that hold together the mature CA subunits in the capsid shell of the mature virion.

The best evidence to date for a major role of the CA domain in determining the organization of Gag molecules within immature particles comes from genetic studies. A large number of CA mutants of RSV, HIV-1 and M-PMV have been found to produce abnormally sized or shaped particles. The deletion of a large portion of CA in RSV results in the release of particles that, although they are of normal density, exhibit moderate to extreme size heterogeneity (Weldon Jr. and Wills 1993). Similar phenotypes have been reported for certain CA mutants of HIV-1 (Dorfman et al. 1994; S.P. Goff and A. Reicin, personal communication). Moreover, mutations at the CA-NC junction of HIV have been reported to cause the formation of anomalously shaped particles or particles of abnormal size (Göttlinger et al. 1989; Kräusslich et al. 1995; Pettit et al. 1994). Deletions in this same region of RSV result in particle size heterogeneity (N.K. Krishna and J.W. Wills, unpublished data). Numerous point mutations within the CA sequence of M-PMV cause the release of filaments and other anomalously shaped particles (Strambio-de-Castilla and Hunter 1992). These alterations to CA have most likely disturbed particle morphology by altering the geometry of packing during budding. Most of these misshapen particles have proven to be either noninfectious or seriously impaired for infectivity, underscoring the importance of appropriate Gag-Gag interactions for the later steps of the replication cycle.

The hypothesis that CA contributes in a major way to the symmetry of packing of Gag proteins during assembly seems reasonable given what is known about the structure of the Gag protein. CA is the largest domain of the Gag protein. Furthermore, the CA sequence is situated in the Gag protein between the membrane-binding domain (M) and the region that mediates RNA interaction (I). Mutations within this central region could, therefore, be expected to have drastically different effects on particle size or shape depending upon the exact nature of the alterations. Mutations that cause relatively minor effects on the structure of the CA domain may subtly alter the packing of Gag molecules into the protein shell and thereby the curvature of the protein-membrane interface. The result may be relatively small effects on the size of the emerging particles. Alterations in the symmetry of packing of Gag molecules could also result in abnormal shapes, such as the long tubes reported in some studies (STRAMBIO-DE-CASTILLIA and HUNTER 1992; GHEYSEN et al. 1989; GÖTTLINGER et al. 1989). Certain mutations have been shown to cause very dramatic effects on size, both in terms of the size distribution and the actual size of the larger particles. This behavior might be explained by a greater degree of disorder in the Gag protein aggregation leading to the release from the cell surface of membrane patches of various sizes (WELDON Jr. and WILLS 1993). Yet other alterations (CRAVEN et al. 1995; REICIN et al. 1995; DORFMAN et al. 1994; TRONO et al. 1989; VON POBLOTZKI et al. 1993; CHAZAL et al. 1994; KRÄUSSLICH et al. 1995) may distort the Gag structure sufficiently to interfere with membrane targeting or to prevent the membrane-binding and RNA interaction domains from working cooperatively. In this case the result would be a budding-negative phenotype. If this scenario is indeed the case, then it is not surprising that removal of most or all of CA could eliminate the potential for interference with budding, as has been observed in HIV-1 and RSV (WELDON Jr. and WILLS 1993; REICIN et al. 1995; SRINIVASAKUMAR et al. 1995).

A clear understanding of how the CA protein and MA domains control protein packing is not yet possible since no three-dimensional structures are available for either the Gag precursor or the free CA protein. Further exploitation of the in vitro assembly methods, coupled with structural studies and genetic studies, will eventually allow a finer dissection of the protein-protein interactions that determine the structure of the nascent particle and the mature capsid.

4.3 Conserved Elements in CA: Role in Gag-Gag Interaction or Core Maturation?

Conserved elements within the CA protein have been implicated in intermolecular interactions. However, whether these elements contribute to such functions during Gag assembly or during the post-budding phase of core maturation is far from clear at this time.

One widely conserved feature of CA proteins is the presence of one or more small peptides separating the CA and NC sequences in Gag. The presence of such spacer peptides has been established in the avian retroviruses (PEPINSKY et al. 1995) and several lentiviruses (TOBIN et al. 1994; HENDERSON et al. 1988a, 1992; ELDER et al. 1993). In fact, MuLV is the only retrovirus that has so far been clearly shown to lack this feature (HENDERSON et al. 1984). Limited sequence similarity between the spacer peptides of HIV and RSV has been noted (PEPINSKY et al. 1995). Where it has been examined, in HIV-1 and RSV, the cleavage at the multiple sites between CA and NC occurs in a temporarily regulated manner with proteolysis occurring first at the SP/NC junction. In HIV-1, the resulting CA/p25 protein is subsequently trimmed at its C-terminus to form p24 (MERVIS et al. 1988; GÖTTLINGER et al. 1989; TRITCH et al. 1991). In the case of RSV and other ASLVs, the first CA species detected in virions (CA+SP or CA1) undergoes further cleavage at three upstream sites after release of the particles from cells, resulting in multiple species of CA (PEPINSKY et al. 1995). It has been suggested that these several species are involved in determining the symmetry of the CA shell in the mature virus (PEPINSKY et al. 1995). The mature RSV CA proteins possess a degree of resistance to detergent extraction that is not characteristic of the immature (CA+SP) species, suggesting that the mature proteins establish interactions with neighboring proteins in the core during the post-budding maturation process (CRAVEN et al. 1993).

Mutations that remove the spacer peptides or alter the flanking cleavage sites in both RSV and HIV-1 have proven, without exception, to have drastic effects on infectivity (CRAVEN et al. 1993; PEPINSKY et al. 1995; GÖTTLINGER et al. 1989; PETTIT et al. 1994; KRÄUSSLICH et al. 1995). The effects on particle assembly have been highly variable, however. RSV deletions that remove different portions of the spacer region all allow efficient particle formation. The cores of particles recovered from the medium are unstable in detergent, indicating that the mutations have disrupted normal core morphogenesis (CRAVEN et al. 1993; PEPINSKY et al. 1995). This phenotype suggests that the sequential processing at the end of CA is important for the establishment of the proper intermolecular bonds during the latest phase of maturation. It is also possible that the spacer peptide itself forms interactions between Gag precursors that (although not essential for budding by RSV) are prerequisite for formation of additional bonds between CA molecules during C-terminal maturation.

The function of the spacer peptide in HIV (also known as p2) may be more complex, since the assembly and release is compromised by many mutations in and near this region (KRÄUSSLICH et al. 1995; PETTIT et al. 1994; ZHAO et al. 1994; CHAZAL et al. 1994; REICIN et al. 1995; HONG and BOULANGER 1993). In one study, deletions of p2 abolished infectivity seemingly without interfering with assembly, although quantitative measurements of release efficiency were not reported (PETTIT et al. 1994). In other studies, deletion of p2 or mutation of the CA-p2 cleavage site resulted in the formation of abnormally shaped particles and drastically reduced the particle release (GÖTTLINGER et al. 1989; KRÄUSSLICH et al. 1995). The discrepancy in the phenotypes of similar mutants from different laboratories is probably due to the different methodologies used. In any case, the HIV-1 results are not inconsistent with the RSV studies. If the p2 peptide does mediate Gag-Gag interactions during particle assembly, then the higher sensitivity of HIV-1 to mutations in this region may reflect a greater importance of such interactions for the budding machinery in HIV-1 than in RSV. Alternatively, HIV-1 Gag may be more sensitive than is the RSV protein to conformational distortion by mutations in this region, causing secondary effects on budding.

Another highly conserved sequence within the CA protein is the major homology region or MHR. Genetic studies of the MHR region of RSV, HIV-1 and M-PMV have found strikingly similar results, indicating that this domain must be playing the same role in each virus (Craven et al. 1995; Mammano et al. 1994; Strambio-de-Castilla and Hunter 1992). In all three cases, conservative changes in the second half of the region allowed particle assembly and release but destroyed virus infectivity. The infectivity defect in these budding-competent mutants does not correlate with defects in RNA, Pol or Env incorporation or inability to initiate reverse transcription in vitro. Rather, the noninfectious but assembly-competent mutants of RSV and HIV-1 exhibit defects in core integrity, indicating that improper intermolecular interactions have been formed during core maturation (Craven et al. 1995; Mammano et al. 1994). Substitutions in the first half of the MHR, in contrast, have proven to prevent assembly in all three viruses. Yet in other mutants these residues can be deleted entirely without compromising budding efficiencies (Craven et al. 1995; Srinivasakumar et al. 1995). Finally, mutations in the MHR of the TY3 element have been found to disturb the structure of the ribonucleoprotein particle (the equivalent of the viral core) and to interfere with transposition (S. Sandmeyer, personal communication).

The best interpretation appears to be that the MHR domain is not an essential part of the budding machinery. The Ty3 particles are not released from the cell at all and so have no need for budding machinery. Rather, the MHR residues seem to be of critical importance for core morphogenesis. It is possible that the MHR residues themselves mediate protein-protein interactions within the core. On the other hand, it may be that interactions between Gag proteins during nascent particle assembly are dependent upon the MHR function and that these interactions are prerequisite for the later steps of morphogenesis that occur after particle release. Many of these uncertainties will be resolved as three-dimensional structural information for both the CA domain of Gag and for the mature CA proteins become available.

5 Perspective

The mapping of functional regions of the Gag protein, as reviewed in this chapter, tends to focus attention on isolated portions of the molecule while often ignoring how these domains influence one another as they act in concert. As a result, this type of analysis certainly understates the complexity of the assembly process. Nevertheless, the studies reviewed here clearly show many more functional similarities than differences among the different virus proteins.

The structural requirements for the budding function of the Gag protein appear to be relatively simple. The Gag protein must have a plasma membrane targeting/binding domain in order to direct assembly to the membrane. Interactions between Gag molecules must be established to hold the protein together, whether through protein-protein interactions or protein-RNA interactions, into a structure that is able to bud through the membrane. Finally, a late function is needed to complete the release of the assembled structure from the cell. Highly regular arrays of proteins are not required for assembly and release to occur. This is indicated by the fact that nonfunctional mutants bearing large deletions can be rescued into particles by intermolecular interactions with functional proteins. Furthermore, the relative insensitivity of the budding machinery to very large deletions in certain parts of the protein, the fact that foreign sequences can be inserted at various places in the Gag protein without interfering with assembly and budding (WELDON et al. 1990; J.W. WILLS, unpublished), and the ease with which budding-competent chimeras (section 2, Fig. 2c) can be created all argue for the existence of a high degree of flexibility in the arrangements of Gag proteins that can be assembled without compromising the budding process.

The formation of an infectious particle, however, demands a much greater degree of precision in assembly. The ability of the viral core to successfully complete the reverse transcription and integration processes requires the proper organization of the protein-RNA complex in the core. The formation of a functional core is in turn dependent upon the proper arrangement of Gag proteins in the immature particle. Exactly how these protein-RNA complexes form is one of the most critical questions of retroviral assembly. The model for RSV and HIV assembly recently proposed by Campbell and Vogt (1995) is founded upon the assumption that the viral genomic RNA is the scaffold on which an array of Gag proteins is built. The protein-coated RNA is proposed to wind itself up. The protein-protein interactions within the assemblage probably determine the curvature of the spherical shell. As the shell grows, the association of the M domains with the membrane causes the growing shell to wrap the membrane around itself. One can also imagine that the Gag proteins bind first to the membrane and then join the RNA strand as it is coiling up at the membrane. Whether the Gag proteins bind to the membrane or the RNA as monomers or as oligomers is unknown at present – either would be consistent with the proposed model (see Fig. 4).

How accurately the story that is being developed for RSV and HIV actually reflects what happens with other retroviruses is still an open question. The fact that the order of the domains in the Gag precursor (MA-CA-NC) is identical in all known retroviruses and the observation that domains from highly divergent retroviruses can function smoothly together to promote budding all suggest that the idea that the RNA scaffold organizes the assembly process applies to all retroviruses. However, the unique features of each virus (such as the shape of the core or the site of assembly in the cell) may be explained by the dynamic interactions between Gag proteins or between Gag and the membrane.

In the case MuLV, for example, evidence suggests that the Gag protein must first interact with the membrane before oligomerizing or binding to RNA. This behaviour may be an indication that Gag-Gag interactions or Gag-RNA interactions in MuLV are not particularly strong, perhaps because of the presence of only one I domain, in which case concentration of the protein on the membrane is needed before the protein can form intermolecular interactions and join the assembly process. Viruses that utilize the B/D-type morphogenetic pathway represent the opposite extreme. In this case, it appears that the Gag-Gag interactions are quite strong and can drive the assembly of the protein-RNA complex into a spherical particle without first associating with the membrane. It may be that the membrane targeting signals in MA are not in the Gag monomers or small oligomers; display of the targeting signal may require a conformational change that occurs upon completion of the entire shell. This model suggests that the point mutations which change the assembly pathway of M-PMV from the D-type to C-type might be acting by exposing the membrane targeting/binding signals prematurely before completion of the intracytoplasmic particle.

It is important to keep in mind that the processes of assembly and budding are functions of the intact Gag protein, not of isolated assembly domains. Further, these processes are dynamic ones requiring cooperative interactions between protein molecules, between Gag and the RNA genome, between Gag and the membrane, and possibly involving host proteins as yet unidentified. The complex interactions involved in budding and maturation must certainly occur in an ordered, stepwise fashion. The formation of bonds between certain viral structural components may be followed by changes in conformation of a protein or modification by proteolysis, leading to stabilization of the interactions and preparation for later steps in assembly. In some cases, proteolytic cleavage of a protein may loosen interactions that were essential during budding but which are not needed, or may even be detrimental, in the mature virion. The greatest challenge will come as we seek methods to study these events as the dynamic processes they really are.

Acknowledgements. Special thanks are due to John Wills for his innumerable contributions to many of the studies described here, for his continuing support and encouragement and for the dynamic interactions that always characterize his laboratory. We thank V. Vogt, S. Goff, D. Rekosh, J.L Darlix, and S. Sandmeyer for their sharing of ideas and unpublished results. The work of the authors has been supported by grants from the National Institutes of Health and by funds from The Pennsylvania State University School of Medicine.

References

Adachi A, Gendelman HE, Koenig S, Folks T, Willey R, Rabson A, Martin MA (1986) Production of acquired immunodeficiency syndrome-associated retrovirus in human and nonhuman cells transfected with an infectious molecular clone. J Virol 59: 284–291
Aldovini A, Young RA (1990) Mutations of RNA and protein sequences involved in human immunodeficiency virus type 1 packaging result in production on noninfectious virus. J Virol 64: 1920–1926
Arcement LJ, Karshin WI, Naso RB, Jamjoon G, Arlinghaus RB (1976) Biosynthesis of Rauscher leukemia viral proteins: presence of p30 and envelope p15 sequences in precursor polypeptides. Virology 69: 763–774
Bennett RP (1994) Molecular interactions required for protease activation and virion assembly by retroviral Gag proteins. PhD dissertation, Pennsylvania State University
Bennett RP, Nelle TD, Wills JW (1993) Functional chimeras of the Rous sarcoma virus and human immunodeficiency virus Gag proteins. J Virol 67: 6487–6498
Bess Jr., Powell PJ, Issaq HJ, Schumack LJ, Grimes MK, Henderson LE, Arthur LO (1992) Tightly bound zinc in human immunodeficiency virus type 1, human leukemia virus type 1 and other retrovirues. J Virol 66: 840–847
Bolognesi DP, Luftig R, Shaper JH (1973) Localization of RNA tumor virus polypeptides. I. Isolation of further virus substructures. Virology 56: 549–564
Bryant M, Ratner L (1990) Myristoylation-dependent replication and assembly of human immunodeficiency virus 1. Proc Natl Acad Sci USA 87: 523–527
Burnette WN, Holladay LA, Mitchell WM (1976) Physical and chemical properties of Moloney murine leukemia virus p30 protein: a major core structural component exhibiting high helicity and self-association. J Mol Biol 107: 131–143
Campbell S, Vogt VM (1995) Self assembly in vitro of purified CA-NC proteins from Rous sarcoma virus and human immunodeficiency virus-1. J Virol 69: 6487–6497
Carrière C, Gay B, Chazal N, Morin N, Boulanger P (1995) Sequence requirements for encapsidation of deletion mutants and chimeras of human immunodeficiency virus type 1 Gag precursor into tetrovirus-like particles. J Virol 69: 2366–2377
Chazal N, Carrière C, Gay B, Boulanger P (1994) Phenotypic characterization of insertion mutants of the human immunodeficiency virus type 1 Gag precursor expressed in recombinant baculovirus-infected cells. J Virol 68: 111–122
Coffin JM (1984) Structure of the retroviral genome. In: Weiss R, Teich N, Varmus H, Coffin J (eds) RNA tumor viruses. Cold Spring Harbor Laboratory, Cold Spring Harbor, NY, pp 261–368
Covey SN (1986) Amino acid sequence homology in gag region of reverse transcribing elements and the coat protein gene of cauliflower mosaic virus. Nucleic Acids Res 14: 623–633
Craven RC, Leure-duPree AE, Erdie CR, Wilson CB, Wills JW (1993) Necessity of the spacer peptide between CA and NC in the Rouse sarcoma virus Gag protein. J Virol 67: 6246–6252
Craven RC, Leure-duPree AE, Weldon RA, Wills JW (1995) Genetic analysis of the major homology region of the Rous sarcoma virus Gag protein. J Virol 69: 4213–4227
Davis NL, Rueckert RR (1972) Properties of a ribonucleoprotein particle isolated from Nonidet P-40-treated Rous sarcoma virus. J Virol 10: 1010–1020
De Rocquigny H, Gabus C, Vincent A, Fournie-Zaluski MC, Roques B, Darlix JL (1992) Viral RNA annealing activities of human immunodeficiency virus type 1 nucleocapsid protein require only peptide domains outside the zinc fingers. Proc Natl Acad Sci USA 89: 6472–6476
Delchambre M, Gheysen D, Thines D, Thiriart C, Jacobs E, Verdin E, Horth M, Burny A, Bex F (1989) The GAG precursor of simian immunodeficiency virus assembles into virus-like particles. EMBO J 8: 2653–2660
Deminie CA, Emerman M (1993) Incorporation of human immunodeficiency virus type 1 Gag proteins into murine leukemia virus virions. J Virol 67: 6499–6506
Dickson C, Eisenman R, Fan H, Hunter E, Teich N (1984) Protein biosynthesis and assembly. In: Weiss R, Teich N, Varmus H, Coffin JM (eds) RNA tumor viruses, vol. 1. Cold Spring Harbor Laboratory, Cold Spring Harbor, NY, pp 513–648
Dorfman T, Bukovsky A, Öhagen A, Höglund S, Göttlinger HG (1994) Functional domains of the capsid protein of human immunodeficiency virus type 1. J Virol 68: 8180–8187
Dupraz P, Spahr P (1992) Specificity of Rous sarcoma virus nucleocapsid protein in genomic RNA packaging. J Virol 66: 4662–4670

Dupraz P, Oertle S, Meric C, Damay P, Spahr PF (1990) Point mutations in the proximal Cys-His box of Rous sarcoma virus nucleocapsid protein. J Virol 64: 4978–4987
Ehrlich LS, Kräusslich HG, Wimmer E, Carter CA (1990) Expression in Escherichia coli and purification of human immunodeficiency virus type 1 capsid protein (p24). AIDS Res Hum Retroviruses 6: 1169–1175
Eisenman RN, Vogt VM, Diggelmann H (1975) Synthesis of avian RNA tumor virus structural proteins. Cold Spring Harbor Symp Quant Biol 39: 1067–1075
Elder JH, Schnölzer M, Hasselkus-Light CS, Henson M, Lerner DA, Phillips TR, Wagaman PC, Kent SBH (1993) Identification of proteolytic processing sites within the Gag and Pol polyproteins of feline immunodeficiency virus. J Virol 67: 1869–1876
Erdie CR, Wills JW (1990) Myristylation of Rous sarcoma virus Gag protein does not prevent replication in avian cells. J Virol 64: 5204–5208
Franke EK, Yuan HEH, Bossolt KL, Goff SP, Luban J (1994) Specificity and sequence requirements for interactions between various retroviral Gag proteins. J Virol 68: 5300–5305
Gelderblom H (1990) Morphogenesis, maturation and fine structure of lentiviruses. In: Pearl LH (ed) Retroviral proteases: maturation and morphogenesis. Stockton, New York, NY, pp 159–180
Gheysen D, Jacobs E, de Foresta F, Thiriart C, Francotte M, Thines D, De Wilde M (1989) Assembly and release of HIV-1 precursor Pr55gag virus-like particles from recombinant baculovirus-infected insect cells. Cell 59: 103–112
Gonzalez SA, Affranchino JL, Gelderblom HR, Burny A (1993) Assembly of the matrix protein of simian immunodeficiency virus-like particles. Virology 194: 548–556
Gorelick RJ, Henderson LE, Hanser JP, Rein A (1988) Point mutations of Moloney murine leukemia virus that fail to package viral RNA: evidence for specific RNA recognition by a "zinc finger-like" protein sequence. Proc Natl Acad Sci USA 85: 8420–8424
Gorelick RJ, Nigida SM, Bess Jr, Arthur LO, Henderson LE, Rein A (1990) Noninfectious human immunodeficiency virus type 1 mutants deficient in genomic RNA. J Virol 64: 3207–3211
Göttlinger HG, Sodroski JG, Haseltine WA (1989) Role of capsid precursor processing and myristoylation in morphogenesis and infectivity of human immunodeficiency virus type 1. Proc Natl Acad Sci USA 86: 5781–5785
Göttlinger HG, Dorfman T, Sodroski JG, Haseltine WA (1991) Effect of mutations affecting the p6 Gag protein on human immunodeficiency virus particle release. proc Natl Acad Sci USA 88: 3195–3199
Hansen MS, Barklis E (1995) Structural interactions between retroviral Gag proteins examined by cysteine cross-linking. J Virol 69: 1150–1159
Hansen LJ, Chalker DL, Sandmeyer SB (1988) Ty3, a yeast retrotransposan associated with tRNA genes, has homology to animal retroviruses. Mol Cell Biol 8: 5245–5256
Henderson LE, Sowder R, Copeland TD, Smythers G, Oroszlan S (1984) Quantitative separation of murine leukemia virus proteins by reversed-phase high-pressure liquid chromatography reveals newly described gag and env cleavage products. J Virol 52: 492–500
Henderson LE, Benveniste RE, Sowder R, Copeland TD, Schultz AM, Oroszlan S (1988a) Molecular characterization of *gag* proteins from simian immunodeficiency virus (SIV_{Mne}). J Virol 62: 2587–2595
Henderson LE, Copeland TD, Sowder RC, Schultz AM, Oroszlan S (1988b) Analysis of proteins and peptides from sucrose banded HTLV-III. In: Bolognesi D (ed) Human retroviruses, cancer and AIDS: approaches to prevention and therapy. Alan R. Liss, New York, NY, pp 135–147
Henderson LE, Bowers MA, Sowder RC, Serabyn SA, Johnson DG, Bess JW Jr., Arthur LO, Bryant DK, Fenselau C (1992) Gag proteins of the highly replicative MN strain of human immunodeficiency virus type 1: posttranslational modifications, proteolytic processings, and complete amino acid sequences. J Virol 66: 1856–1865
Hong SS, Boulanger P (1993) Assembly-defective point mutants of the human immunodeficiency virus type 1 Gag precursor phenotypically expressed in recombinant baculovirus-infected cells. J Virol 67: 2787–2798
Hoshikawa N, Kojima A, Yasuda A, Takayashiki E, Masuko S, Chiba J, Sata T, Kurata T (1991) Role of the *gag* and *pol* genes of human immunodeficiency virus in the morphogenesis and maturation of retrovirus-like particles expressed by recombinant vaccinia virus: an ultrastructural study. J Gen Virol 72: 2517
Jowett JB, Hockley DJ, Nermut MV, Jones IM (1992) Distinct singnals in human immunodeficiency virus type 1 Pr55 necessary for RNA binding and particle formation [published erratum appears in J Gen Virol 1993 May; 74(5): 943]. J Gen Virol 73: 3079–3086

Karpel RL, Henderson LE, Oroszlan S (1987) Interactions of retroviral structural proteins with single-stranded nucleic acids. J Biol Chem 262: 4961–4967
Kawakami T, Sherman L, Dahlberg J, Gazit A, Yaniv A, Tronick SR, Aaronson SA (1987) Nucleotide sequence analysis of equine infectious anemia virus proviral DNA. Virology 158: 300–312
Kräusslich H-G, Facke M, Heuser A-M, Konvalinka J, Zentgraf H (1995) The spacer peptide between human immunodeficiency virus capsid and nucleocapsid proteins is essential for ordered assembly and viral infectivity. J Virol 69: 3407–3419
Luftig RB, Yoshinaka Y (1978) Rauscher leukemia virus populations enruched for "immature" virions contain increased amounts of P70, the *gag* gene product. J Virol 25: 416–421
Luo L, Li Y, Dales S, Yong Kang C (1994) Mapping of functional domains for HIV-1 *gag* assembly into virus-like particles. Virology 205: 496–502
Mammano F, Ohagen A, Hoglund S, Göttlinger HG (1994) Role of the major homology region of human immunodeficiency virus type 1 in virion morphogenesis. J Virol 68: 4927–4936
Massiah MA, Starich MR, Paschall C, Summers MF, Christensen AM, Sundquist WI (1994) Three-dimensional structure of the human immunodeficiency virus type 1 matrix protein. J Mol Biol 244: 198–223
Matthews S, Barlow P, Boyd J, Russell R, Mills H, Cunningham M, Meyers N, Burns NC, Kingsman A, Campbell I (1994) Structural similarity between the p17 matrix protein of HIV-1 and interferon-gamma. Nature (London) 370: 666–668
Maurer B, Bannert H, Darai G, Flugel RM (1988) Analysis of the primary structure of the long terminal repeat and the *gag* and *pol* genes of the human spumaretrovirus. J Virol 62: 1590–1597
McLaughlin S, Aderam A (1995) The myristoyl-electrostatic switch: a modulator of reversible protein-membrane interactions. Trends Biochem Sci 20: 272–276
Mergener K, Facke M, Welker R, Brinkmann V, Gelderblom HR, Kräusslich HG (1992) Analysis of HIV particle formation using transient expression of subviral constructs in mammalian cells. Virology 186: 25–39
Méric C, Spahr P (1986) Rous sarcoma virus nucleic acid-binding protein p12 is necessary for viral 70S RNA dimer formation and packaging. J Virol 60: 450–459
Méric, C, Goff SP (1989) Characterization of Moloney murine leukemia virus mutants with single-amino-acid substitutions in the Cys-His box of the nucleocapsid protein. J Virol 63: 1558–1568
Méric C, Gouilloud E, Spahr P (1988) Mutations in Rous sarcoma virus nucleocapsid protein p12 (NC): deletions of Cys-His boxes. J Virol 62: 3328–3333
Mervis RJ, Ahmad N, Lillehoj EP, Raum MG, Salazar FHR, Chan HW, Venkatesan S (1988) The *gag* gene products of human immunodeficiency virus type 1: alignment within the *gag* open reading frame, identification of posttranslational modifications, and evidence for alternative *gag* precursors. J Virol 62: 3993–4002
Morellet N, Jullian N, De Rocquigny H, Maigret B, Darlix JL, Roques BP (1992) Determination of the structure of the nucleocapsid protein NCp7 from the human immunodeficiency virus type 1 by 1H NMR. EMRO J 11: 3059–3065
Nelle TD, Wills JW (1996) A large region within the Rous sarcoma virus matrix protein is dispensable for budding and infectivity. J Virol 70: 2269–2276
Ottoman M, Gabus C, Darlix J (1995) The central globular domain of the nucleocapsid protein of human immunodeficiency virus type 1 is critical for virion structure and infectivity. J Virol 69: 1778–1784
Overton HA, Fuji Y, Prince IR, Jones IM (1989) The protease and *gag* gene products of the human immunodeficiency virus: authentic cleavage and post-translational modification in an insect cell expression system. Virology 170: 107–116
Parent LJ, Bennett RP, Craven RC, Nelle TD, Krishna NK, Bowzard JB, Wilson CB, Puffer BA, Montelaro RC, Wills JW (1995) Positionally independent and exchangeable late budding functions of the Rousy sarcoma virus and human immunodeficiency virus Gag proteins. J Virol 69: 5455–5460
Parent LJ, Wilson CB, Resh MD, Wills JW (1996) Evidence for a second function of the MA sequence in the Rous sarcoma virus Gag protein. J Virol 70: 1016–1026
Patarca R, Haseltine WA (1985) A major retroviral core protein related to EPA and TIMP. Nature (London) 318: 390
Paxton W, Connor RI, Landau NR (1993) Incorporation of Vpr into human immunodeficiency virus type 1 virions: requirement for the p6 region of Gag and mutational analysis. J Virol 67: 7229–7237
Peitzsch RM, McLaughlin S (1993) Binding of acylated peptides and fatty acids to phospholipid vesicles: pertinence to myristoylated proteins. Biochemistry 32: 10436–10443

Pepinsky RB (1983) Localization of lipid-protein and protein-protein interactions within the murine retrovirus *gag* precursor by a novel peptide-mapping technique. J Biol Chem 258: 11299–11235
Pepinsky RB, Vogt VM (1984) Fine-structure analyses of lipid-protein and protein-protein interactions of Gag protein p19 of the avian sarcoma and leukemia viruses by cyanogen bromide mapping. J Virol 52: 145–153
Pepinsky RB, Cappiello D, Wilkowski C, Vogt VM (1980) Chemical crosslinking of proteins in avian sarcoma and leukemia viruses, Virology 102: 205–210
Pepinsky RB, Papayannopoulos IA, Chow EP, Krishna NK, Craven RC, Vogt VM (1995) Differential proteolytic processing leads to multiple forms of the CA protein in Avian sarcoma and leukemia viruses. J Virol (in press)
Pettit SC, Moody MD, Wehbie RS, Kaplan AH, Nantermet PV, Klein CA, Swanstrom R (1994) The p2 domain of human immunodeficiency virus type 1 Gag regulates sequential proteolytic processing and is required to produce fully infectious virions. J Virol 68: 8017–8027
Prats AC, Sarih L, Gabus C, Litvak S, Keith G, Darlix J (1988) Small finger protein of avian and murine retroviruses has nucleic acid annealing activity and positions the replication primer tRNA onto genomic RNA. EMBO J 7: 1777–1783
Reicin AS, Paik S, Berkowitz RD, Luban J, Lowy I, Goff SP (1995) Linker insertion mutations in the human immunodeficiency virus type 1 *gag* gene: effects on virion particle assembly, release and infectivity. J Virol 69: 642–650
Rein A, McClure MR, Rice NR, Luftig RB, Schultz AM (1986) Myristylation site in Pr65gag is essential for virus particle formation by Moloney murine leukemia virus. Proc Natl Acad Sci USA 83: 7246–7250
Resh MD (1994) Myristylation and palmitylation of Src family members: the fats of the matter. Cell 76: 411–413
Rhee SS, Hunter E (1987) Myristylation is required for intracellular transport but not for assembly of D-type retrovirus capsids. J Virol 61: 1045–1053
Royer M, Cerutti M, Gay B, Hong SS, Devauchelle G, Boulanger P (1991) Functional domains of HIV-1 Gag-polyprotein expressed in baculovirus-infected cells. Virology 184: 417–422
Sakalian M, Wills JW, Vogt VM (1994) Efficiency and selectivity of RNA packaging by Rous sarcoma virus Gag deletion mutants. J Virol 68: 5969–5981
Schultz AM, Rein A (1989) Unmyristylated Moloney murine leukemia virus Pr65gag is excluded from virus assembly and maturation events. J Virol 63: 2370–2373
Schawartz DE, Tizard R, Gilbert W (1983) Nucleotide sequence of Rous sarcoma virus. Cell 32: 853–869
Shioda T, Shibuta H (1990) Production of human immunodeficiency virus (HIV)-like particles from cells infected with recombinant vaccinia viruses carrying the *gag* gene of HIV. Virology 175: 139–148
Smith AJ, Cho MI, Hammarskjold ML, Rekosh D (1990) Humain immunodeficiency virus type 1 Pr55gag and Pr160gag-pol expressed from a simian virus 40 late replacement vector are efficiently processed and assembled into viruslike particles. J Virol 64: 2743–2750
Sonigo P, Alizon M, Staskus K, Klatzmann D, Cole S, Danos O, Retzel E, Tiollais P, Haase A, Wain-Hobson S (1985) Nucleotide sequence of the visna lentivirus: relationship to the AIDS virus. Cell 42: 369–382
Srinivasakumar N, Hammarskjold M-L, Rekosh D (1995) Characterization of deletion mutations in the capsid region of HIV-1 that effect particle formation and Gag-Pol precursor incorporation. J Virol 69: 6106–6114
Stewart L, Schatz G, Vogt VM (1990) Properties of avian retrovirus particles defective in viral protease. J Virol 64: 5076–5092
Strambio-de-Castillia C, Hunter E (1992) Mutational analysis of the major homology region of Mason-Pfizer monkey virus by use of saturation mutagenesis. J Virol 66: 7021–7032
Stromberg K, Hurley NE, Davis NL, Rueckert RR, Fleissner E (1974) Structural studies of avian myeloblastosis virus: comparision of polypetides in virion and core component by dedecyl sulfate-polyarcylamide gel electrophoresis. J Virol 13: 513–528
Stys D, Blaha I, Strop P (1993) Structural and functional studies in vitro on the p6 protein from the HIV-1 *gag* open reading frame. Biochim Biophys Acta 1182: 157–161
Teich N (1984) Taxonomy of retroviruses. In: Weiss R, Teich N, Varmus H, Coffin J (eds) RNA tumor viruses, vol. 1. Cold Spring Harbor Laboratory, Cold Spring Harbor, NY, pp 25–207
Tobin GJ, Sowder RC, Fabris D, Hu MY, Battles JK, Fenselau C, Henderson LE, Gonda MA (1994) Amino acid sequence analysis of the proteolytic cleavage products of the bovine immunodeficiency virus Gag precursor polypeptide. J Virol 68: 7620–7627

Tritch RJ, Cheng YE, Yin FH, Erickson-Viitanen S (1991) Mutagenesis of protease cleavage sites in the human immunodeficiency virus type 1 Gag polyprotein. J Virol 65: 922–930
Trono D, Feinberg MB, Baltimore D (1989) HIV-1 Gag mutants can dominantly interfere with the replication of wild-type virus. Cell 59: 113–120
Verderame MF, Nelle TD, Wills JW (1996) The membrane-binding domain of the Rous sarcoma virus Gag protein. J Virol 70: 2664–2668
Vogt VM, Eisenman R (1973) Identification of a large polypetide precursor of avian oncornavirus proteins. Proc Natl Acad Sci USA 70: 1734–1738
Vogt VM, Eisenman R, Diggelmann H (1975) Generation of avian myeloblastosis virus structural proteins by proteolytic cleavage of a precursor polyprotein. J Mol Bios 96: 471–493
Vogt VM, Bruckenstein DA, Bell AP (1982) Avain sarcoma virus *gag* precursor polypetide is not processed in mammalian cells. J Virol 44: 725–730
von Poblotzki A, Wagner R, Niedrig M, Wanner G, Wolf H, Modrow S (1993) Identification of a region in the Pr55^{gag}-polyprotein essential for HIV-1 particle formation. Virology 193: 981–985
Voynow SL, Coffin JM (1985) Truncated *gag*-related proteins are produced by large delection mutants of Rous sarcoma virus and form virus particles. J Virol 55: 79–85
Wagner R, Deml L, Fliesbach H, Wanner G, Wolf H (1994) Assembly and extracellular release of chimeric HIV-1 Pr55gag retrovirus-like particles. Virology 200: 162–175
Wang CT, Barklis E (1993) Assembly, processing, and infectivity of human immunodeficiency virus type 1 Gag mutants. J Virol 67: 4264–4273
Weldon RA Jr., Wills JW (1993) Characterization of a small (25-kilodalton) derivative of the Rous sarcoma virus Gag protein competent for particle release. J Virol 67: 5550–5561
Weldon RA Jr., Erdie CR, Oliver MG, Wills JW (1990) Incorporation of chimeric Gag proteins into retroviral particles. J Virol 64: 4169–4179
Wills JW, Craven RC (1991) Form, function, and use of retroviral Gag proteins [editorial]. AIDS 5: 639–654
Wills JW, Craven RC, Achacoso JA (1989) Creation and expression of myristylated forms of Rous sarcoma virus Gag protein in mammalian cells. J Virol 63: 4331–4343
Wills JW, Craven RC, Weldon RA Jr., Nelle TD, Erdie CR (1991) Suppression of retroviral MA deletions by the amino-terminal membrane-binding domain of p60src. J Virol 65: 3804–3812
Wills JW, Cameron CE, Wilson CB, Xiang Y, Bennett RP, Leis J (1994) An assembly domain of Rous sarcoma virus Gag protein required late in budding. J Virol 68: 6605–6618
Witte ON, Baltimore D (1978) Relationship of retrovirus polyprotein cleavages to virion maturation studied with temperature-sensitive murine leukemia virus mutants. J Virol 26: 750–761
Yu X, Yuan X, Matsuda Z, Lee T-H, Essex M (1992) The matrix protein of human immunodeficiency virus type 1 is required for incorporation of viral enveloped protein into mature virions. J Virol 66: 4966–4971
Zhao Y, Jones IM, Hockley DJ, Nermut MV, Roy P (1994) Complementation of human immunodeficiency virus (HIV-1) Gag particle formation. Virology 199: 403–408
Zhou W, Parent LJ, Wills JW, Resh MD (1994) Identification of membrane-binding domain within the amino-terminal region of human immunodeficiency virus type 1 Gag protein which interacts with acidic phospholipids. J Virol 68: 2556–2569

Proteolytic Processing and Particle Maturation

V.M. VOGT

1 Introduction and Historical Overview

Proteolytic processing of the structural and enzymatic proteins of viruses is a common phenomenon (KRÄUSSLICH and WIMMER 1988; DOUGHERTY and SEMLER 1993). Well-studied examples include picornaviruses (such as poliovirus), alphaviruses (such as Sindbis virus), adenoviruses, plant viruses (such as potyviruses), and bacteriophage (such as T4). Cleavages of viral polypeptides have diverse functions, but they can be viewed most simply as serving to drive a reaction in one direction, since hydrolysis of the peptide bond in most situations is irreversible. In particular, cleavages often play roles in virion morphogenesis; in the most extreme cases complete degradation of a "scaffolding protein" is needed for the final infectious virus particle to be formed from an immature particle. In retroviruses all three major virion proteins, Gag, Pol, and Env, are proteolytically processed. For Gag and Pol it is the virus-encoded protease (PR) that accomplishes these cleavages, while for Env it is a host-encoded protease found in the Golgi apparatus. However, in some viruses the very C-terminus of Env also is processed by PR. In this chapter only PR-mediated processing is

Section of Biochemistry, Molecular and Cell Biology, Biotechnology Building, Cornell University, Ithaca, NY 14853, USA

discussed, with a focus on the regulation of processing and the consequences of processing. Related topics such as PR structure, mechanism of catalysis, action of protease inhibitors, comparative aspects of retroviral proteases, and specificity of cleavage are treated perfunctorily. Many aspects of proteolytic processing in retroviruses have been reviewed elsewhere (KRÄUSSLICH and WIMMER 1988; SKALKA 1989; FITZGERALD and SPRINGER 1991; WILLS and CRAVEN 1991; DEBOUCK 1992; WLODAWER and ERICKSON 1993; KATZ and SKALKA 1994; RINGE 1994; TOMASSELLI and HEINDRIKSON 1994).

An outline of the genetic organization of the structural genes of prototypic retroviruses (Fig. 1) serves as a reference point for discussions of proteolytic processing. For assembly and maturation probably the most studied viruses are Moloney murine leukemia virus (Mo-MLV), belonging to the mammalian C-type virus genus; Rous sarcoma virus (RSV), belonging to the avian sarcoma and leukemia virus (ASLV) genus; human immunodeficiency virus type 1 (HIV1), belonging to the lentivirus genus; and Mason-Pfizer monkey virus (MPMV), belonging to the D-type virus genus. The discussion in this chapter is concentrated on these model systems. In the sections that follow, "MLV" is used as a general term to refer to the murine leukemia viruses. Similarly, "ASLV" is used to refer to Rous sarcoma virus, avian leukosis virus (ALV), as well as myeloblastosis-associated virus (MAV), which are all very closely related in sequence and hence can be considered strains of the same species. "HIV" is used to refer to HIV1, which despite its name is much more distantly related to HIV2 than the ASLVs are to each other or the MLVs are to each other.

The genomes of all retroviruses contain the genes *gag, pro, pol,* and *env*; for the majority of retroviruses that have no accessory genes in fact these are the only genes. *gag* encodes the internal structural proteins of the virion: matrix (MA), capsid (CA), nucleocapsid (NC), and other proteins designated by their molecular weight because their function is not known. *pro* encodes the viral protease (PR). *pol* encodes the viral enzymes reverse transcriptase (RT), RNaseH (which usually is considered as part of RT), integrase (IN), and in a few viruses an additional enzyme, dUTPase (DU). *env* encodes the two envelope glycoproteins, the surface protein (SU) and transmembrane protein (TM), which are dispensable for formation of virus particles, although they are essential for infectivity of the particles. With few exceptions the protein products of *pro* and *pol* are derived by ribosomal frameshifting, or in viruses closely related to MLV, by suppression of the *gag* termination codon. Thus only a fraction of the ribosomes translating *gag* proceed on into *pro* and *pol*. However, this coding segment by convention has been given the status of a distinct gene because of the variety of ways in which it may be expressed. In most retroviruses *pro* is simply part of *pol* and thus has neither an initiation codon nor a termination codon (as in MLV and HIV). In a few retroviruses it stands in a reading frame separate from both *gag* and *pol* [as in MPMV, mouse mammary tumor virus (MMTV), and human T-cell leukemia virus (HTLV)]. In these cases *pro* is expressed by ribosomal frameshifting as a Gag-Pro protein, and a second frame-shifting event is needed for expression of *pol*. In a few viruses *pro* is the distal part of *gag*

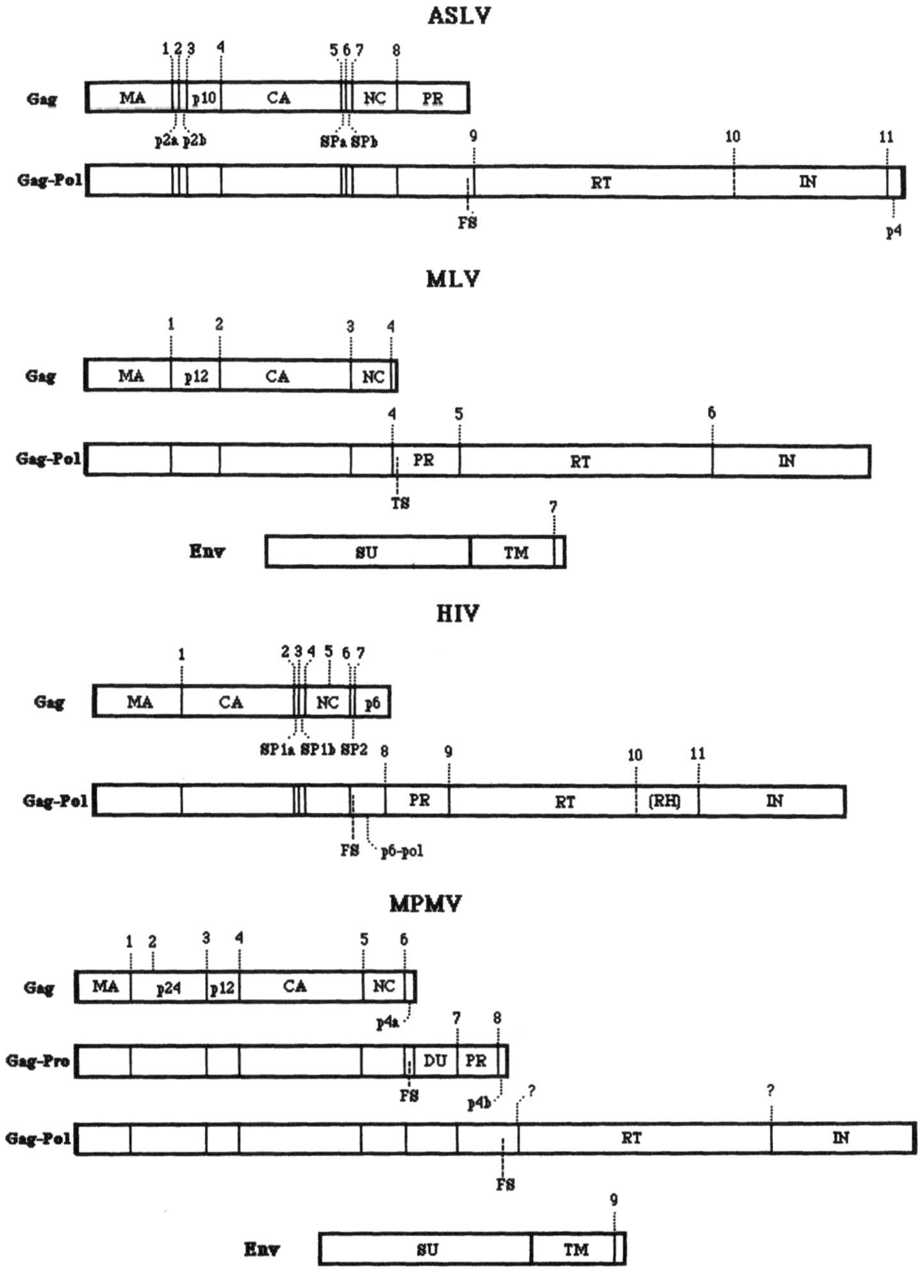

Fig. 1. Schematic representation of cleavage sites for prototypic retroviral proteases. The Gag, Gag-Pol, Gag-Pro, and Env proteins of four retroviruses are shown diagrammatically, with PR cleavage sites marked with *dotted lines above the rectangle that represents the protein*. Names of the mature proteins derived by cleavage are given *within the rectangle or underneath for small peptides*. *FS*, site of frame shifting; *TS*, site for termination suppression. *Parentheses* around RH signify that this domain is proteolytically separated from the RT domain only in the enzymatically inactive p51 polypeptide (see text)

(as in ASLV). A fundamental consequence of this genetic organization is that the retroviral enzymes all are translated with the same N-terminal segment of polypeptide, namely Gag[1]. The evolutionary logic of this scheme is that the same protein-protein interactions that underly assembly of the virion also are used for incorporation of the enzymes needed for replication and maturation.

The Gag, Gag-Pro, and Gag-Pro-Pol proteins often are called polyproteins or precursors, because they give rise to the several mature proteins found in the infectious virus. (For the purposes of this review, the more common term Gag-Pol is used to stand for Gag-Pro-Pol.) In the simplest model for the structure of the Gag and Gag-Pol precursors, each consists of a series of independently folded domains, like beads on a string. The viral protease acts to sever the connections between the domains, which then become the mature proteins. Whether all the domains are really separated in this way, or some are intertwined, is not known. Some domains with common functions are found in all retroviruses, for example MA, CA, and NC in Gag and RT and IN in Gag-Pol. Others are characteristic of the genus. For example, the segment between MA and CA may give rise to one mature protein (p12 in MLV), more than one protein (p2a, p2b, and p10 in ASLV), or no proteins (in HIV). As discussed below, the exact timing of proteolytic processing is uncertain. In any case, the cleavages that lead to the mature proteins incorporated into virions are late events, probably occurring in the last stages of budding as the virion is being readied for release from the cell. Preventing cleavage, e.g. by mutation of PR, does not prevent assembly. But the resulting particles, which are called "immature," are of different morphology and are not infectious. Immature particles also have a number of other properties that may shed light on the assembly process, since they represent a static picture of assembly in its last stage. For example, they are distinctly more stable to weak detergents than are mature particles. Teleologically, one can think of proteolytic maturation of Gag and Gag-Pol proteins as readying the virion for a new round of infection. In this view maturation destabilizes the virus core, facilitating its dissociation in the newly infected cell and perhaps facilitating reverse transcription of the genome. At least in some retrovirus systems, maturation also leads to activation of the fusion potential of the Env proteins, and to activation of reverse transcriptase activity. No doubt many more subtle functions of maturation have yet to be described.

Proteolytic cleavage of retroviral proteins was discovered over 20 years ago, in the ASLV system (Vogt and Eisenman 1973). The use of pulse-chase labeling with 35S-methionine combined with SDS polyacrylamide gel electrophoresis (SDS PAGE) and immune precipitation with antiserum directed against mature virus proteins allowed the broad outline of processing to be discerned. The full-

[1]Conventionally the viral genes are designated with three lower case letters in italics (for example *gag*) while the protein products are designated with the same three letters in normal type and with the first letter capitalized (for example Gag). For historical reasons this nomenclature usually is not applied to protease, which is called PR rather than Pro. Nevertheless, in viruses like MPMV the precursor for PR is called Gag-Pro rather than Gag-PR.

length Gag precursor, named Pr76 (or later, $Pr76^{Gag}$) after its apparent molecular weight and its status as a precursor, was shown to be processed and incorporated into virions with a half-life of somewhat less than an hour, a number similar to that found subsequently for $Pr65^{Gag}$ in the major other model retrovirus system at that time, MLV. Attempts to position the mature proteins on the precursor, based on inhibition of translation initiation, were only partially successful, and the accurate placement of mature proteins had to await the development of DNA sequencing. The Gag-Pol protein, $Pr180^{Gag\text{-}Pol}$ in the avian viruses, was found soon thereafter (OPPERMANN et al. 1977), long before the discovery of ribosomal frameshifting clarified how this protein was synthesized in a fixed ca. 1:20 ratio to Gag (JACKS and VARMUS 1988). The first indication that the enzyme responsible for Gag cleavage is a virion protein was the observation in ASLV that specific proteolytic activity could be recovered from an SDS polyacrylamide gel, from a band corresponding to a Gag protein (VON DER HELM 1977). The assignment of enzymatic activity to what was considered to be a structural protein was supported later by more careful purification studies. The MLV protease activity proved more difficult to isolate as a homogeneous protein species, which was explained in part by the differences in genetic organization of the ASLV and MLV genomes, the former producing about 20 times as much PR as the latter. More recent milestones in the understanding of proteolytic maturation have come from biochemical and physical studies of the protease. Analysis of the conserved amino acid residues in the PR of diverse retroviruses led to the conclusion that all PRs belong to the aspartic family of cellular proteases (TOH et al. 1985). The three-dimensional structure of ASLV PR and HIV1 PR (NAVIA et al. 1989; JASKÓLSKI et al. 1990; WLODAWER et al. 1989; LAPATTO et al 1989), inferred from X-ray crystallography, provided a detailed view of this enzyme. The potential therapeutic importance of HIV PR inhibitors greatly spurred interest in this protein, and a large number of three-dimensional structures with and without various inhibitors have been worked out. The importance of these fundamental studies is underscored by the ongoing clinical trials of PR inhibitors to treat AIDS.

The earliest view of the function of proteolysis was that it played a direct role in assembly of the budding virus (see, e.g. review by EISENMAN and VOGT 1978). Thus it was an important insight that virus particles could be formed and released in the absence of proteolysis. Perhaps the first definitive proof that uncleaved Gag itself could efficiently form a virus-like particle came from a PR-defective mutant of MLV, Gazdar murine sarcoma virus (PINTER and DEHARVEN 1979). Cells expressing this mutant were found to release particles consisting of a single protein, $Pr65^{Gag}$. The same conclusion also was suggested from earlier studies on wild-type MLV (LUFTIG and YOSHINAKA 1978), which showed that a fraction of virus particles remained immature by EM and a similar fraction of the Gag protein remained uncleaved. This was the first indication that it was cleavage of Gag that triggers the dramatic morphological changes now called maturation. While the conclusion that proteolysis directly causes maturation has been accepted for many years, recent data challenge some aspects of this view. Indeed the conclusion that proteolysis plays no role in virion assembly and budding also has been challenged, at least for the HIV system. The data underlying these revisionist ideas are discussed in the sections below.

2 Structure and Specificity of Retroviral Proteases

Retroviral PRs are aspartic proteases. This class of enzymes is characterized by an active site containing two aspartate residues, each located in a conserved stretch of residues in a separate polypeptide or polypeptide domain. The numerous cellular aspartic proteases are encoded in a single gene as two homologous domains with very similar sequence. The polypeptide folds into what could be called a pseudodimer, with two lobes that come together to create the active site. Well-studied examples of cellular aspartic proteases include pepsin, renin, chymosin, penicillopepsin, and cathepsins D and E (reviewed in DAVIES 1990). Most of these normally function at acid pH. By contrast to these cellular proteases, retroviral PRs are true dimers, with two identical subunits held together by noncovalent interactions. The monomers are somewhat smaller than one lobe of a typical cellular aspartic protease, ranging in size from about 100 to 130 amino acid residues. Retroviral PRs are much less active proteases than most of their distant cousins encoded in the genomes of cells. Critical for the understanding of PRs is that the active site is between the two subunits, and thus only dimers can possess enzymatic activity. Despite their structural and functional similarities, PRs from different genera of retroviruses are quite di-

```
aa#R:30                      64              108
RSV: vyiTaLLDSGADiTiis  ...  ihGIGGgip  ...  ilGRDcLqglg ...
MLV: qpvTfLVDTGAqhsvlt  ...  vqGatGskn  ...  llGRDlLtklk ...
MPMV:kmfTgLIDTGADvTiik  ...  lrGIGqsnn  ...  lwGRDlLsqmk ...
HIV: klkeaLLDTGADdTvle  ...  igGIGGfik  ...  iiGRnlLtqig ...
aa#H:18                      47              84
```

Fig. 2. Major sequence similarities of prototypic retroviral proteases. The data are taken from RAO et al. (1991) with the MPMV PR sequence added from SONIGO et al. (1986). *Uppercase letters* indicate conserved residues identical in these four viruses. For HIV the sequence of the BH10 strain is shown, for MLV the Moloney strain, and for RSV the Prague C strain. The active site aspartate residue is *underlined*. The residue numbering is shown both for the RSV PR (*upper line*) and for the HIV PR (*lower line*)

vergent in sequence, with only about one-third of the residues being identical in pairwise comparisons between viruses of different genera. The identical residues are largely clustered in three short blocks of sequences (RAO et al. 1991) (Fig. 2). As expected, these blocks of sequences also contain many of the amino acid residues that cannot be mutated without resulting in loss of enzymatic activity (LOEB et al. 1989).

The three-dimensional structure of PR has been reviewed elsewhere (WLODAWER and ERICKSON 1993), and only the most salient features are presented here (Fig. 3). The two polypeptides form a symmetrical bilobal structure with a cleft or groove in which the substrate binds. As inferred from crystal structures with bound inhibitors (MILLER et al. 1989; WEBER et al. 1989; reviewed in RINGE 1994), the peptide substrate lies in the groove in an extended conformation as if it were in a β-sheet, with side chains pointing alternately in opposite directions. It is interesting that an intrinsically symmetrical protein binds to, and acts on, a substrate that is intrinsically asymmetrical; thus the enzyme/substrate complex also is asymmetrical. From modeling of inhibitors, it appears that the enzyme contacts seven or eight residues in the substrate. Peptides of seven residues generally are the smallest that are actively cleaved, with four residues required upstream and three downstream of the scissile bond. By convention the amino acid residues at the cleavage site are called, from N- to C-terminus of the substrate, P4 to P1 and P1′ to P4′ (Fig. 4), the scissile bond

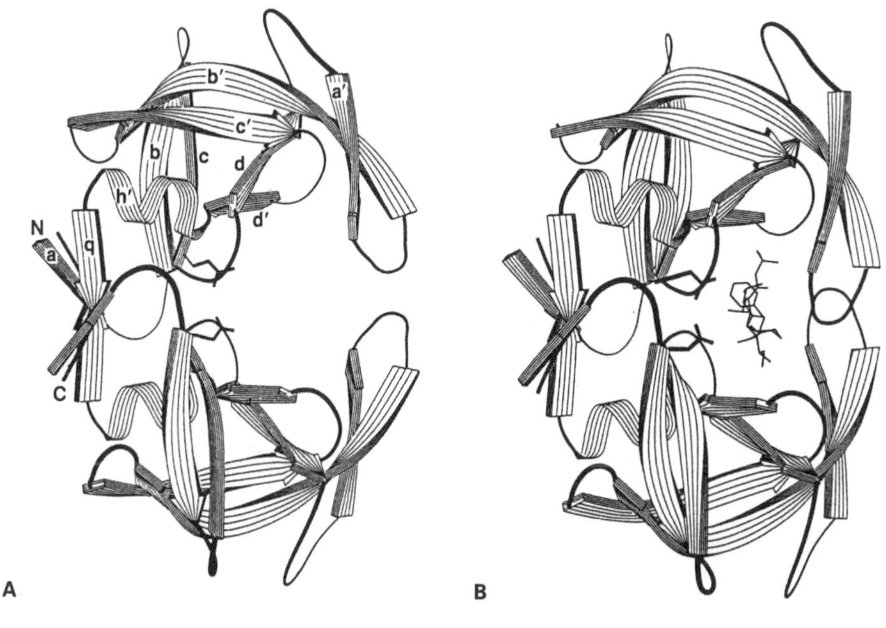

Fig. 3. Ribbon diagram of HIV1 protease. *Diagram* shows the three-dimensional structure of the PR dimer, with and without bound inhibitor. From WLODAWER and ERICKSON (1993)

being between P1 and P1′. In the groove the eight residues make contact with the enzyme, both through peptide backbone and side chains. The latter interact with the enzyme at locations referred to as "subsites," which are named by analogy S4 - S1 and S1′ to S4′. These sites alternate between the two subunits. Thus P4, P2, P1′, and P3′ contact mainly the one subunit, while P3, P1, P2′, and P4′ contact mainly the other. Positioned over the cleft are two segments of polypeptide called the "flaps." The flaps, whose motion has been tracked directly by nuclear magnetic resonance (NMR) techniques (NICHOLSON et al. 1995), clamp down on the substrate once it is bound. In contrast to the viral proteases, cellular aspartic proteases have only a single flap.

Enzymatic function of the protease relies on formation of the dimer and, of course, on the proper conditions. Perhaps the most key interactions for holding the dimer together are provided by the short four-stranded antiparallel β-sheet that contains the N- and C-terminal ends of the two subunits (Fig. 3). Disruption of these interactions, for example in the case of HIV by truncation due to self-digestion (ROSÉ et al. 1993) or competition by peptides (ZHANG et al. 1991; BABÉ

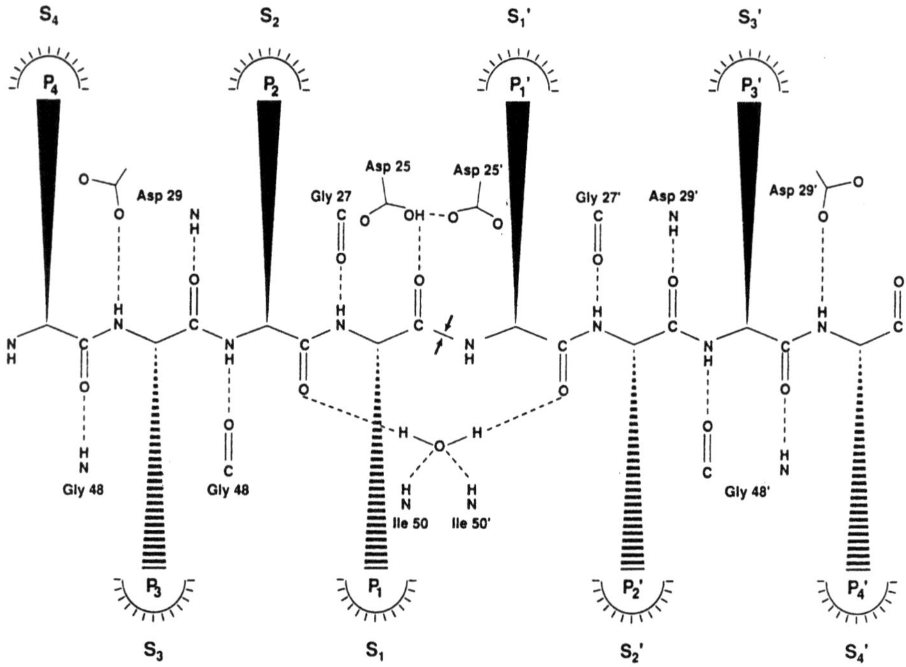

Fig. 4. Schematic representation of substrate bound to HIV PR. The backbone and side chains for an eight-residue peptide are shown diagrammatically in the extended β-sheet form in the cleft of PR. The peptide bond to be cleaved is between P1 and P1′, *shown by the two arrows*. The two subunits of PR and the subsites interacting with the side chains of the substrate are indicated, as well as some of the contact residues in PR. Asp 25 and Asp 25′ are the Asp residues at the active site. One subunit is above the plane of the paper and the other below. (From WLODAWER and ERICKSON 1993)

et al. 1992), leads to loss of enzymatic activity. Published estimates of the dissociation constant governing subunit interactions are in the nanomolar range (DARKE et al. 1994; reviewed in DARKE 1994). Consistent with their assignment to the class of aspartic proteases, retroviral PRs all function best at mildly acid pH values, typically from pH 4 to 6. However, significant activity also is observed up to neutrality, at which processing reactions presumably are carried out in vivo. When acting on peptide substrates in vitro, all PRs work most efficiently in high salt, and in some studies up to 2*M* NaCl has been used. The high ionic strength increases the strength of binding to the substrate, rather than the catalytic rate itself, and thus probably the enhancement of cleavage is due to the known strengthening of hydrophobic interactions by salt. Curiously, HIV-1 PR is abnormal in that it prefers low salt when the substrates are polyproteins, even though it is like other PRs in preferring high salt with peptide substrates (KONVALINKA et al. 1995a).

The specificity of retroviral PRs has been the subject of numerous investigations, with information being generated by two methods. One is simply the compilation of sequences at bona fide cleavage sites in Gag and Pol and in nonviral proteins found to be substrates for PR (PETTIT et al. 1991; POORMAN et al. 1991; reviewed in TOMASSELLI and HEINDRIKSON 1994). If a large enough data set is available, analysis can lead to rules that at least qualitatively define the potential of a sequence to be recognized by PR. When applied to the collection of diverse retroviruses, of course, this approach obscures any genus-specific differences among the PRs. The other is the systematic testing of short synthetic peptides for ability to be cleaved (for example GRIFFITHS et al. 1992; GRINDE et al. 1992; TÖZSÉR et al. 1992; reviewed in DUNN et al. 1994). This extensively used method has the advantage that quantitative information can be obtained, namely the kinetic parameters that give information about binding affinity of the substrate and the catalytic efficiency once the substrate is bound. Both methods have been augmented by construction of specific alterations of known cleavage sites, by in vitro mutagenesis or peptide chemistry. For interpretation of the data generated, it is important to remember that neither of these approaches can lead to a strong prediction that PR will cleave at a particular sequence found in a given polypeptide, because the secondary and tertiary folding may preclude access to the sequence. The substrate must be in an extended conformation, at least transiently, to fit into the active site and be hydrolyzed. It has been known since the discovery of PR as a viral enzyme that if nonviral polypeptides are partially unfolded, for example with SDS, otherwise cryptic PR cleavage sites can be recognized.

In broad outline, the specificity of retroviral PRs can be summarized as follows. The amino acid recognition sequences all are relatively hydrophobic, as is apparent by inspection of the collection of natural sites for ASLV, HIV, MLV, and M-PMV (Table 1). In general the requirements for residues upstream of the scissile bond are different from those downstream. From systematic studies it is clear that the most pronounced preferences are for the P1 position, which

Table 1. Cleavage sites in prototypic viruses

Junction	Site number	Sequence	Gag/Pol residue number[a]	Comments
RSV (PrC)				
Gag				
MA-p2a	1	TSCY HCG	155	Late cleavage, sometimes incomplete
p2a-p2b	2	IGCN CAT	164	inferred from in vitro clv[b]
p2b-p10	3	PYVG SGL	177	
p10-CA	4	VVAM PVV	239	
CA-SPa	5	QGIA AAM	476	Ca. two-thirds of CA ends here[c]
SPa-SPb	6	AAAM SSA	479	Ca. one-third of CA ends here[c]
SPb-NC	7	PLIM AVV	488	
NC-PR	8	PAVS LAM	577	
Gag-Pol				
-RT	9	ATVL TVA	708	
RT-IN	10	FQAY PLR	1280	
IN-p4	11	PLFA GIS	1566	
MLV (Mo)				
Gag				
MA-p12	1	SSLY PAL	131	
p12-CA	2	SQAF PLR	215	
CA-NC	3	SKLL ATV	478	
NC-SP	4	TSLL TLD	534	
Gag-Pol				
NC-PR	4	TSLL TLD	534	
PR-RT	5	LQVL TLN	659	
RT-IN	6	STLL IEN	1330	
Env				
TM-	7	VQAL VLT	–	
HIV (NL4-3)				
Gag				
MA-CA	1	SQNY PIV	132	
CA-SP1a	2	ARVL AEA	363	
SP1a-SP1b	3	AEAM SQV	367	
SP1b-NC	4	ATIM IQK	377	
(NC)	5	VKCF NCG	392	"Early" cleavage?[d]
NC-SP2	6	RQAN FLG	432	
SP2-p6	7	PGNF LQS	448	
Gag-Pol				
-PR	8	SFSF PQI	488	
PR-RT	9	TLNF PIS	587	
RT-(RH)	10	AETF YVD	1027	p66/p51 cleavage site[e]
(RH)-IN	11	RKVL FLD	1147	
M-PMV				
Gag				
MA-p24	1	PQVM AAV	100	
(p24)	2	FPVL LTA	161	Partial cleavage: "p16"[f]
p24-p12	3	PTVM AVV	216	
p12-CA	4	KDIF PVT	299	
CA-NC	5	GLAM AAA	525	
NC-p4	6	KQAY GAV	621	
Gag-Pro				
DU-PR	7	SDIY WVQ	759	Inferred from in vitro clv[g]
PR-p4	8	MMCS PND	866	Inferred from in vitro clv[g]
Env:				
TM-	9	QVHY HRL	–	

[a] Residue number is the amino acid in P1, with the initiating methionine counted as number 1.
[b] From PEPINSKY et al. (1996)
[c] From PEPINSKY et al. (1995).
[d] Based on cleavage of peptides (from WONDRAK et al. 1994).
[e] Since the HIV RT is a heterodimer, one-half of the polypeptides cleaved at this position.
[f] From SONIGO et al. (1986).
[g] From HRUSKOVÁ-HEIDINGSFELDOVA et al. (1995).

typically is occupied by an amino acid with a bulky side chain, and in natural sites never by an amino acid with a β-branched aliphatic side chain. In P1′ bulky side chains also are preferred. At least for some model peptides, certain residues also are not permitted in the P3, P2, P1′, and P2′ positions, with restriction at the P2′ position being especially stringent for HIV PR. Some residues stimulate cleavage at particular positions, for example glutamate in P2′ for HIV PR. In no case for any retroviral PR is only a single amino acid residue tolerated at a particular position in the substrate. The sequence specificities of PRs from different retrovirus genera are not identical, and as expected the more divergent the PR the more dissimilar the cleavage site specificity. More closely related PRs, like those of HIV1 and HIV2, still are distinguishable in their preference for substrates.

A particularly common motif in cleavage sites in all retroviruses is the sequence Tyr/Pro or Phe/Pro at the scissile bond. This "aromatic/proline" sequence is found almost invariably at the N-terminus of CA proteins, and in addition at scattered other locations in Gag or Pol (Table 1). The common placement of this sequence at one end of CA has led to the idea that it may have some particular significance for processing, but it is equally possible that the conservation of this site reflects structural requirements that have nothing to do with cleavage per se. The significance of the differences in cleavage site sequences is unknown. It is clear that some naturally occurring sites are "optimal" while others are distinctly "suboptimal." As defined by model peptides, cleavage rates for natural sites can range over more than 20-fold. Conceivably the suboptimal sites evolved because it is important that these be cleaved later, thus facilitating the correct temporal order of cleavage. However, it is important to note that cleavage efficiency of Gag protein does not necessarily mirror cleavage efficiency of peptides. Indeed given that the "consensus" target sequence for retroviral proteases is so weak, one might argue that in natural substrates a primary determinant for cleavage is secondary and tertiary structure. A useful though certainly oversimplified view of processing in vivo is that PR simply finds stretched amino acids that are hydrophobic and both accessible and flexible. Apparently such stretches mostly are located between the separately folded domains of Gag and Gag-Pol.

More specific algorithms for predicting what defines a cleavage site have been developed, which are reasonably accurate in their predictions for peptides. These algorithms are generated by scoring the frequency with which a particular amino acid is found at a particular site (e.g. P1) in a collection of substrates, and then assigning a quantitative value to this frequency (Poorman et al. 1991). Since the number of natural cleavage sites is limited, usually artificial sites in nonviral proteins are used as well. The assumption in this treatment is that the seven or eight sites act independently in determining the "goodness" with which a segment of polypeptide can fit into the PR cleft and be hydrolyzed. Recently more sophisticated algorithms have been developed, which take into account possible indirect interactions of amino acid residues as they lie in the PR cleft (Chou 1993). For example, the residues at P2 and P1′ have contacts in the subsites S2 and S1′, which are next to each other in the same subunit (Fig. 4), and thus may influence each other. All of these predictive methods suffer from the inherent problem that they are based on qualitatively scoring of protein or peptide substrates as "cleavable" or "not cleavable" by the enzyme. In reality of course there is a wide range in efficiency of cleavage, covering more than two orders of magnitude.

Models of the three-dimensional structure of HIV PR with bound inhibitors show directly which amino acid residues form the subsites in which the peptide substrate docks. Comparisons between the HIV and ASLV PR structures thus have allowed predictions about which of the subsite residues account for the different specificities of the two PRs. Some of these predictions have been tested experimentally by genetic engineering of ASLV PR (KONVALINKA et al. 1992; GRINDE et al. 1992; CAMERON et al. 1994). Several key changes in subsite residues together were found to make the specificity of the avian enzyme resemble that of HIV PR on some peptide substrates. However, as might be expected, the cleavage efficiencies of the different engineered ASLV mutant PRs, as measured on a panel of peptides, only partially mimicked that of HIV PR. Thus not only the obviously different subsite residues contribute to the differences in specificity of these two PRs. This conclusion also is consistent with the observation that even residues in the flaps play a role in specificity (MOODY et al. 1995). While the information generated by this type of in vitro mutagenesis, to make one enzyme more similar to another, is useful qualitatively, it is difficult to evaluate in more quantitative terms, and predictions of activity of such mutant enzymes on polypeptide substrates are uncertain. For example, a mutant ASLV PR resembling HIV PR in its cleavage of some peptides failed to show any differences in cleavage of natural Gag polyprotein substrates (KONVALINKA et al. 1995a).

The best-studied retroviral PRs, those of HIV and ASLV, are markedly different in their catalytic efficiencies, with ASLV PR being roughly tenfold less active on optimal peptide substrates. This difference might be rationalized by an evolutionary argument based on the genetic structure of ASLV: since the avian PR is unusual in being encoded as the C-terminal domain of Gag, it is synthesized and incorporated into virions at ca. 20-fold higher levels than is HIV PR, and thus a lower specific activity is needed to accomplish the Gag and Pol cleavages. It is provocative that ASLV PR also is one of the few examples of retroviral proteases with the active site sequence Asp-Ser-Gly as compared with the more usual Asp-Thr-Gly. At least for these two PRs, apparently the presence of the Ser residue signifies lower specific protease activity, since a Thr to Ser change in HIV PR reduces its activity under defined conditions (KONVALINKA et al. 1995b; ROSÉ et al. 1995). Other residues clearly also contribute to catalytic potential, however. Despite the attenuation in activity, the Thr to Ser mutant PR still is able to provide full proteolytic function to a replicating virus. However, when expressed in *E. coli* or in mammalian cells, it is no longer toxic and apparently no longer cleaves at least some of the host protein substrates (KONVALINKA et al. 1995b). Detailed enzymological comparisons of other retroviral PRs with Ser/Thr changes at the active site have not been reported.

3 Inhibition of Protease

One of the consequences of the AIDS epidemic has been the search for anti-HIV drugs. The three viral enzymes, reverse transcriptase, integrase, and protease, make ideal targets for pharmaceutical approaches to intervention in this disease, since each plays an essential role in the virus life cycle, and thus blocking enzyme function halts the spread of the virus. Each also has key features that distinguish it from related cellular enzymes. To be useful, inhibitors must be effective in the nanomolar range or lower. They must permeate cells, be bioavailable and nontoxic, and not be rapidly degraded. Several strategies have been used to find such inhibitors of HIV PR (reviewed in DEBOUCK 1992; WLODAWER and ERICKSON 1993). One is the random screening of libraries of compounds for ability to inhibit PR or to inhibit virus replication. Another is modification of compounds that are known to inhibit an important cellular aspartic protease, human renin. In the latter case, the knowledge that PR cleaves well when proline occupies the P1′ position facilitated development of specific inhibitors, since few cellular proteases cut in front of a proline residue.

In the classical approach to design a protease inhibitor (reviewed in DUNN et al. 1994), a nonhydrolyzable analogue of a dipeptide is used as the central part of the compound. This is built into a larger oligopeptide, or molecule that mimics such a peptide, with structural features like those of substrates known to bind tightly in the PR cleft. The numerous detailed studies of substrate specificity of HIV PR provide a foundation for the latter task. Various chemical configurations have been used in the design of non-hydrolyzable analogues resembling the scissile peptide bond. These include (in approximate order of effectiveness) hydroxyethylenes or dihydroxyethylenes, difluoroketones, phosphinates, or reduced amides. Another structure used to replace the cleaved bond is found in statin, an unusual amino acid derived from pepstatin, a specific inhibitor of aspartic proteases isolated from species of Actinomyces.

In view of the inherently poor stability and bioavailability of peptide-based inhibitors, many groups have found or designed novel peptide-like compounds by structure-based drug design, i.e. by using the high-resolution three-dimensional structure of PR with and without bound inhibitors. More than 100 structures, most unpublished, have been determined from crystals of this protein (WLODAWER and ERICKSON 1993). An example of such an approach is the design of twofold or pseudo-twofold symmetrical PR inhibitors, based on the knowledge that retroviral PRs exhibit twofold symmetry (for example, ABDEL-MEGUID et al. 1993). Another approach is to search the databases for known compounds with steric complementarity to the active site and binding cleft (for example, DESJARLAIS et al. 1990).

A number of PR inhibitors have been found or developed that satisfy the criteria for possible use as pharmaceuticals. It is important to remember that while such drugs will prevent infection of new cells, they will not hinder continued production of virus particles from cells infected previously. Patients with AIDS have high levels of viremia by the time the disease is full blown, with on the order of 10^8 HIV particles in fluids in the body. PR inhibitors rapidly decrease the levels of virus detected in the plasma, as measured by amplification of viral sequences from the RNA in particles (Wei et al. 1995; Ho et al. 1995). Virus titers may drop several orders of magnitude, with some resurgence of CD4 cell numbers. The rapidity of this decrease, which is evident already after a few days of treatment, implies that most of the virus detected is from newly infected cells. Thus there is a dynamic equilibrium between infection of cells and the death of these cells. However, just as for RT inhibitors, the virus titer returns to high levels after weeks, with mutants resistant to the anti-PR drugs dominating the population. From cell culture studies, these mutants have changes in the amino acid sequence of PR at selected sites, which may or may not differ among apparently dissimilar inhibitors (Otto et al. 1993; Jacobsen et al. 1992). It seems likely that selection of drug-resistant mutants will be the limiting factor in the utility of all anti-PR pharmaceuticals, but the utility of this type of therapy has not been fully evaluated. One might expect that in combination with other inhibitors, for example of RT and IN, PR inhibitors will be able to stymie an HIV infection, at least if it is in an early stage.

4 Regulation of Proteolytic Processing

When does PR-mediated proteolysis occur? The exact answer to this fundamental question is not known. Some conflicting data and interpretations have been reported, reflecting in part differences among retroviruses and the cells they infect and in part imprecision of the techniques available to address these issues experimentally. However, it is widely agreed that in all viruses most cleavages take place late in assembly, probably immediately before the completion of budding and release of the virus particle. For the majority of retroviruses mutations in Gag that block assembly at an early stage also prevent cleavage. The best examples of this class are mutations that prevent transport to or stable association with the plasma membrane. For the mammalian viruses this can be achieved by preventing myristylation of the N-terminal Gly residue of Gag (Rein et al. 1986; Rhee and Hunter 1987). For ASLV Gag, which is not myristylated, small deletions at or near at the N-terminus yield the same phenotype (Wills et al. 1991). HIV seems to be an important exception to this rule, since significant cleavage of Gag also occurs in a compartment that is interpreted to be the cytoplasm, both in wild-type (for example, Kaplan and Swanstrom 1991) and to a variable extent in mutants in which Gag is unable to associate with the membrane (Göttlinger et al. 1989; Bryant and Ratner 1990). In this sense the HIV PR can be considered less highly regulated than other retroviral PRs, which is also consistent with the cleavage of host proteins (see below). By contrast, mutations in Gag that interfere with late stages in assembly show variable and sometimes high levels of processed Gag proteins "in" the cell (WILLS et al. 1994; PARENT et al. 1995). A recurring difficulty in all studies dealing with the site and timing of processing should be noted: there is a lack of clear criteria to distinguish the last stages in budding. Particles that are fully released into the medium are easy to identify. But the Gag proteins in particles that are partially formed, in particles that are completely formed but with viral membrane still continuous with plasma membrane, and in particles that are completely formed with viral membrane distinct from plasma membrane but still somehow tethered to it – all will appear to be "intracellular" upon biochemical analysis.

Thin-section electron microscopy provides a view of the virion at all stages of budding. As discussed below, immature particles formed in the absence of any proteolysis are distinctively different from mature particles with fully cleaved Gag proteins. It is noteworthy that by EM all retroviruses observed to be budding are immature in morphology, up to and including the last stage of assembly when they appear fully formed and ready for release into the medium. These observations may be taken to suggest that processing actually occurs immediately after release of the particle. Although this hypothesis is difficult to exclude rigorously, the genetic results from late assembly mutants argue against it. The possibility that morphological changes in the viral core do not follow immediately upon cleavage makes it difficult to draw a direct correlation between appearance in EM and extent of processing.

While PR-mediated processing of retroviral proteins takes place at a distinct stage in assembly, at least in most viruses processing plays no causal role in assembly. It has been demonstrated repeatedly that inactivating mutations in PR do not prevent assembly and budding (CRAWFORD and GOFF 1985; KATOH et al. 1985; KOHL et al. 1988; WILLS et al. 1989; STEWART et al. 1990; SOMMERFELT et al. 1992). However, recent evidence has been presented that HIV may be an exception to this generalization (KAPLAN et al. 1994b). At least in some cell types PR inhibitors slow the release of particles, and a virus expressing a mutant PR with reduced protease activity also shows slower budding, with accumulation of particles that look as if they are "tethered" to the cell membrane. The generality of these findings for other HIV strains and infected cell types has not yet been established. Investigating if this conclusion holds in other retrovirus systems will require the development of suitable cell-permeant inhibitors for other PRs. Assuming that PR action is timed to occur late in the assembly process but is not required for that process, one may speculate that there is only a window in time in which processing can occur in order for the infectious particle to be produced. As discussed below, it is known that premature activation of proteolysis, induced for example by generating a linked PR dimer, abrogates assembly. What would happen if PR activation were forcibly delayed until after assembly and budding were complete? At present this question cannot be addressed precisely. Membrane-permeable PR inhibitors cannot be removed easily once incorporated into the virion. Although temperature sensitive mutants in HIV PR have been characterized (MANCHESTER et al. 1994; KOTTLER et al. 1995), the mutant PR may not be able to be reversibly inactivated in a tight enough manner to allow study of delayed function of PR.

Whether proteolysis takes place in the last stages of budding or after release of the virus particle, the timing of this process must be controlled. The need for timing is underscored by the examples of B-type and D-type viruses, in which assembly and budding are distinct and uncoupled reactions. In these viruses, the best studied being M-PMV, the viral core is preassembled in the cytoplasm from uncleaved Gag and Gag-Pol proteins. Only after transport of the core to the plasma membrane and subsequent envelopment by the membrane does cleavage occur (RHEE and HUNTER 1987). The mechanisms underlying regulation of proteolysis are unknown, but diverse experimental findings suggest the importance of PR dimerization. For example, when two *pro* genes are fused to code for a "linked dimer," with the two PR moieties held together by a flexible linker peptide, expression of the resulting protein leads to premature cleavage of Gag and Gag-Pol and thereby abrogates assembly, both in the HIV (KRÄUSSLICH 1991) and ASLV (BURSTEIN et al. 1991) systems. These results demonstrate directly that only the uncleaved Gag protein, and not the mature proteins derived therefrom, has the ability to organize the assembly process. They also demonstrate that dimerization of PR is a possible point of control; forcing dimerization by covalent linkage prematurely activates its proteolytic activity. According to one model that assigns a regulatory role to PR dimerization, it is simply the high concentration of PR in the nascent virion that promotes dimerization, thus leading to activation of enzymatic activity. However, in M-PMV this model cannot account for the delay of processing until envelopment. Nor can it account in general for activation of processing late in budding, since the local concentration of PR presumably would be just as high when the bud is just starting to form.

It remains a matter of speculation how dimerization of PR could be regulated, because the structure of the PR domain as it is embedded in the Gag-Pol or Gag precursors is not known. The crystal structures of several mature PRs imply that the four-stranded β-sheet tying up N- and C-terminal ends of both subunits is important for holding the dimer together, since about half of the intersubunit contact area is in this structure. Consistent with the importance of the β-sheet, removal of several N-terminal amino acid residues reduces or abolishes protease activity, in HIV (ROSÉ et al. 1993) and by inference also in ASLV (SCHATZ et al. 1996). Even removal of a single N-terminal residue is highly deleterious for activity (PICHOVA et al. 1992). Thus it is easy to imagine that sequences flanking the PR domain could interact with abutting sequences in the PR domain itself, and thereby alter dimerization potential. The several clues that support this possibility are discussed below.

The structure and function of cellular aspartic proteases suggest general models for regulation of retroviral PRs. The cellular enzymes are synthesized as zymogens, with N-terminal extensions of some dozens of amino acid residues. Proteolytic removal of these N-terminal extensions in most cases is required for activation of the full enzymatic activity (DAVIES 1990). For example, pepsinogen differs from pepsin by a 44 amino acid segment, which is cleaved out autocatalytically when the proenzyme is exposed to acid pH. In this case basic residues in this segment fold into the active site, thereby blocking it (JAMES and SIELECKI 1986). One could thus speculate that in retroviruses the segment of polypeptide upstream of the PR domain serves a similar function. Several results support this notion. ASLV cleavage site mutations engineered to block cleavage between NC and PR in vivo lead to global processing defects, with no mature Gag or Pol proteins formed in avian or mammalian cells (OERTLE and SPAHR 1990; BURSTEIN et al. 1992). In these mutants limited cleavage – up to 50% of the Gag and Gag-Pol molecules – does occur, but only at two sites: at the junction between NC and PR domains in Gag (BURSTEIN et al. 1992; SCHATZ et al. 1996), and at this junction plus the junction between PR and RT in Gag-Pol (STEWART and VOGT 1994). Because of the mutation, cleavage between NC and PR is incorrect, leading to a PR moiety that is apparently inactive, since it is missing the normal first three amino acid residues (SCHATZ et al. 1996). Cleavage at the same site also takes place in wild type virus in a fraction of Gag molecules (PEPINSKY et al. 1996). Alterations at the C-terminal end of PR also lead to slowed or blocked processing (BENNETT et al. 1991). Analogous mutations at the N-terminus of PR in HIV also have major effects on processing, blocking or greatly slowing the formation of mature Gag proteins in transfected cells (ZYBARTH et al. 1994). These data support the hypothesis that proteolytic liberation of the PR domain is essential for it to carry out the several Gag and Gag-Pol cleavages.

Studies with PR-related proteins expressed in *E. coli* or in vitro support the hypothesis that PR itself and the PR domain embedded in other sequences are not necessarily folded identically. ASLV NC-PR, purified from *E. coli* after denaturation and renaturation, has extremely little proteolytic activity (SELLOS-MOURA and VOGT 1996). This protein remains monomeric under the conditions tested, providing an explanation for its lack of activity. However, after prolonged incubation slow "auto"-cleavage between the two domains is observed, giving rise to a fully active PR. A simple interpretation of these observations is that the association constant governing NC-PR dimerization is much lower than for PR itself because the upstream NC sequences somehow perturb dimer contacts. One can hypothesize that the same holds for the PR domains of the ASLV Gag protein. Another precedent for alterations in folding of the ASLV PR domain that are induced by neighboring sequences is provided by Gag-Pol. The PR domain in Gag-Pol apparently is inactive, being neither necessary nor sufficient for processing of Gag and Gag-Pol proteins in virions (STEWART and VOGT 1991; CRAVEN et al. 1991); a contrary conclusion (OERTLE et al. 1992) might be due to an adventitious mutation that was noted. Finally, when ASLV Gag-Pol or various related truncated proteins are incubated in vitro with PR, no cleavage is observed between the PR domain and the RT domain, i.e. between Gag and Pol. This is the only cleavage site in Gag or Pol that is not processed in vitro, apparently because the local sequence environment causes it to fold into a coiled coil (STEWART and VOGT 1994). All of these observations support the potential of PR to fold in alternative ways, but of course their relevance in vivo remains to be ascertained.

In the HIV system a variety of PR-containing proteins have been assessed for their enzymatic activity in vitro. These include PRs extended at the N- and/or C-termini with viral sequences, in some cases fused to nonviral proteins for ease of purification and assay. An example is a construct consisting of the *E. coli* maltose-binding protein (MBP) fused to HIV sequences just upstream of the N-terminus of PR, and also including a short stretch of viral sequences downstream of PR (LOUIS et al. 1994). As isolated from *E. coli* in an aggregated form, this protein has little protease activity. When the aggregates are dissociated in urea and renatured, the protein slowly cleaves itself at the correct N-terminus of PR. As measured on an exogenous substrate, this single cleavage leads to a more than 100-fold activation of protease activity, consistent with the notion that activity can be modulated by upstream sequences. However, another study found no difference in specific activity in vitro between mature HIV PR and a PR derivative carrying a 25 amino acid N-terminal extension, in the absence of fusion to another protein (CO et al. 1994). In this case the PR was a mutant with reduced catalytic activity, chosen because it lengthens the half-life of the extended protein in *E. coli*, thus allowing its purification. Comparison of the results in these two studies suggests that the foreign MBP sequences or the C-terminal extension may play a role in modulating enzyme activity, perhaps indirectly during refolding. Interpretations of all results with purified PRs expressed in *E. coli* are made difficult by the need to denature and renature the protein, because of its insolubility or aggregation.

In HIV the role of flanking sequences in enzymatic activity and the dimerization potential of PR also has been studied with translation in vitro in a rabbit reticulocyte lysate (ZYBARTH and CARTER 1995). Although the radioactively labeled products made by in vitro translation comprise only a tiny fraction of the total protein, this system has potential advantages in that chaperonins and other host factors present may contribute to the proper folding of the protein. PR extended from 14 to more than 70 residues to its N-terminus shows little or no autoproteolytic activity. However, still longer extended versions, resulting in the inclusion of the entire NC domain, are highly active in this assay. A qualitative assay for dimerizaton in the same system, based on binding of labeled proteins to an immobilized, six-histidine-tagged PR derivative, indicates that the inactive forms of PR fail to dimerize. These results suggest an interplay between negatively and positively acting elements upstream of the PR domain.

In sum, diverse data suggest that the PR domain when embedded in its precursor does not have the full enzymatic activity of the mature protein whose crystal structure is known. It is important to emphasize, however, that not all studies have come to this conclusion. Some HIV PR fusion proteins are active (BOUTELJE et al. 1990; VALVERDE et al. 1992; CO et al. 1994; PHYLIP et al. 1992), and the PR domain also is reported to be active in a normal precursor form in *E. coli* (KOTLER et al. 1992). Because of the diversity and complexity of systems in which PR regulation has been addressed, it is perhaps not surprising that the published results are not uniform. Nevertheless, it remains likely that the activity of the PR domain in Gag or Gag-Pol is not identical to that of PR itself. Differences in polypeptide folding – perhaps only subtle differences – represent the most plausible mechanism for this sort of control.

The notion that the PR domain in Gag or Gag-Pol is not fully active does not itself explain the timing of processing, but simply implies another layer of control. What events trigger activation? A priori it seems plausible that a cellular protease might act as a trigger, proteolytically liberating a small amount of the PR domain, which then can dimerize and function in a positive feedback loop to release more PR from the polypeptide precursor. Such a hypothetical protease might not cleave at the same sites as PR itself. While this model seems attractive in view of the abundance of proteolytic cascades in biological systems, the available evidence does not support it. For example, when PR activity is abolished by mutation of the active site, the budded particles do not show evidence of any cleaved Gag or Gag-Pol products. However, few molecules processed in this way would escape detection and yet might be sufficient to initiate a cascade. In the absence of data to the contrary, it seems most probable that the first cleavages in fact are carried out by the PR domain functioning as a dimer. The local structure of two dimerized PR domains is a matter of speculation, but at least at the active site presumably it is very similar to that of the mature PR. On the other hand, the interface contacts holding the domains together might be fundamentally different than those in the mature PR. Generally it has been assumed that the dimerized PR domains must cleave "in *trans*," i.e. act on a neighboring Gag or Gag-Pol molecule in the budding virion. Cleavage "in *cis*" seems unlikely because of the geometry of the active site seen in the crystal structure of PR: the dog cannot bite off his own tail (Fig. 3). That is, the groove that binds the substrate as well as the active site are on one side of the dimer, while the N- and C-terminal ends both are on the other side. These ideas notwithstanding, kinetic evidence was found to favor cleavage in *cis*, at least in vitro for certain "mini-precursors" of PR (Co et al. 1994; LOUIS et al. 1994). Molecular modeling suggests that one of the two N-terminal strands in the β-sheet could peal back, without disrupting the remaining three strands, and slide around into the active site. Since this disruption of the β-sheet would be transient, the stability of the dimer might not be compromised. Whether cleavage in *cis* actually occurs in the virion is a difficult and important question that remains to be addressed.

None of the other speculations about primary signals that might trigger proteolysis has been supported experimentally. It is unlikely that local acidification of the budding virion plays any role in regulation of protease activity, since all known proton pumps in fact pump in the opposite direction. One might imagine that the virus particle could sense when its connection to the cytoplasm is severed because of the inevitable influx of ions such as Ca^{2+}, which is at a concentration in the medium that is at least 1000 times higher than that in the cell. Given the ubiquitous nature of Ca^{2+} as a messenger, this seems an attractive model. But in fact evidence argues against such a role in the case of the ASLV (Vogt et al. 1992). Similarly, one might imagine that the oxidizing environment of the medium or the hydrolysis of ATP trapped during budding is sensed by the virion. Perhaps protein phosphorylation or dephosphorylation by a plasma membrane-bound kinase or phosphatase provides the signal. This would be consistent with the observation that morphological maturation does not occur properly when budding is redirected to a different membrane compartment (see below). These hypotheses remain to be tested.

5 Processing of Gag

As measured by abundance, the Gag protein is the major target of proteolytic processing in retroviruses. The ca. 2000 molecules of Gag that form the structure of the immature core each are ultimately cleaved at about a half dozen locations, giving rise to the major mature proteins as well as some smaller peptides. The sites for these cleavages in the viruses used as prototypes in this review are shown in Fig. 1 and Table 1. As has been known from pulse-chase experiments since the time processing was discovered, some sites are cleaved before others. Several factors make the significance of the apparent ordering of processing difficult to assess. Since budding cannot be synchronized, the Gag and Gag-Pol molecules under scrutiny may reside in virions that are at different stages of assembly and maturation. Some molecules, for example ones cleaved prematurely, may represent dead end products not incorporated into virions. Nevertheless, in overview the many published studies suggest that for most cleavage sites there is no obligatory requirement for order of cleavage, in the sense that processing at one site must wait for processing to be completed at another. Rather, cleavage appears to be a stochastic process whose rate at each site is governed by the intrinsic susceptibility of that site to being attacked by PR. Not only the primary sequence, but also the secondary and tertiary structure around the site no doubt play key roles in determining this rate. This picture is consistent with the findings that processing of in vitro-translated Gag takes place correctly, and with similar relative efficiencies, in the presence of exogenous protease (von der Helm 1977; Kräusslich et al. 1988; Erickson-Viitanen et al. 1989; Tritch et al. 1991; Pettit et al. 1994). Processing of immature particles with exogenous PR, in presence of detergents to remove the lipid envelope, also leads to the correct Gag products (Stewart et al. 1990; Konvalinka et al. 1995a). Thus the structure of the immature virion and the position of the PR domain in it does not seem to be of critical importance for the bulk of the processing. In fact, despite the differences in their "preferred target sequences", retroviral proteases in vitro cleave heterologous Gag molecules in immature cores of unrelated viruses, and many if not most of these cleavages resemble those made by the homologous PR (KONVALINKA et al. 1995a).

The exceptions to the generalization that processing at one site does not depend on processing at another are the cleavages leading to release of PR. By pulse-chase kinetics, in ASLV the site at the N-terminus of the PR domain in Gag is the first to be cut (VOGT et al. 1975), and probably the same holds for the Gag-Pol protein (EISENMAN et al. 1980). For the retroviruses in which the PR domain is part of Gag-Pol as in HIV, the relative timing of the cleavages flanking the PR domain compared with cleavages of Gag is not known. As discussed above, both in ASLV and in HIV excision of PR appears to be a prerequisite for subsequent steps, plausibly because the enzymatic activity of PR is activated in this way and/or because only the freed PR can diffuse through the virion to gain access to the other sites.

Among the last processing steps in ASLV and HIV are those that occur at the C-terminus of CA. In both of these viruses the CA domain is separated from the downstream NC domain by a "spacer" peptide, here called "SP" in ASLV and "SP1" in HIV (Fig. 1). The spacer is 14 amino acid residues long in HIV (with some variation among different strains) and 12 residues in ASLV. The processing events in these otherwise very different viruses are remarkably similar: The first cleavage is at the downstream site, thus generating the mature N-terminus of NC (MERVIS et al. 1988; BENNETT et al. 1991; CRAVEN et al. 1993). The sites only a few residues upstream become targets for PR later. In both virus types there is an additional cleavage site in the spacer segment itself, dividing it into segments *a* and *b* (Fig. 1). The additional site in HIV is 4 residues (see KRÄUSSLICH et al. 1995) and in ASLV 3 residues (PEPINSKY et al. 1995) from the N-terminus of the spacer. The final mature C-terminus of ASLV CA was originally misassigned to this latter position, and the correct placement of processing sites has been worked out only recently. In fact, the population of CA molecules in mature ASLV is heterogeneous, with some ending at one site and some at the other. The spacer peptide in HIV and ASLV appears to play an important role in the infectious virus. Deletion of this segment of Gag renders the viruses noninfectious (CRAVEN et al. 1993; KRÄUSSLICH et al. 1995; Pepinsky et al. 1995). In HIV these particles show severe aberrations in budding from COS cells (KRÄUSSLICH et al. 1995), though perhaps not in HeLa cells (PETTIT et al. 1994). In ASLV deletions in the spacer lead to heterogeneously sized particles (N.K.KRISHNA and J.W.WILLS, unpublished observations). Evidence from detergent sensitivity of ASLV suggests that the final CA cleavage somehow stabilizes the viral core. For most retroviruses, cores from mature particles are disrupted by nonionic detergents while cores from immature particles are stable (PEPINSKY 1983; STEWART et al. 1990). Nevertheless, in ASLV at least a portion of the population of CA molecules can be obtained in a particulate form under defined conditions. In mutants lacking all or parts of the 12-residue spacer segment, this partial detergent resistance is lost (CRAVEN et al. 1993; PEPINSKY et al. 1995).

Several examples of Gag cleavages with special properties have been described. A drastic alteration of the HIV CA-SP1*a* cleavage site originally was found to inhibit budding (GÖTTLINGER et al. 1989). However, in view of the importance of the spacer peptide for correct assembly and budding, this phenotype most likely is not a consequence of the processing defect per se, but rather of alteration of a sequence that is critical for other reasons. This interpretation is consistent with the highly aberrant, tubular morphology of the mutant virus particles originally seen to be in the process of assembling. Studies of wild-type and mutated forms of the HIV Gag protein synthesized in vitro suggest that cleavage at the SP1*b*-NC cleavage site, which normally takes place first, somehow reduces the efficiency of processing at the upstream CA-SP1*a* site (PETTIT et al. 1994). These observations have been interpreted to mean that the 14-residue "tail" at the end of CA somehow slows the further processing that leads to its release. However, the relevance of these results for processing in vivo is uncertain, since both in HIV (KRÄUSSLICH et al. 1995) and ASLV (CRAVEN et al. 1993; PEPINSKY et al. 1995a), when the equivalent 14-or 12-residue spacer, respectively, is deleted, processing in vivo at the single remaining site is slowed rather than accelerated, virions showing considerable uncleaved CA-NC protein.

In ASLV alterations in and around the PPPY sequence in p2b, which is found in most retroviruses though not lentiviruses, lead to increased amounts of a Gag species missing the PR domain, as well as to an inhibition of particle budding that appears to be somewhat dependent on cell type (BOWLES et al. 1994; WILLS et al. 1994). In one study (BOWLES et al. 1994) this result was interpreted as evidence that in wild-type virus this portion of Gag may play a direct role in inhibiting premature cleavage of Gag. A peptide that models the p2*b*-p10 junction is poorly cleaved in vitro though it binds well to PR, and thus can act as an inhibitor (CAMERON et al. 1992), which led to the speculation that this cleavage site might modulate PR activity, perhaps by blocking the active site. However, the fact that the sequences in question are part of an important "assembly domain" that somehow facilitates the final release of the completed virion (PARENT et al. 1995) makes it seem likely, as in the HIV example above, that mutations in this region affect proteolysis indirectly.

In HIV, one cleavage near the end of HIV Gag is special in a different way. Processing in the C-terminal p15 domain, which occurs late in assembly, separates the NC from the p6 domains. In most HIV strains PR cleaves at two closely spaced sites between these domains; the short segment of polypeptide between the sites here is termed SP2 (Fig. 1). In vitro PR-mediated cleavage between NC and p6 is dependent on the presence of RNA (SHENG and ERICKSON-VIITANEN 1994), a property not found for any other PR-mediated processing events. Mutation of the basic residues between the two Cys-His motifs in NC abolishes the RNA dependence, suggesting that it is the binding of NC to RNA that modulates accessibility of the cleavage site to PR. Thus processing may be tied to encapsidation or condensation of the genomic RNA. The biological significance of this phenomenon in vivo remains to be explored.

Finally, while proteolytic processing of Gag is generally attributed exclusively to the PR, recent analyses of ASLV Gag protein imply that cellular enzymes also play a role in the genesis of the final mature products (PEPINSKY et al. 1995b).

Most of p10 and a fration of CA are missing the C-terminal methionine residue predicted by the known PR-mediated cleavage event at this site, implying the existence of a carboxypeptidase activity. Also, the p2b peptide as it is found in virions as well as minor, truncated forms of MA is most easily explained by invoking the action of other proteases. Whether any of these processing events plays a role in the life cycle of the virus is uncertain.

6 Consequences of Gag Cleavages

Processing of the Gag protein leads to morphological maturation of the virus. This conclusion is derived from the properties of PR-defective viruses (KATOH et al. 1985; PENG et al. 1989; STEWART et al. 1990), from the correlation between uncleaved Gag molecules and immature particles in wild-type MLV (YOSHINAKA and LUFTIG 1977; LUFTIG and YOSHINAKA 1978), and from the effects of HIV PR inhibitors (KAPLAN et al. 1993; SOMMERFELT et al. 1992). By thin-section electron microscopy, the electron-dense ring underneath the lipid envelope in immature particles disappears, concomitantly with the appearance of an electron-dense core located approximately in the center of the virion. The morphology of budding and mature HIV has been especially well studied (HÖGLUND et al. 1992; NERMUT et al. 1994; reviewed in GELDERBLOM 1991). The final shape of the mature core is dependent on the virus genus. The cores of ASLV and MLV are central and round, while the cores of HIV and other lentiviruses are cone shaped and appear to touch the envelope at one point. In other retroviruses cores may be round and eccentric (MMTV), or central and bar shaped (M-PMV). Although rigorous proof is lacking, it is widely accepted that the outside layer of the visible mature core (which is useful to designate as the "capsid") is made of CA protein. Thus the presumption is that the CA proteins of different viruses have the propensity to form these different shapes. However, probably this is a too simplistic view, and other portions of Gag, in particular the adjoining NC domain, may also play a role in core morphology.

Exactly how does proteolytic processing lead to the collapse or condensation of the ring of Gag molecules to yield the mature core? Definitive answers to this question are not available, in large part because the budding process itself, with which maturation is somehow intimately associated as a late step, is poorly understood. For many purposes immature virions released from cells expressing PR-defective mutants can be considered to have the same fundamental structure as wild-type virions in the process of budding. Thus one can imagine that the electron-dense crescent-shaped shell in the bud grows until it is a sphere, at which time the membrane surrounding it fuses in a pinching off process, and the particle is released. In fact, however, budded PR-defective particles do not all have uniformly round immature cores. In all viruses where such particles have been examined by thin sectioning, many of the cores appear incompletely closed, as if the process of assembly had been halted when the growing core reached the neck or stalk at which the virion is attached to the cell prior to release. Since the closure of the core is not visible in all planes of sectioning, it may be that all of the cores in immature particles are incompletely closed.

In general terms it is clear that in the budding virion the MA domain, by itself binding to the membrane, holds the rest of the Gag molecule close to the membrane as well. Thus one might imagine that breaking the polypeptide backbone between the MA and CA domains would be the primary event leading to core condensation. One of the first tests of this idea was carried out by mutagenesis of HIV Gag, which conveniently has only a single cleavage site between MA and CA. Alteration of this site led to noninfectious particles with an abnormal morphology: in many of the resulting virions the core appeared to remain attached to the viral membrane (GÖTTLINGER et al. 1989). In later studies, partial inhibition of processing by limiting concentrations of an HIV PR inhibitor also was observed to give rise to aberrant morphologies, with partially condensed cores some of which remained juxtaposed to the membrane (KAPLAN et al. 1993). Even after only modest reduction in processing, with less than 50% of the Gag cleavage events blocked, the budded virions by thin-section EM were almost all abnormal. These results are consistent with a pivotal role for the cleavage between MA and CA, but they do not allow assignment of a one-to-one correspondence between morphological alteration and any cleavage.

Some studies have called into question the simple equation that a particular set of processing events leads directly and invariably to morphological maturation. When large portions of the HIV MA domain are deleted, most of the Gag protein for unknown reasons is redirected to the endoplasmic reticulum, leading to budding of the virus into this internal space (FÄCKE et al. 1993). While the small numbers of mutant particles that still bud from the plasma membrane look morphologically mature, the large numbers of mutant particles that bud into the ER look immature. Nevertheless, in the lysed cells the extent of Gag cleavage in this mutant was found to be similar to that in wild-type virus. Thus differences in proteolysis appear not to explain the different morphologies of budded particles in the ER and in the medium. Intracisternal A particles (IAPs), expressed from an endogenous retrovirus-like element in rodent cells, also bud into the ER and are immature by EM (KUFF and LUEDERS 1988). IAPs as isolated from cells consist of uncleaved Gag protein, notwithstanding the fact that they have a PR domain that is active when this Gag protein is artificially targeted to the plasma membrane (WELKER et al. 1995). A second and striking discordance between cleavage and maturation is provided by certain HIV mutants with alterations in NC that compromise the ability of this domain of Gag to correctly interact with RNA (BERKOWITZ et al. 1995; S. ERICKSON-VIITANEN and S. PETTIT, unpublished observations). These mutants remain grossly immature in morphology despite nearly normal levels of Gag cleavage. These several studies point to the possibilities that the site of budding may play a determining role in activation of PR, and that events other than proteolysis, in particular NC-RNA interacton, may be critical for proper morphological maturation.

Implicit in the morphological change precipitated by cleavage of Gag is the condensation of the RNA. Little information is available on the location of the RNA in immature particles. It is presumed to be attached at multiple points to the NC domains of the Gag molecules that form the electron-dense ring seen in EM, and perhaps also to loosely fill the "hollow" center of the particle. Upon liberation by cleavage, the NC protein becomes tightly bound to the RNA. Given the stoichiometry of RNA (ca. 15–20 kB / virion) and Gag (ca. 2000 molecules / virion; STROMBERG et al. 1974) and the size of the NC-binding site (ca. 6 nucleotides), the RNA probably becomes entirely coated by NC. This ribonucleoprotein complex then presumably is squeezed into the middle of the core, by the NC-RNA interaction itself, by the formation of the capsid, or by both. Cross-linking studies in ASLV suggest that in the immature virion the contacts between the NC domain and the RNA are quite different than those in the mature particle, being either "looser" or less numerous (STEWART et al. 1990). This observation has not been explained in biochemical terms.

The genomic RNA itself undergoes a dramatic change in structure as a consequence of maturation. In the immature particle the RNA occurs as a variable mixture of monomers and dimers (KORB et al. 1976; STEWART et al. 1990; OERTLE and SPAHR 1990; FU and REIN 1993; FU et al. 1994;), while in the mature particle it has the well-known dimeric structure. Even that fraction of the RNA that is observed to be dimeric in the immature particle is held together more loosely, as evidenced by its slower mobility in electrophoresis and its lower melting temperature. These observations are consistent with the properties of NC in vitro. This small basic protein strongly promotes annealing of complementary sequences (PRATS et al. 1988), probably by simply lowering the activation energy in the annealing reaction. Thus in the mature virion NC probably allows a sequence to more efficiently "search" for a complementary or partially complementary sequence on the same or on the other RNA, allowing the RNA to gradually settle into a form with maximal base pairing. It has not been established if this RNA maturation is linked directly to the morphological maturation seen by EM.

7 Processing of Pol

Retrovirus *pol* genes encode at least three enzymatic activities – reverse transcriptase (RT), RNaseH, and integrase (IN) (reviewed in KATZ and SKALKA 1994). The Pol protein is processed in different ways depending on the genus of virus (Fig. 1). In each case at least some IN is proteolytically released as a separate entity and some RNaseH remains attached to the upstream RT domain. In ASLV reverse transcriptase is a heterodimer composed of α- and β-subunits, with α containing the sequences for RT and RNaseH, and β containing in addition the sequence of IN. Processing also removes from the C-terminus a short segment of polypeptide that is dispensable for virus replication (KATZ and SKALKA 1988). When purified or expressed individually, both α and β manifest RT activity (HIZI et al. 1977; SOLTIS and SKALKA 1988), although the longer β-subunit is more active and more stable. β also has the same endonuclease activity manifested by the mature IN. In MLV reverse transcriptase isolated as a monomeric polypeptide consisting of the RT plus RNaseH domains is active. However, this protein appears to dimerize when presented with a template (TELESNITZKY and GOFF 1993).

In HIV the reverse transcriptase also is a heterodimer, composed of a p66 subunit and a p51 subunit. The former contains the RT and RNaseH domains, while the latter contains only the RT domain. As in ASLV, in qualitative terms both p66 and p51 are active enzymes under defined conditions when expressed and assayed individually, although p51 is more feeble. However, in the natural heterodimer only the p66 subunit possesses enzymatic activity. These conclusions were derived originally from the properties of heterodimers selectively reconstituted so that one or the other subunit was mutated at the active site (LEGRICE et al. 1991; HOSTOMSKY et al. 1992). The three-dimensional structure of HIV RT with bound inhibitor (KOHLSTAEDT et al. 1992) or with a bound DNA oligonucleotide (JACOBO-MOLINA et al. 1993) provides an explanation of these findings, showing that p51 folds somewhat differently than p66. By several in vitro assays homodimers of p66 possess similar enzymatic activities to those of the native heterodimer, but the homodimer is considerably less stable (RESTLE et al. 1990). The genesis of the p51 subunit in vivo also has not been established definitively. PR-mediated cleavage of p66 monomers could occur first, followed by dimer formation, or alternatively the p66 homodimer could be the substrate for PR. In vitro the two subunits can be separated and reassociated readily, and hence in vivo monomeric and dimeric forms of both subunits probably are in equilibrium. Reformation of an active enzyme in vitro appears to involve two steps, rapid formation of a dimer that lacks enzymatic activity, followed by a much slower isomerization to recreate the active enzyme (DIVITA et al. 1995).

Like the structures of the mature reverse transcriptases, the functional consequences of Pol processing appear to vary from one virus to another. In ASLV the natural processing steps activate RT activity by over 100-fold, as measured on detergent-treated, PR-defective virions with exogenous assays using homopolymer polyA template and oligodT primer (STEWART et al. 1990). In teleological terms it is tempting to speculate that RT activity in a retrovirus-producing cell has to be restrained in order to protect the cell from the potentially damaging consequences of rampant reverse transcription of cellular RNAs. For viruses like ASLV that show little proteolysis of Gag before budding, processing of the Gag-Pol polypeptide would appear to be an ideal way to insure that activation of RT activity occurs only in the virion. In ASLV the low RT activity measured for Gag-Pol is not a result of inaccessibility of the RT domain in the viral core, since Gag-Pol precursor polypeptides expressed in insect cells also show extremely weak activity (STEWART and VOGT 1993). From the properties of a variety of ASLV RT fusion proteins with upstream Gag or downstream Pol sequences, it appears that the PR domain immediately abutting the RT domain somehow is responsible for inhibiting activity.

By contrast to ASLV, in MLV and HIV processing appears to play a more minor role or no role in activation of enzymatic activity. Immature MLV virions carrying a deletion in PR show activity similar to their wild-type counterparts (CRAWFORD and GOFF 1985). For HIV there are conflicting data about the extent of activation of RT activity by proteolytic processing, but according to most accounts Gag-Pol in PR-defective virions shows only a modest decrease in activity, perhaps five fold relative to the natural p66/p51 heterodimer (for example PENG et al. 1991). Consistent with studies on PR mutants, virions treated with PR inhibitors also show substantial RT activity (KAPLAN et al. 1994a). Published studies have not addressed the question of whether processing alters the IN activity in any of these viruses.

In most cases where it has been studied, digestion of Pol products in vitro with PR mimics most of the processing events seen in vivo. A notable exception is the processing that activates ASLV RT, inferred to be cleavage between the PR and RT domains. In vitro both in immature cores and in isolated polypeptides the spacer sequence between *gag* and *pol*, including the last residues of PR and the first residues of RT, is refractory to cleavage under all conditions tested (STEWART and VOGT 1994). Paradoxically, in vivo the first Gag-Pol processing step may occur here. A model to account for these findings is based on the predicted ability of this region to form a coiled coil. The model postulates that somehow the structure of the immature virion holds this stretch of polypeptide in Gag-Pol in an extended and intrinsically unfavorable conformation, ready to be the first Gag-Pol target for the activated PR. On the other hand, if under artificial conditions processing begins elsewhere, the segment of polypeptide spanning Gag and Pol is allowed to collapse into a PR-resistant coiled coil. The theme in this model, the ability of a viral polypeptide sequence to fold into alternative structures depending on the context, may well apply to other Gag and Pol domains. The different folding of the p51 and p66 subunits of the HIV heterodimeric RT is an example of this theme, as is perhaps the slow isomerization step upon reformation of the dimer in vitro noted above. The still hypothetical alternative folding of the PR domain as it is embedded in the Gag-Pol precursor may turn out to be another. In the absence of crystal structures, for example of combined domains of Gag or Gag-Pol, it will remain difficult to precisely define folding alternatives in these cases.

In all retroviruses where appropriate mutations have been created and tested, preventing the expression of an active PR results in the accumulation of the unprocessed Gag-Pro-Pol precursor polypeptide in cells and in virions, of ca. 180–200 kDa. An exception to this finding is provided by the spumavirus HSRV (KONVALINKA et al. 1995c). In this case an active site PR mutation resulted in the disappearance of polypeptides presumed to correspond to the mature RT and IN, with an increase in the amounts of a polypeptide of ca. 120 kDa carrying antigenic determinant for PR, RNaseH, and IN. However, there was no evidence of a larger protein. In this retrovirus system a splicing event recently has been demonstrated to the underlying mechanism for Pol expression (Yu et al. 1996). How the resulting Pol protein becomes incorporated into virions remains to be established.

8 Other Processed Events Mediated by PR

In some retroviruses a portion of the C-terminal cytoplasmic tail of the Env TM component is removed by PR, as a late step in maturation after the virion has budded from the cell. This reaction was first described in MLV as the processing of what was called p15E to yield p12E. PR removes the last 16 amino acid residues of TM, leaving a protein with about 16 residues extending past the transmembrane anchor domain. Only recently has the important functional role of this cleavage come to be appreciated (Rein et al. 1994). When the C-terminal segment that is normally proteolytically removed is deleted, cells infected by the resulting mutant virus readily undergo fusion with each other, a phenomenon not observed with wild-type virus. Thus the tail of TM is inferred to restrain the fusion activity of the SU/TM Env complex until this protein has left the cell in virus particles. Consistent with this picture, mutation of the cleavage site leads to virus particles that contain normal amounts of SU and TM, but are poorly infectious. In MPMV the Env TM protein also is trimmed by PR (Brody et al. 1992), and similar conclusions have been reported for the function of this processing (Brody et al. 1994). Equine infectious anemia virus (EIAV), a lentivirus distantly related to HIV, appears to have a PR cleavage site in the cytoplasmic domain of TM as well (Rice et al. 1990). It is well known that for all retrovirus Env proteins that have been studied, the Golgi cleavage to generate SU and TM is the primary signal to activate fusion potential, as it is in the best-studied of all viral membrane proteins, HA of influenza virus. Hence the additional, PR-mediated cleavage represents a second level of control. Given the difficulties of accurate C-terminal analysis, it seems possible that Env processing by PR in other retroviruses also occurs but remains to be discovered. It will be experimentally challenging to unravel the mechanisms underlying regulation of fusion by retroviral Env proteins.

For retroviruses in which some Gag and Gag-Pol processing takes place in the cytoplasm, uncoupled from assembly and budding, it is not surprising that host proteins also are targets for PR. Even in the absence of knowledge about particular substrates, it is clear from their toxicity that retroviral proteases cleave host proteins. HIV PR is the best studied in this regard. Expression of the *pro* gene is lethal both in procaryotic and eucaryotic cells, and cells can be spared of the toxic effects of expression if PR inhibitors are added to the culture (Kräusslich 1992). Increasing numbers of proteins have been found to be targets of PR in vivo (Poorman et al. 1991). For example, the intermediate filament proteins desmin and vimentin are cleaved to measurable extents in HIV-infected cells, as they are in vitro (SHOEMAN et al. 1990). Other cytoskeletal proteins and potentially important cellular proteins (TOMASSELLI et al. 1991; RIVIERE et al. 1991; WALLIN et al. 1991; AINSZTEIN and PURICH 1992; ZHANG et al. 1995), as well as the viral regulatory protein Nef (FREUND et al. 1994), also are substrates for retroviral PRs in vitro. However, it is uncertain if any of these events has relevance to the infectious cycle or contributes to the cytopathic effects associated with infection. There are multiple ways in which infection by HIV can lead to cytotoxicity, and PR expression is but one of these. Recent molecular studies argue that the toxic effects of the PR do not represent an essential aspect of infection, but rather a by-product. Mutation of the active site threonine residue to serine in HIV PR leads to a protease that is less active in vitro but still is able to provide all the functions needed for the infectious life cycle (ROSÉ et al. 1995; KONVALINKA et al. 1995b). However, this mutant PR is no longer toxic when expressed by itself in cells, and cleavage of cellular intermediate filament proteins upon infection also no longer occurs (KONVALINKA et al. 1995b). As yet it is unclear if PR-mediated toxicity is important in AIDS.

It had been assumed until recently that the action of PR is restricted to the late stages in budding, namely to those events discussed in the sections above. In lentiviruses this dogma has been challenged by experiments implicating PR early in the infection process, i.e. the time when the virus enters a cell. It was originally observed for EIAV that incubation of viral cores isolated after detergent treatment led to the PR-mediated cleavage of the NC protein resident in the cores (ROBERTS et al. 1991). The selectivity and efficiency of this cleavage suggested that the same event might occur upon infection, and thus play an important role in the virus life cycle. This hypothesis has been addressed in HIV by use of PR inhibitors early in infection. Treatment of cells with PR inhibitors before infection, or similar treatment of the virus, was reported to cause marked diminution of infectivity, while treatment 30 min after infection showed no effect (NAGY et al. 1994). Some step in the synthesis or stabilization of the viral cDNA appeared to be affected by the inhibitor, leading to the speculation that the NC cleavage is the relevant target. However, it cannot be excluded that other proteins, for example host proteins, play a role. Furthermore, data suggesting no role of HIV PR early in infection also have been published (JACOBSEN et al. 1992). More work will be needed establish the generality of PR action in the infection process and to pinpoint the mechanism of its action.

9 Summary and Prognosis

The broad outlines of proteolytic processing and maturation in retroviruses were worked out in the first decade after the discovery of these processes, by the mid 1980s. In the following decade and extending to the present, the mechanics of processing were elucidated, driven to a large degree by biochemical and physical studies of PR and the intense interest in possible anti-HIV therapies. What are the frontiers for this field in the next decade? It is likely that ongoing and future work will lead to understanding of exactly how cleavages lead to biochemical changes in Gag and Pol proteins, for example the nucleic acid binding activity of NC and the enzymatic activities of RT, IN, and PR itself. These advances probably will continue to rely heavily on studies of proteins expressed in *E. coli* or other systems. The vexing problem of how proteolysis is activated is likely to yield to the combined efforts of biochemical and genetic analyses, perhaps with the aid of reconstitution studies of particles assembled with membranes in vitro. The fundamental and still enigmatic problems of structure will be addressed: How are the Gag and Gag-Pol proteins put together to make an immature virion, and what changes occur upon maturation? Biophysical analyses are likely to be important contributors to these efforts, including crystallography and NMR of mature Gag proteins and their processing intermediates, as well as cryo-electron microscopy of viral and subviral particles made in vitro and in vivo. Finally, it seems likely that a fuller understanding will emerge of the timing of proteolysis and the role of proteolysis in the final pinching off of the virus, if indeed there is such a role. What drives the final membrane fusion event that has to occur in particle release has remained mysterious, not only for retroviruses but for all enveloped viruses. In vitro assembly systems, perhaps based on permeabilized cells or crude extracts plus membranes, may be expected to be contributors in this area.

Without doubt unanticipated aspects of proteolysis and maturation will be uncovered in the next decade. More host proteins cleaved by PR, at least in lentivirus infection, will be identified, perhaps including proteins that play a role in virus assembly or egress in the early stages of infection. It seems possible that the action of PR could help explain how the bud can push out of the cell despite the presence of subcortical cytoskeletal elements underneath the plasma membrane. As happens repeatedly in biological research, less-studied "maverick" systems may contribute important insights. Examples for proteolysis and maturation could include some of the retrotransposon cousins of retroviruses, and the still poorly studied spumaviruses, among which only the primate species have been characterized in molecular terms. Given the very low levels of mature Gag and Pol proteins found associated with spumavirus infection, it may be that proteolysis in this system plays a different role or is regulated differently, as is suggested by the report that spumavirus particles contain high levels of viral DNA (Yu et al.1996).

References

Abdel-Meguid SS, Zhao B, Murthy KHM, Winborne E, Choi JK, Desjarlais RL, Minnich MD, Culp JS, Debouck D, Tomaszek TA Jr, et al (1993) Inhibition of human immunodeficiency virus-1 protease by a c-2- symmetric phosphinate. Synthesis and crystallographic analysis. Biochemistry 32: 7972–7980

Ainsztein AM, Purich DL (1992) Cleavage of bovine brain microtubule associated protein-2 by human immunodeficiency virus proteinase. J Neurochem 59: 874–880

Babé LM, Rosé J, Craik CS (1992) Synthetic "interface" peptides alter dimeric assembly of the HIV 1 and 2 proteases. Protein Sci 1: 1244–1253

Bennett RP, Rhee S, Craven RC, Hunter E, Wills JW (1991) Amino acids encoded downstream of Gag are not required by Rous sarcoma virus protease during Gag-mediated assembly. J Virol 65: 272–280

Berkowitz RD, Ohagen A, Höglund S, Goff SP (1995) Retroviral nucleocapsid domains mediate the specific recognition of genomic viral RNAs by chimeric Gag polyproteins during RNA packaging in vivo. J Virol 69: 6445–6456

Boutelje J, Karlstrom AR, Hartmanis MGN, Holmgren E, Sjogren A, Levine RL (1990) Human immunodeficiency viral protease is catalytically active as a fusion protein: characterization of the fusion and native enzymes produced in *Escherichia coli*. Arch Biochem Biophys 283: 141–149

Bowles N, Bonnet D, Mulhauser F, Spahr PF (1994) Site-directed mutagenesis of the P2 region of the Rous sarcoma virus Gag gene: effects on Gag polyprotein processing. Virology 203: 20–28

Brody BA, Rhee SS, Sommerfelt MA, Hunter E (1992) A viral protease-mediated cleavage of the transmembrane glycoprotein of Mason-Pfizer monkey virus can be suppressed by mutations within the matrix protein. Proc Natl Acad Sci USA 89: 3443–3447

Brody BA, Rhee SS, Hunter E (1994) Postassembly cleavage of a retroviral glycoprotein cytoplasmic domain removes a necessary incorporation signal and activates fusion activity. J Virol 68: 4620–4627

Bryant M, Ratner L (1990) Myristoylation-dependent replication and assembly of human immunodeficiency virus 1. Proc Natl Acad Sci USA 87: 523–527

Burstein H, Bizub D, Skalka AM (1991) Assembly and processing of avian retroviral gag polyproteins containing linked protease dimers. J Virol 65: 6165–6172

Burstein H, Bizyb D, Kotler M, Schatz G, Vogt VM, Skalka AM (1992) Processing of avian retroviral Gag polyprotein precursors is blocked by a mutation in the NC-PR cleavage site. J Virol 66: 1781–1785

Cameron CE, Grinde B, Jentoft J, Leis J, Weber IT, Copeland TD, Wlodawer A (1992) Mechanism of inhibition of the retroviral protease by a Rous sarcoma virus peptide substrate representing the cleavage site between the Gag P2 and P10 proteins. J Biol Chem 267: 23735–23741

Cameron CE, Ridky TW, Shulenin S, Leis J, Weber IT, Copeland T, Wlodawer A, Burstein H, Bizub-Bender D, Skalka AM (1994) Mutational analysis of the substrate binding pocket of the Rous sarcoma virus and human immunodeficiency virus-1 protease. J Biol Chem 269: 11170–11177

Chou JJ (1993) A vectorized sequence-coupling model for predicting HIV protease cleavage sites in proteins. J Biol Chem 268: 16938–16948

Co E, Koelsch G, Lin Y, Ido E, Hartsuck JA, Tang J (1994) Proteolytic processing mechanisms of a miniprecursor of the aspartic protease of human immunodeficiency virus type 1. Biochemistry 33: 1248–1254

Craven RC, Bennett RP, Wills JW (1991) Role of the avian retroviral protease in the activation of reverse transcriptase during virion assembly. J Virol 65: 6205–6217

Craven RC, Leure-duPree AE, Erdie CR, Wilson CB, Wills JW (1993) Necessity of the spacer peptide between CA and NC in the Rous sarcoma virus Gag protein. J Virol 67: 6246–6252

Crawford S, Goff SP (1985) A deletion mutation in the 5′ part of the pol gene of Moloney murine leukemia virus blocks proteolytic processing of the gag and pol polyproteins. J Virol 53: 899–907

Darke PL (1994) Stability of dimeric retroviral proteases. In: Kuo LC, Shafer JA (eds) Retroviral proteases. Academic, New York, pp 104–127 (Methods in enzymology, vol 241)

Darke PL, Jordan SP, Hall DL, Zugay JA, Shafer JA, Kuo LC (1994) Dissociation and association of the HIV-1 protease dimer subunits: equilibria and rates. Biochemistry 33: 98–105

Davies DR (1990) The structure and function of the aspartic proteinases. Annu Rev Biophys Chem 19: 189–215

Debouck C (1992) The HIV-1 protease as a therapeutic target of AIDS. AIDS Res Hum Retroviruses 8: 153–164

Desjarlais RL, Seibel GL, Kuntz ID, Furth PS, Alvarez JC, Ortiz de Montellano PR, Decamp DL, Babé LM, Craik CS (1990) Structure-based design of nonpeptide inhibitors specific for the human immunodeficiency virus 1 protease. Proc Natl Acad Sci USA 87: 6644–6648

Divita G, Rittinger K, Geourjon C, Deléage G, Goody RS (1995) Dimerization kinetics of HIV-1 and HIV-2 reverse transcriptase: a two step process. J Mol Biol 245: 508–521

Dougherty WG, Semler BL (1993) Expression of virus-encoded proteinases: functional and structural similarities with cellular enzymes. Microbiol Rev 57: 781–822

Dunn DM, Gustchina A, Wlodawer A, Kay J (1994). Subsite preferences of retroviral proteases. In: Kuo LC, Shafer JA (eds) Retroviral proteases. Academic, New York, pp 254–278 (Methods in enzymology, vol 241)

Eisenman R, Vogt VM (1978) The biosynthesis of oncovirus proteins. Biochim Biophys Acta (Cancer Rev) 473: 187–239

Eisenman RN, Mason WS, Linial M (1980) Synthesis and processing of polymerase proteins of wild type and mutant avian retroviruses. J Virol 36: 89–104

Erickson-Viitanen S, Manfredi J, Viitanen P, Tribe DE, Tritch R, Hutchison CAIII, Loeb DD, Swanstrom R (1989) Cleavage of HIV-1 Gag polyprotein synthesized in-vitro. Sequential cleavage by the viral protease. AIDS Res Hum Retroviruses 5: 577–592

Fäcke M, Janetzko A, Shoeman RL, Kräusslich HG (1993) A large deletion in the matrix domain of the human immunodeficiency virus *Gag* gene redirects virus particle assembly from the plasma membrane to the endoplasmic reticulum. J Virol 67: 4972–4980

Freund J, Kellner R, Konvalinka J, Wolber V, Kräusslich H-G, Kalbitzer HR (1994) A possible regulation of negative factor (Nef) activity of human immunodeficiency virus type 1 by the viral protease. Eur J Biochem 223: 589–593

Fitzgerald PMD, Springer JE (1991) Structure and function of retroviral proteases. Annu Rev Biophys Chem 20: 299–320

Fu W, Rein A (1993) Maturation of dimeric viral RNA of Moloney murine leukemia virus. J Virol 67: 5443–5449

Fu W, Gorelick RJ, Rein A (1994) Characterization of human immunodeficiency virus type 1 dimeric RNA from wild-type and protease-defective virions. J Virol 68: 5013–5018

Gelderblom HR (1991) Assembly and morphology of HIV: potential effect of structure on viral function. AIDS 5: 617–638

Göttlinger HG, Sodroski JG, Haseltine WA (1989) Role of capsid presursor processing and myristoylation in morphogenesis and infectivity of human immunodeficiency virus type 1. Proc Natl Acad Sci USA 86: 5781–5785

Griffiths JT, Phylip LH, Konvalinka J, Strop P, Gustchina A, Wlodawer A, Davenport RJ, Briggs R, Dunn BM, Kay J (1992) Different requirements for productive interaction between the active site of HIV-1 proteinase and substrates containing hydrophobic or aromatic pro cleavage sites. Biochemistry 31: 5193–5200

Grinde B, Cameron CE, Leis J, Weber IT, Wlodawer A, Burstein H, Skalka AM (1992) Analysis of substrate interactions of the Rous sarcoma virus wild type and mutant proteases and human immunodeficiency virus 1 protease using a set of systematically altered peptide substrates. J Biol Chem 267: 9491–9498

Hizi A, Leis JP, Joklik WK (1977) RNA-dependent DNA polymerase of avian sarcoma virus B77. I. Isolation and partial characterization of the α, β_2, and $\alpha\beta$ enzyme forms. J Biol Chem 252: 2290–2295

Ho D, Neumann AU, Perelson AS, Chen W, Leonard JM, Markowitz M (1995) Rapid turnover of plasma virons and CD4 lymphocytes in HIV-1 infection. Nature 373: 123–126

Höglund S, Öfverstedt L-G, Nilsson A, Lundquist P, Gelderblom H, Özel M, Skoglund U (1992) Spatial visualization of the maturing HIV-1 core and its linkage to the envelope. AIDS Res Hum Retroviruses 8: 1–7

Hostomsky Z, Hostomska Z, Fu T-B, Taylor J (1992) Reverse transcriptase of human immunodeficiency virus type 1: functionality of subunits of the heterodimer in DNA synthesis. J. Virol 66: 3179–3182

Hrusková-Heidingsfeldová O, Andreansky M, Fabry M, Blaha I, Strop P, Hunter E (1995) Cloning, bacterial expression and characterization of the Mason-Pfizer monkey virus proteinase. J Biol Chem 270: 15053–15058

Jacks T, Varmus HE (1985) Expression of the Rous sarcoma virus pol gene by ribosomal frameshifting. Science 230: 1237–1242

Jacobo-Molina A, Ding J, Nanni R, Clark AD, Lu X, Tantillo C, Williams RL, Kamer G, Ferris AL, Clark P, Hizi A, Hughes SH, Arnold E (1993) Crystal structure of human immunodeficiency virus type 1 reverse transcriptase complexed with double stranded DNA at 3.0 Å resolution shows bent DNA. Proc Natl Acad Sci USA 90: 6320–6324

Jacobsen H, Ahlborn-Laake L, Gugel R, Mous J (1992) Progression of early steps of human immunodeficiency virus type 1 replication in the presence of an inhibitor of viral protease. J Virol 66: 5087–5091

James MNG, Sielecki AR (1986) Molecular structure of an aspartic proteinase zymogen, porcine pepsinogen, at 1.8 Å resolution. Nature 319: 33–38.
Jaskólski M, Miller M, Rao JKM, Leis J, Wlodawer A (1990) Structure of the aspartic protease from Rous sarcoma retrovirus refined at 1 Å resolution. Biochemistry 29: 5889–5898
Kaplan AH, Swanstrom R (1991) Human immunodeficiency virus type 1 Gag proteins are processed in two cellular compartments. Proc Natl Acad Sci USA 88: 4528–4532
Kaplan AH, Zack JA, Knigge M, Paul DA, Kempf DJ, Norbeck DW, Swanstrom R (1993) Partial inhibition of the human immunodeficiency virus type 1 protease results in aberrant virus assembly and the formation of noninfectious particles. J Virol 67: 4050–4055
Kaplan AH, Krogstad P, Kempf DJ, Norbeck DW, Swanstrom R (1994a) Human immunodeficiency virus type 1 virions composed of unprocessed Gag and Gag-Pol precursors are capable of reverse transcribing viral genomic RNA. Antimicrobial Agents Chemotherapy 38: 2929–2933
Kaplan AH, Manchester M, Swanstrom R (1994b) The activity of the protease of human immunodeficiency virus type 1 is initiated at the membrane of infected cells before the release of viral proteins and is required for release to occur with maximum efficiency. J Virol 68: 6782–6786
Katoh I, Yoshinaka Y, Rein A, Shjibuya M, Adaka T, Oroszlan S (1985). Murine leukemia virus maturation: protease region required for conversion from "immature" to "mature" core form and for virus infectivity. Virology 145: 280–292
Katz R A, Skalka AM (1988) A C-terminal domain in the avian sarcoma-leukosis virus pol gene product is not essential for viral replication. J Virol 62: 528–533
Katz RA, Skalka AM (1994) The retroviral enzymes. In: Richardson CC (ed) Annual Review of Biochemistry, vol. 63. Annual Reviews, Palo Alto, pp 133–173
Kohl NE, Emini EA, Schleif WA, Davis LJ, Heimbach JC, Dixon RAF, Scolnick EM, Sigal IS (1988) Active human immunodeficiency virus protease is required for viral infectivity. Proc Natl Acad Sci USA 85: 4686–4690
Kohlstaedt LA, Wang J, Friedman JM, Rice PA, Steitz TA (1992) Crystal structure at 3.5 Å resolution of HIV-1 reverse transcriptase complexed with an inhibitor. Science 256: 1783–1790
Konvalinka J, Horejsi M, Andreansky M, Novek P, Pichova I, Blaha I, Fabry M, Sedlacek J, Foundling S, Strop P (1992) An engineered retroviral proteinase from myeloblastosis associated virus acquires pH dependence and substrate specificity of the HIV-1 proteinase. EMBO J 11: 1141–1144
Konvalinka J, Heuser AM, Hruskova-Heidingsfeldova O, Vogt VM, Sedlacek J, Strop P, Kräusslich HG (1995a) Proteolytic processing of particle-associated retroviral polyproteins by homologous and heterologous viral proteinases. Eur J Biochem 228: 191–198
Konvalinka J, Litterst MA, Welker R, Rippmann R, Heuser A-M, Kräusslich HG (1995b) HIV-1 protease (PR) active site mutation causes reduced PR activity and loss of PR-mediated cytotoxicity without apparent effect on virus maturation and infectivity J Virol 69: 7180–7186
Konvalinka J, Loechelt M, Zentgraf H, Fluegel RM, Kräusslich H-G (1995c) Active foamy virus proteinase is required for virus infectivity but not for formation of a Pol polyprotein. J Virol 69: 7264–7268
Korb J, Trávnícek M, Ríman J (1976) The oncornavirus maturation process: quantitative correlation between morphological changes and conversion of genomic virion RNA. Intervirology 7: 211–224
Kotler M, Arad G, Hughes SH (1992) Human immunodeficiency virus type 1 Gag-protease fusion proteins are enzymatically active. J Virol 66: 6781–6783
Kottler H, Weber J, Konvalinka J, Kräusslich H-G (1995) A mutation in HIV-1 proteinase gives rise to a virus temperature-sensitive for polyprotein processing and for viral infectivity (manuscript in preparation)
Kräusslich HG (1991) Human immunodeficiency virus proteinase dimer as a component of the viral polyprotein prevents particle assembly and viral infectivity. Proc Natl Acad Sci USA 88: 3213–3217
Kräusslich HG (1992) Specific inhibitor of human immunodeficiency virus proteinase prevents the cytotoxic effects of a single-chain proteinase dimer and restores particle formation. J Virol 66: 567–572
Kräusslich HG, Wimmer E (1988) Viral proteinases. Annu Rev Biochem 57: 701–754
Kräusslich HG, Schneider H, Zybarth G, Carter CA, Wimmer E (1988) Processing of *in vitro*-synthesized Gag precursor proteins of human immunodeficiency virus HIV type 1 by HIV proteinase generated in *Escherichia coli*. J Virol 62: 4393–4397
Kräusslich HG , Ingraham RH, Skoog MT, Wimmer E, Pallai PV, Carter (1989) Activity of purified biosynthetic proteinase of human immunodeficiency virus on natural substrates and synthetic peptides. Proc Nat Acad Sci USA 86: 807–811

Kräusslich HG, Fäcke M, Heuser AM, Konvalinka J, Zentgraf W (1995) The spacer peptide between human immunodeficiency virus capsid and nucleocapsid proteins is essential for ordered assembly and viral infectivity. J Virol 69: 3407–3419
Kuff EL, Lueders KK (1988) The intracisternal A particle gene family: structure and functional aspects. Adv Cancer Res 51: 183–276
Lapatto R, Blundell T, Hemmings A, Overington J, Wilderspin A, Wood S, Merson JR, Whittle PJ, Danley DE, Geoghegan KF, Hawryklik SJ, Lee SE, Scheld KG, Hobart PM (1989) X-ray analysis of HIV-1 proteinase at 2.7 Å resolution confirms structural homology among retroviral enzymes. Nature 342: 299–302
LeGrice SFJ, Naas T, Wohlgensinger B, Schatz O (1991) Subunit-selective mutagenesis indicates minimal polymerase activity in heterodimer associated p51 of HIV-1 reverse transcriptase. EMBO J 10: 3905–3911
Loeb DD, Swanstrom RS, Everitt L, Manchester M, Stamper SE, Hutchison CA III (1989) Complete mutagenesis of the HIV-1 protease. Nature 340: 397–400
Louis JM, Nashed NT, Parris KD, Kimmel AR, Jerina DM (1994) Kinetics and mechanism of autoprocessing of human immunodeficiency virus type 1 protease from an analog of the Gag-Pol polyprotein. Proc Natl Acad Sci USA 91: 7970–7974
Luftig RB, Yoshinaka Y (1978) Rauscher leukemia virus populations enriched for "immature" virions contain increased amounts of P70, the gag gene product. J Virol 25: 416–421
Manchester M, Everitt L, Loeb DD, Hutchison CAIII, Swanstrom R (1994) Identification of temperature-sensitive mutants of the human immunodeficiency virus type 1 protease through saturation mutagenesis: amino acid side chain requirements for temperature sensitivity. J Biol Chem 269: 7689–7695
Mervis RJ, Ahmad N, Lillehoj EP, Raum MG, Salazar FHR, Chan HW, Venkatesan S (1988) The gag gene products of human immunodeficiency virus type 1. Alignment within the gag open reading frame, identification of posttranslational modifications, and evidence for alternative gag precursors. J Virol 62: 3993–4002
Miller M, Schneider J, Sathyanarayana BK, Toth MV, Marchall GR, Clawson L, Selk L, Kent SBH, Wlodawer A (1989) Structure of complex of synthetic HIV-1 protease with a substrate based inhibitor at 2.3 Å resolution. Science 246: 1149–1152
Moody MD, Pettit SC, Shao W, Everitt L, Loeb DD, Hutchison CA III, Swanstrom R (1995) A side chain at position 48 of the human immunodeficiency virus type-1 protease flap provides an additional specificity determinant. Virology 207: 475–485
Nagy K, Young M, Baboonian C, Merson J, Whittle P, Oroszlan S (1994) Antiviral activity of human immunodeficiency virus type 1 protease inhibitors in a single cycle of infection: evidence for a role of protease in the early phase. J Virol 68: 757–765
Navia MA, Fitzgerald PMD, McKeever BM, Leu C-T, Heimbach JC, Herber WK, Sigal IS, Darke PL, Springer JP (1989) Three-dimensional structure of aspartyl protease from human immunodeficiency virus HIV-1. Nature 337: 615–620
Nermut MV, Hockley DJ, Jowett JBM, Jones IM, Garreau M, Thomas D (1994) Fullerene-like organization of HIV gag protein shell in virus-like particles produced by recombinant baculovirus. Virology 198: 288–296
Nicholson LK, Yamazaki T, Torchia DA, Grzesiek S, Bax A, Stahl SJ, Kaufman JD, Wingfield PT, Lam PYS, Jadhav PK, Hodge CN, Domaille PJ, Chang CH (1995) Flexibility and function in HIV-1 protease. Nature Struct Biol 2: 274–280
Oertle S, Spahr P–F (1990) Role of the Gag polyprotein precursor in packaging and maturation of Rous sarcoma virus genomic RNA. J Virol 64: 5757–5763
Oertle S, Bowles N, Spahr P-F (1992) Complementation studies with RSV Gag and Gag-Pol polyprotein mutants. J Virol 66: 3873–3878
Oppermann H, Bishop JM, Varmus HE, Levintow L (1977) A joint product of the genes gag and pol of avian sarcoma virus: a possible precursor of reverse transcriptase Cell 12: 993–1005
Otto MJ, Garber S, Winslow DL, Reid CD, Aldrich P, Jadhar PK, Patterson CE, Hodge CN, Cheng Y-SE (1993) In vitro isolation and identification of human immunodeficiency virus (HIV) variant with reduced sensitivity to C-2 symmetrical inhibitors of HIV type 1 protease. Proc Natl Acad Sci USA 90: 7543–7547
Parent LJ, Bennett RP, Craven RC, Nelle TD, Krishna NK, Bowzard JB, Wilson CB, Puffer BA, Montelaro RC, Wills JW (1995) Positionally independent and exchangeable late budding functions of the RSV and HIV Gag proteins. J Virol 69

Peng C, Ho BK, Chang TW, Chang NT (1989) Role of human immunodeficiency virus type 1-specific protease in core protein maturation and viral infectivity. J Virol 63: 2550–2556
Peng C, Chang NT, Chang TW (1991) Identification and characterization of human immunodeficiency virus type 1 gag-pol fusion protein in transfected mammalian cells. J Virol 65: 2751–2756
Pepinsky RB (1983) Localization of lipid-protein and protein-protein interactions within the murine retrovirus gag precursor by a novel peptide mapping technique. J Biol Chem 258: 11229–11235
Pepinsky RB, Papayannopoulos IA, Chow EP, Krishna NK, Craven RC, Vogt VM (1995) Differential proteolytic processing leads to multiple forms of the CA protein of avian sarcoma and leukemia viruses. J Virol 69: 6430–6438
Pepinsky RB, Papayannopoulos IA, Campbell S, Vogt VM (1996) Analysis of Rous sarcoma virus Gag proteins by mass spectrometry indicates trimming by host exopeptidase. J Virol 70: (in press)
Pettit SC, Simsic J, Loeb DD, Everitt L, Hutchison CA III, Swanstrom R (1991) Analysis of retroviral protease cleavage sites reveals two types of cleavage sites and the structural requirements of the P1 amino acid. J Biol Chem 266: 14539–14547
Pettit SC, Moody MD, Wehbie RS, Kaplan AH, Nantermet PV, Klein CA, Swanstrom R (1994) The p2 domain of human immunodeficiency virus type 1 Gag regulates sequential proteolytic processing and is required to produce fully infectious virions. J Virol 68: 8017–8027
Pichova I, Strop P, Sedlacek J, Kapralek F, Benes V, Travnicek M, Pavlickova L, Soujcek M, Kostka V, Foundling S (1992) Isolation and biochemical characterization and crystallizaton of the p15Gag proteinase of myeloblastosis associated virus expressed in E. coli. Int J Biochem 24: 235–242
Pinter A, DeHarven E (1979) Protein composition of a defective murine sarcoma virus particle possessing the enveloped Type-A morphology. Virology 99: 103–110
Phylip LH, Mills JS, Parten BF, Dunn BM, Kay J (1992) Intrinsic activity of precursor forms of HIV-1 proteinase. FEBS Lett 314: 449–454
Poorman RA, Tomasselli AG, Heinrikson RL, Kézdy FJ (1991) A cumulative specificity model for proteases from human immunodeficiency virus types 1 and 2, inferred from statistical analysis of an extended substrate data base. J Biol Chem 266: 14554–14561
Prats A-C, Sarin L, Gabus C, Litvak S, Keith G, Darlix J-L (1988) Small finger protein of avian and murine retroviruses has nucleic acid annealing activity and positions the replication primer tRNA onto genomic RNA. EMBO J. 7: 1777–1783
Rao JKM, Erickson JW, Wlodawer A (1991) Structural and evolutionary relationships between retroviral and eucaryotic aspartic proteinases. Biochemistry 30: 4663–4671
Rein A, McClure MR, Rice NR, Luftig RB, Schultz AM (1986) Myristylation site in Pr65gag is essential for virus particle formation by Moloney murine leukemia virus. Proc Nat Acad Sci USA 83: 7246–7250
Rein A, Mirro J, Haynes JG, Ernst SM, Nagashima K (1994) Function of the cytoplasmic domain of a retroviral transmembrane protein: p15E-p12E cleavage activates the membrane fusion capability of the murine leukemia virus Env protein. J Virol 68: 1773–1781
Restle T, Mueller B, Goody RS (1990) Dimerization of human immunodeficiency virus type 1 reverse transcriptase. J Biol Chem 265: 8986–8988
Rhee SS, Hunter E (1987) Myristylation is required for intracellular transport but not for assembly of D-type retrovirus capsids. J Virol 61: 1046–1053
Rice NR, Henderson LE, Sowder RC, Copeland TD, Oroszlan S, Edwards JF (1990) Synthesis and processing of the transmembrane envelope protein of equine infectious anemia virus. J. Virol 64: 3770–3778
Ringe D (1994) X-ray structures of retroviral proteases and their inhibitor-bound complexes. In: Kuo LC, Shafer JA (eds) Retroviral proteases. Academic, New York, pp 157–177 (Methods in enzymology, vol 241)
Riviere Y, Blank R, Kourilsky P, Israel I (1991) Processing of the precursor of NK-kappa B by the HIV-1 protease during acute infection. Nature 350: 625–626
Roberts MM, Copeland TD, Oroszlan S (1991) *In situ* processing of a retroviral nucleocapsid protein by the viral proteinase. Protein Eng 4: 695–700
Rosé JR, Salto R, Craik CS (1993) Regulation of autoproteolysis of the HIV-1 and HIV-2 proteases with engineered amino acid substitutions. J Biol Chem 268: 11939–11945
Rosé JR, Babé LM, Craik CS (1995) Defining the level of human immunodeficieincy virus type 1 (HIV-1) protease activity required for HIV1 particle maturation and infectivity. J Virol 69: 2751–2758
Schatz G, Pichova I, Vogt VM (1996) Cleavage and miscleavage at the mutant and wild type NC-PR junction in avian sarcoma and leukemia viruses (manuscript in preparation)

Sellos-Moura M, Vogt VM (1995) Proteolytic activity of the NC-PR fragment of avian sarcoma and leukemia virus Gag protein purified from E. coli. Virology (in press)
Sheng N, Erickson-Viitanen S (1994) Cleavage of p15 protein in vitro by human immunodeficiency virus type 1 protease is RNA dependent. J Virol 68: 6207–6214
Shoeman RL, Honer B, Stoller TJ, Kesselmeier C, Miedel MC, Traub P, Graves MC (1990) Human immunodeficiency virus-1 protease cleaves the intermediate filament proteins vimentin, desmin, and glial fibrillary acidic protein. Proc Natl Acad Sci USA 87: 6636–6340
Skalka AM (1989) Retroviral proteases: a first glimpse at the anatomy of a processing machine. Cell 56: 911–913
Soltis DA, Skalka AM (1988) The α and β chains of avian retrovirus reverse transcriptase independently expressed in E. coli: characterization of enzymatic activities. Proc Natl Acad Sci USA 85: 3372–3376
Sonigo P, Barker C, Hunter E, Wain-Hobson S (1986) Nucleotide sequence of Mason-Pfizer monkey virus, an immunosuppressive type D retrovirous. Cell 45: 373–386
Sommerfelt MA, Petteway JR, Dreyer GB, Hunter E (1992) Effect of retroviral proteinase inhibitors on Mason-Pfizer monkey virus maturation and transmembrane protein cleavage. J Virol 66: 4220–4227
Stewart L, Vogt VM (1991) Trans-acting viral protease is necessary and sufficient for activation of avian leukosis virus reverse transcriptase. J Virol 65: 6218–6231
Stewart L, Vogt VM (1993) Reverse transcriptase and protease activities of avian leukosis virus Gag-Pol fusion proteins expressed in insect cells. J Virol 67: 7582–7596
Stewart L, Vogt VM (1994) Proteolytic cleavage at the Gag-Pol junction in avian leukosis viruses: differences in vitro and in vivo. Virology 204: 45–59
Stewart L, Schatz G, Vogt VM (l990) Properties of avian retrovirus particles defective in viral protease. J Virol 64: 5076–5092
Stromberg K, Hurley NE, Davis NL, Rueckert RR, Fleissner E (1974) Structural studies of avian myeloblastosis virus: comparison of polypeptides in virion and core component by dodecyl sulfate-polyacrylamide gel electrophoresis. J Virol 13: 513–528
Telesnitsky A, Goff SP (1993) RNase H domain mutations affect the interaction between Moloney murine leukemia virus reverse transcriptase and its primer-template. Proc Natl Acad Sci USA 90: 1276–1280
Toh H, Ono M, Saigo K, Miyata T (1985) Retroviral protease-like sequence in the yeast transposon Ty1. Nature 315: 691–692
Tomasselli AG, Heindrikson RL (1994) Specificity in retroviral protease: an analysis of viral and non-viral protein substrates. In: Kuo LC, Shafer JA (eds) Retroviral proteases. Academic, New York, pp 279–301 (Methods in enzymology, vol 241)
Tomasselli AG, Hui JO, Adams L, Chosay J, Lowery D, Greenberg B, Yem A, Deibel MR, Zurcher-Neely H, Heinrikson RL (1991) Actin, troponin c, Alzheimer amyloid precursor protein, and pro-interleukin 1-beta as substrates of the protease from human immunodeficiency virus. J Biol Chem 266: 14548–14553.
Tözsér, Weber IT, Gustchina A, Bláha In, Copeland TD, Louis JM, Oroszlan S (1992) Kinetic and modeling studies of S3-S3′, subsites of HIV proteinases. Biochemistry 31: 4793–4800
Tritch RJ, Cheng YS, Yin FH, Erickson-Viitanen S (1991) Mutagenesis of protease cleavage sites in the human immunodeficiency virus type 1 gag polyprotein. J Virol 65: 922–930
Valverde V, Lemay P, Masson JM, Gay B, Boulanger P (1992) Autoprocessing of the human immunodeficiency virus type 1 protease precursor expressed in Escherichia coli from a synthetic gene. J Gen Virol 73: 639–651
Vogt VM, Eisenman R (1973) Identification of a large polypeptide precursor of avian oncornavirus proteins. Proc Natl Acad Sci USA 70: 1734–1738
Vogt VM, Eisenman R, Diggelmann H (1975) Generation of avian myeloblastosis virus structural proteins by proteolytic cleavage of a precursor polypeptide. J Mol Biol 96: 471–493
Vogt VM, Burstein H, Skalka AM (1992) Proteolysis in the maturation of avian retroviruses does not require calcium. Virology 189: 771–774
von der Helm K (1977) Cleavage of Rous sarcoma viral polypeptide precursor into internal structural proteins in vitro involves viral protein p15. Proc Natl Acad Sci USA 74: 911–915
Wallin M, Deinum J, Goobar L, Danielson UH (1991) Proteolytic cleavage of microtubule associated proteins by retroviral proteinases. J Gen Virol 71: 1985–1992
Weber IT, Miller M, Jaskolski M, Leis J, Skalka AM, Wlodawer A (1989) Molecular modeling of the HIV-1 protease and its substrate binding site. Science 243: 928–931

Wei X, Ghosh SK, Taylor ME, Johnson VA, Emini EA, Deutsch P, Lifson JD, Bonhoeffer S, Nowak M, Hahn BH, Saag MS, Shaw GM (1995) Viral dynamics in human immunodeficiency virus type 1 infection. Nature 373: 117–122

Welker R, Janetzko A, Heuser A-M, Kräusslich H-G (1995) Targeting of intracisternal A-type particle Gag polyprotein to the plasma membrane leads to release of extracellular particles and induces polyprotein processing (submitted for publication)

Wills JW, Craven RC (1991) Form function and use of retroviral Gag proteins. AIDS (Phil) 5: 639–654

Wills JW, Craven RC, Achacoso JA (1989) Creation and expression of myristylated forms of Rous sarcom virus Gag protein in mammalian cells. J Virol 55: 79–85

Wills JW, Craven RC, Weldon RA Jr, Nelle TD, Erdie CR (1991) Suppression of retroviral MA deletions by the amino-terminal membrane binding domain of $p60^{src}$. J Virol 65: 3804–3812

Wills JW, Cameron CE, Wilson CB, Xiang Y, Bennett RP, Leis J (1994) An assembly domain of the Rous sarcoma virus Gag protein required late in budding. J Virol 68: 6605–6618

Wlodawer A, Erickson JW (1993) Structure-based inhibitors of HIV-1 protease. In: Richardson CC (ed) Annual review of biochemistry, vol 62, Annual Reviews, Palo Alto, pp 543–585

Wlodawer A, Miller M, Jaskólski M, Sathyanarayana BK, Baldwin E, Weber IT, Selk LM, Clawson L, Schneider J, Kent SBH (1989) Conserved folding in retroviral proteases: crystal structure of a synthetic HIV1 protease. Science 245: 616–621

Wondrak EM, Sakaguchi K, Rice WG, Kun E, Kimmel AR, Louis JM (1994) Removal of zinc is required for processing of the mature nucleocapsid protein of human immunodeficiency virus, type1, by the viral protease. J Biol Chem 269: 21948–21950

Yoshinaka Y, Luftig RB (1977) Murine leukemia virus morphogenesis: cleavage of P70 in vitro can be accompanied by a shift from a concentrically coiled internal strand ("immature") to a collapsed ("mature") form of the virus core. Proc Natl Acad Sci USA 74: 3446–3450

Yu SF, Baldwin DN, Gwynn SR, Yendapalli S, Linial ML (1996) Human foamy virus replication: A pathway distinct from that of retroviruses and hepadnaviruses. Science 271: 1579–1582

Zhang ZY, Poorman RA, Maggiora LL, Heinrikson RL, Kezdy FJ (1991) Dissociative inhibition of dimeric enzymes: kinetic characterization of the inhibition of HIV-1 protease by its carboxyl terminal tetrapeptide. J Biol Chem 266:15591–15594

Zhang D, Zhang N, Wick MM, Byrn RA (1995) HIV type 1 protease activation of NF-kappa-B within T lymphoid cells. AIDS Res and Hum Retroviruses 11: 223–230

Zybarth G, Carter C (1995) Domains upstream of the protease (PR) in human immunodeficiency virus type 1 Gag-Pol influence PR autoprocessing. J Virol 69: 3878–3884

Zybarth G, Kräusslich HG, Partin K, Carter C (1994) Proteolytic activity of novel human immunodeficiency virus type 1 proteinase proteins from a precursor with a blocking mutation at the N terminus of the PR domain. J Virol 68: 240–250

Maturation and Assembly of Retroviral Glycoproteins

D. EINFELD

1 Introduction

Retroviruses infect cells after binding to a diverse set of specific cell surface receptors via their envelope glycoproteins (Env) (reviewed in (WEISS 1993). Retroviral *env* genes are translated into a precursor protein that enters the secretory pathway of the host cell. During the process of being transported to the plasma membrane, the site at which the glycoproteins become incorporated into virions, the precursor undergoes maturation to a bipartite complex composed of the N-terminal, external surface subunit (SU), and a membrane-spanning C-terminal protein (TM). The receptor specificity of the complex is determined in large part, if not exclusively, by the SU domain, which retains receptor binding activity outside of the complex with TM (DELARCO and TODARO 1976; LASKY et

Genvec Inc., 12111 Parklawn Dr., Rockville, MD 20852, USA

al. 1987). The diversity of receptor specificities within some groups of retroviruses has facilitated the identification of regions of the SU proteins that contribute to receptor binding. The heterogeneity of SU proteins that is apparent from *env* sequences of different retroviruses suggests that the structural details of Env-receptor interactions may vary among these viruses. In addition, the sequences or domains that have been implicated in receptor binding differ in their positions within SU. The gp70 of the murine leukemia viruses (MuLVs) appears to be organized into N-terminal and C-terminal structural domains with the receptor-binding determinants in the N-terminal region (BATTINI et al. 1992; HEARD and DANOS 1991; MORGAN et al. 1993; OTT and REIN 1992), while receptor specificity of avian sarcoma and leukemia virus (ASLV) Env is dependent on sequence in the middle third of SU (BOVA et al. 1986, 1988; DORNER and COFFIN 1986; DORNER et al. 1985). Receptor binding by gp120 of human immunodeficiency virus (HIV-1) involves a major determinant in the C-terminal third of the protein together with discrete segments from other regions of the protein (CORDONNIER et al. 1989; DOWBENKO et al. 1988; LASKY et al. 1987; OLSHEVSKY et al. 1990; POLLARD et al. 1992). Whether these differences extend to the structures of the SU proteins or only indicate variety in the interaction of a relatively conserved structure with different receptors is unclear.

On the other hand, common features among TM proteins are apparent from comparison of their sequences (PATARCA and HASELTINE 1984). A hydrophobic sequence at or near the N-terminus of TM is positioned external to the membrane and plays a critical role in membrane fusion during virus entry. TM proteins also share two or three closely spaced cysteine residues within their ectodomains. The external domains of TM proteins are relatively conserved in sequence within virus groups while their sizes are similar among different retroviruses. In addition to their contribution to fusion, this region of TM proteins also interacts with a potentially diverse group of SU proteins to anchor them to the virus. Although the relative conservation among TM ectodomains fits the general functions attributed to this protein, many aspects of Env function that are essential for virus entry following receptor binding may depend on coordinate involvement of both SU and TM.

This chapter considers the process whereby the initial *env* translation product is converted to the mature virion-associated complex. An attempt is made to recognize general features of this process without ignoring differences which are significant and may be related to the diversity of receptors that have been selected by retroviruses. Other aspects of Env function and structure have been discussed in previous reviews (HUNTER and SWANSTROM 1990; WEISS 1993). Where less is known about Env proteins, particularly in the very early stages of maturation, information obtained in other systems has been incorporated. In particular, studies involving the well-characterized influenza virus hemagglutinin (HA) and glycoprotein G of vesicular stomatitis virus (VSV) have contributed significantly to our understanding of cellular processes involved in protein maturation and transport. The latter proteins are also functional analogs of Env proteins and the structural organization of HA at least superficially resembles that of Env. This material has been included to provide additional context for discussing the maturation of Env proteins.

2 Insertion of the *env* Product into the Endoplasmic Reticulum

Like many cellular membrane proteins, Env precursors are synthesized with an N-terminal signal sequence which allows them to be recognized by the signal recognition particle and begin translocation across the membrane of the endoplasmic reticulum (ER). As a result of translocation, the precursor spans the membrane with its N-terminus in the lumen of the ER and its C-terminus in the cytoplasm. Mutated Env proteins expressed without this signal sequence remained in the cytoplasm and were degraded (WILLS et al. 1984). The signal sequences of retroviral Env proteins exhibit the general pattern recognized for membrane proteins, which consists of a positively charged N-terminal region, a central hydrophobic sequence, and a more polar C-terminal region that terminates with an uncharged, small side chain residue at the site of cleavage (VON HEIJNE 1985, 1986). While most signal peptides are 15–30 amino acids in length, those of some Env proteins are much longer, such as the 62 amino acid signal of Rous sarcoma virus (RSV). The genetic organization of retroviruses is such that signal peptides overlap other open reading frames. A consequence of this overlap is that part of the sequence of the signal peptide may be conserved to maintain the function of other gene products. In ASLV the splice site of the *env* mRNA results in Gag and Env having the same six N-terminal residues. The next 37 amino acids of the signal peptide overlap *pol* in a different reading frame while the hydrophobic core of the sequence occurs after the overlap. The signal sequence of HIV-1 is not as long as that of RSV, but it is unusual for the high number of charged residues, including two in the relatively hydrophobic region. The ability of this sequence to direct translocation in vitro, however, was not abolished by removal of the positive charges or the introduction of an additional charged residue in the middle of the hydrophobic core (ELLERBROK et al. 1992).

The signal sequence functions to target proteins to the ER and may also contribute to the activity of the translocation apparatus which enables proteins to cross the membrane. The ability of signal peptides to open channels in bacterial membranes might be shared by N-terminal signals interacting with similar channels that function in the ER membrane (SIMON and BLOBEL 1991, 1992). During translocation the signal peptide appears to interact with the translocation apparatus, which in its simplest form may be composed of only the two chains of the SRP receptor and the multi-chain sec61p complex (CROWLEY et al. 1993; GÖRLICH and RAPOPORT 1993). The influence of the signal peptide on Env maturation presumably ends with its proteolytic removal by signal peptidase in

ASLV	LL**K**GLLLGLVVILLLVVCLPCLLQXXXXXX**RK**MIN
MuLV	FN**R**SPWFTTLISTIMGPLIVLLLILLFGPCILN**R**LVQF
FeLV	FN**K**SPWFTTLISSIMGPLLILLLILLFGPCILN**R**LVQF
M-PMV	LI**D**NPFWTSFHGFLPYVMPLLGPLLCLLLVLSFGPIIFN**K**LMT
REV-A	LY**D**NPLWNGLNGFLPYLLPSLGPLFGLQLFLTLGPCI**RK**TLT
MMTV	NPL**D**WTQYFIFIGVGALLLVIVLMIFPIVFQCLA**K**SL**D**
HIV-1	**K**WASLWNWFNITNWLWYI**K**LFIMIVGGLVGL**R**IVFAVLSIVN**R**
SIV	**K**YIQYGVYIVVGVILL**R**IVIYIVQMLA**R**L**RK**

Fig. 1. Membrane-spanning sequences of retroviral TM proteins. The proposed membrane-spanning domains are *underlined*. Charged residues flanking or internal to these sequences are *highlighted*. In the ASLV sequence *X* indicates residues that are highly variable among subgroups

the ER. In general these sequences are thought to be degraded soon after cleavage, although the VP7 glycoprotein of rotavirus is retained in the ER through interaction with its N-terminal signal sequence following signal peptidase cleavage (Poruchynsky and Atkinson 1988; Stirzaker and Both 1989). This interaction in turn could protect the signal sequence from degradation. A major histocompatibility complex (MHC) class I epitope comprising the hydrophobic core of a functional signal sequence was presented at the cell surface in cells that lacked the transporter for transferring peptides into the ER from the cytoplasm (Guéguen et al. 1994). This observation fits with degradation of signal sequences within the ER.

Translocation of Env proteins into the ER is brought to a halt by the hydrophobic sequence that ultimately becomes embedded in the membrane. The membrane-spanning domains (MSDs) of retroviruses range from 27 amino acids for RSV to 41 amino acids for HIV-1 (Fig. 1). These hydrophobic sequences are generally flanked by charged residues and are enriched for the highly hydrophobic residues valine, leucine, and isoleucine. Evidence for the membrane-anchoring function of these sequences has been provided by the secretion of truncated proteins from which they are absent, but the precise boundaries of these domains has not been defined. The MSD of HIV-1 is unusual for the presence of two basic residues that divide the sequence into three segments of similar lengths. The long HIV-1 sequence is also distinguished by the presence of several tryptophans. HIV-1 proteins truncated at the Arg residue within the putative MSD were still stably anchored in the membrane and transported, while truncation four residues upstream resulted in a protein blocked in transport out of the ER which may or may not have been membrane anchored (Owens et al. 1994).Further shortening of the MSD to the internal Lys residue resulted in secretion of Env (Owens et al. 1994). Truncation of an MuLV Env just prior to the cysteine in the MSD yielded a protein that appeared to be transported (Granowitz et al. 1991). In addition, deletion of the GPCILNRL sequence at the C-terminus of the MSD did not have an effect on membrane anchoring as this protein appeared to be incorporated into virus. Hydrophobic residues C-terminal to the deletion may have compensated for the shortening of the MSD in this mutant. These results fit with earlier experiments involving deletions in MSD sequences. Progressive shortening of MSD sequences was found to lead first to membrane-anchored proteins that had a defect in transport, and further shortening resulted in secretion (ADAMS and ROSE 1985; DOYLE et al. 1986).

The mechanism by which the MSD stops translocation is not clear. The translocation apparatus may assemble or be activated as the signal sequence initially interacts with the membrane. This protein channel would provide a less formidable energy barrier for translocation of polar residues through the membrane. Entry of the MSD into the translocation pore, however, may alter or disrupt the pore so that charged amino acids are no longer tolerated in the membrane and a stable association of the MSD with the lipid environment ensues. It has also been proposed that the hydrophobic MSD is able to move through the wall of the translocation pore into the lipid environment (SINGER et al. 1987). Analyses of the association of translocating proteins and the translocation apparatus by cross-linking indicate that the MSD remained associated with this complex until translation was completed (THRIFT et al. 1991). This observation suggests that the interruption of translocation upon entry of the MSD into the membrane is not due to dissociation of the translocation channel or movement of this sequence out of the pore and into the lipid bilayer; although the cross-linking approach might not detect subtle but important alterations. Rather, a ribosome-dependent association between the translocation apparatus and the nascent chain is maintained. It does appear that there is a specific interaction between Sec61p and ribosomes (GÖRLICH et al. 1992; HIGH et al. 1991), and that association with the ribosome helps to maintain the translocation apparatus as an open protein channel (SIMON and BLOBEL 1991). This contribution of the ribosome might result in an enhanced stop-translocation activity of hydrophobic sequences when placed near the C-terminus of proteins since these sequences would encounter a less stable or active translocation complex due to release of the ribosome at the time they enter the membrane. The observation that an MSD of suboptimal length resulted in a transport-defective, membrane-anchored protein indicates that the shorter MSD had not lost its stop-translocation activity, but may have interfered with transport due to an effect on flanking residues normally located outside the membrane.

Retroviral glycoproteins contain an additional hydrophobic segment present in the extracellular domain of TM that is active in fusion. Viral fusion sequences must be translocated across the ER membrane despite similarity to stop-translocation sequences, and this probably limits their hydrophobicity. The length of hydrophobic fusion sequences is similar to that of stop-translocation sequences, ranging from 26 to 32 for most retroviruses (Fig. 2). In general the fusion peptides contain fewer highly hydrophobic residues and no more than three of these occur in a row. These sequences are also enriched for glycines and alanines. The fusion peptide of the paramyxovirus SV5 fusion protein can function as a

ASLV	^{14}QLWGPTA**R**IFASILAPGVAAAQAL**RE**^{39}
MuLV	^{1}**E**PVSLTLALLLGGLTMGGIAAGVGTGTTALVAT**K**^{34}
FeLV	^{1}**E**PISLTVALMLGGLTVGGIAAGVGTGT**K**ALL**E**^{32}
M-PMV	^{1}AIQLIPLFVGLGITTAVSTGAAGLGVSITQYT**K**^{33}
REV-A	^{1}AVQFIPLLVGLGITGATLAGGTGLGVSVHTYH**K**^{33}
MMTV	^{1}FVAAIILGIISALIAIITSFAVATTALV**KE**^{30}
HIV-1	^{1}AVGIGALFLGFLGAAGSTMGA*SMTLTVQA**R**^{31}
SIV/HIV-2	^{1}GVFVLGFLGFLATAGSAMGAASLTLSAQS**R**TL^{32}

Fig. 2. Fusion sequences of retroviral TM proteins. The hydrophobic fusion sequences are *underlined* while flanking charged residues are *highlighted*. The amino acid position within the TM protein is indicated for both the N- and C-terminal residues of the displayed sequence. In the HIV-1 sequence * indicates a position which is occupied by Arg in some isolates while other isolates have Val, Thr, Ala, or Gly

membrane anchor when it is placed at the carboxy terminus of a secreted protein (Davis and Hsu 1986; Paterson and Lamb 1987). Restoration of the five Arg residues normally present at the N-terminus of this sequence also disrupted its membrane-anchoring activity (Paterson and Lamb 1987).

Stable anchoring in the membrane is thought to occur once the hydrophobic MSD encounters the membrane. Interestingly the TM proteins of MuLV have a potential site for N-linked glycosylation at the putative lumenal border of the MSD. This site is not modified by oligosaccharide addition, presumably because of its position at the membrane. The fact that this site is not utilized supports the view that the transition from a translocating to a stably integrated protein is tight. This contrasts, however, with the scenario described for the T-cell receptor α-chain, where stable membrane anchoring is dependent on association with other chains of the receptor complex (Shin et al. 1993).

3 Maturation of Env as an ER Membrane Protein

Membrane anchoring of the Env protein places the flanking sequences in distinct environments, with the N-terminal, extracytoplasmic domain exposed to the lumen of the ER while the cytoplasmic domain is accessible to the cytosol as well as factors on the external surface of the ER membrane.

3.1 Folding and Oligomerization of the ER Lumenal Domain of Env

Entry of the Env polypeptide into the lumen of the ER exposes the polypeptide chain to modification through glycosylation, disulfide bond formation, and potential interaction with several factors that may influence its folding and assembly into oligomeric complexes. These aspects of protein maturation in vivo have been the focus of considerable research and of recent reviews (DOMS et al. 1993; GETHING and SAMBROOK 1992).

3.1.1 Glycosylation in the ER

N-Glycosylation involves the attachment of a glucose$_3$-mannnose$_9$-*N*–acetylglucosamine$_2$ oligosaccharide to asparagine residues of the nascent polypeptide as it enters the ER. The fact that addition of this large hydrophilic complex essentially requires only the presence of the Asn in an Asn-X-Ser/Thr motif means that this process could cause a dramatic change in the propensity of the surrounding sequence during folding. In addition, the structure of these oligosaccharides is subject to modification during passage of glycoproteins through the ER and the Golgi complex. The SU subunits of Env proteins are often heavily N-glycosylated with 20–26 sites in HIV, as many as 16 sites in ASLV and 7–10 sites for MuLV. On the other hand, only four sites are present in mouse mammary tumor virus (MMTV) SU proteins. The TM domains are less heavily glycosylated with four sites on HIV gp41, two sites for ASLV and MMTV, and a single site for D-type viruses and the related reticuloendotheliosis-associated virus (REV-A). In the case of MuLV and the related viruses, feline leukemia virus (FeLV) and gibbon ape leukemia virus (G$_a$LV), TM is not glycosylated despite conservation of a potential site located just N-terminal to the membrane-spanning sequence. In general, most of the potential sites within Env proteins are modified.

Blocking the addition of N-linked oligosaccharides with tunicamycin resulted in apparently misfolded Env precursors that were not transported (PAL et al. 1989a; PINTER et al. 1984; WILLS et al. 1984). Deoxynojirimycin inhibition of ER glucosidases, which perform the initial trimming of the oligosaccharide, was reported to cause a decrease in the surface expression of MuLV Env and accumulation of precursor proteins (PINTER et al. 1984). Transport and processing of gp160 was also affected by this inhibitor but the extent of this effect varied (FENOUILLET and GLUCKMAN 1991; GRUTERS et al. 1987; PAL et al. 1989b; RATNER et al. 1991).In the case of RSV, on the other hand, this inhibitor had no effect on virus infectivity or cleavage of the precursor, but processing was relatively inefficient in the absence of the inhibitor as well (BOSCH and SCHWARZ 1984). Thus while addition of oligosaccharides appears essential for proper folding of Env proteins, these proteins differ in the extent to which their transport and function is affected by less dramatic alterations in the structure of the oligosaccharide.

Characterization of individual oligosaccharides on Env proteins indicated that these structures are a mixed population of minimally processed high-mannose chains and complex chains of varied structure (GEYER et al. 1988; MIZOUCHI et al. 1990). The oligosaccharides attached to individual asparagine residues can also be mixtures (GEYER et al. 1990). Since the presence or absence of oligosaccharides can affect folding, modification of individual oligosaccharides is likely to be influenced by protein structure, e.g., by controlling the access of modifying enzymes. Diversity in the oligosaccharide chains present at some individual sites, however, indicates that Env structure, and function, tolerates some variation.

The contribution of glycosylation to Env structure and function has also been examined through mutation of glycosylation sites. In MuLV SU, only one site appears essential for intracellular transport and processing of Env (FELKNER and ROTH 1992; KAYMAN et al. 1991). Mutation of a second site, in the N-terminal, receptor-binding region, gave rise to a temperature-sensitive phenotype in certain cell types where a defect in transport occurred (FELKNER and ROTH 1992). The N-terminal fragment of ecotropic gp70, which retains receptor binding in the form of a truncated, secreted molecule, contains two glycosylation sites (HEARD and DANOS 1991). Mutation of both of these sites blocked the secretion of this fragment, either through an effect on transport or stability (BATTINI et al. 1994). But the mutated N-terminal fragment retained receptor-binding activity. Unlike the full-length protein, therefore, maturation of this fragment was dependent on these glycosylation sites. Each of the 24 glycosylation sites in an HIV isolate was mutated separately without disrupting the transport and cleavage of the precursor (LEE et al. 1992a).Mutation, individually, of each of the four conserved N-linked glycosylation sites on HIV-1 TM also had no effect on processing or intracellular transport of Env (DEDERA et al. 1992; LEE et al. 1992b). Collective mutation of three of these sites resulted in appearance of a TM protein of the same apparent molecular weight as deglycosylated gp41, but processing was reduced to about 10% compared to the wild type (FENOUILLET et al. 1993).

It appears that few of the N-linked glycosylation sites are absolutely required for maturation of Env proteins, especially in regions where multiple sites are present. On the other hand, altered or absent oligosaccharides can have an impact on functions of mature Env proteins that are critical for infectivity. For the influenza HA and the VSV G protein, maturation and transport requires only that a minimum number of glycosylation sites be utilized and no single site appears indispensable (GALLAGHER et al. 1992; MACHAMER et al. 1985; ROBERTS et al. 1993). The variation in glycosylation sites among Env proteins of the same group is consistent with a positionally flexible requirement for glycosylation. The oligosaccharides may facilitate folding by preventing aggregation of nascent proteins, possibly by enhancing their solubility. Apart from this direct effect of the oligosaccharides on the polypeptide chain, these oligosaccharides may also be important for the interaction of Env proteins with factors in the ER such as the chaperone calnexin. As discussed below, N-linked oligosaccharides affect the interaction of glycoproteins with this chaperone. The function of mature Env proteins may in some cases have developed a dependence on particular oligosaccharides, but retention of partial activity by deglycosylated molecules suggests that alteration of these structures is also tolerated (FENOUILLET et al. 1990). The variable regions in HIV-1 Env proteins have been noted to exhibit a bias toward codons that favor the appearance of N- and O-linked glycosylation sites (BOSCH et al. 1994). In this way the virus may exploit a flexible dependence of the maturation of its Env proteins on these modifications to enhance its evasion of the immune response.

3.1.2 Disulfide Formation

As in the case of N-glycosylation, formation of disulfide bonds can occur very early as the protein enters the lumen of the ER. Native disulfides, retained in the fully folded protein, and nonnative disulfides can form if the involved cysteine residues come into the correct proximity during folding. These bonds may stabilize folding intermediates as well as the native structure. Env proteins possess multiple cysteine residues and these are generally conserved within virus groups. HIV-1 Env variants retain 18 cysteine residues within gp120 and HIV-2 and simian immunodeficiency virus (SIV) have 20 in their SU domains. The SU proteins of ASLV isolates which differ in their receptor specificities have a common set of 14 cysteines. Sixteen cysteines are conserved among MuLV SU sequences, while the N-terminal variable region has four additional cysteines in ecotropic viruses. A common feature of the lumenal segment of TM sequences is a set of two or three closely spaced cysteines (PATARCA and HASELTINE 1984; SCHULZ et al. 1992). The lentiviruses, including HIV and SIV, have a pair of cysteines separated by five to seven amino acids, as does MMTV. A third cysteine, adjacent to the C-terminal member of the pair is present in other C-type retroviruses and primate D-types (SCHULZ et al. 1992).

Disulfides have been mapped in both MuLV and HIV-1 SU proteins. Following release from virus by repeated freezing and thawing the SU protein of MuLV contains no free sulfhydryls (LINDER et al. 1992). Comparison of the disulfides in an ecotropic and a polytropic gp70 confirmed that the four additional cysteines in the former protein form disulfides internally while the remaining cysteines form identical disulfide linkages for the two proteins (LINDER et al. 1992, 1994). All of the disulfides were internal to either the N- or C-terminal domains of the protein, which are separated by a proline-rich sequence. This fits with earlier analyses, employing proteolytic fragmentation, which suggested that the N- and C-terminal domains of gp70 constitute structurally independent folding domains (PINTER and HONNEN 1983, 1984). Recombinant gp120 expressed in the absence of gp41 has also been characterized by disulfide mapping (LEONARD et al. 1990). All 18 of the cysteine residues form disulfides which, as in the case of MuLV, span relatively short distances in the primary sequence. The identified disulfide loops correlate with structural features such as the extensively analyzed V3 loop.

Disulfide linkage between SU and TM has been observed in MuLV proteins (LEAMNSON et al. 1977; PINTER and FLEISSNER 1977, 1979; PINTER et al. 1978; WITTE et al. 1977). Unoxidized cysteines could be detected in gp70 isolated from cells in the presence of an alkylating agent (GLINIAK et al. 1991) and isomerization of an interchain disulfide with one or more free cysteines was proposed to account for the lability of the disulfide link between gp70 and p15E (GLINIAK et al. 1991). One of the disulfide loops in gp70 involves the very short sequence ECWLCL, which might isomerize with little effect on overall structure, but this motif is also present, as DCWLCL, in D-type Env proteins that do not form intersubunit disulfides. The only cysteines in the ectodomain of p15E are the three present in the conserved cysteine cluster that is also present in TM proteins which do not form detectable disulfide links to SU. The SU and TM proteins of ASLV have a stable disulfide linkage, but the cysteine residues involved have not been identified. Since there are 14 cysteines in the SU domain, formation of a disulfide between SU and TM must involve more than one SU cysteine or result in an unpaired cysteine in this domain. Another possibility is that an intrasubunit disulfide in either SU or TM forms a structure that interlocks with the other subunit. The ASLV TM protein differs from other TM proteins in having two additional cysteines within its N-terminal third. ASLV Env does not have the ECWLCL motif but has immediately adjacent cysteines flanked by glycines in a similar position within SU. The structural features that mediate and favor disulfide linkage between SU and TM remain unclear.

Mutation of individual cysteine residues in membrane proteins generally results in transport defects (DOMS et al. 1993). An MuLV Env protein with a Cys to Arg mutation in the C-terminal region of its SU protein was not only defective for transport out of the ER, but also inhibited the transport of coexpressed wild-type ecotropic Env (MATANO et al. 1993). Cysteine substitutions in the HIV-1 SU protein resulted in the accumulation of the gp160 precursor but some of the mutants exhibited significant processing to gp120/gp41 and cell surface expression (TSCHACHLER et al. 1990). Mutation of either or both of the cysteine residues in the ectodomain of gp41 disrupted transport and inhibited cleavage of the precursor (DEDERA et al. 1992; SYU et al. 1991).

Studies involving the VSV G protein and the HA of influenza virus have provided insight into the process of disulfide formation. Despite the presence of 12 cysteines both of these proteins acquire their apparent native disulfide structure with a $t_{1/2}$ of about 5 min (BRAAKMAN et al. 1991; DE SILVA et al. 1990). This rate does vary somewhat with the conditions of expression, indicating the influence of cellular factors (BRAAKMAN et al. 1991). Two prominent folding intermediates of HA with incomplete disulfides have been detected during maturation and these same intermediates appeared during refolding of HA that had been reduced in cells exposed briefly to dithiothreitol (DTT) (BRAAKMAN et al. 1992a). These experiments suggest that the disulfides which form are determined by the primary sequence and the local structure of the protein rather than the temporal order of appearance of the cysteine residues in the lumen of the ER during translocation. Similar implications can be drawn from analyses of cysteine mutants of HA (SEGAL et al. 1992).Intermediates with incomplete disulfides have also been identified for VSV G protein (MACHAMER et al. 1990; MACHAMER and ROSE 1988). Reduced, incompletely folded G protein that accumulated in the presence of DTT was able to fold into its native structure following removal of the reducing agent (DE SILVA et al. 1993). Although removal of DTT caused formation of disulfide-linked aggregates of G protein which are not detected during synthesis of G protein in untreated cells, formation of these aberrant disulfides did not compromise the ability of these proteins to fold efficiently into their native structure. These results suggest that formation of nonnative disulfides in itself does not inhibit protein folding. But folding may be prolonged when nonnative disulfide structures are relatively stable.

A mixture of disulfide isomers of the MuLV Env precursor has been reported to resolve to a more homogeneous structure prior to transport out of the ER (GLINIAK et al. 1991). It is not clear to what extent the apparent disulfide heterogeneity of this precursor reflects nonnative versus incomplete disulfide formation. Intermolecular disulfides of gp160 detected under nonreducing conditions were interpreted as not being intermediates in folding of native proteins since no intermolecular disulfides were detected involving gp41 or gp120 and disulfide complexes of gp160 did not chase into the mature products (OWENS and COMPANS 1990).

Protein disulfide isomerase (PDI) is a major component of the ER. This protein has two thioredoxin-like active sites able to catalyze disulfide exchange but has also come to be recognized as a multi-functional protein (FREEDMAN 1989; NOIVA and LENNARZ 1992). The importance of PDI for formation of native disulfides within the ER is supported by in vitro studies with microsomes depleted of the enzyme (BULLEID and FREEDMAN 1988). In addition, the intracellular transport of a disulfide-containing protein was affected in yeast having a PDI mutated in the thiol/disulfide-active site (LAMANTIA and LENNARZ 1993). The ability of PDI to catalyze disulfide exchange may be critical for rearrangement of nonnative disulfides in folding intermediates that would otherwise accumulate by virtue of their stability (WEISSMAN and KIM 1993). The susceptibility of the native disulfide form of HA to reduction in DTT-treated cells is dependent on some factor present in the intact ER (TATU et al. 1993) which might be PDI. Once HA trimerizes, however, it is no longer susceptible to reduction under these conditions (TATU et al. 1993).

PDI has also been found to have broad specificity for binding to peptides regardless of the presence or absence of cysteine residues (MORJANA and GILBERT 1991; NOIVA et al. 1991), and a peptide-binding domain has been identified in an acidic region near the C-terminus of mammalian PDI (NOIVA et al. 1993). It associates with nascent and misfolded proteins in the ER (CHESSLER and BYERS 1992; PERSSON and PETTERSSON 1991; ROTH and PIERCE 1987) and will also bind denatured proteins in vitro (OTSU et al. 1994).This binding activity fits proposals that PDI functions as a chaperone (NOIVA and LENNARZ 1992). The fact that PDI facilitates in vitro refolding of glyceraldehyde-3-phosphate dehydrogenase, a protein that has an active site thiol but lacks disulfide bonds, and that this effect is not inhibited by DTT (CAI et al. 1994) suggests that PDI might have chaperone activity apart from any involvement in disulfide interactions. While PDI is encoded by an essential gene in yeast, mutation of the thiol groups to render the protein defective in disulfide exchange is not lethal, indicating the importance of additonal functions of this protein (LAMANTIA and LENNARZ 1993). The proposed chaperone-like activity of PDI appears to depend on both the peptide-binding site and the thiol/disulfide- active sites (PUIG et al. 1994). Given the interdependence of disulfide formation and folding, the potential of PDI to interact with the polypeptide as well as sulfhydryl groups suggests it could make a significant contribution to maturation of Env proteins.

3.1.3 Interaction with Chaperones

Two functions have been identified for resident ER proteins that are involved in folding of nascent polypeptides. One is the catalysis of particular steps in folding, such as isomerization of disulfide bonds; the other is interaction with newly translocated proteins at various stages of folding to prevent misfolding, aggregation, improper assembly, and exit of intermediates from the ER (reviewed in GETHING and SAMBROOK 1992). The latter functions are attributed to chaperone proteins, and the possible contribution of PDI in this area has been discussed above. Here the contributions of two other ER proteins, BiP and calnexin, will be considered.

BiP, a member of the hsp70 family of heat shock proteins, is known to bind many nascent proteins in the ER, and hydrolysis of ATP by BiP allows dissociation of these complexes. Different functions, driven by hydrolysis of ATP, have been proposed for BiP, ranging from induction of a conformational change in the bound protein to a cyclical occupation and release of binding sites(GETHING and SAMBROOK 1992; HUBBARD and SANDER 1991; ROTHMAN 1989) In the latter scenario release of the binding site frees it to form other associations either through folding or assembly into a multichain complex, but if the binding site is retained BiP binds to it again. The mechanism by which BiP might prevent misfolding while allowing formation of native structure is not fully understood.It has also been suggested that BiP could aid in the dissolution of misfolded complexes as in the ATP-dependent dissociation of aggregates of HA or VSV G which form in cells exposed to DTT and ATP-depletion (BRAAKMAN et al. 1992b; DE SILVA et al. 1993).

In peptide-binding assays BiP exhibited a preference for a seven-residue site enriched in hydrophobic, nonaromatic amino acids (FLYNN et al. 1991), although charged and hydrophilic residues were tolerated (FLYNN et al. 1989). A probe of BiP-binding specificity using peptides presented on bacteriophage particles led to identification of a heptameric motif with large hydrophobic residues at positions 1, 3, 5, and 7 (BLOND-ELGUINDI et al. 1993). The relative tolerance at the other positions of the heptamer, with the exception of position 2, where tryptophan was favored, indicates that BiP does not recognize substrate based only on hydrophobicity. The specificity of BiP suggests that binding sites may be hydrophobic regions of proteins that are exposed in nascent polypeptides but become sequestered in the compact, native protein or oligomeric complex. BiP-binding sites have been characterized in some proteins. Deletion of the BiP-binding sequence from immunoglobulin heavy-chain abrogates retention of the unassembled protein in the ER (HENDERSHOT et al. 1987; POLLOK et al. 1987). For the hemagglutinin-neuraminidase (HN) of SV5, however, deletion mutants lacking the BiP-binding activity of the wild-type protein were still retained in the ER (NG et al. 1992). This indicates that there are other mechanisms for ER retention, possibly by interaction with other chaperones.

BiP appears to bind proteins very early upon entry into the ER. The BiP dependence of protein translocation into the ER may involve interaction of this chaperone with the incoming protein as well as with components of the translocation machinery (MÖSCH et al. 1992; NICCHITTA and BLOBEL 1993; SANDERS et al. 1992; VOGEL et al. 1990).Interactions between BiP and nascent proteins are expected to have a range of affinities so that the detectable complexes only include the higher affinity binding sites which continue to be expressed as the protein folds. Time course studies of protein maturation are consistent with BiP interaction occurring very early in the folding of nascent proteins (KIM et al. 1992; MACHAMER et al. 1990). Association of HIV-1 Env with BiP has been detected by its coprecipitation with a BiP antibody (EARL et al. 1991). Peak levels of the complex were observed after a 15 -min pulse or an additional 15 -min chase. The capacity to bind CD4 was acquired by gp160 in a time interval that overlapped with these complexes while dissociation from BiP coincided with the appearance of gp160 dimers (EARL et al. 1991). Since only a fraction of gp160 could be detected in association with BiP it was not possible to directly relate BiP binding or release to these changes in the Env protein.

A second chaperone protein identified in the ER is calnexin, which unlike BiP is an integral membrane protein. The role of calnexin in protein maturation has been the subject of recent reviews (BERGERON et al. 1994; HAMMOND and HELENIUS 1993). While association of Env proteins with calnexin has not been reported, the ability of calnexin to bind N-linked glycoproteins and the dependence of Env maturation on glycosylation suggest that this resident of the ER could impact the assembly and maturation of Env proteins.

Association of proteins with calnexin is prolonged when these proteins misfold through incorporation of amino acid analogs while tunicamycin, which can also affect folding by blocking the addition of N-linked oligosaccharides, interferes with calnexin binding (OU et al. 1993). Studies involving glycosylation inhibitors and mutant cell lines indicate that calnexin interacts with N-linked glycoproteins at the stage when their oligosaccharides have been trimmed to a monoglucosylated form (HAMMOND et al. 1994; HAMMOND and HELENIUS 1993; KEARSE et al. 1994). Binding of calnexin to oligosaccharide chains, however, has not been demonstrated and the dependence of binding on the state of the oligosaccharide chain has been proposed to involve UDP-glucose:glycoprotein glucosyltransferase (HAMMOND et al. 1994; HAMMOND and HELENIUS 1993). Calnexin has been implicated in the ER retention of unassembled components of some multi-chain complexes (Jackson et al. 1994; Rajagopalan et al. 1994). In the case of flu HA and VSV G proteins the kinetics of dissociation from calnexin and of trimer formation are very similar (Hammond et al. 1994). Unlike VSV G, however, maturation of HA is not affected when its interaction with calnexin is blocked by the glucosidase inhibitors castanospermine or deoxynojirimycin (Hammond et al. 1994; Hammond and Helenius 1994a). Dependence of folding or assembly on calnexin may be limited by structural features of the nascent protein and redundancy of function among chaperones. Calnexin appears to interact with proteins at a later stage than BiP but the functional consequences of these interactions are not clear (Hammond and Helenius 1994a). Another ER resident protein representing a potential chaperone is GRP94, which also appears to interact with newly synthesized proteins at a later stage than BiP (Melnick et al. 1994).

The contribution of chaperones to the folding and maturation of Env proteins has not been studied extensively. Appreciation of the unique environment the ER provides for protein folding has come from the identification of cellular factors that contribute to this process. As the functions of these factors become more fully characterized, additional insight into the process whereby nascent proteins acquire their mature structure can be anticipated. Extrapolation from other systems provides a useful context for looking at Env maturation but unique features of Env proteins may be uncovered by further characterization.

3.1.4 Oligomerization

As the ectodomains of Env proteins fold in the lumen of the ER, these molecules are assembled into oligomeric complexes. Initial analyses of cross-linked MuLV glycoproteins on virions or infected cells indicated the presence of oligomeric complexes composed of multiple SU and TM subunits (Pinter et al. 1978; Takemoto, et al. 1978). Cross-linked oligomers containing SU and TM proteins were also observed in ASLV and MMTV (Gebhardt et al. 1984; Pepinsky et al. 1980; Racevskis and Sarkar 1980). Sedimentation gradient fractionation of lysates from pulse-chase experiments allowed the formation of ASLV Env oligomers to be monitored and indicated that assembly occurred in the ER (Einfeld and Hunter 1988). The Env protein of ASLV forms a noncovalent oligomer, probably a trimer, that is transported out of the ER and is ultimately incorporated into virus (Einfeld and Hunter 1988). The Env proteins from Moloney MuLV-infected cells exhibited complexes on sucrose gradients that appeared to be trimers (Kamps et al. 1991), while Friend MuLV, upon crosslinking, presented a mixture of complexes thought to include dimers, trimers, and tetramers (Tucker et al. 1991). The Env proteins of HIV-1, HIV-2 and SIV formed stable dimers detectable on sucrose gradients (Chakrabarti et al. 1990; Earl et al. 1990; Schawaller et al. 1989) and dimers of the HIV-2 and SIV precursors have also been detected directly on protein gels (Rey et al. 1989). These dimers have been identified as maturation intermediates based on their appearance prior to cleavage of the precursor and by their incorporation of [^{3}H]fucose (Earl et al. 1991;

REY et al. 1989). These complexes, however, are not stable following cleavage to SU and TM. The structure of the mature HIV/SIV Env complex has been probed by cross-linking, which yielded a mixture of dimers and larger complexes which might be trimers and tetramers (CHAKRABARTI et al. 1990; EARL et al. 1990; REY et al. 1989; SCHAWALLER et al. 1989).

The ER lumenal domain of Env proteins appears to be sufficient for oligomerization. Neither the cytoplasmic nor membrane-spanning sequences are essential for assembly of the Env glycoproteins of ASLV and MuLV (EINFELD and HUNTER 1988; TUCKER et al. 1991). An HIV-1 Env protein truncated after amino acid 129 of gp41, and consequently lacking the cytoplasmic and membrane-spanning sequences together with about 25 amino acids normally present outside the membrane, formed oligomers as efficiently as the full-length protein (EARL et al. 1991). Transport and oligomerization of glycophosphatidylinositol-anchored HIV-1 Env (SALZWEDEL et al. 1993; WEISS and WHITE 1993) and RSV Env (GILBERT et al. 1993) confirms that the membrane-spanning and cytoplasmic domains are dispensable for assembly.

The TM subunits appear to be the major site for oligomeric interactions. Under reducing conditions that disrupt the disulfide link between SU and TM proteins of RSV, the TM subunit remained in an oligomer although SU dissociated from the complex (EINFELD and HUNTER 1988). The RSV TM protein also retained the ability to oligomerize when it was expressed in the absence of SU (EINFELD and HUNTER 1994). Oligomers of TM could also be detected after the HIV-1, HIV-2, and SIV Env proteins were cleaved (CHAKRABARTI et al. 1990; EARL et al. 1990; PINTER et al. 1989; REY et al. 1989, 1990) and the TM protein of caprine arthritis-encephalitis virus formed a prominent dimer in addition to larger complexes (MCGUIRE et al. 1992). These observations do not rule out a contribution by the SU domain to assembly, and there are reports of complexes composed of SU proteins (MORIKAWA et al. 1992; OWENS and COMPANS 1990; POLLARD et al. 1992; RACEVSKIS and SARKAR 1980; TUCKER et al. 1991; WEISS et al. 1990). The majority of monoclonal antibodies that exhibit specificity for oligomeric rather than monomeric gp160, however, were reported to bind epitopes in the TM domain (BRODER et al. 1994). In the case of RSV, independent expression of the SU domain resulted in no detectable oligomers, and comparison of the wild-type Env with the independently expressed TM protein suggested that the presence of SU in the former resulted in slower assembly of oligomers (EINFELD and HUNTER 1994).

The TM proteins of retroviruses, despite variation in sequence and in the length of cytoplasmic domains, have ectodomains of similar size. The length of the sequence between the N-terminus of TM and putative membrane-spanning sequences ranges from 119 to 166 amino acids (Table 1). The cluster of two or three cysteine residues is found at a conserved position between amino acids 81 and 99. The ASLV protein has an additional pair of cysteines between this cluster and its N-terminus while CaEV has an additional cysteine C-terminal of this motif. With the exception of the MuLV protein, one or more N-glycosylation sites are present just downstream of the cysteines. Conserved among TM proteins of

Table 1. Conserved features of retroviral TM proteins. Listed are the number of residues in the TM ectodomains and the positions relative to the N-terminus of cysteines and of asparagines present in the N-glycosylation motif N-X-S/T

Virus	Ectodomain	Cysteines	N-X-S/T
ASLV	149	90, 97, 98	52, 100
MuLV	135	86, 93, 94	(98)[a]
D-type	119	81, 88, 89	93
MMTV	164	86, 94	42, 101
HIV-1	155	88, 94	101, 106, 115, 127
SIV/HIV-2	164	86, 92	100, 109, 125
CAEV	166	89, 96, 117	41, 109, 116, 132
EIAV	162	91, 99	39, 46, 106, 113

[a]The position of an aspartic acid residue found in the context of D-X-S/T is indicated for these nonglycosylated proteins.
ASLV, avian sarcoma and leukemia virus; MuLV, murine leukemia virus; MMTV, mouse mammary tumor virus; HIV-1, human immunodeficiency virus; SIV, simian immunodeficiency virus; CAEV, caprine arthritis encephalitis virus; EIAV, equine infectious anemia virus.

MuLV-related viruses, however, is a D-H-T/S motif starting at position 98. The lentiviruses have multiple glycosylation sites in this region while more N-terminally positioned glycosylation sites are a variable feature of TM proteins.

Coexpression of the HIV-1 Env protein with the Env protein of SIV or HIV-2 led to the appearance of mixed heterodimers that represented about 20% of all Env dimers in these cells (Doms et al. 1990). Formation of these dimers was thought to involve the ectodomains of the TM proteins since these are the most highly conserved in sequence and this region had been implicated in earlier studies mentioned above. The HIV-1 Env truncated after amino acid 129 of TM that oligomerized and was secreted (Earl et al. 1991) contained the cysteine motif and the glycosylation sites in TM (see Table 1). Separate mutation of each of the glycosylation sites at 106, 115, and 127, and of the two cysteine residues, did not disrupt assembly or cleavage of the truncated precursor (Earl and Moss 1993). The lack of effect by the cysteine mutations is striking in light of their conservation and indicates that if a disulfide is formed between these residues this structure is not essential for oligomer formation. Maturation of the full-length protein, however, was dependent on these cysteines (Dedera et al. 1992; Syu et al. 1991). Thus a structure involving these cysteine residues may be more critical in the context of the complete ectodomain or a membrane-anchored protein. Deletion of this entire region by truncation at residue 68 of TM resulted in a loss of assembly and processing. A truncated RSV protein containing 20 foreign amino acids appended to residue 102 of TM, and thus retaining the cysteines and glycosylation site, did not form oligomers and was blocked in transport (Einfeld and Hunter 1988). Truncation after residue 105 of TM yielded an RSV protein that was also impaired in the formation of stable oligomers and not efficiently transported (D. Einfeld and E. Hunter, unpublished data). But inclusion of the N-terminal 120 amino acids of TM allowed truncated proteins to assemble and be transported out of the cell. This suggests that sequence C-

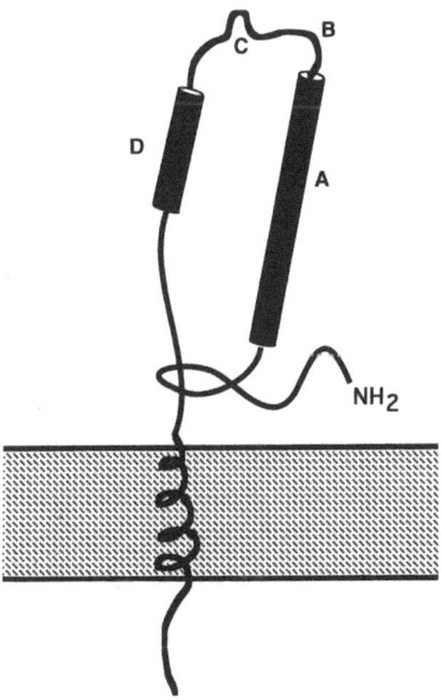

Fig. 3. Schematic diagram of a retroviral TM protein. Features of the ectodomain include: *A* a potential helical region containing a hepted repeat, *B* a region of high sequence similarly among oncoviruses, *C* proximal cysteine residues, and *D* a second potential helical region of variable length and position. This model is based on similarities of TM proteins to the influenza HA2 molecule (GALLAHER et al. 1989)

terminal of the cysteines and the adjacent glycosylation site contributes to formation of a stable oligomer by the RSV protein.

A prominent feature lying between the fusion sequence and the cysteine cluster of TM proteins is a potential α-helix with amphipathic content (Fig. 3) (CHAMBERS et al. 1990; DELWART and MOSALIOS 1990; GALLAHER et al. 1989). The amphipathicity of these sequences has the form of a heptad repeat with hydrophobic residues at positions 1 and 4 so that folding of these sequences into helices would give structures with distinct hydrophobic and polar faces. Spectroscopic analysis of the gp160 precursor and cleaved gp120 of HIV-1 indicated that the TM domain has a relatively high α-helical content, consistent with the predicted structure (DECROLY et al. 1993). In the case of HIV-1, the heptad repeat lies between residues 35 and 77 of TM and has primarily Leu, Ile, and Val at the hydrophobic positions. This region includes a motif reminiscent of a leucine zipper, Ile-X_6-Leu-X_6-Ile-X_6-Ile-X_6-Leu, which is very highly conserved. SIV and HIV-2 have a hydrophobic repeat of essentially the same length at residues 33–75 and identity of amino acids with HIV-1 at 8 of the 13 hydrophobic positions. Heptad repeat sequences span residues 36–78 of MuLV, 34–73 of primate D-type, and, if allowance is made for an extra residue near the center, residues 36–78 of MMTV. The potential helical region of RSV TM at residues 30–85 comprises a sequence with similarity to a heptad repeat although the predicted hydrophobic positions, 1 and 4, are occupied by small side-chain and polar residues in addition to Ile, Leu, and Val. This helix would also have a cysteine residue and an N-linked glycosylation site at adjacent positions on one of its polar faces.

Mutation of the middle Ile in the putative leucine zipper of HIV to Ala, Gly, Ser, Asp, or Glu had no apparent effect on oligomer formation or on the transport and processing of gp160 (DUBAY et al. 1992a). Substitution of a proline at this position or for other members of the set of Ile and Leu residues also had no apparent effect on oligomerization of Env proteins and did not block processing of the precursor (CHEN et al. 1993). The heptad repeat sequence exhibited characteristics of an α-helix assembled into a coiled-coil when analyzed in peptide form and this structure was disrupted by substitutions at the central Ile which had no effect on gp160 oligomerization (WILD et al. 1992,1994). These results argue against this region of the TM protein being a major determinant of oligomerization through formation of a leucine zipper. In the case of RSV, an Env protein truncated after residue 88 and containing the heptad repeat sequence formed an oligomer but was defective in transport (D. Einfeld and E. Hunter, unpublished). Oligomerization of the RSV protein was also quite tolerant of deletions that shorten the helical region at 30–85 and only a deletion that would shift the putative polar and hydrophobic faces out of alignment was blocked in oligomer formation. As noted above, the available evidence suggests that sequences C-terminal of the cysteine cluster are required for optimal assembly and intracellular transport. Since mutations of the leucine zipper in the HIV-1 protein did have an inhibitory effect on fusion (CHEN et al. 1993; DUBAY et al. 1992a), the leucine zipper/coiled-coil of Env proteins might be critical for formation of an alternative conformation involved in fusion (CARR and KIM 1993) or in assembly of a multimer of Env oligomers in a fusion pore (FREED et al. 1992). These results do not rule out a contribution of the heptad repeat to the initial assembly of the oligomer. Complete deletion of the heptad repeat region of the RSV protein resulted in an apparently monomeric protein that was blocked in transport (D. Einfeld and E. Hunter, unpublished).

Analysis of the maturation of flu HA and VSV G indicated that trimerization occurred after the monomers reached their fully oxidized form (DE SILVA et al. 1993; TATU et al. 1993). These results suggest that interactions between monomers do not have a critical influence on folding. This fits with the observation that while trimerization rate is dependent on expression level, folding rate is not (BRAAKMAN et al. 1991). This contrasts with the hemagglutinin-neuraminidase of human parainfluenza virus which appears to assemble oligomers before folding is complete(COLLINS and MOTET 1991). Folding intermediates of Env proteins have not been characterized so it is not clear to what extent folding and oligomerization are interdependent or separable. Dimerization of HIV-1 Env appears to be a later step kinetically than folding into a structure competent to bind CD4 (EARL et al. 1991). Receptor binding by Env proteins, however, may not be dependent on complete folding. Some transport-defective Env proteins are able to confer resistance to superinfection by homologous retroviruses, a resistance thought to involve receptor binding by these proteins (DELWART and PANGANIBAN 1989; MATANO et al. 1993). On the other hand, some MSD mutants of MuLV did not protect against superinfection even though they were processed (GRANOWITZ et al. 1991). CD4 binding has been observed also for modified gp120 molecules that may not fold like the wild-type protein (MORIKAWA et al. 1992; POLLARD et al. 1992).

3.1.5 SU/TM Interactions

Folding of Env precursors includes formation of stable interactions between the SU and TM domains so that upon cleavage the two proteins remain associated in a functional complex. A possible cysteine loop in TM proteins is thought to be important for these interactions (Fig. 3) (SCHULZ et al. 1992). The occurrence of disulfide links between the SU and TM proteins of MuLV supports the view that this region comes into close proximity to SU since these are the only cysteines of p15E which could form disulfides. At the same time, however, it suggests that the structure of this region of TM may not be limited to a disulfide loop. But the conservation of this motif among TM proteins argues for a common function. The C-terminus of SU has also been proposed to interact with TM, possibly through formation of a pocket that accommodates the cysteine loop of TM (SCHULZ et al. 1992). This again is consistent with the disulfide link in MuLV proteins which involves the C-terminal domain of gp70 (PINTER and HONNEN 1984).

Overlapping the C-terminus of the heptad repeat motif and just N-terminal to the cysteine cluster is a sequence that is highly conserved among the TM proteins of D-type and C-type oncoviruses (Fig. 3). Deletion of the M-PMV sequence, EVVLQNRRGLD, resulted in shedding of the SU protein from cells, while cell-associated mature proteins could not be detected (BRODY and HUNTER 1992). Failure to detect the modified TM product suggests that it was destabilized, either as a result of the deletion or its dissociation from SU. The majority of point mutations introduced into the QNRRGLDL motif did not affect processing or incorporation into viruses but did affect the infectivity of these viruses (BRODY et al. 1994a). More dramatic changes, such as introduction of a glycosylation site or a proline, however, did disrupt SU-TM association. Whether the defect in infectivity for the other viruses involved SU-TM interaction or some other function of this region is unclear.

Linker insertion mutagenesis of the MuLV *env* led to isolation of two temperature-sensitive (ts) mutants and two viruses defective in a syncytia assay (GRAY and ROTH1993). These mutants exhibited a decreased level of Env proteins on virions, and for the ts mutants this defect was magnified at the nonpermissive temperature. The mutations affected stable association between SU and TM, resulting in shedding of SU from the cell surface. The insertions in the ts mutants were located in the proline-rich region of SU while the others had insertions in the C-terminal domain of SU.

These mutational analyses identified modifications that do not disrupt the folding of Env that is a prerequisite for transport but do affect the stable association between SU and TM. One might argue that the fact these proteins were transported makes it less likely that the mutations caused global effects on Env structure. But much remains to be learned about the interactions that form a stable SU/TM complex.

Expression of p15E was shown to complement a glycophosphatidylinositol-anchored MuLV Env protein in formation of viruses capable of transferring a marker gene to receptor-bearing cells (RAGHEB and ANDERSON 1994). Similar complementation did not occur when p15E was coexpressed with an Env molecule lacking the extracellular domain of TM or a lipid-anchored gp70. Complementation with the lipid-anchored form of Env might be accounted for by interactions between the two molecules via their TM ectodomains. Complexes that might have formed on these viruses were not examined. The lack of complementation with the two membrane-anchored forms of SU could indicate an inability of SU and TM to associate when expressed separately. Experiments involving coexpression of the RSV SU and TM proteins did not detect complexes between the two proteins (D. Einfeld and E. Hunter, unpublished).

3.2 Role of the Cytoplasmic Domain in Env Maturation

The assembly of retroviral glycoproteins into transport competent structures is generally not dependent on their cytoplasmic domains. Truncations in the cytoplasmic domains of the Env proteins of RSV (PEREZ et al. 1987), MuLV (GRANOWITZ et al. 1991), HIV-1 (DUBAY et al. 1992a; GABUZDA et al. 1992; YU et al. 1993), SIV (JOHNSTON et al. 1993; SPIES and COMPANS 1994; ZINGLER and LITTMAN 1993), and M-PMV (BRODY et al. 1994b) did not inhibit their movement through the secretory pathway. Some truncations and deletions within the cytoplasmic domains of HIV and SIV caused the proteins to become unstable (DUBAY et al. 1992a; GABUZDA et al. 1992; SPIES and COMPANS 1994). It is not clear whether this degradation was the result of an aberrant structure in the cytoplasmic domain or from an effect of this domain on other regions of the protein. An interesting observation from cytoplasmic domain truncations is that for some Env proteins fusion activity was enhanced (BRODY et al. 1994b; MULLIGAN et al. 1992; REIN et al. 1994; RITTER et al. 1993) (see Chap. 4). The basis of this effect is unknown and in some cases exhibited a cell-type dependence (MULLIGAN et al. 1992). In the case of SIV, a truncation of intermediate length resulted in increased syncytial activity that was lost in a more extensive truncation (SPIES and COMPANS 1994). While the conformation of the cytoplasmic domain may be important for interactions with other factors that influence Env function, truncation of the cytoplasmic domain of SIV Env was also reported to affect the conformation of the ectodomain (SPIES et al. 1994). This altered conformation, detected as a difference in susceptibility of the TM protein to biotinylation at the cell surface, could involve a subtle alteration in the association between monomers which changes the accessibility of a reactive site. Similar effects by some altered cytoplasmic domains might interfere with maturation of the Env protein or cause it to be degraded.

The cytoplasmic domains of lentiviral Env proteins have the potential to fold into more intricate structures than those of other Env proteins, which are about one-fourth as long. In addition, the TM proteins of HIV have a sequence at their C-terminus that is predicted to form a pair of amphipathic helices (VENABLE et al. 1989). This region of TM may interact with the membrane in an association that is critical for infection (DUBAY et al. 1992b; HAFFAR et al. 1991). Peptides corresponding to this region have also been found to interact with membranes and result in the formation of pore structures (CHERNOMORDIK et al. 1994; GAWRISCH et al. 1993; SRINIVAS et al. 1992). On the other hand, fusion proteins composed of the cytoplasmic tail of EIAV Env protein appended to Tat have been detected (BEISEL et al. 1993). If this class of protein is common among lentiviruses and important for replication, this would provide an alternative explanation for the extended cytoplasmic domains of lentiviral Env proteins.

4 Transport out of the ER

The ER has been recognized to exert a form of quality control in that proteins which misfold due to mutation, incorporation of certain amino acid analogs, or inhibition of glycosylation are retained in this organelle. In addition, transport of polypeptides that assemble into multichain complexes is generally restricted to those molecules that are assembled in complexes while the individual unassembled proteins do not move out of the ER. Acquisition of resistance to endoglycosidase H (endoH), which results from oligosaccharide processing in the cis/medial compartment of the Golgi complex, provides a mark to distinguish those proteins which have been transported out of the ER. Interactions with the chaperone proteins discussed above can mediate retention of proteins in the ER. But where folding or assembly intermediates can be detected in complexes with these chaperones, dissociation of these complexes is not always the rate-limiting step for conversion of the proteins to an endo-H resistant form (OU et al. 1993).

There is a significant lag between the oligomerization of HIV-1 Env proteins ($t_{1/2}$ of 30 min) and the appearance of endoH resistance ($t_{1/2}$ of 80 min) (EARL et al. 1991). Oligomerization of the RSV Env protein appears to be slower with a $t_{1/2}$ of more than 60 min (EINFELD and HUNTER 1988). The TM protein of RSV expressed alone assembles into an apparent trimer more rapidly than the full-length protein ($t_{1/2}$ of 20 min), but conversion of this protein to endoH resistance is still relatively slow ($t_{1/2}$ of 50 min) (EINFELD and HUNTER 1994). By contrast the lag between trimer formation and arrival in the Golgi is on the order of 5–10 min for flu HA (TATU et al. 1993). This suggests that either the acquisition of an oligomeric structure by the Env proteins is not sufficient to meet the quality control of the ER or that their movement of the ER is slower. What distinguishes Env proteins from more rapidly maturing proteins has not been determined, but the slower kinetics for Env proteins may be useful for exploring these aspects of transport.

Progression of Env proteins beyond the ER is dependent on their entry into budding structures that allow them to move to the Golgi apparatus. The intermediate compartment through which proteins pass on their way to the Golgi is distinguished from other components of the secretory pathway by distinct markers. Isolation of a cellular fraction containing this compartment indicated an absence of factors that are residents of the ER, such as BiP and PDI, or of the *cis*-Golgi cisternae (SCHWEIZER et al. 1991). An earlier study that isolated a vesicle fraction involved in ER-to-Golgi transport reported that the rate of entry of secretory proteins into these vesicles reflected the rates at which the individual proteins became endoH resistant in intact cells (LODISH et al. 1987). Together with more recent evidence (KRIJNSE-LOCKER et al. 1994), this suggests that the lag in conversion to endoH resistance for Env proteins is determined by their entry into the intermediate compartment.

The temperature-sensitive mutant ts045 of the VSV G protein is defective for oligomerization and transport at the restrictive temperature, forming incomplete and aberrant disulfides (MACHAMER and ROSE 1988). When protein synthesis inhibitors were added at the nonpermissive temperature, and cells were chased for relatively long periods, the distribution of this protein was found by immunofluorescence to include dispersed spots in addition to the ER (HAMMOND and HELENIUS 1994b). These spots stained positively for intermediate compartment markers but were negative for ER markers, with the exception of BiP. This overlap between BiP and intermediate compartment markers was observed only in VSV G expressing cells. The VSV G protein did not become endoH resistant but appeared to be recycled back to the ER upon reaching the Golgi complex. These results indicate that movement of a protein out of the ER does not preclude its retention and return to the ER although this may be dependent on a persistent interaction with BiP which recycles to the ER.

An RSV Env mutant, X1, which has a 14 amino acid deletion near its N-terminus, exhibited a distribution in immunofluorescence that extended beyond the ER (HARDWICK et al. 1986). While some of the protein colocalized with a region stained by a Golgi marker, no modification by Golgi components was detected. Although it was defective in transport, this mutant was able to form oligomers (EINFELD and HUNTER 1988). Understanding the basis of the transport defect of this protein, in addition to those of assembly-competent mutants having deletions in the TM domain, could provide insight into regulation of protein transport through the intermediate compartment.

5 Modification Within the Golgi Apparatus

5.1 Glycosylation

Upon entry into the Golgi apparatus, Env proteins are exposed to a set of glycosidases and transferases that can direct restructuring of their N-linked oligosaccharide chains. While some of these chains may persist in the high-mannose form found among ER localized proteins, others are trimmed by mannosidases and become resistant to cleavage by endoH. The latter chains can be modified by addition of *N*-acetylglucosamine, galactose, fucose, and sialic acid to yield a variety of complex structures. The structure of complex oligosaccharides indicates the ordered process by which these modifications occur. But the relative distribution of the enzymes that catalyze these modifications within the multiple compartments of the Golgi apparatus is less clear (MELLMAN and SIMONS 1992).

The Env proteins of MuLV and related viruses are also modified by addition of O-linked oligosaccharides (PINTER and HONNEN 1988) These structures were detected on a maturation intermediate, Pr90, distinguished by its retarded mobility on SDS-PAGE gels relative to the major precursor Pr80. The Thr and Ser residues to which these oligosaccharides are attached lie in the proline-rich region of SU (GEYER et al. 1990). Whereas Pr80 was completely sensitive to endoH, most of the oligosaccharide chains of Pr90 were resistant (PINTER and HONNEN 1988). O-Glycosylation also has been reported for the HIV-1 Env protein (BERNSTEIN et al. 1994; HANSEN et al. 1992).

The MuLV Pr90 intermediate was also altered by neuraminidase treatment (PINTER and HONNEN 1988), pointing to the presence of sialic acid, which is thought to be a late modification in the Golgi. These characteristics of the oligosaccharides of Pr90 are shared by gp70, the SU protein arising from cleavage within the Golgi. An endoH-resistant precursor of increased size has been observed also for MMTV (COREY and STALLCUP 1990) and REV (TSAI and OROSZLAN 1988). No such intermediate has been detected for RSV. The RSV Env precursor, Pr95, remains endoH sensitive and no conversion to a larger species has been detected. When cleavage of Pr95 is blocked, however, a species of Env appears that exhibits an increased apparent molecular weight as a result of modifications within the Golgi complex (PEREZ and HUNTER 1987). Thus the failure to detect a larger precursor of RSV is not due to the absence of modification of the oligosaccharide chains on the RSV protein. A maturation intermediate representing a larger molecular weight form of the HIV precursor gp160 is not apparent, but an increase in the apparent molecular weight of HIV gp160 in the absence of cleavage is also less obvious, due at least in part to less efficient cleavage and poorer resolution (BOSCH and PAWLITA 1990). Detection of intermediates in processing in the Golgi, as at any other point during transport to the cell surface, is dependent on the efficiency of the individual steps involved in this process. For example, the presence of a particular oligosaccharide only on the mature products SU or TM and its absence from the precursor does not indicate that this modification must occur after cleavage, but only that if it occurs before cleavage there is no detectable lag between the two events. The apparent difference between those Env proteins that exhibit a larger precursor intermediate and the RSV protein suggests that cleavage of the former is delayed, due either to a less efficient progress of the protein to the site of cleavage or poorer recognition of the cleavage site in the less mature precursor. Interestingly, mutation of the one glycosylation site of MuLV Env which affected transport resulted in accumulation of the Pr90 precursor which was not cleaved (KAYMAN et al. 1991).

5.2 Cleavage of the Env Precursor

One or more cellular proteases cleave Env precursors into the SU and TM products during transit of the Golgi complex. Env proteins blocked at earlier stages of intracellular transport are not cleaved (HARDWICK et al. 1986; WILLEY et al. 1988; WILLS et al. 1984) and biosynthetic labeling of the oligosaccharides of Env proteins has given results consistent with this picture although this approach only allows identification of stages beyond which cleavage must occur (BOSCH et al. 1982; NG et al. 1982; STEIN and ENGLEMAN 1990; TSAI and OROSZLAN 1988). The presence of sialic acid on some uncleaved precursors suggests that cleavage occurs late in the Golgi complex (BEDGOOD and STALLCUP 1992). But the fact that BFA promoted cleavage of the MuLV precursor (KAYMAN et al. 1991) indicates that processing activity resides within the component of the Golgi complex that can be recycled to the ER under these conditions, i.e., the medial Golgi.

The peptide bond of the precursor that is cleaved has been inferred from N-terminal sequencing of TM (HUNTER et al. 1983; VERONESE et al. 1985) and lies immediately C-terminal to a pair of basic residues within the consensus motif Arg-X-Arg/Lys-Arg. Positions in this consensus sequence are designated P4 through P1, moving from the N-terminus to the C-terminus. Arg is conserved among Env proteins at the P1 position while Lys predominates at P2. FeLV isolates and SRV-2 have Arg at P2 while some HIV-2 isolates are unusual with Thr at P2 (HIV-2_{ROD} and HIV-2_{NIHZ}). Exceptions from Arg at P4 are limited to Lys, which occurs in M-PMV and in some isolates of MuLV. Wide divergence, on the other hand, is seen at the P3 position, but within retrovirus groups this residue is relatively conserved. MMTV has Ala at this position while MuLV has His and Tyr. FeLV isolates have large hydrophobic residues (Phe, Leu) but ASLV and HIV-1 have the oppositely charged residues Arg and Glu, respectively. The Glu residue is highly conserved among diverse HIV-1 isolates, but HIV-2 has Asn, His, or Pro at this position and Asn predominates for SIV isolates. The protease that recognizes this sequence appears to tolerate a variety of amino acids at the P3 position, but the more limited variation within individual Env proteins suggests the side chain of this residue may be critical for optimal recognition of the cleavage site in the context of different proteins. Alternatively, there may be subtle differences between proteases of different species or different cell types.

Mutation of the P2 Lys to Glu inhibited cleavage of the RSV Env but residual cleavage occurred that exhibited a different inhibitor sensitivity from cleavage of the wild-type protein (PEREZ and HUNTER 1987). Combination of this P2 mutation with a Ser-to-Arg substitution at P4 resulted in a more complete block of cleavage (DONG et al. 1992b). That this represented a specific effect on recognition by processing enzymes was supported by the similar sensitivities of wild-type and mutant proteins to specific cleavage by trypsin (DONG et al. 1992b). Mutation of the P2 Lys of HIV, to Asn, caused a reduction in cleavage similar to that seen for RSV (BOSCH and PAWLITA 1990), but substitution of Ile or Glu at this position exhibited no inhibition of cleavage (FREED et al. 1989; DUBAY et al. 1995). Substitution of nonbasic residues at P2 and P4 had an additive effect resulting in complete inhibition of cleavage (BOSCH and PAWLITA 1990). Substitution of a Lys for the P1 Arg caused a dramatic reduction in cleavage of the MuLV precursor (FREED and RISSER 1987) while the same mutation in HIV-1 gp160 had little effect (FREED et al. 1989).

The consensus sequence at the SU/TM cleavage site of retroviral glycoproteins closely resembles the canonical recognition sequence of the cellular proprotein processing enzyme, furin (HATSUZAWA et al. 1990; HOSAKA et al. 1991; KORNER et al. 1991). This enzyme belongs to a family of serine proteases distinguished by their similarity to the yeast KEX2/kexin endoprotease with catalytic domains homologous to bacterial subtilisins (reviewed in BARR 1991; NAGAI 1993; STEINER et al. 1992). While some of these proteases are expressed only in select cell types and within the regulated secretory pathway, furin, PACE4, and PC6/PC5 are found in a wide variety of cells (HATSUZAWA et al. 1990; KIEFER et al. 1991; LUSSON et al. 1993; SCHALKEN et al. 1987). Members of the former, more restricted group appear to recognize dibasic sequences, while the latter proteases recognize the consensus sequence R-X-K/R-R. Furin is able to cleave some dibasic sequences (BRESLIN et al. 1993) and earlier analyses focused on this specificity. But the presence of a basic residue at the P4 position is essential for cleavage of some substrates (HOSAKA et al. 1991; MOLLOY et al. 1992). PACE4 and PC6/PC5 have a similar sequence specificity but are somewhat more restricted than furin in the range of substrates they can cleave (CREEMERS et al. 1993; REHEMTULLA et al. 1993). While PC5/PC6 is widely distributed in its expression, very little is detected in some tissues such as liver (LUSSON et al. 1993), where furin expression is relatively high.

The ability of furin to cleave viral glycoprotein precursors has been examined by coexpression of the two proteins even though furin is expressed in most cells and processing of Env proteins occurs in the absence of exogenous furin. One approach has been to express furin from vaccinia virus and analyze cleavage at a later point in infection when the endogeous protease activity of the cell is shut down (BRESNAHAN et al. 1990 ; STIENEKE-GROBER et al. 1992). Under these conditions furin has been shown to cleave the HIV-1 precursor, gp160 (HALLENBERGER et al. 1992). Cleavage of gp160 by furin was also observed when both were expressed in insect cells using baculovirus, although additional cleavage occurred in the V3 loop (MORIKAWA et al. 1993). The level of expression of furin using vaccinia is much higher than in uninfected cells and overexpression of furin can result in cleavage of proteins that are not cleaved, or are only poorly cleaved, by the endogenous proteases which are thought to include furin (BRESNAHAN et al. 1990; SMEEKENS et al. 1992). Analysis of mutations in the R-K-R-K-K-R cleavage sequence of HA from avian influenza viruses indicated that overexpressed furin had a broader specificity than the endogenous protease of CV-1 cells (HORIMOTO et al. 1994; WALKER et al. 1994). The specificity of the endogenous protease more closely resembled that of purified furin in vitro. Clearly the expression level of the protease has a marked effect on its apparent specificity.

Evidence for the role of furin in cleavage of gp160 also comes from the use of inhibitors. Peptidyl chloromethylketones linked to an R-X-K-R peptide are effective inhibitors of furin and inhibited cleavage of gp160 by endogenous protease when added to cells (HALLENBERGER et al. 1992). While cleavage was inhibited in cells treated with compounds containing the peptides R-E-K-R or R-A-K-R, R-A-I-R was not effective. The negative result with the R-A-I-R chloromethylketone is interesting in light of evidence that the minimal furin recognition sequence is R-X-X-R (MOLLOY et al. 1992). But a similar R-E-I-R compound was 1000-fold less effective in inhibition of HA cleavage by purified furin compared to one having an R-E-K-R peptide (STIENEKE-GROBER et al. 1992). As noted above, mutations indicated that basic residues at both P4 and P2 are essential for optimal gp160 cleavage (BOSCH and PAWLITA 1990). The inhibition of gp160 cleavage in the presence of these inhibitors, then, is consistent with the involvement of furin but still leaves open the possibility for other related proteases. A second inhibitor of furin in vitro, a modified α_1-antitrypsin with R-I-P-R at its reactive site, was also able to block gp160 processing when coexpressed with the latter (ANDERSON et al. 1993).

Cleavage of gp160 occurred in furin-defective LoVo cells which are unable to process a number of precursor proteins (OHNISHI et al. 1994; TAKAHASHI et al. 1993). Although the product of the mutant furin gene could not be excluded as the mediator of gp160 cleavage in these cells, it is striking that the Newcastle disease virus fusion glycoprotein precursor, F_0, was not cleaved in these cells despite the close resemblance of its cleavage site, R-Q-K-R, to that of gp160 (OHNISHI et al. 1994). This result suggests that more than one protease may be capable of cleaving gp160 and that whatever protease cleaved gp160 in these cells is able to distinguish its cleavage site from that of the F_0 protein. Of the other subtilisin-like proteases, PC5/6 has been shown to recognize cleavage site mutants of the avian influenza virus HA that were also cleaved by endogenous proteases whereas PACE4, with little activity against wild-type HA, failed to cleave any of the mutants (HORIMOTO et al. 1994). PACE4, however, is able to cleave some proteins with the R-X-K-R sequence, such as pro-von Willebrand factor (CREEMERS et al. 1993; REHEMTULLA et al. 1993).

An analysis of membrane fractions from a T-cell line resulted in isolation of a protease activity capable of cleaving gp160, but unlike furin this apparent ER/Golgi activity was not Ca^{2+} dependent (KIDO et al. 1993). Other chromatographic fractions, including a Ca^{2+}-dependent activity, reportedly could cleave a dibasic fluorogenic peptide but exhibited no activity on gp160.The dibasic protease PC1, which is not dependent on a basic residue at P4, has also been shown to cleave gp160 in vitro (DECROLY et al. 1994) and this enzyme has been detected in $CD4^+$ lymphocytes where furin expression is relatively low. But the ability of PC1 to cleave other substrates in vivo varies with cell type (BENJANNET et al. 1992).

The HIV-1 Env precursor has a second potential cleavage site sequence R/K-X-K-R-R just upstream of the site at the N-terminus of gp41. Individual mutation of the basic residues in this site did not disrupt cleavage whereas replacement of the complete set with nonbasic residues resulted in very little cleavage of gp160 (Bosch and Pawlita 1990). In the context of a modified first site, mutation of one or two residues at the second site can block cleavage but it still appears the first site is where cleavage occurs. Using an antibody to the C-terminus of gp120, however, cleavage at the upstream site has been detected to occur at a rate of about 10% (Fenouillet and Gluckman 1992; Morikawa et al. 1993). This second site does not fit the consensus for a furin cleavage site. If the last Arg of this motif constitutes P1, this site would lack a basic residue at P4 and the presence of basic residues at P3 and P5 would not be expected to enhance cleavage (Watanabe et al. 1992, 1993). In addition, the large hydrophobic residue C-terminal to P1, at P1', could inhibit cleavage (Oda et al. 1991). On the other hand, if cleavage occurred between the two Arg residues, a basic residue would be present at P4 and a favorable hydrophobic residue at P2'. The susceptibility of a site with Arg at P1' is not clear but this residue is often polar and is also predicted to be oriented away from the enzyme so that there is tolerance at this position (Siezen et al. 1994).

In the case of many precursors cleaved after multiple basic residues, these residues are subsequently removed from the N-terminal cleavage product. Trimming of basic residues from the C-terminus of the NDV F protein and avian influenza HA cleavage products has been observed (Garten et al. 1982;Gorman et al. 1990). The cleavage product of a prohormone precursor cleaved in nonendocrine mammalian cells by introduction of the yeast Kex2p protease was detected in a form that retained the dibasic residues of the cleavage site at its C-terminus (Thomas et al. 1988). Removal of the basic residues could be achieved when the yeast Kex1p carboxypeptidase, which is specific for basic residues, was coexpressed (Thomas et al. 1990). This result suggests that removal of C-terminal basic residues is not mediated by a ubiquitous enzyme. Whether C-terminal basic residues are removed from SU proteins and the identity of the carboxypeptidase that could mediate this have not been elucidated.

6 Sorting at the TGN and Transport to the Plasma Membrane

From the trans-Golgi network (TGN) retroviral glycoproteins are targeted to the basolateral domain of the plasma membrane in polarized epithelial cells (Roth et al. 1983). Proteins localized to either the basolateral or apical membranes may be sorted into separate vesicles in the TGN for transport to their separate domains of the plasma membrane, as has been observed for VSV G protein, flu HA, and cellular proteins in polarized Madin-Darby canine kidney cells (LE BIVIC et al. 1990b; PFEIFFER et al. 1985; RINDLER et al. 1984, 1985; WANDINGER-NESS et al. 1990). Alternatively, both classes of proteins may be transported to the basolateral surface with apical proteins then being redirected to that membrane domain, as was observed in cell fractionation studies of hepatocytes (BARTLES et al. 1987). Both pathways appear to be used in intestinal epithelial cells, depending on the individual apical protein (LE BIVIC et al. 1990a; MATTER et al. 1990).

Some basolaterally targeted proteins can be redirected to the apical membrane by deletions or mutations in their cytoplasmic domains (CASANOVA et al. 1991; HUNZIKER et al. 1991; PRILL et al. 1993; YOKODE et al. 1992). Appending these cytoplasmic domains to apically targeted proteins can redirect the latter to the basolateral membrane (CASANOVA et al. 1991; PRILL et al. 1993; THOMAS et al. 1993). Mutational analysis suggests that signals for basolateral targeting and for endocytosis may be similar (BREWER and ROTH 1991) but these appear to involve distinct regions on wild type proteins (DARGEMONT et al. 1993; PRILL et al. 1993). Features of Env proteins that cause polarized sorting to the basolateral membrane may not be localized to the cytoplasmic domain, since a truncated MuLV Env protein lacking the entire cytoplasmic domain was still sorted to the basolateral surface (KILPATRICK et al. 1987)

Env proteins are incorporated into budding capsids after their arrival at the plasma membrane. The presence of Env proteins at the cell surface may be dynamic, both in terms of internalization and recycling to the cell surface and with regard to lateral movement within the membrane. Both of these are likely to impact incorporation. While rapid internalization would be detrimental to incorporation, specific inclusion of Env proteins in budding viruses might be enhanced by movement of these proteins to specialized structures of the plasma membrane.

7 Incorporation of Env Proteins into Budding Virions

Assembly of infectious virus requires the presence of Env proteins in the plasma membrane that is acquired as capsids bud from the cell. Since the budding of capsids does not require expression of Env proteins for the release of particles, a mechanism must exist to insure the enrichment of Env proteins at the site of budding. While a specific, direct interaction between Env proteins and the budding capsid could allow an efficient assembly of infectious virus, evidence for such an interaction is incomplete. Incorporation of heterologous membrane proteins into retroviruses can occur. The selective inclusion of Env proteins and the relative exclusion of cellular proteins may be achieved through other mechanisms than a direct interaction of the Env protein and a component of the capsid.

The fact that retroviruses can incorporate membrane proteins other than the product of their own *env* gene was recognized first in the formation of pseudotypes during mixed infection or complementation of Env-defective viruses (WEISS 1993; ZAVADA 1982). These studies indicated that retroviruses could not only incorporate other viral envelope proteins but also that these proteins could confer infectivity. Later analyses have provided more quantitative information on the relative incorporation of heterologous proteins. Foreign proteins that can be incorporated by ASLV capsids include Env proteins of MuLV-related viruses (LEVY 1977; WEISS and WONG 1977), Sindbis virus glycoproteins (ZAVADOVA et al. 1977), VSV G protein (WEISS et al. 1977), and influenza HA (DONG et al. 1992a). The incorporation of HA into ASLV particles was readily detected in pelleted virus at levels similar to Env when the glycoproteins were coexpressed with Gag using the vaccinia/T7 system (DONG et al. 1992a). Thus incorporation of HA was not dependent on co-expression of Env proteins. Replacing the membrane-spanning and cytoplasmic domains of HA with those of RSV Env did not enhance its incorporation into Gag particles, which would be expected if a specific signal made incorporation of Env more efficient. Expression of HA from the viral genome in place of *env* confirmed this incorporation during virus assembly and resulted in the generation of infectious virus.

Among other retroviruses, capsids of MuLV can incorporate the VSV G protein (EMI et al. 1991; LIVINGSTON et al. 1976) as well as the human T-cell leukemia virus (HTLV)-1 Env protein (VILE et al. 1991; WILSON et al. 1989). Based on the transfer of a marker gene, the efficiency of incorporation of VSV G in the absence of Env appeared similar to that of MuLV Env and was not specifically inhibited in the presence of Env (EMI et al. 1991). On the other hand, incorporation of HTLV-1 was much lower than the MuLV or related GaLV Env proteins (WILSON et al. 1989). Mixed infections between HIV and VSV or herpes simplex virus led to production of HIV cores that exhibited infectivity indicative of the presence of the heterologous envelope glycoproteins (ZHU et al. 1990). Formation of infectious particles consisting of HIV-1 capsids and an MuLV Env protein was not dependent on the presence of HIV-1 Env. The infectivity of these particles, assayed by transfer of a marker gene, was similar to particles containing the HIV protein (PAGE et al. 1990). HIV capsids also efficiently incorporated the HTLV glycoprotein in the presence or absence of gp160 to produce infectious particles (LANDAU et al. 1991).

Incorporation of foreign proteins is not limited to viral glycoproteins. Gradient-purified HIV-1, HIV-2, and SIV virions grown in H9 cells were found to contain both β_2-microglobulin and HLA class II molecules at levels similar to those expected for Env (ARTHUR et al. 1992). The presence of these molecules was reported by others using different methods (HOXIE et al. 1987; SCHOLS et al. 1992). Incorporation of class II molecules was specific for those encoded by HLA-DR, since HLA-DQ and HLA-DP could not be detected despite production of viruses in cells that expressed all of these. A series of other antibodies gave negative results in these studies. The levels of these proteins observed in virions points to an efficient incorporation. But the nature of this association and whether Env proteins contribute to their presence has not been determined. MuLV has been reported to specifically incorporate Thy-1(CALAFAT et al. 1983). No association between Env and Thy-1 was observed in this study when the individual proteins were capped using monoclonal antibodies. The proteins detected in the above studies are expected to be encountered by virus during the course of natural infections. Introduction of the human CD4 gene into avian cells infected with ASLV resulted in incorporation of this protein into virions (YOUNG et al. 1990).

Mutational analysis of the RSV Env protein indicated that the cytoplasmic tail did not contribute to virus incorporation since mutants lacking this domain exhibited wild-type levels of glycoprotein and infectivity (PEREZ et al. 1987). The membrane-spanning sequence retained by this mutant might still mediate an interaction with capsids. In fact a deletion of three amino acids from this domain resulted in a protein defective for incorporation, despite the similarity in its processing and stability to the wild-type protein (DONG and HUNTER 1993). A decrease in incorporation could result from decreased stability of the protein at the cell surface so that the amount of Env available for incorporation is reduced. Alternatively, the distribution of the protein in the plasma membrane may be affected so that it is not concentrated in regions of virus budding due to altered interaction with cellular factors or an effect of the deletion on the mobility of the protein within the membrane. If the deletion has a direct effect on specific interaction between Env and Gag, this must be reconciled with the lack of specificity that allows incorporation of foreign proteins.

The contribution of the cytoplasmic tail of the HIV Env protein to its incorporation is less clear. Analyses of a series of deletions in the cytoplasmic domain led to the conclusion that the cytoplasmic domain was not required for incorporation (GABUZDA et al. 1992; WILK et al. 1992). One of the truncation mutants, however, was defective for incorporation (WILK et al. 1992) while in a different experimental system a reduction in incorporation was observed for a series of proteins terminating at positions throughout the cytoplasmic domain (DUBAY et al. 1992b; YU et al. 1993). In contrast, reported truncations of the SIV protein do not impair incorporation and may actually enhance it (JOHNSTON et al. 1993; ZINGLER and LITTMAN 1993).

A mutant MuLV Env protein lacking the cytoplasmic domain was not incorporated into virions but the processing and transport of this protein were not characterized (GRANOWITZ et al. 1991). A truncation within the cytoplasmic domain of the MuLV protein was introduced to mimic the cleavage in this region by the viral protease following assembly (REIN et al. 1994). This protein was incorporated into virions, although at reduced levels. The efficiency of its incorporation is not clear in the absence of information on its expression at the cell surface.

In the case of the D-type M-PMV TM protein, truncation of its 38 amino acid cytoplasmic domain to 15 amino acids resulted in a loss of infectivity and a marked reduction in the incorporation of Env into virus (BRODY et al. 1994a).

Truncation of the cytoplasmic domain to 27 amino acids gave a protein with an intermediate phenotype for both infectivity and incorporation. Since the levels of the wild-type and the mutant proteins at the cell surface were similar, the differences in incorporation were attributed to a function of the mutated region as a signal for incorporation. Incorporation at intermediate levels would reflect inefficient recognition of a modified signal or might indicate a level of non-signal-mediated incorporation. On the other hand, alteration of the Env protein or its interactions with cellular factors, as suggested by its increased fusogenicity, might impede its incorporation into virions. Dependence of incorporation of the M-PMV Env protein on the cytoplasmic domain could also reflect a difference in the assembly mechanism for D type viruses, where the Env protein encounters Gag as an assembled capsid.

In summary, while incorporation of the RSV Env protein is not dependent on the cytoplasmic domain, a deletion in the MSD affected incorporation. But it remains to be seen if this deletion directly impacts an interaction with Gag. Truncations in the cytoplasmic domain of the M-PMV protein and, at least in some circumstances, the HIV-1 Env affected incorporation. Reduced incorporation for all truncations beyond a certain point in the cytoplasmic domain argues against an inhibitory effect of an altered structure in this domain. But specific signals have not been characterized and the loss of incorporation through truncation appears to be progressive.

The component of the capsid that is expected to interact most closely with Env protein is the matrix protein (MA) (CARDIFF et al. 1978; GEBHARDT et al. 1984; MONTELARO et al. 1978). There is evidence that MA mutations or sequence differences can affect glycoprotein incorporation into virions (CASELLA and PANGANIBAN 1993; DORFMAN et al. 1994; YU et al. 1992). Deletions in HIV-1 MA protein which did not interfere with capsid assembly and release did exhibit a defect for incorporation of Env proteins (YU et al. 1992). Analysis of a series of point mutations and small deletions in HIV-1 MA indicated that only mutations near the C-terminus failed to affect incorporation of Env (DORFMAN et al. 1994). Thus it is possible that the conformation of the MA protein is critical for incorporation. Since visna virus capsids having a substitution of the HIV sequence in place of their own MA acquired the ability to efficiently incorporate the HIV-1 Env (DORFMAN et al. 1994), it would appear that a specific interaction contributes to incorporation.

Reports that the polarized budding of HIV capsids is dependent on Env expression also point to a specific interaction between Env and Gag (LODGE et al. 1994; OWENS et al. 1991). While the yield of capsids budding from the apical and basal surfaces of polarized cells is roughly equal in the absence of Env, expression of the basolaterally targeted Env protein results in budding of capsids almost exclusively from the basolateral surface. A dose-response analysis indicated that even at a 1:10 ratio of Env to Gag more than 90% of capsids budded from the basolateral surface (LODGE et al. 1994). The ability of Env to alter the budding of capsids was lost by truncations near its C-terminus (LODGE et al. 1994). This effect of Env is likely to be important for understanding how retroviruses incorporate glycoproteins. The membrane association of Gag that initiates budding might be enhanced by the presence of Env. But since the production of capsids appears higher in the absence of Env (LODGE et al. 1994) the possibility exists that Gag budding is regulated in the presence of Env. It remains to be seen if the polarized transport of the truncated Env proteins was altered or if the deletion disrupted a motif recognized by Gag. Deletion of the cytoplasmic domain of MuLV Env did not affect its polarized sorting (KILPATRICK et al. 1987). Nevertheless, this system may prove fruitful for understanding the process of incorporation.

8 Perspectives

The identification and characterization of molecular components of the secretory pathway continues to advance a detailed understanding of the processes involved in the intracellular maturation and transport of membrane proteins. The complement of this understanding is an appreciation of the features of transported proteins that interact with these components. As critical as the receptor binding and fusogenicity of Env proteins is to the replication of retroviruses, the initial challenge is a productive interaction with the secretory pathway, allowing transport of the translated *env* product to the cell surface. The first requirement is acquisition of the poorly defined quality of transport competence which allows the protein to progress beyond the ER. While general cellular processes are involved, an ability to specifically interfere with the assembly of Env complexes could prove useful for inhibiting virus replication. A second critical step is cleavage of the precursor to produce a molecule that can be fusogenic. Apart from the sequence at the cleavage site, other structural features of Env are likely required for efficient cleavage. Knowing the proteases involved and their other functions in the cell is essential for attempting to block this event. The third major step in the Env transport pathway is incorporation of the glycoprotein into virus. The stringency of this process is called into question by the possibility of non-specific incorporation, but the Env protein must still avoid the fate of the majority of cellular proteins which are excluded from virions. Provided that specific interactions are involved, intervention with this process, occurring at the cell surface, would represent an additional target for reducing virus production.

Since retroviruses make use of a major system for protein maturation and transport in cells, analysis of the interaction of Env proteins with components of this pathway may contribute to an understanding of the functions of these components. For example, the relative inefficiency with which Env proteins are able to be transported out of the ER could make them useful tools for exploring aspects of transport competence. On the other hand, the possibility that incorporation of Env proteins into budding capsids is dependent on targeting and mobility within specialized plasma membrane structures suggests that the analysis of this process could provide insights into the dynamic structure of this cellular compartment.

References

Adams GA, Rose JK (1985) Structural requirements of a membrane-spanning domain for protein anchoring and cell surface transport. Cell 41: 1007–1015

Anderson ED, Thomas L, Hayflick JS, Thomas G (1993) Inhibition of HIV-1 gp160-dependent membrane fusion by a furin-directed α1-antitrypsin variant. J Biol Chem 268: 24887–24891

Arthur LO, Bess JW, Sowder RC, Benveniste RE, Mann DL, Chermann J-C, Henderson LE (1992) Cellular proteins bound to immunodeficiency viruses:implications for pathogenesis and vaccines. Science 258: 1935–1938

Barr PJ (1991) Mammalian subtilisins: the long-sought dibasic processing endoproteases. Cell 66: 1–3

Bartles JR, Feracci HM, Stieger B, Hubbard AL (1987) Biogenesis of the rat hepatocyte plasma membrane in vivo: comparison of the pathways taken by apical and basolateral proteins using subcellular fractionation. J Cell Biol 105: 1241–1251

Battini J-L, Heard JM, Danos O (1992) Receptor choice determinants in the envelope glycoproteins of amphotropic, xenotropic, and polytropic murine leukemia viruses. J Virol 66: 1468–1475

Battini J-L, Kayman SC, Pinter A, Heard JM, Danos O (1994) Role of N-linked glycosylation in the activity of the Friend murine leukemia virus SU protein receptor-binding domain. Virology 202: 496–499

Bedgood RM, Stallcup MR (1992) A novel intermediate in processing of murine leukemia virus envelope glycoproteins. Proteolytic cleavage in the late Golgi region. J Biol Chem 267: 7060–7065

Beisel CE, Edwards JF, Dunn LL, Rice NR (1993) Analysis of multiple mRNAs from pathogenic equine infectious anemia virus (EIAV) in an acutely infected horse reveals a novel protein, Ttm, derived from the carboxy terminus of the EIAV transmembrane protein. J Virol 67: 832–842

Benjannet S, Reudelhuber T, Mercure C, Rondeau N, Chretien M, Seidah NG (1992) Proprotein conversion is determined by a multiplicity of factors including convertase processing, substrate specificity, and intracellular environment. Cell type-specific processing of human prorenin by the convertase PC1. J Biol Chem 267: 11417–11423

Bergeron JJ, Brenner MB, Thomas DY, Williams DB (1994) Calnexin: a membrane-bound chaperone of the endoplasmic reticulum. Trends Biochem Sci 19: 124–128

Bernstein HB, Tucker SP, Hunter E, Schutzbach JS, Compans RW (1994) Human immunodeficiency virus type 1 envelope glycoprotein is modified by O-linked oligosaccharides. J Virol 68: 463–468

Blond-Elguindi S, Cwirla SE, Dover WJ, Lipshutz RJ, Sprang SR, Sambrook JF, Gething M-JH (1993) Affinity panning of a library of peptides displayed on bacteriophages reveals the binding specificity of BiP. Cell 75: 717–728

Bosch JV, Schwarz RT (1984) Processing of gPr92env, the precursor of the glycoproteins of Rous sarcoma virus: use of inhibitors of oligosaccharide trimming and glycoprotein transport. Virology 132: 95–109

Bosch JV, Schwartz RT, Ziemiecki A, Friis RR (1982) Oligosaccharide modifications and the site of processing of gPr92env, the precursor for the viral glycoproteins of Rous sarcoma virus. Virology 119: 122–132

Bosch ML, Andeweg AC, Schipper R, Kenter M (1994) Insertion of N-linked glycosylation sites in the variable regions of the human immunodeficiency virus type 1 surface glycoprotein through AAT triplet reiteration. J Virol 68: 7566–7569

Bosch V, Pawlita M (1990) Mutational analysis of the human immunodeficiency virus type 1 env gene product proteolytic cleavage site. J Virol 64: 2337–2344

Bova CA, Manfredi JP, Swanstrom R (1986) env genes of avian retroviruses: nucleotide sequence and molecular recombinants define host range determinants. Virology 152: 343–354

Bova CA, Olsen JC, Swanstrom R (1988) The avian retrovirus env gene family: molecular analysis of host range and antigenic variants. J Virol 62: 75–83

Braakman I, Hoover-Litty H, Wagner KR, Helenius A (1991) Folding of influenza hemagglutinin in the endoplasmic reticulum. J Cell Biol 114: 401–411
Braakman I, Helenius J, Helenius A (1992a) Manipulating disulfide bond formation and protein folding in the endoplasmic reticulum. EMBO J 11: 1717–1722
Braakman I, Helenius J, Helenius A (1992b) Role of ATP and disulphide bonds during protein folding in the endoplasmic reticulum. Nature 356: 260–262
Breslin MB, Lindberg I, Benjannet S, Mathis JP, Lazure C, Seidah NG (1993) Differential processing of proenkephalin by prohormone convertases 1(3) and 2 and furin. J Biol Chem 268: 27084–27093
Bresnahan PA, Leduc R, Thomas L, Thorner J, Gibson HL, Brake HL, Barr PJ, Thomas G (1990) Human fur gene encodes a yeast KEX2-like endoprotease that cleaves pro-β-NGF in vivo. J Cell Biol 111: 2851–2859
Brewer CB, Roth MG (1991) A single amino acid change in the cytoplasmic domain alters the polarized delivery of influenza virus hemagglutinin. J Cell Biol 114: 413–421
Broder CC, Earl PL, Long D, Abedon ST, Moss B, Doms RW (1994) Antigenic implications of human immunodeficiency virus type 1 envelope quaternary structure: oligomer-specific and -sensitive monoclonal antibodies. Proc Natl Acad Sci USA 91: 11699–11703
Brody BA, Hunter E (1992) Mutations within the env gene of Mason-Pfizer monkey virus: effects on protein transport and SU-TM association. J Virol 66: 3466–3475
Brody BA, Kimball MG, Hunter E (1994a) Mutations within the transmembrane glycoprotein of Mason-Pfizer monkey virus: loss of SU-TM association and effects on infectivity. Virology 202: 673–683
Brody BA, Rhee SS, Hunter E (1994b) Postassembly cleavage of a retroviral glycoprotein cytoplasmic domain removes a necessary incorporation signal and activates fusion activity. J Virol 68: 4620–4627
Bulleid NJ, Freedman RB (1988) Defective co-translational formation of disulphide bonds in protein disulphide-isomerase deficient microsomes. Nature 335: 649–651
Cai H, Wang CC, Tsou CL (1994) Chaperone-like activity of protein disulfide isomerase in the refolding of a protein with no disulfide bonds. J Biol Chem 269: 24550–24552
Calafat J, Hansen H, Demant P, Hilgers J, Zavada J (1983) Specific selection of host cell glycoproteins during assembly of murine leukemia virus and vesicular stomatitis virus: presence of Thy-1 glycoprotein and absence of H-2, Pgp-1 and T200 glycoproteins on the envelopes of these virus particles. J Gen Virol 64: 1241–1253
Cardiff RD, Puentes MJ, Young LJT, Smith GH, Teramoto YA, Altrock BW, Pratt TS (1978) Serological and biochemical characterization of the mouse mammary tumor virus with localization of p10. Virology 85: 157–167
Carr CM, Kim PS (1993) A spring-loaded mechanism for the conformational change of influenza hemagglutinin. Cell 73: 823–832
Casanova JE, Apodaca G, Mostov KE (1991) An autonomous signal for basolateral sorting in the cytoplasmic domain of the polymeric immunoglobulin receptor. Cell 66: 65–75
Casella CR, Panganiban AT (1993) The matrix region is responsible for the differential ability of two retroviruses to function as helpers for vector propagation. Virology 192: 458–464
Chakrabarti S, Mizukami T, Franchini G, Moss B (1990) Synthesis, oligomerization, and biological activity of the human immunodeficiency virus type 2 envelope glycoprotein expressed by a recombinant vaccinia virus. Virology 178: 134–142
Chambers P, Pringle CR, Easton AJ (1990) Heptad repeat sequences are located adjacent to hydrophobic regions in several types of virus fusion glycoproteins. J Gen Virol 71: 3075–3080
Chen SS-L, Lee C-N, Lee W-R, McIntosh K, Lee T-H (1993) Mutational analysis of the leucine zipper-like motif of the human immunodeficiency virus type 1 envelope transmembrane glycoprotein. J Virol 67: 3615–3619
Chernomordik L, Chanturiya AN, Suss-Toby E, Nora E, Zimmerberg J (1994) An amphipathic peptide from the C-terminal region of the human immunodeficiency virus envelope glycoprotein causes pore formation in membranes. J Virol 68: 7115–7123
Chessler SD, Byers PH (1992) Defective folding and stable association with protein disulfide isomerase/prolyl hydroxylase of type I procollagen with a deletion in the proα2(I) chain that preserves the Gly-X-Y repeat pattern. J Biol Chem 267: 7751–7757
Collins PL, Motet G (1991) Homooligomerization of the hemagglutinin-neuraminidase glycoprotein of human parainfluenza virus type 3 occurs before acquisition of correct intramolecular disulfide bonds and mature immunoreactivity. J Virol 65: 2362–2371
Cordonnier A, Montagnier L, Emerman M (1989) Single amino-acid changes in HIV envelope affect viral tropism and receptor binding. Nature 340: 571–574

Corey JL, Stallcup MR (1990) The order of processing events in mouse mammary tumor virus envelope protein maturation: implications for the location of glucocorticoid-regulated step. Cell Regul 1: 531–541
Creemers JW, Kormelink PJ, Roebroek AJ, Nakayama K, Van de Ven WJ (1993) Proprotein processing activity and cleavage site selectivity of the Kex2-like endoprotease PACE4. FEBS Lett 336: 65–69
Crowley KS, Reinhart GD, Johnson AE (1993) The signal sequence moves through a ribosomal tunnel into a noncytoplasmic aqueous environment at the ER membrane early in translocation. Cell 73: 1101–1115
Dargemont C, Le Bivic A, Rothenberger S, Iacopetta B, Kühn LC (1993) The internalization signal and the phosphorylation site of transferrin receptor are distinct from the main basolateral sorting information. EMBO J 12: 1713–1721
Davis NG, Hsu M-C (1986) The fusion-related hydrophobic domain of Sendai F protein can be moved through the cytoplasmic membrane of *Escherichia coli.* Proc Natl Acad Sci USA 83: 5091–5095
Decroly E, Cornet B, Martin I, Ruysshaert J-M, Vandenbranden M (1993) Secondary structure of gp160 and gp120 envelope glycoproteins of human immunodeficiency virus type 1: a Fourier transform infrared spectroscopy study. J Virol 67: 3552–3560
Decroly E, Vandenbranden M, Ruysschaert JM, Cogniaux J, Jacob GS, Howard SC, Marshall G, Kompelli A, Basak A, Jean F et al (1994) The convertases furin and PC1 can both cleave the human immunodeficiency virus (HIV)-1 envelope glycoprotein gp160 into gp120 (HIV-1 SU) and gp41 (HIV-I TM). J Biol Chem 269: 12240–12247
Dedera D, Gu R, Ratner L (1992) Conserved cysteine residues in the human immunodeficiency virus type 1 transmembrane envelope protein are essential for precursor envelope cleavage. J Virol 66: 1207–1209
DeLarco JE, Todaro GJ (1976) Membrane receptors for murine leukemia viruses: characterization using the purified viral envelope glycoprotein gp71. Cell 8: 365–371
Delwart EL, Mosalios G (1990) Retroviral envelope glycoproteins contain a "leucine zipper"-like repeat. AIDS Res Hum Retroviruses 6: 703–706
Delwart EL, Panganiban AT (1989) Role of reticuloendotheliosis virus envelope glycoprotein in superinfection interference. J Virol 63: 273–280
de Silva A, Balch WE, Helenius A (1990) Quality control in the endoplasmic reticulum: folding and misfolding of vesicular stomatitis virus G protein in cells and in vitro. J Cell Biol 111: 857–866
de Silva A, Braakman I, Helenius A (1993) Posttranslational folding of vesicular stomatitis virus G protein in the ER: involvement of noncovalent and covalent complexes. J Cell Biol 120: 647–655
Doms RW, Earl PL, Chakrabarti S, Moss B (1990) Human immunodeficiency virus types 1 and 2 and simian immunodeficiency virus env proteins possess a functionally conserved assembly domain. J Virol 64: 3537–3540
Doms RW, Lamb RA, Rose JK, Helenius A (1993) Folding and assembly of viral membrane proteins. Virology 193: 545–562
Dong J, Hunter E (1993) Analysis of retroviral assembly using a vaccinia/T7-polymerase complementation system. Virology 194: 192–199
Dong J, Roth MG, Hunter E (1992a) A chimeric avian retrovirus containing the influenza virus hemagglutinin gene has an expanded host range. J Virol 66: 7374–7382
Dong JY, Dubay JW, Perez LG, Hunter E (1992b) Mutations within the proteolytic cleavage site of the Rous sarcoma virus glycoprotein define a requirement for dibasic residues for intracellular cleavage. J Virol 66: 865–874
Dorfman T, Mammano F, Haseltine WA, Gottlinger HG (1994) Role of the matrix protein in the virion association of the human immunodeficiency virus type 1 envelope glycoprotein. J Virol 68: 1689–1696
Dorner AJ, Coffin JM (1986) Determinants for receptor interaction and cell killing on the avian retrovirus glycoprotein gp85. Cell 45: 365–374
Dorner AJ, Stoye JP, Coffin JM (1985) Molecular basis of host range variation in avian retroviruses. J Virol 53: 32–39
Dowbenko D, Nakamura G, Fennie C, Shimasaki C, Riddle L, Harris R, Gregory T, Lasky L (1988) Epitope mapping of the human immunodeficiency virus type 1 gp120 with monoclonal antibodies. J Virol 62: 4703–4711
Doyle C, Sambrook J, Gething M-J (1986) Analysis of progressive deletions of the transmembrane and cytoplasmic domains of influenza hemagglutinin. J Cell Biol 103: 1193–1204
Dubay JW, Roberts SJ, Brody B, Hunter E (1992a) Mutations in the leucine zipper of the human immunodeficiency virus type 1 affect fusion and infectivity. J Virol 66: 4748–4756

Dubay JW, Roberts SJ, Hahn BH, Hunter E (1992b) Truncation of the human immunodeficiency virus type 1 glycoprotein cytoplasmic domain blocks virus infectivity. J Virol 66: 6616–6625
Earl PL, Moss B (1993) Mutational analysis of the assembly domain of the HIV-1 envelope glycoprotein. AIDS Res Hum Retroviruses 9: 589–594
Earl PL, Doms RW, Moss B (1990) Oligomeric structure of the human immunodeficiency virus type 1 envelope glycoprotein. Proc Natl Acad Sci USA 87: 648–652
Earl PL, Moss B, Doms R (1991) Folding, interaction with GRP78-BiP, assembly, and transport of the human immunodeficiency virus type 1 envelope protein. J Virol 65: 2047–2055
Einfeld D, Hunter E (1988) Oligomeric structure of a prototype retrovirus glycoprotein. Proc Natl Acad Sci USA 85: 8688–8692
Einfeld DA, Hunter E (1994) Expression of the TM protein of Rous sarcoma virus in the absence of SU shows that this domain is capable of oligomerization and intracellular transport. J Virol 68: 2513–2520
Ellerbrok H, D'Auriol L, Vaquero C, Sitbon M (1992) Functional tolerance of the human immunodeficiency virus type 1 envelope signal peptide to mutations in the amino-terminal and hydrophobic regions. J Virol 66: 5114–5118
Emi N, Friedmann T, Yee JK (1991) Pseudotype formation of murine leukemia virus with the G protein of vesicular stomatitis virus. J Virol 65: 1202–1207
Felkner RH, Roth MJ (1992) Mutational analysis of the N-linked glycosylation sites of the SU envelope protein of Moloney murine leukemia virus. J Virol 66: 4258–4264
Fenouillet E, Gluckman JC (1991) Effect of a glucosidase inhibitor on the bioactivity and immunoreactivity of human immunodeficiency virus type 1 envelope glycoprotein. J Gen Virol 72: 1919–1926
Fenouillet E, Gluckman JC (1992) Immunological analysis of human immunodeficiency virus type 1 envelope glycoprotein proteolytic cleavage. Virology 187: 825–828
Fenouillet E, Gluckman JC, Bahraoui E (1990) Role of N-linked glycans of envelope glycoproteins in infectivity of human immunodeficiency virus type 1. J Virol 64: 2841–2848
Fenouillet E, Jones I, Powel B, Schmitt D, Kieny MP, Gluckman JC (1993) Functional role of the glycan cluster of the human immunodeficiency virus type 1 transmembrane glycoprotein (gp41) ectodomain. J Virol 67: 150–160
Flynn GC, Chappell TG, Rothman JE (1989) Peptide binding and release by proteins implicated as catalysts of protein assembly. Science 245: 385–390
Flynn GC, Pohl J, Flocco MT, Rothman JE (1991) Peptide-binding specificity of the molecular chaperone BiP. Nature 353: 726–730
Freed EO, Risser R (1987) The role of envelope glycoprotein processing in murine leukemia virus infection. J Virol 61: 2852–2856
Freed EO, Myers DJ, Risser R (1989) Mutational analysis of the cleavage sequence of the human immunodeficiency virus type 1 envelope glycoprotein precursor gp160. J Virol 63: 4670–4675
Freed EO, Delwart EL, Buchschacher GLJ, Panganiban AT (1992) A mutation in the human immunodeficiency virus type 1 transmembrane glycoprotein gp41 dominantly interferes with fusion and infectivity. Proc Natl Acad Sci USA 89: 70–74
Freedman RB (1989) Protein disulfide isomerase: multiple roles in the modification of nascent secretory proteins. Cell 57: 1069–1072
Gabuzda DH, Lever A, Terwilliger E, Sodroski J (1992) Effects of deletions in the cytoplasmic domain on biological functions of human immunodeficiency virus type 1 glycoproteins. J Virol 66: 3306–3315
Gallagher PJ, Henneberry JM, Sambrook JF, Gething M-JH (1992) Glycosylation requirements for intracellular transport and function of the hemagglutinin of influenza virus. J Virol 66: 7136–7145
Gallaher WR, Ball JM, Garry RF, Griffin MC, Montelaro RC (1989) A general model for the transmembrane proteins of HIV and other retroviruses. AIDS Res Hum Retroviruses 5: 431–440
Garten W, Liner D, Rott R, Klenk H-D (1982) The cleavage site of the hemagglutinin of fowl plague virus. Virology 122: 186–190
Gawrisch K, Han KH, Yang JS, Bergelson LD, Ferretti JA (1993) Interaction of peptide fragment 828–848 of the envelope glycoprotein of human immunodeficiency virus type 1 with lipid bilayers. Biochemistry 32: 3112–3118
Gebhardt A, Bosch JV, Ziemiecki A, Friis R (1984) Rous sarcoma virus p19 and gp35 can be chemically crosslinked to high molecular weight complexes. J Mol Biol 174: 297–317
Gething M-J, Sambrook J (1992) Protein folding in the cell. Nature 355: 33–45
Geyer H, Holschbach C, Hunsmann G, Schneider J (1988) Carbohydrates of human immunodeficiency virus. Structures of oligosaccharides linked to the envelope glycoprotein 120. J Biol Chem 263: 11760–11767

Geyer R, Dabrowski J, Dabrowski U, Linder D, Schlüter M, Schott H-H, Stirm S (1990) Oligosaccharides at individual glycosylation sites in glycoprotein 71 of Friend murine leukemia virus. Eur J Biochem 187: 95–110

Gilbert JM, Hernandez LD, Chernov-Rogan T, White JM (1993) Generation of a water-soluble oligomeric ectodomain of the Rous sarcoma virus envelope glycoprotein. J Virol 67: 6889–6892

Gliniak BC, Kozak SL, Jones RT, Kabat D (1991) Disulfide bonding controls the processing of retroviral envelope glycoproteins. J Biol Chem 266: 22991–22997

Görlich D, Rapoport TA (1993) Protein translocation into proteoliposomes reconstituted from purified components of the endoplasmic reticulum membrane. Cell 75: 615–630

Görlich D, Prehn S, Hartmann E, Kalies K-U, Rapoport TA (1992) A mammalian homolog of SEC61p and SECYp is associated with ribosomes and nascent polypeptides during translocation. Cell 71: 489–503

Gorman JJ, Corino GL, Selleck PW (1990) Comparison of the positions and efficiency of cleavage activation of fusion protein precursors of virulent and avirulent strains of Newcastle disease virus: insights into the specificities of activating proteases. Virology 177: 339–351

Granowitz C, Colicelli J, Goff SP (1991) Analysis of mutations in the envelope gene of Moloney murine leukemia virus: separation of infectivity from superinfection resistance. Virology 183: 545–554

Gray KD, Roth M (1993) Mutational analysis of the envelope gene of Moloney murine leukemia virus. J Virol 67: 3489–3496

Gruters RA, Neefjes JJ, Tersmette M, De Goede REV, Huisman G, Miedema R, Ploegh HL (1987) Interference with HIV-induced syncytium formation and viral infectivity by inhibitors of trimming glucosidase. Nature 330: 74–77

Guguen M, Biddison WE, Long EO (1994) T cell recognition of an HLA-A2 restricted epitope derived from a cleaved signal sequence. J Exp Med 180: 1989–1994

Haffar OK, Dowbenko DJ, Berman PW (1991) The cytoplasmic tail of HIV-1 gp160 contains regions that associate with cellular membranes. Virology 180: 439–441

Hallenberger S, Bosch V, Angliker H, Shaw E, Klenk HD, Garten W (1992) Inhibition of furin-mediated cleavage activation of HIV-1 glycoprotein gp160. Nature 360: 358–361

Hammond C, Helenius A (1993) A chaperone with a sweet tooth. Curr Biol 3: 884–886

Hammond C, Helenius A (1994a) Folding of VSV G protein: sequential interaction with BiP and calnexin. Science 266: 456–458

Hammond C, Helenius A (1994b) Quality control in the secretory pathway: retention of a misfolded viral membrane glycoprotein involves cycling between the ER, intermediate compartment, and Golgi apparatus. J Cell Biol 126: 41–52

Hammond C, Braakman I, Helenius A (1994) Role of N-linked oligosaccharide recognition, glucose trimming, and calnexin in glycoprotein folding and quality control. Proc Natl Acad Sci USA 91: 913–917

Hansen JE, Clausen H, Hu SL, Nielsen JO, Olofsson S (1992) An O-linked carbohydrate neutralization epitope of HIV-1 gp 120 is expressed by HIV-1 env gene recombinant vaccinia virus. Arch Virol 126: 11–20

Hardwick JM, Shaw KES, Wills JW, Hunter E (1986) Amino-terminal deletion mutants of the Rous sarcoma virus glycoprotein do not block signal peptide cleavage but can block intracellular transport. J Cell Biol 103: 829–838

Hatsuzawa K, Hosaka M, Nakagawa T, Nagase M, Shoda A, Murakami K, Nakayama K (1990) Structure and expression of mouse furin, a yeast Kex2-related protease: lack of processing of coexpressed prorenin in GH_4C_1 cells. J Biol Chem 265: 22075–22078

Heard JM, Danos O (1991) An amino-terminal fragment of the Friend leukemia virus envelope glycoprotein binds the ecotropic receptor. J Virol 65: 4026–4032

Hendershot L, Bole D, Köhler G, Kearney JF (1987) Assembly and secretion of heavy chains that do not associate posttranslationally with immunoglobulin heavy chain-binding protein. J Cell Biol 104: 761–767

High S, Görlich D, Wiedmann M, Rapoport TA, Dobberstein B (1991) The identification of proteins in the proximity of signal-anchor sequences during targeting to and insertion into the membrane of the ER. J Cell Biol 113: 35–44

Horimoto T, Nakayama K, Smeekens SP, Kawaoka Y (1994) Proprotein-processing endoproteases PC6 and furin both activate hemagglutinin of virulent avian influenza viruses. J Virol 68: 6074–6078

Hosaka M, Nagahama M, Kim W-S, Watanabe T, Hatsuzawa K, Ikemizu J, Murakami K, Nakayama K (1991) Arg-X-Lys/Arg-Arg motif as a signal for precursor cleavage catalyzed by furin within the constitutive secretory pathway. J Biol Chem 266: 12127–12130

Hoxie JA, Fitzharris TP, Youngbar PR, Matthews DM, Rackowski JL, Radka SF (1987) Nonrandom association of cellular antigens with HTLV-III virions. Hum Immunol 18: 39–52

Hubbard TJP, Sander C (1991) The role of heat-shock and chaperone proteins in protein folding: possible molecular mechanisms. Protein Eng 4: 711–717

Hunter E, Swanstrom R (1990) Retrovirus envelope glycoproteins. In: Swanstrom R, Vogt PK (eds) Retroviruses. Strategies to replication. Springer, Berlin Heidelberg New York, pp 187–253 (Current topics in microbiology and immunology, vol 157)

Hunter E, Hill E, Hardwick M, Bhown A, Schwartz DE, Tizard R (1983) Complete sequence of the Rous sarcoma virus env gene: identification of structural and functional regions of its product. J Virol 46: 920–936

Hunziker W, Harter C, Matter K, Mellman I (1991) Basolateral sorting in MDCK cells requires a distinct cytoplasmic domain determinant. Cell 66: 907–920

Jackson MR, Cohen-Doyle MF, Peterson PA, Williams DB (1994) Regulation of MHC class I transport by the molecular chaperone, calnexin (p88, IP90). Science 263: 384–387

Johnston PB, Dubay JW, Hunter E (1993) Truncations of the simian immunodeficiency virus transmembrane protein confer expanded host range by removing a block to virus entry into cells. J Virol 67: 3077–3086

Kamps CA, Lin YC, Wong PKY (1991) Oligomerization and transport of the envelope protein of Moloney murine leukemia virus-TB and of ts–1, a neurovirulent temperature-sensitive mutant of MoMuLV-TB. Virology 184: 687–694

Kayman SC, Kopelman R, Projan S, Kinney DM, Pinter A (1991) Mutational analysis of N-linked glycosylation sites of Friend murine leukemia virus envelope protein. J Virol 65: 5323–5332

Kearse KP, Williams DB, Singer A (1994) Persistence of glucose residues on core oligosaccharides prevents association of TCRα and TCRβ proteins with calnexin and results specifically in accelerated degradation of nascent TCRα proteins within the endoplasmic reticulum. EMBO J 13: 3678–3686

Kido H, Kamoshita K, Fukutomi A, Katunuma N (1993) Processing protease for gp160 human immunodeficiency virus type I envelope glycoprotein precursor in human $T4^+$ lymphocytes. J Biol Chem 268: 13406–13413

Kiefer MC, Tucker JE, Joh R, Landsberg KE, Saltman D, Barr PJ (1991) Identification of a second human subtilisin-like protease gene in the fes/fps region of chromosome 15. DNA Cell Biol 10: 757–769

Kilpatrick DR, Srinivas SR, Stephens EB, Compans RW (1987) Effects of deletion of the cytoplasmic domain upon surface expression and membrane stability of a viral envelope glycoprotein. J Biol Chem 262: 16116–16121

Kim PS, Bole D, Arvan P (1992) Transient aggregation of nascent thyroglobulin in the endoplasmic reticulum: relationship to the molecular chaperone, BiP. J Cell Biol 118: 541–549

Korner J, Chun J, O'Bryan L, Axel R (1991) Prohormone processing in *Xenopus* oocytes: characterization of cleavage signals and cleavage enzymes. Proc Natl Acad Sci USA 88: 11393–11397

Krijnse-Locker J, Ericsson M, Rottier PJM, Griffiths G (1994) Characterization of the budding compartment of mouse hepatitis virus: evidence that transport from the RER to the Golgi complex requires only one vesicular transport step. J Cell Biol 124: 55–70

LaMantia M, Lennarz WJ (1993) The essential function of yeast protein disulfide isomerase does not reside in its isomerase activity. Cell 74: 899–908

Landau NR, Page KA, Littman DR (1991) Pseudotyping with human T-cell leukemia virus type I broadens the human immunodeficiency virus host range. J Virol 65: 162–169

Lasky LA, Nakamura G, Smith DH, Fennie C, Shimasaki DH, Patzer E, Berman P, Gregory T, Capon DJ (1987) Delineation of a region of the human immunodeficiency virus type 1 gp120 glycoprotein critical for interaction with CD4. Cell 50: 975–985

Le Bivic A, Quaroni A, Nichols B, Rodriguez-Boulan E (1990a) Biogenetic pathways of plasma membrane proteins in Caco-2, a human intestinal epithelial cell line. J Cell Biol 111: 1351–1361

Le Bivic A, Sambury Y, Mostov K, Rodriguez-Boulan E (1990b) Vectorial targeting of an endogenous apical membrane sialoglycoprotein and uvomorulin in MDCK cells. J Cell Biol 110: 1533–1539

Leamnson RN, Shander MHN, Halpern MS (1977) A structural protein complex in Moloney leukemia virus. Virology 76: 437–439

Lee W-R, Syu W-J, Du B, Matsuda M, Tan S, Wolf A, Lee T-H, Essex M (1992a) Nonrandom distribution of gp120 N-linked glycosylation sites important for infectivity of human immunodeficiency virus type 1. Proc Natl Acad Sci USA 89: 2212–2217

Lee W-R, Yu X-F, Syu W-J, Essex M, Lee T-H (1992b) Mutational analysis of conserved N-linked glycosylation sites of human immunodeficiency virus type 1 gp41. J Virol 66: 1799–1803
Leonard CK, Spellman MW, Riddle L, Harris RJ, Thomas JN, Gregory TJ (1990) Assignment of intrachain disulfide bonds and characterization of potential glycosylation sites of the type 1 recombinant human immunodeficiency virus envelope glycoprotein (gp120) expressed in Chinese hamster ovary cells. J Biol Chem 265: 10373–10382
Levy JA (1977) Murine xenotropic type C viruses. III. Phenotypic mixing with avian leukosis and sarcoma viruses. Virology 77: 811–825
Linder M, Linder D, Hahnen J, Schott HH, Stirm S (1992) Localization of the intrachain disulfide bonds of the envelope glycoprotein 71 from Friend murine leukemia virus. Eur J Biochem 203: 65–73
Linder M, Wenzel V, Linder D, Stirm S (1994) Structural elements in glycoprotein 70 from polytropic Friend mink cell focus-inducing virus and glycoprotein 71 from ecotropic Friend murine leukemia virus, as defined by disulfide-bonding pattern and limited proteolysis. J Virol 68: 5133–5141
Livingston DM, Howard T, Spence C (1976) Identification of infectious virions which are vesicular stomatitis virus pseudotypes of murine type C virus. Virology 70: 432–439
Lodge R, Gottlinger H, Gabuzda D, Cohen EA, Lemay G (1994) The intracytoplasmic domain of gp41 mediates polarized budding of human immunodeficiency virus type 1 in MDCK cells. J Virol 68: 4857–4861
Lodish HF, Kong N, Hirani S, Rasmussen J (1987) A vesicular intermediate in the transport of hepatoma secretory proteins from the rough endoplasmic reticulum to the Golgi complex. J Cell Biol 104: 221–230
Lusson J, Vieau D, Hamelin J, Day R, Chretien M, Seidah NG (1993) cDNA structure of the mouse and rat subtilisin/kexin-like PC5: a candidate proprotein convertase expressed in endocrine and nonendocrine cells. Proc Natl Acad Sci USA 90: 6691–6695
Machamer CE, Rose JK (1988) Vesicular stomatitis virus G proteins with altered glycosylation sites display temperature-sensitive intracellular transport and are subject to aberrant intermolecular disulfide bonding. J Biol Chem 263: 5955–5960
Machamer CE, Florkiewicz RZ, Rose JK (1985) A single N-linked oligosaccharide at either of the two normal sites is sufficient for transport of vesicular stomatitis virus G protein to the cell surface. Mol Cell Biol 5: 3074–3083
Machamer CE, Doms RW, Bole DG, Helenius A, Rose JK (1990) Heavy chain binding protein recognizes incompletely disulfide-bonded forms of vesicular stomatitis virus G protein. J Biol Chem 265: 6879–6883
Matano T, Odawara T, Ohshima M, Yoshikura H, Iwamoto A (1993) trans-dominant interference with virus infection at two different stages by a mutant envelope protein of Friend murine leukemia virus. J Virol 67: 2026–2033
Matter K, Brauchbar M, Bucher K, Hauri H-P (1990) Sorting of endogenous plasma membrane proteins occurs from two sites in cultured human intestinal epithelial cells (Caco-2). Cell 60: 429–437
McGuire TC, Knowles DPJ, Davis WC, Brassfield AL, Stem TA, Cheevers WP (1992) Transmembrane protein oligomers of caprine arthritis-encephalitis lentivirus are immunodominant in goats with progressive arthritis. J Virol 66: 3247–3250
Mellman I, Simons K (1992) The Golgi complex: in vitro veritas? Cell 68: 829–840
Melnick J, Dul JL, Argon Y (1994) Sequential interaction of the chaperones BiP and GRP94 with immunoglobulin chains in the endoplasmic reticulum. Nature 370: 373–375
Mizouchi T, Matthews TJ, Kato M, Hamako J, Titani K, Solomon J, Feizi T (1990) Diversity of oligosaccharide structures on the envelope glycoprotein of human immunodeficiency virus 1 from the lymphoblastoid cell line H9: presence of complex-type oligosaccharides with bisecting N-acetylglucosamine residues. J Biol Chem 265: 8519–8524
Molloy SS, Bresnahan PA, Leppla SH, Klimpel KR, Thomas G (1992) Human furin is a calcium-dependent serine endoprotease that recognizes the sequence Arg-X-X-Arg and efficiently cleaves anthrax toxin protective antigen. J Biol Chem 267: 16396–16402
Montelaro RC, Sullivan SJ, Bolognesi DP (1978) An analysis of type-C retrovirus polypeptides and their associations in the virion. Virology 84: 19–31
Morgan RA, Nussbaum O, Muenchau DD, Shu L, Couture L, Anderson WF (1993) Analysis of the functional and host range-determining regions of the murine ecotropic and amphotropic retrovirus envelope proteins. J Virol 67: 4712–4721
Morikawa Y, Moore JP, Fenouillet E, Jones IM (1992) Complementation of human immunodeficiency virus glycoprotein mutations in trans. J Gen Virol 73: 1907–1913

Morikawa Y, Barsov E, Jones I (1993) Legitimate and illegitimate cleavage of human immunodeficiency virus glycoproteins by furin. J Virol 67: 3601–3604
Morjana NA, Gilbert HF (1991) Effect of protein and peptide inhibitors on the activity of protein disulfide isomerase. Biochemistry 30: 4985–4990
Mulligan MJ, Yamshchikov GV, Ritter GD, Gao F, Jin MJ, Nail CD, Hahn BH, Compans RC (1992) Cytoplasmic domain truncation enhances fusion activity by the exterior glycoprotein complex of human immunodeficiency virus type 2 in selected cell types. J Virol 66: 3971–3975
Müsch A, Wiedmann M, Rapoport TA (1992) Yeast Sec proteins interact with polypeptides traversing the endoplasmic reticulum membrane. Cell 69: 343–352
Nagai Y (1993) Protease-dependent virus tropism and pathogenicity. Trends Microbiol 1: 81–87
Ng DT, Watowich SS, Lamb RA (1992) Analysis in vivo of GRP78-BiP/substrate interactions and their role in induction of the GRP78-BiP gene. Mol Biol Cell 3: 143–155
Ng VL, Wood TG, Arlinghaus RB (1982) Processing of the env gene products of MuLV. J Gen Virol 59: 329–343
Nicchitta CV, Blobel G (1993) Lumenal proteins of the mammalian endoplasmic reticulum are required to complete protein translocation. Cell 73: 989–998
Noiva R, Lennarz WJ (1992) Protein disulfide isomerase: a multifunctional protein resident in the lumen of the endoplasmic reticulum. J Biol Chem 267: 3553–3556
Noiva R, Kimura H, Roos J, Lennarz WJ (1991) Peptide binding by protein disulfide isomerase, a resident protein of the endoplasmic reticulum. J Biol Chem 266: 19645–19649
Noiva R, Freedman RB, Lennarz WJ (1993) Peptide binding to protein disulfide isomerase occurs at a site distinct from the active site. J Biol Chem 268: 19210–19217
Oda K, Ikeda M, Tsuji E, Sohda M, Takami N, Misumi Y, Ikehara Y (1991) Sequence requirements for proteolytic cleavage of precursors with paired basic amino acids. Biochem Biophys Res Commun 179: 1181–1186
Ohnishi Y, Shioda T, Nakayama K, Iwata S, Gotoh B, Hamaguchi M, Nagai Y (1994) A furin-defective cell line is able to process correctly the gp160 of human immunodeficiency virus type 1. J Virol 68: 4075–4079
Olshevsky U, Helseth E, Furman C, Li J, Haseltine W, Sodroski J (1990) Identification of individual human immunodeficiency virus type 1 gp120 amino acids important for CD4 receptor binding. J Virol 64: 5701–5707
Otsu M, Omura F, Yoshimori T, Kikuchi M (1994) Protein disulfide isomerase associates with misfolded human lysozyme in vitro. J Biol Chem 269: 6874–6877
Ott D, Rein A (1992) Basis for receptor specificity of nonecotropic murine leukemia virus surface glycoprotein $gp70^{SU}$. J Virol 66: 4632–4638
Ou WJ, Cameron PH, Thomas DY, Bergeron JJ (1993) Association of folding intermediates of glycoproteins with calnexin during protein maturation. Nature 364: 771–776
Owens RJ, Compans RW (1990) The human immunodeficiency virus type 1 envelope glycoprotein precursor acquires aberrant intermolecular disulfide bonds that may prevent normal proteolytic processing. Virology 179: 827–833
Owens RJ, Dubay JW, Hunter E, Compans RW (1991) Human immunodeficiency virus envelope protein determines the site of virus release in polarized epithelial cells. Proc Natl Acad Sci USA 88: 3987–3991
Owens RJ, Burke C, Rose JK (1994) Mutations in the membrane-spanning domain of the human immunodeficiency virus envelope glycoprotein that affect fusion activity. J Virol 68: 570–574
Page KA, Landau NR, Littman DR (1990) Construction and use of a human immunodeficiency virus vector for analysis of virus infectivity. J Virol 64: 5270–5276
Pal R, Hoke GM, Sarngadharan MG (1989a) Role of the oligosaccharides in the processing and maturation of envelope glycoproteins of human immunodeficiency virus type 1. Proc Natl Acad Sci USA 86: 3384–3388
Pal R, Kalyanaraman VS, Hoke GM, Sarngadharan MG (1989b) Processing and secretion of envelope glycoproteins of human immunodeficiency virus type 1 in the presence of trimming glucosidase inhibitor deoxynojirimycin. Intervirology 30: 27–35
Patarca R, Haseltine WA (1984) Similarities among retrovirus proteins. Nature 312: 496
Paterson RG, Lamb RA (1987) Ability of the hydrophobic fusion-related external domain of a paramyxovirus F protein to act as a membrane anchor. Cell 48: 441–452
Pepinsky RB, Cappiello D, Wilkowski C, Vogt VM (1980) Chemical crosslinking of proteins in avian sarcoma and leukemia viruses. Virology 102: 205–210

Perez LG, Hunter E (1987) Mutations within the proteolytic cleavage site of the Rous sarcoma virus glycoprotein that block processing to gp85 and gp37. J Virol 61: 1609–1614
Perez LG, Davis GL, Hunter E (1987) Mutants of the Rous sarcoma virus envelope glycoprotein that lack the transmembrane anchor and/or cytoplasmic domains: analysis of intracellular transport and assembly into virions. J Virol 61: 2981–2988
Persson R, Pettersson RF (1991) Formation and intracellular transport of a heterodimeric viral spike protein complex. J Cell Biol 112: 257–266
Pfeiffer S, Fuller SD, Simons K (1985) Intracellular sorting and basolateral appearance of the G protein of vesicular stomatitis virus in Madin-Darby canine kidney cells. J Cell Biol 101: 470–476
Pinter A, Fleissner E (1977) The presence of disulfide-linked gp70-p15(E) complexes in AKR MuLV. Virology 88: 222–227
Pinter A, Fleissner E (1979) Characterization of oligomeric complexes of murine and feline leukemia virus envelope and core components formed upon crosslinking. J Virol 30: 157–165
Pinter A, Honnen WJ (1983) Comparison of structural domains of gp70s of ecotropic AKV and its dualtropic recombinant MCF-247. Virology 129: 40–50
Pinter A, Honnen WJ (1984) Characterization of structural and immunological properties of specific domains of Friend ecotropic and dualtropic murine leukemia virus gp70s. J Virol 49: 452–458
Pinter A, Honnen WJ (1988) O-linked glycosylation of retroviral envelope gene products. J Virol 62: 1016–1021
Pinter A, Lieman-Hurwitz J, Fleissner E (1978) The nature of the association between the murine leukemia virus envelope proteins. Virology 91: 345–351
Pinter A, Honnen WJ, Li JS (1984) Studies with inhibitors of oligosaccharide processing indicate a functional role for complex sugars in the transport and proteolysis of Friend mink cell focus-inducing murine leukemia virus envelope proteins. Virology 136: 196–210
Pinter A, Honnen WJ, Tilley SA, Bona C, Zaghouani H, Gorny MK, Zolla-Pazner S (1989) Oligomeric structure of gp41, the transmembrane protein of human immunodeficiency virus type 1. J Virol 63: 2674–2679
Pollard SR, Rosa MD, Rosa JJ, Wiley DC (1992) Truncated variants of gp120 bind CD4 with high affinity and suggest a minimum CD4 binding region. EMBO J 11: 585–591
Pollok BA, Anker R, Eldridge P, Hendershot L, Levitt D (1987) Molecular basis of the cell-surface expression of immunoglobulin μ chain without light chain in human B lymphocytes. Proc Natl Acad Sci USA 84: 9199–9203
Poruchynsky MS, Atkinson PH (1988) Primary sequence domains required for retention of rotavirus VP7 in the endoplasmic reticulum. J Cell Biol 107: 1697–1706
Prill V, Lehmann L, von Figura K, Peters C (1993) The cytoplasmic tail of lysosomal acid phosphatase contains overlapping but distinct signals for basolateral sorting and rapid internalization in polarized MDCK cells. EMBO J 12: 2181–2193
Puig A, Lyles MM, Noiva R, Gilbert HF (1994) The role of the thiol/disulfide centers and peptide binding site in the chaperone and anti-chaperone activities of protein disulfide isomerase. J Biol Chem 269: 19128–19135
Racevskis J, Sarkar NH (1980) Murine mammary tumor virus structural protein interactions: formation of oligomeric complexes with cleavable cross-linking agents. J Virol 35: 937–948
Ragheb JA, Anderson WF (1994) Uncoupled expression of Moloney murine leukemia virus envelope polypeptides SU and TM: a functional analysis of the role of TM domains in viral entry. J Virol 68: 3207–3219
Rajagopalan S, Xu Y, Brenner MB (1994) Retention of unassembled components of integral membrane proteins by calnexin. Science 263: 387–390
Ratner L, Vander Heyden N, Dedera D (1991) Inhibition of HIV and SIV infectivity by blockade of α-glucosidase activity. Virology 181: 180–192
Rehemtulla A, Barr PJ, Rhodes CJ, Kaufman RJ (1993) PACE4 is a member of the mammalian propeptidase family that has overlapping but not identical substrate specificity to PACE. Biochemistry 32: 11586–11590
Rein A, Mirro J, Haynes JG, Ernst SM, Nagashima K (1994) Function of the cytoplasmic domain of a retroviral transmembrane protein: p15E-p2E cleavage activates the membrane fusion capability of the murine leukemia virus Env protein. J Virol 68: 1773–1781
Rey M-A, Krust B, Laurent AG, Montagnier L, Hovanessian AG (1989) Characterization of human immunodeficiency virus type 2 envelope glycoproteins: dimerization of the glycoprotein precursor during processing. J Virol 63: 647–658

Rey M-A, Laurent AG, McClure J, Krust B, Montagnier L, Hovanessian AG (1990) Transmembrane envelope glycoproteins of human immunodeficiency virus type 2 and simian immunodeficiency virus SIV-mac exist as homodimers. J Virol 64: 922–926
Rindler MJ, Ivanov IE, Plesker H, Rodriguez-Boulan E, Sabatini DD (1984) Viral glycoproteins destined for apical or basolateral plasma membrane domains traverse the same Golgi apparatus during their intracellular transport in doubly infected Madin-Darby canine kidney cells. J Cell Biol 98: 1304–1319
Rindler MJ, Ivanov IE, Plesker H, Sabatini DD (1985) Polarized delivery of viral glycoproteins to the apical and basolateral plasma membranes of Madin-Darby canine kidney cells infected with temperature sensitive virus. J Cell Biol 100: 136–151
Ritter GD Jr, Mulligan MJ, Lydy SL, Compans RW (1993) Cell fusion activity of the simian immunodeficiency virus envelope protein is modulated by the intracytoplasmic domain. Virology 197: 255–264
Roberts PC, Garten W, Klenk H-D (1993) Role of conserved glycosylation sites in maturation and transport of influenza A virus hemagglutinin. J Virol 67: 3048–3060
Roth MG, Srinivas RV, Compans RW (1983) Basolateral maturation of retroviruses in polarized epithelial cells. J Virol 45: 1065–1073
Roth RA, Pierce SB (1987) In vivo cross-linking of protein disulfide isomerase to immunoglobulins. Biochemistry 26: 4179–4182
Rothman JE (1989) Polypeptide chain binding proteins: catalysts of protein folding and related processes. Cell 59: 591–601
Salzwedel K, Johnston PB, Roberts SJ, Dubay JW, Hunter E (1993) Expression and characterization of glycophospholipid-anchored human immunodeficiency virus type 1 envelope glycoproteins. J Virol 67: 5279–5288
Sanders SL, Whitfield KM, Vogel JP, Rose MD, Schekman RW (1992) Sec61p and BiP directly facilitate polypeptide translocation into the ER. Cell 69: 353–365
Schalken JA, Roebroek AJM, Oomen PPCA, Wagenaar SS, Debruyne FMJ, Bloemers HPJ, Van de Ven WJM (1987) fur gene expression as a discriminating marker for small cell and nonsmall cell lung carcinomas. J Clin Invest 80: 1545–1549
Schawaller M, Smith GE, Skehel JJ, Wiley DC (1989) Studies with crosslinking reagents on the oligomeric structure of the env glycoprotein of HIV. Virology 172: 367–369
Schols D, Pauwels R, Desmyter J, De Clerq E (1992) Presence of class I histocompatibility DR proteins on the envelope of human immunodeficiency virus demonstrated by FACS analysis. Virology 189: 374–376
Schulz TF, Jameson BA, Lopalco L, Siccardi AG, Weiss RA, Moore JP (1992) Conserved structural features in the interaction between retroviral surface and transmembrane proteins? AIDS Res Human Retrovirus 8: 1571–1580
Schweizer A, Matter K, Ketcham CM, Hauri H-P (1991) The isolated ER-Golgi intermediate compartment exhibits properties that are different from ER and cis-Golgi. J Cell Biol 113: 45–54
Segal MS, Bye JM, Sambrook J, Gething M-J (1992) Disulfide bond formation during the folding of influenza virus hemagglutinin. J Cell Biol 118: 227–244
Shin J, Lee S, Strominger JL (1993) Translocation of TCRα chains into the lumen of the endoplasmic reticulum and their degradation. Science 259: 1901–1904
Siezen RJ, Creemers JWM, Van de Ven WJM (1994) Homology modelling of the catalytic domain of human furin – a model for the eukaryotic subtilisin-like proprotein convertases. Eur J Biochem 222: 255–266
Simon SM, Blobel G (1991) A protein-conducting channel in the endoplasmic reticulum. Cell 65: 371–380
Simon SM, Blobel G (1992) Signal peptides open protein-conducting channels in *E. coli*. Cell 69: 677–684
Singer SJ, Maher PA, Yaffe MP (1987) On the transfer of integral proteins into membranes. Proc Natl Acad Sci USA 84: 1960–1964
Smeekens SP, Montag AG, Thomas G, Albiges-Rizo C, Carroll R, Benig M, Phillips LA, Martin S, Ohagi S, Gardner P, et al. (1992) Proinsulin processing by the subtilisin-related proprotein convertases furin, PC2, and PC3. Proc Natl Acad Sci USA 89: 8822–8826
Spies CP, Compans RW (1994) Effects of cytoplasmic domain length on cell surface expression and syncytium-forming capacity of the simian immunodeficiency virus envelope glycoprotein. Virology 203: 8–19
Spies CP, Ritter GD Jr, Mulligan MJ, Compans RW (1994) Truncation of the cytoplasmic domain of the simian immunodeficiency virus envelope glycoprotein alters the conformation of the external domain. J Virol 68: 585–591

Srinivas SK, Srinivas RV, Anantharamaiah GM, Segrest JP, Compans RW (1992) Membrane interactions of synthetic peptides corresponding to amphipathic helical segments of the human immunodeficiency virus type-1 envelope glycoprotein. J Biol Chem 267: 7121–7127
Stein BS, Engleman EG (1990) Intracellular processing of the gp160 envelope precursor. J Biol Chem 265: 2640–2649
Steiner DF, Smeekens SP, Ohagi S, Chan SJ (1992) The new enzymology of precursor processing endoproteases. J Biol Chem 267: 23435–23438
Stieneke-Grober A, Vey M, Angliker H, Shaw E, Thomas G, Roberts C, Klenk HD, Garten W (1992) Influenza virus hemagglutinin with multibasic cleavage site is activated by furin, a subtilisin-like endoprotease. EMBO J 11: 2407–2414
Stirzaker SC, Both GW (1989) The signal peptide of the rotavirus glycoprotein VP7 is essential for its retention in the ER as an integral membrane protein. Cell 56: 741–747
Syu W-J, Lee W-R, Du B, Yu Q-C, Essex M, Lee T-H (1991) Role of conserved gp41 cysteine residues in the processing of human immunodeficiency virus envelope precursor and viral infectivity. J Virol 65: 6349–6352
Takahashi S, Kasai K, Hatsuzawa K, Kitamura N, Misumi Y, Ikehara Y, Murakami K, Nakayama K (1993) A mutation of furin causes the lack of precursor-processing activity in human colon carcinoma LoVo cells. Biochem Biophys Res Commun 195: 1019–1026
Takemoto LJ, Fox CF, Jensen FC, Elder JH, Lerner RA (1978) Nearest-neighbor interactions of the major RNA tumor virus glycoprotein on murine cell surfaces. Proc Natl Acad Sci USA 75: 3644–3648
Tatu U, Braakman I, Helenius A (1993) Membrane glycoprotein folding, oligomerization and intracellular transport: effects of dithiothreitol in living cells. EMBO J 12: 2151–2157
Thomas DC, Brewer CB, Roth MG (1993) Vesicular stomatitis virus glycoprotein contains a dominant cytoplasmic basolateral sorting signal critically dependent upon a tyrosine. J Biol Chem 268: 3313–3320
Thomas G, Thorne BA, Thomas L, Allen RG, Hruby DE, Fuller R, Thorner J (1988) Yeast KEX2 endopeptidase correctly cleaves a neuroendocrine prohormone in mammalian cells. Science 241: 226–230
Thomas L, Cooper A, Bussey H, Thomas G (1990) Yeast KEX1 protease cleaves a prohormone processing intermediate in mammalian cells. J Biol Chem 265: 10821–10824
Thrift RN, Andrews DW, Walter P, Johnson AE (1991) A nascent membrane protein is located adjacent to ER membrane proteins throughout its integration and translation. J Cell Biol 112: 809–821
Tsai W-P, Oroszlan S (1988) Novel glycosylation pathways of retroviral envelope proteins identified with avian reticuloendotheliosis virus. J Virol 62: 3167–3174
Tschachler E, Buchow H, Gallo RC, Reitz MS Jr (1990) Functional contribution of cysteine residues to the human immunodeficiency virus type I envelope. J Virol 64: 2250–2259
Tucker SP, Srinivas RV, Compans RW (1991) Molecular domains involved in oligomerization of the Friend murine leukemia virus envelope glycoprotein. Virology 185: 710–720
Venable RM, Pastor RR, Brooks BR, Carson FW (1989) Theoretically determined three-dimensional structures for amphipathic segments of the HIV-1 gp41 envelope glycoprotein. AIDS Res Hum Retroviruses 5: 7–22
Veronese FM, DeVico AL, Copeland TD, Oroszlan S, Gallo RC, Sarngadharan MG (1985) Characterization of gp41 as the transmembrane protein encoded by the HTLV-III/LAV envelope gene. Science 229: 1402–1405
Vile RG, Schulz TF, Danos OF, Collins MKL, Weiss RA (1991) A murine cell line producing HTLV-1 pseudotype virions carrying a selectable marker. Virology 180: 420–424
Vogel JP, Misra LM, Rose MD (1990) Loss of BiP/GRP78 function blocks translocation of secretory proteins in yeast. J Cell Biol 110: 1885–1895
von Heijne G (1985) Signal sequences: the limits of variation. J Mol Biol 184: 99–105
von Heijne G (1986) Towards a comparative anatomy of N-terminal topogenic sequences. J Mol Biol 189: 239–242
Walker JA, Molloy SS, Thomas G, Sakaguchi T, Yoshida T, Chambers TM, Kawaoka Y (1994) Sequence specificity of furin, a proprotein-processing endoprotease, for the hemagglutinin of a virulent avian influenza virus. J Virol 68: 1213–1218
Wandinger-Ness A, Bennett MK, Antony C, Simons K (1990) Distinct transport vesicles mediate the delivery of plasma membrane proteins to the apical and basolateral domains of MDCK cells. J Cell Biol 111: 987–1000

Watanabe T, Nakagawa T, Ikemizu J, Nagahama M, Murakami K, Nakayama K (1992) Sequence requirements for precursor cleavage within the constitutive secretory pathway. J Biol Chem 267: 8270–8274
Watanabe T, Murakami K, Nakayama K (1993) Positional and additive effects of basic amino acids on processing of precursor proteins within the constitutive secretory pathway. FEBS Lett 320: 215–218
Weiss CD, White JM (1993) Characterization of stable Chinese hamster ovary cells expressing wild-type, secreted, and glycosylphosphatidylinositol-anchored human immunodeficiency virus type 1 envelope glycoprotein. J Virol 67: 7060–7066
Weiss CD, Levy JA, White JM (1990) Oligomeric organization of gp120 on infectious human immunodeficiency virus type 1 particles. J Virol 64: 5674–5677
Weiss RA (1993) Cellular receptors and viral glycoproteins involved in retrovirus entry. In Levy J A.(ed) The retroviridae, vol 2. Plenum, New York, pp 1–108
Weiss RA, Wong AL (1977) Phenotypic mixing between avian and mammalian RNA tumor viruses: I. Envelope pseudotypes of Rous sarcoma virus. Virology 76: 826–834
Weiss RA, Boettiger D, Murphy HM (1977) Pseudotypes of avian sarcoma viruses with the envelope properties of vesicular stomatitis virus. Virology 76: 808–825
Weissman JS, Kim PS (1993) Efficient catalysis of disulphide bond rearrangements by protein disulphide isomerase. Nature 365: 185–188
Wild C, Oas T, McDanal C, Bolognesi D, Matthews T (1992) A synthetic peptide inhibitor of human immunodeficiency virus replication: correlation between solution structure and viral inhibition. Proc Natl Acad Sci USA 89: 10537–10541
Wild C, Dubay JW, Greenwell T, Baird T Jr, Oas TG, McDanal C, Hunter E, Matthews T (1994) Propensity for a leucine zipper-like domain of human immunodeficiency virus type 1 gp41 to form oligomers correlates with a role in virus-induced fusion rather than assembly of the glycoprotein complex. Proc Natl Acad Sci USA 91: 12676–12680
Wilk T, Pfeiffer T, Bosch V (1992) Retained in vitro infectivity and cytopathogenicity of HIV-1 despite truncation of the C-terminal tail of the env gene product. Virology 189: 167–177
Willey RL, Bonifacino JS, Potts BJ, Martin MA, Klausner RD (1988) Biosynthesis, cleavage, and degradation of the human immunodeficiency syndrome virus 1 envelope glycoprotein gp160. Proc Natl Acad Sci USA 85: 9580–9590
Wills JW, Srinivas RV, Hunter E (1984) Mutations of the Rous sarcoma virus env gene that affect the transport and subcellular location of the glycoprotein products. J Cell Biol 99: 2011–2023
Wilson C, Reitz MS, Okayama H, Eiden MV (1989) Formation of infectious hybrid virions with gibbon ape leukemia virus and human T-cell leukemia virus retroviral envelope glycoproteins and the gag and pol proteins of Moloney murine leukemia virus. J Virol 63: 2374–2378
Witte ON, Tsulamoto-Adey A, Weissman IL (1977) Cellular maturation of oncornavirus glycoproteins: topological arrangement of precursor and product forms in cellular membranes. Virology 76: 539–553
Yokode M, Pathak RK, Hammer RE, Brown MS, Goldstein JL, Anderson RGW (1992) Cytoplasmic sequence required for basolateral targeting of LDL receptor in livers of transgenic mice. J Cell Biol 117: 39–46
Young JA, Bates P, Willert K, Varmus HE (1990) Efficient incorporation of human CD4 into avian leukosis virus particles. Science 250: 1421–1423
Yu X, Yuan X, Matsuda Z, Lee T-H, Essex M (1992) The matrix protein of human immunodeficiency virus type 1 is required for incorporation of viral envelope protein into mature virions. J Virol 66: 4966–4971
Yu X, Xin Y, McLane MF, Lee T-H, Essex M (1993) Mutations in the cytoplasmic domain of human immunodeficiency virus type 1 transmembrane protein impair the incorporation of Env proteins into mature virions. J Virol 67: 213–221
Zavada J (1982) The pseudotypic paradox. J Gen Virol 63: 15–24
Zavadova Z, Zavada J, Weiss RA (1977) Unilateral phenotypic mixing of envelope antigens between togaviruses and vesicular stomatitis or avian RNA tumor virus. J Gen Virol 37: 557–567
Zhu Z, Chen SSL, Huang AS (1990) Phenotypic mixing between human immunodeficiency virus and vesicular stomatitis virus or herpes simplex virus. J AIDS 3: 215–219
Zingler K, Littman DR (1993) Truncation of the cytoplasmic domain of the simian immunodeficiency virus envelope glycoprotein increases Env incorporation into particles and fusogenicity and infectivity. J Virol 66: 2824–2831

RNA Packaging

R. BERKOWITZ[1], J. FISHER[2], and S.P. GOFF[2]

1 Introduction

Wild-type retrovirus particles contain a variety of different RNA molecules. Most particles contain a homodimer of the viral genome, an 8- to 10-kilobase terminally redundant RNA containing a 5′ guanosine cap and a 3′ polyadenosine tail

[1]Gladstone Institute for Virus Research, University of California, PO Box 914100, San Francisco, CA 94110–9100, USA
[2]Departments of Microbiology and Biochemistry and Molecular Biophysics, Howard Hughes Medical Institute, Columbia University, NY 10032, USA

(Coffin 1984). In addition, each particle contains multiple copies of cellular 7S L RNA (Bishop et al. 1970; Erikson et al. 1973; Ullu et al. 1982; Walker et al. 1974), 5S ribosomal RNA (Faras et al. 1973; Sawyer and Dahlberg 1973; Taylor et al. 1975) and approximately 30 copies of 4S cellular tRNA molecules (Erikson et al. 1973; Mak et al. 1994; Sawyer and Dahlberg 1973; Walker et al. 1974). Occasionally, retrovirus particles have been reported to contain subgenomic spliced viral RNAs (Katz et al. 1986; Luban and Goff 1994; Martin et al. 1986; Stacey 1979; Svoboda et al. 1986), cellular 18S and 28S rRNA (Waters and Mullin 1977), and cellular mRNAs (Adkins and Hunter 1981; Ikawa et al. 1974) as well. However, the viral genomic RNA dimer is the dominant virion RNA, usually accounting for over 50% of the virion nucleic acid by weight. The incorporation of the genomic RNA into the assembling virion is an essential step in the retroviral life cycle, as particles lacking the viral genome are noninfectious.

One can imagine several possible mechanisms through which a virus particle acquires its genomic RNA. One mechanism involves the random diffusion of the RNA into the assembling virion. However, since the viral genomic RNA typically accounts for less than 1% of the total RNA in the cytoplasm the genomic RNA must be selectively enriched during packaging. This enrichment could occur if packaging was coupled to translation, since the genomic RNA serves as the template for synthesis of the virus capsid proteins. Although this mechanism may be used by the hepadnaviruses (Bartenschlager et al. 1990; Hirsch et al. 1990), there is no evidence for it in retroviral RNA packaging. Another potential mechanism involves the specific recognition and binding of the genomic RNA molecule by the Gag and Gag/Pol precursor polyproteins which comprise the assembling virion. In fact, in the last 15 years researchers have obtained much evidence for such an interaction in vivo, through the identification of necessary "packaging elements" in the viral RNA and RNA-binding sites in the Gag polyprotein. By a combination of mutational and structural analyses, we are learning that the interaction between the viral RNA and the Gag polyprotein is similar in nature to many of the RNA-protein interactions of the eukaryotic cell.

2 Packaging Elements in the RNA

What is the signal in the viral genomic RNA that is recognized by the Gag polyprotein? In several nonretroviral viruses which selectively encapsidate their single-stranded RNA genomes, a small (~100 nt) region containing one or more stem-loop structures functions as the signal (Fujimura et al. 1990; Schlesinger et al. 1994). In retroviruses, the signal appears to contain stem-loop structures as well, but may span regions of several hundred nucleotides and may be noncontiguous in the viral genome. The segments of the retrovirus genome, though varying in length, are constant in their position within the genome. The order of segments, from 5′ to 3′, is R, U5, the primer-binding site (PBS), the 5′ untranslated region, the coding region, the 3′ untranslated region, U3, and R. The U3, R, and U5 regions are defined by the events of reverse transcription and are juxtaposed during this process to create the long terminal repeat (LTR) at each edge of the DNA provirus. The term "leader" refers to the noncoding sequences at the 5′ end of the genome; although some reports in the literature use this term to describe only the region between the PBS and the *gag* start codon, in this review the leader RNA is meant to describe the entire 5′ untranslated region, including the 5′ LTR, PBS, and the region leading up to the *gag* start codon.

Precise identification of the packaging signal has been problematic for several reasons. First, retroviral packaging signals are complex, in that they appear to contain multiple elements which act in redundant, additive, or synergistic fashions. Second, the ability of an element to be recognized for packaging is dependent on its conformation; mutations in both the element and its neighboring sequences can disrupt packaging. Third, it is difficult to analyze mutations in some portions of the viral RNA because of the role those portions play in the synthesis of viral proteins. One solution to this problem is to analyze the mutations in the setting of a retroviral vector, while providing the proteins *in trans* from a separate construct. For a review of retroviral vectors, see Boris-Lawrie and Temin (1994).

It is through the mutation of retroviral genomes and the analysis of retroviral packaging vectors that the preliminary elucidation of retroviral packaging signals has occurred. In the following sections, we review the mutational studies that have served to define the packaging signals in the RNAs of the major retroviral genera.

2.1 Avian Retroviruses

The first characterized virus which had a defect in genomic RNA packaging was a Rous sarcoma virus (RSV) mutant isolated from the infected quail cell line SE21Q1b. This mutant packaged less than 1% of the normal levels of viral RNA (Linial et al. 1978), due to a deletion of 179 nucleotides near the 5′ end of the genome, including the entire PBS (Anderson et al. 1992). Several years later, another spontaneous RSV mutant with a packaging defect, TK15, was described; these particles contained only 0.5% of the normal levels of genomic RNA (Koyama et al. 1984). The lesion in the TK15 variant is a 237-nt deletion in the leader, beginning immediately downstream of the PBS (Nishizawa et al. 1985). It was reasoned that the TK15 and SE21Q1b mutations destroyed a specific packaging signal in the viral genomic RNA, leading to the production of virus particles which rarely contained viral RNA.

Targeted mutational analysis of the viral genome has also been used to demonstrate the presence of the packaging signal in the leader. Deletion of 153 nucleotides from the avian leukosis virus (ALV) leader, including the 3′ half of the PBS, rendered the virus completely noninfectious, though virion RNA levels were not analyzed (STOKER and BISSELL 1988). Deletion analysis of the avian sarcoma virus (ASV) leader demonstrated a role in packaging for sequences within a 31-nt region which itself lies in the region missing from the SE21Q1b and TK15 genomes (KATZ et al. 1986). In contrast, small deletions in the U5 region (COBRINIK et al. 1988,1991) and large deletions in the coding region (NORTON and COFFIN 1985; STOKER and BISSELL 1988) do not affect RNA packaging .

Sequences at the 3′ end of the genome may also be involved in packaging. Deletion of both 115-nt direct repeats (DRs) flanking the v-*src* gene of RSV completely blocked viral RNA packaging, while deletion of either repeat individually had no effect on packaging (SORGE et al. 1983). A more recent study demonstrated that deletion of both DRs reduced packaging only tenfold (ARONOFF and LINIAL 1991), indicating that the DRs may serve as auxiliary packaging elements. However, very recent experiments suggest that the DRs function in achieving the high-level cytoplasmic expression (R. Ogert and K. Beemon, personal communication) or translation (L. Zhang and C.M. Stoltzfus, personal communication) of unspliced RSV RNA, and in some contexts are completely dispensable for RNA packaging (L. Zhang and C.M. Stoltzfus, personal communication).

What are the structures within the leader that form the packaging signal? Recently, ARONOFF et al. (1993) demonstrated that a 270-nt block of RSV RNA, beginning immediately after the PBS and ending at the splice donor, can mediate the efficient packaging of a non-LTR-based vector. This RSV segment contains sequences missing from the SE21Q1b and TK15 genomes, including the 31 nucleotides implicated in packaging by KATZ et al. (1986). Eight of these 31 nucleotides are thought to form the 3′ side of a well-conserved stem, termed the O3 stem (HACKETT et al. 1992). Disruptive mutation of either side of the O3 stem reduced packaging 20- to 40-fold, while restoration of the stem by combination of the two compensatory stem mutations restored the efficiency of packaging to 30% of the wild-type levels (KNIGHT et al. 1994). Recently the minimal block of RSV RNA competent for packaging has been narrowed even further to a 160-nt segment including the O3 stem and its loop, as well as 4 nucleotides on the 5′ side and 81 nucleotides on the 3′ side of the stem-loop (A. Yeo and M. Linial, personal communication).

The described packaging elements are present in the spliced, subgenomic RNAs as well as the genomic RNA. Since avian retrovirus genomic RNA is packaged with 10- to 20-fold higher efficiency than spliced RNAs (KATZ et al. 1986), there must be a way to discriminate between these RNAs. Splicing per se does not inhibit packaging (ARONOFF et al. 1993), but the juxtaposition of the leader and envelope sequences in the spliced RNA may alter the conformation of the packaging signal. Alternatively, it is possible that the viral intron contains additional packaging elements. Along these lines, Pugatsch and Stacey demonstrated that deletion of sequences from a putative stem-loop structure 140 nucleotides downstream of the splice donor of RSV appeared to impair packaging of the genome by several orders of magnitude (PUGATSCH and STACEY

1983). However, several other studies using retroviral vectors did not detect packaging elements in this region (LIPSICK et al. 1986; NORTON and COFFIN 1985; STOKER and BISSELL 1988). Perhaps Pugatsch and Stacey's deletion altered the conformation of the packaging element in the leader.

In summary, recent studies have better defined the packaging element in the 5′ end of the genome. Formation of the O3 stem is required, perhaps to bind directly to the Gag polyprotein or to help sequences in the O3 loop bind to Gag. In addition, other stem-loop structures in the leader may be required for packaging (HACKETT et al. 1992; KNIGHT et al. 1994). A thorough structural analysis of the leader, combined with mutagenesis of the stem-loops, should help to further identify the packaging element(s). The role of the DRs in packaging should perhaps be re-evaluated; if a packaging element is found, experiments could be performed to ascertain whether the element interacts with the packaging element in the leader or whether the two elements bind to Gag independently. Lastly, the mechanism accounting for the preferential encapsidation of the genomic RNA is completely unknown.

2.2 Murine Retroviruses

In the case of murine retroviruses, deletion of 351 nucleotides near the 5′ end of the Moloney murine leukemia virus (M-MuLV) RNA caused more than a 100-fold decrease in genomic RNA packaging (MANN et al. 1983). This packaging element, located between nt #215 and 565, was termed Ψ. In the amphotropic murine virus 4070, deletion of 280 nucleotides from the leader also blocked RNA packaging (SORGE et al. 1984).

Murine retroviral LTR-based vectors extending through the Ψ region are encapsidated by murine virions. However, the efficiency of packaging increases dramatically when the vectors also include the remaining 55 nucleotides of the leader and the 5′ 407 nucleotides of the *gag* gene (ARMENTANO et al. 1987; BENDER et al. 1987), termed collectively the gag^{+} region. The increase in packaging is due to the presence of the extra RNA sequences per se and not due to translation of the truncated Gag protein (BENDER et al. 1987). Adam and Miller showed that a non-LTR-based vector, containing the composite, 813-nt Ψ-gag^{+} region, was packaged by murine virus particles as efficiently as the wild-type murine retrovirus genome (ADAM and MILLER 1988).

One explanation for the augmentation of packaging mediated by the extra sequences is that the gag^{+} region itself contains packaging elements. However, no one has been able to demonstrate that the gag^{+} region region by itself can mediate packaging of a heterologous RNA. Another hypothesis is that the gag^{+} region does not contain packaging elements, but instead augments Ψ-mediated packaging by stabilizing the packaging elements in the Ψ region. It is known that the efficacy of the Ψ region is dependent on its neighboring sequences; translocation of the Ψ region to the 3′ viral region of an M-MuLV vector resulted in a fivefold decrease in marker transduction (MANN and BALTIMORE 1985). Work with HIV-1 indicates that placement of heterologous sequences immediately downstream of the packaging signal can interfere with packaging, but not if viral sequences separate the heterologous sequences from the packaging signal (see Sect. 2.4). In addition, Adam and Miller found that the Moloney murine sarcoma virus (M-MuSV) Ψ region, but not the M-MuLV Ψ region, could, by itself, mediate the packaging of a non-LTR-based vector with 50% of the efficiency of wild-type viral RNA (Adam and Miller 1988). Thus, it is possible that the Ψ region is the only region of the murine retrovirus genome which contains packaging elements.

Analysis of the M-MuLV leader led to the surprising finding that mutants with deletions in the 5′ portion of the U5 region, which is upstream of the Ψ region, exhibited 25- to 100-fold reductions in RNA packaging (Murphy and Goff 1989). Since addition of this region to Adam and Miller's packageable Ψ-gag^{+} vector had no effect on its efficiency of packaging (Adam and Miller 1988), it is unlikely that the U5 region normally contributes to packaging. Instead, the deletion may have caused conformational changes in the downstream Ψ region, destabilizing the packaging signal.

Can the Ψ region be further narrowed? Early efforts to define the packaging elements within the Ψ region showed that sequences from nt #400 to 484 and from nt #537 to 602 can be deleted from the M-MuLV Ψ region without affecting virus replication (Schwartzberg et al. 1983). Mutation of sequences near nt #348 severely reduced the packaging of Adam and Miller's Ψ vector (Adam and Miller 1988). Analysis of the conservation and covariation of sequences in this region of the murine and related C-type retroviruses, coupled with computer-generated RNA-folding predictions, indicated the possibility that two highly conserved, extremely stable stem-loop structures, each containing the four nucleotides GACG as its loop, surround nt #348 (Konings et al. 1992). This potential GACG double hairpin, nts #310–374, is immediately followed by a third potential stem-loop structure, nts #382–399, which exhibits marked conservation of a very stable 5-bp stem but little conservation of its 8-nt loop. Chemical modification analysis of M-MuLV genomic RNA (Alford et al. 1991) and in-vitro-synthesized Ψ RNA (Tounekti et al. 1992) supported the existence of these three potential stem-loops and indicated that the portion of the leader 5′ to the stem-loops, especially nts #215–280, lacked stable secondary structure. High-resolution compensatory mutagenesis of the M-MuLV leader indicates that the two GACG stem-loops act in a synergistic fashion to mediate RNA packaging, while the third potential stem-loop may not be involved (J. Fisher and S.P. Goff, unpublished observations).

Thus, given the results of the structural and mutational analyses, it is possible that the GACG double hairpin may serve as the entire MuLV packaging signal. Since two adjacent GACG stem-loops (with minimum stem lengths of 5 base pairs) are not found in vertebrate or viral gene banks (Konings et al. 1992), a packaging signal composed of these two stem-loops would ensure the exclusivity of viral genome packaging. However, recent studies on the endogenous murine virus-like 30S (VL30) retrotransposon indicated that the GACG double hairpin may not be the only packaging element recognized by assembling MuLV virions (TORRENT et al. 1994a, b).

A significant difference between the murine and avian retroviral packaging systems is that while avian virions contain a significant amount of *env* mRNA (KATZ et al. 1986), murine virions contain very little. The basis for this difference probably lies in the relative placement of the splice donor (SD) and the packaging elements. In the avian viruses, the SD is located downstream of the *gag* start codon, so that both the genomic RNA and spliced mRNAs contain the entire 5′ leader and its packaging elements. In the murine viruses, the SD is located several hundred nucleotides upstream of the *gag* start codon, so that the spliced *env* mRNA lacks much of the leader, including the Ψ region. This arrangement makes the discrimination against the packaging of spliced RNAs in the murine retroviruses straightforward to understand.

In summary, the packaging signal of the murine retroviruses appears to be contained within one small region of the genome. The GACG double hairpin motif appears to be a critical element of the signal, but there may be additional elements nearby which constitute independent parts of the signal or which stabilize the double hairpin. To assess the relative contributions of the GACG double hairpin or any other structure to the packaging of murine retrovirus RNA, non-LTR-based vectors containing each structure as the sole retrovirus sequence could be constructed and compared to the viral genomic RNA for their relative packaging efficiencies. In addition, subtle point mutations in the RNA and the Gag polyprotein may help identify the exact molecular interactions between these species that occur during encapsidation of the RNA.

2.3 Reticuloendotheliosis Viruses

Reticuloendotheliosis viruses (REVs), including spleen necrosis virus (SNV), are viruses which infect avian cells but whose genomes most closely resemble those of the murine retroviruses. By cotransfection of cells with an SNV LTR-based vector and wild-type REV, it was demonstrated that a 185-bp deletion in the region between the splice donor and the *gag* start codon in the vector blocked packaging of the vector RNA and reduced transduction by over 10^4-fold (WATANABE and TEMIN 1982). This region was termed E, and contains the GACG double hairpin identified in the murine retroviral leaders (KONINGS et al. 1992).

A later study demonstrated that the 5′ and 3′ portions of the E region can, on their own, mediate a moderate level of packaging, indicating that the packaging signal within the E region may be composed of several discrete packaging elements (EMBRETSON and TEMIN 1987). High-resolution mutagenesis of the E region revealed that the GACG double hairpin comprises one of the REV packaging elements (YANG and TEMIN 1994). Both stems were required for efficient packaging, with both the primary and secondary structures of each stem mediating its activity. In contrast, only one of the two GACG tetraloops was required for packaging. A second, weaker packaging element may reside slightly downstream of the GACG double hairpin, in the 3′ portion of the E region, as linker insertion mutations in this 24-nt region caused moderate decreases in packaging (YANG and TEMIN 1994).

Last, and perhaps most remarkably, it was found that the REV GACG double hairpin could be replaced by the M-MuLV GACG double hairpin, and that the entire REV E region could be replaced by the M-MuSV Ψ region, without any effect on REV replication (YANG and TEMIN 1994). Thus, the REVs appear to use the same packaging signal as the murine retroviruses, i.e., the GACG double hairpin, and perhaps a small region downstream of the double hairpin.

2.4 Human Immunodeficiency Virus Type 1

Due to the location of the Ψ region in the murine retroviruses and REVs, most of the mutational analyses of the HIV-1 packaging signal have focused on the 45-nt region in between the splice donor and the *gag* start codon. Several labs have found that deletions of approximately 20 nucleotides in this region cause greater than 50-fold reductions in RNA packaging (ALDOVINI and YOUNG 1990; LEVER et al. 1989; VICENZI et al. 1994). However, other deletion mutants in this region exhibit modest or insignificant reductions in packaging (CLAVEL and ORENSTEIN 1990; KIM et al. 1994; LUBAN and GOFF 1994; VICENZI et al. 1994) (H. Göttlinger, M. McBride, and A. Panganiban, personal communications). The differences in packaging may be partially attributable to differences in the virion-production systems. A 19-nt deletion mutant which was at least 50-fold impaired in RNA packaging when virions were produced by the Jurkat T-cell line (LEVER et al. 1989) was only fivefold impaired when produced by Cos cells (LUBAN and GOFF 1994), twofold impaired when produced in SW480 cells (KIM et al. 1994), threefold impaired when produced in MT4 cells (KIM et al. 1994), and only two- to fourfold impaired when produced by the Hela cell line (H. Göttlinger, personal communication). In addition, other mutants have been found to exhibit larger packaging defects in Jurkat cells than Cos cells (A.M.L. Lever, personal communication; see Sect. 2.5.2). The reason for the relatively stringent RNA packaging exhibited by Jurkat cells (recently quanitated by KAYE and LEVER 1996) is unknown, but in theory could be due to relatively low levels of HIV-1 RNA expression.

In addition, sequences upstream of the major splice donor and sequences in the *gag* coding region appear to be involved in RNA packaging. Replacement of 13 nucleotides by a heterologous linker 36 nt upstream of the SD decreased packaging six- to sevenfold and delayed viral growth (KIM et al. 1994). Viral growth was similarly delayed when 63 nucleotides were replaced by the same linker 11 nt upstream of the SD. Deletion of the 5′ 250 nucleotides of the genome has been found to reduce RNA packaging fivefold (M.McBride and A. Panganiban, personal communication). Deletion of the 5′-most 14 (R. Berkowitz and S.P. Goff, unpublished observations) or 40 (LUBAN and GOFF 1994) nucleotides of *gag* reduced the packaging of an HIV-1 vector fivefold. Like the case with the MuLV vectors, inclusion of *gag* sequences in HIV-1 vectors has been found to increase vector packaging efficiencies, although the gains are modest (Buchschacher and Panganiban 1992; Parolin et al. 1994).

Remarkably, one group has found that insertion of a 46-nt portion of the NL4-3 genome, including 39 nucleotides upstream of the *gag* start codon and 7 nucleotides of the *gag* coding region, into a vaccinia vector rendered that vector packageable by HIV-1 virus-like particles produced by a recombinant vaccinia virus (Hayashi et al. 1992). The evaluation of the vector's packaging efficiency relative to wild-type HIV-1 was not performed. The authors argue, based on predictions from a computer RNA-folding program, that the 46-nt HIV-1 segment exists as a stem-loop structure (an extended version of stem-loop #3 in Fig. 1), which serves as the packaging signal. However, no direct evidence for the existence of this stem-loop structure in vitro or in vivo was presented.

Biochemical evidence for the existence of the central portion of this stem-loop, as well as several other stem-loops, has been supplied by three separate structural analyses of the HIV-1 leader (Baudin et al. 1993; Clever et al. 1995; Harrison and Lever 1992). Although each of these analyses has yielded dif-

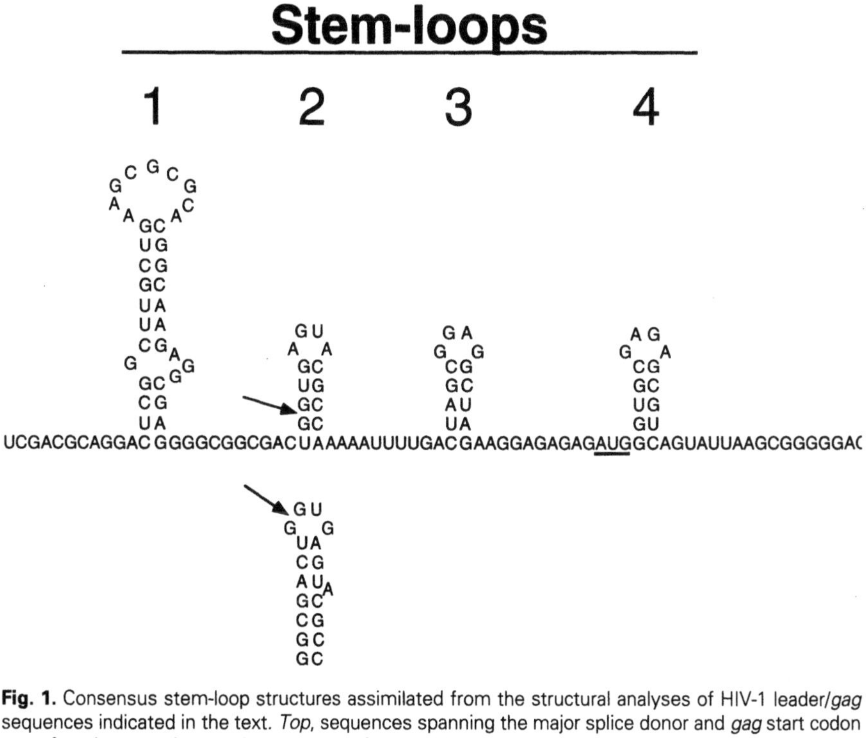

Fig. 1. Consensus stem-loop structures assimilated from the structural analyses of HIV-1 leader/*gag* sequences indicated in the text. *Top*, sequences spanning the major splice donor and *gag* start codon were found to contain stem-loops at up to four positions, *numbered at the top*. Stem-loops #1, 3, and 4 varied in different studies only in the length of the stem, not in the identity of the loop. Stem-loop #2 varied in the identity of the stem and loop, with the alternative structure depicted underneath. The splice donor cleavage site in stem-loop #2 is *indicated with an arrow*, and the *gag* start codon next to stem-loop #4 is *underlined*

ferent proposed stem-loop structures, careful inspection of the proposed structures indicates that this region of the leader has the potential to form only four stem-loops, with stem-loop #2 possessing two possible conformations (Fig. 1). All of the proposed structural models assert the existence of the central portion of the Hayashi stem-loop (#3); the models differ only in which of the other three structures are present and to what extent the intervening sequences contribute to extending the stem-loops. Interestingly, the sequence of the loops of stem-loops #2–4, assuming that stem-loop #2 exists in the alternate conformation, is GGUG, GGAG, and GAGA; the consensus sequence, GPuXPu, also includes the GACG sequence of the murine/REV stem-loops. In addition, recently reported strains of HIV-1 contain different sequences in the stems of stem-loops #1 and 3 but still maintain the stem structures, strongly implicating the existence and conservation of these two stem-loops (G. Harrison, personal communication).

Do these stem-loops mediate encapsidation of the viral genomic RNA? Disruptive mutation of either side of the stem of stem-loop #3 was found to significantly impair virus replication in Jurkat cells, while combination of the two mutations restored wild-type kinetics of replication (G. Harrison and A.M.L. Lever, personal communication). Disruptive mutation of stem-loop #1 and stem-loop #3 individually reduced RNA packaging minimally, but together reduced packaging fivefold in HeLa cells; compensatory restoration of both stems restored packaging to wild-type efficiency (M. McBride and A. Panganiban, personal communication). Disruptive mutation of stem-loop #2 had no effect on packaging. In addition, we recently found that disruptive mutations of stem-loop #4 did not impair the packaging of an HIV-1 vector by wild-type virions produced in Cos cells, although deletion of the stem-loop reduced packaging fivefold (R. Berkowitz and S.P. Goff, unpublished observations).

Sequences in the 3′ portion of the HIV-1 genome may be involved in RNA packaging. In one study, several vectors were encapsidated efficiently only when they contained a 1.1-kb fragment of the *env* gene, indicating that this *env* fragment contains packaging elements (Richardson et al. 1993). However, placement of any one of several marker genes next to the 5′ viral sequences was found to impair packaging efficiency 15- to 30-fold. Recent results from this group indicate that the *env* fragment does not contain packaging elements, but instead augments vector packaging by overcoming the inhibitory effects of the marker gene (Kaye et al. 1995).

Three groups have turned to in vitro binding assays to probe the interaction of RNAs containing the stem-loops with Gag proteins. Using northwestern blot or gel mobility shift assays, HIV-1 Gag polyproteins and NC proteins were found to bind with specificity to several non-overlapping segments of the leader and *gag* coding region, indicating the presence of multiple sequence-specific binding elements in the 5′ portion of the RNA genome (Berkowitz and Goff 1994; Berkowitz et al. 1993; Clever et al. 1995; Luban and Goff 1991). Binding assays of the stem-loops, either alone or in combinations, revealed that each stem-loop bound to the Gag polyprotein or NC protein weakly, with high-affinity binding occurring only when the stem-loops and their flanking regions were linked (BERKOWITZ and GOFF 1994; CLEVER et al. 1995). A 47-nt RNA lying upstream of the stem-loops was also found to bind to the proteins with specificity (CLEVER et al. 1995).

Another study of in vitro HIV-1 RNA-Gag protein interactions involved gel mobility shift assays using peptides of the NC domain and a 44-nt RNA comprising stem-loops #2 and 3 (SAKAGUCHI et al. 1993). Chemical modification analysis of this RNA indicated that both structures were formed, with stem-loop #2 existing in the alternative conformation. The authors claim, based on 4-nt substitution mutations in the stems or loops of each structure, that stem-loop #2, but not stem-loop #3 or the linker region between the two stem-loops, was involved in binding of the RNA to the full 55 amino acid NC peptide. However, the authors presented no evidence that the interaction between their RNAs and the peptide was sequence specific. Since NC is known to possess a weak nonspecific RNA-binding activity in addition to its strong Ψ-binding activity (see Sect. 3), it is possible that the authors measured nonspecific binding between the RNAs and the peptide. Even if the binding between the 44-nt RNAs and the NC peptide was specific, however, close inspection of the data reveals that mutation of stem-loop #2 did not reduce the extent of binding, but only changed the position of the shifted bands. Mutation of stem-loop #3, however, did reduce the extent of binding five- to tenfold.

In sum, the HIV-1 packaging signal appears to be different than the avian and murine/REV packaging signals. The ALV and MuLV/REV signals are compact, consisting of one or two adjacent stem-loop structures. Mutations in the stem-loops cause large defects in packaging, while mutations outside of the stem-loops are not disruptive. Packaging vectors can be generated with less than 160 nucleotides of viral sequences. In HIV-1, mutations throughout the leader and *gag* cause reductions in packaging, but in most laboratory cell lines the reductions are not severe. Increasing the number of mutations increases the severity of the packaging defect. These results, in combination with in vitro binding results, indicate that the HIV-1 packaging signal is composed of multiple dispersed elements which act in a roughly additive fashion. It appears that stem-loops #1 and 3 are each packaging elements. In addition, the data indicate that a packaging element is likely to lie upstream of stem-loop #1. Sequences downstream of stem-loop #3 are also involved in packaging, but we do not yet know whether this region contains a packaging element(s) or instead acts to stabilize the upstream stem-loops. The identification of the other elements that contribute to the HIV-1 packaging signal are possible studies for the future.

With the exception of the Hayashi study, no one has been able to generate an HIV-1 packaging vector containing less than 1 kb of viral sequences (BERKOWITZ et al. 1995b). Although the data could be interpreted that the HIV-1 packaging signal is small but extremely sensitive to its context, it is also possible that the packaging signal is extremely dispersed, and requires certain elements for its function. Candidate regions are the 5′ edge of the genome and portions of the *pol* and *env* coding regions.

2.5 Other Retroviruses

2.5.1 Bovine Leukemia Virus

RNA gel mobility shift assays have been used to demonstrate that the bovine leukemia virus (BLV) NC protein binds to several BLV RNA segments in vitro, while the BLV MA cleavage precursor p15 binds only to the RNA segment containing the leader and 5′ *gag* sequences (KATOH et al. 1991). As seen with the HIV-1 in vitro binding assay (BERKOWITZ and GOFF 1994), at least four non-overlapping fragments, including sequences from the U5 region to the 5′ part of the *gag* coding region, each mediate a low level of protein binding; when the fragments are linked, p15-binding levels increase incrementally (KATOH et al. 1993).

Recently, the construction of a BLV-packaging vector has been described (MANSKY et al. 1995). Deletion of a 148-nt segment, including the 3′ 77 nucleotides of the leader and the 5′ 71 nucleotides of the *gag* coding region, caused a 50-fold reduction in the level of vector RNA packaged by helper virus. Deletion of a 137-nt segment in the CA domain of *gag* caused a sevenfold reduction in vector packaging. This packaging element appeared to be discontinuous with the primary packaging element further upstream, since deletion of the sequences in between these elements did not affect vector packaging.

2.5.2 Simian Immunodeficiency Virus and HIV-2

A simian immunodeficiency virus (SIV) LTR-based vector, containing the *hyg*^{r} gene in place of the viral coding regions, was found to be packaged by a co-transfected HIV-1 helper virus (RIZVI and PANGANIBAN 1993). The 92 nucleotides immediately following the 5′ LTR were critical for packaging, since a vector containing this region was packaged while the same vector lacking this region was not. Packaging of SIV vectors by SIV virions has not been reported.

Deletions between the SD and *gag* of HIV-2 have been analyzed for their effects on RNA packaging and viral replication (E. McCann and A.M.L. Lever, personal communication). Virus particles shed from transfected Cos cells contained wild-type levels of the deleted RNAs. However, when these particles were used to infect Jurkat cells, the mutant virus replicated slower than wild type, and virus particles collected 12 days after infection contained fivefold less RNA than wild-type particles.

2.5.3 Mason-Pfizer Monkey Virus

A Mason-Pfizer monkey virus (MPMV) LTR-based vector containing the 5′ 989 nucleotides of the genome, including 495 nucleotides of the *gag* coding region, was found to be packaged efficiently by MPMV helper virus (VILE et al. 1992). Conversely, the same vector lacking the *gag* coding sequences and the 3′ 124 nucleotides of the 130-nt SD-*gag* spacer region was not packaged.

Structural analysis of the MPMV leader indicated the presence of several stem-loop structures, including an extremely stable stem-loop containing a 5-bp stem and a 7-nt loop upstream of the SD, and a small stem-loop containing the *gag* start codon (HARRISON et al. 1995). Recently, a small deletion upstream of the splice donor was found to impair RNA packaging and virus infectivity, as well as transduction of an MPMV vector (F. Guesdon, S. Rhee, E. Hunter, and A.M.L. Lever, personal communication). Since the deletion included the 5′ stem-loop, this stem-loop may function as a packaging element.

3 Role of the Gag Polyprotein in RNA Packaging

The viral Gag polyprotein is sufficient to specify the assembly and release of virion particles from cells. Further, virus-like particles generated in the absence of the *pol* and *env* genes have been found to selectively package viral RNA (KAYE and LEVER 1996; OERTLE and SPAHR 1990; SAKALIAN et al. 1994; SHIELDS et al. 1978), indicating that the Gag polyprotein contains all of the determinants for the specific recognition of the packaging signal on the viral RNA. Genetic analysis of the Gag polyprotein has unequivocally determined that the nucleocapsid (NC) domain is intimately involved in viral RNA packaging, and several studies strongly suggest that NC is the sole *trans* factor involved in packaging. There are very few reports of mutations outside of NC affecting RNA packaging. One report demonstrated that a deletion in the RSV protease domain impaired packaging (OERTLE and SPAHR 1990); however, this result has not been confirmed in the same or other laboratories (SAKALIAN et al. 1994).

3.1 Structure of the NC Protein

The NC domain of the Gag polyprotein has several distinguishing characteristics. First, it always lies C-terminal to the matrix and capsid domains, occupying either the ultimate or penultimate C-terminal position in the Gag polyprotein. A second characteristic of NC is the existence of a conserved motif of 14 residues, termed the Cys-His box, which possesses the canonical sequence **Cys-Xaa$_2$-Cys-Xaa$_4$-His-Xaa$_4$-Cys** (Fig. 2). This motif is found once or twice in all retroviral NCs with the exception of the spumaviruses, and can even be found among distantly related retro-elements such as the *Drosophila* copia element, yeast Ty transposons, and nonretroviral human nucleic acid binding proteins. The cysteines and the histidine are absolutely conserved, as is a glycine immediately N-terminal to the histidine, and an aromatic residue is usually located in either the first or second variable position. The positions of the two Cys-His boxes of HIV-1, avian, and other retroviruses define subdomains of the NC domain; we will refer to these subdomains, from N- to C-terminus, as the N-terminal subdomain, box

Virus	N-terminal subdomain	Cys-His box	C-terminal subdomain
AKV MuLV	ATVVSGQRQDRQGGERRRPQLDKDQ	CAYCKEKGHWAKDC	PKKPRGPRGPRPQTSLL
cas-br-e MuLV	ATVVSGQKQDRQGGERRRPQLDKDQ	CAYCKEKGHWAKDC	PKKPRGPRGPRPQTSLL
Moloney MuLV	ATVVSGQKQDRQGGERRRSQLDRDQ	CAYCKEKGHWAKDC	PKKPRGPRGPRPQTSLL
Rauscher MuLV	ATVVSGQRQDRQGGERRRPQLDRDQ	CAYCKEKGHWAKDC	PKKPRGPRGPRPQASLL
Friend MuLV pvc-211	ATVVSGQRQDRQGGERRRPQLDHDQ	CAYCKEKGHWARDC	PKKPRGPRGPRPQASLL
BaEV m7	AAVVTEKRAGKSGETRRRPKVDKDQ	CAYCKERGHWIKDC	PKRPRDQKKPAPVL
FeLV	ATVVAQNRDKDREENKLGDQRKIPLGKDQ	CAYCKEKGHWVRDC	PKRPRKKPANSTLL
GALV	AAVVSREGSTGRQTGNLSNQAKKTPRDGRPPLDKDQ	CAYCKEKGHWAREC	PRKKHVREAKVLALDN
SNV ptw44	LAQESRAERGSKKTPPGKGRPPLGKNQ	CAYCKEEGHWKKNC	PKLVSGAAPVL

Virus	N-terminal subdomain	Box 1	Linker subdomain	Box 2	C-terminal subdomain
MPMV	AAAFSGQTVKDFLNNKNKEKGGC	CFKCGKKGHFAKNC	HEHAHNNAEPKVPGL	CPRCKRGKHWANEC	KSKTDNQGNPIPPHQGNGWRGQPQAPKQAY
SRV1	AAAFSGQTVKDFLNNKNKEKGGC	CFKCGRKGHFAKNC	HEHIHNNSETKAPGL	CPRCKRGKHWANEC	KSKTDSQGNPLPPHQGNGLRGQPQAPKQAY
MMTV c3h	AAAMRGQKUSTFVKQTYGGGKGGQGSKGPV	CFSCGKTGHIKRDC	KEEKGSKRAPPGL	CPRCKKGYHWKSEC	KSKFDKDGNPLPPLETNAENSKNL
BIV 127	ASQTSGPEDGRR	CYGCGKTGHLKRNC	KQQK	CYHCGKPGHQARNC	RSKNGKCSSAPYGQRSQPQNNF
EIAV wyoming	QTGLAGPFKGGALKGGPLKAAQT	CYNCGKPGHLSSQC	RAPKV	CFKCKQPGHFSKQC	RSVPKNGKQGAQGRPQKQTF
CAEV cork	AQALRPGKGKGNGQPQR	CYNCGKPGHQARQC	RQGII	CHNCGKRGHMQKEC	RGKRDIRGKQQGNGRRGIRCCPSAPPME
Visna kv1772	AQALRPQGKAGQKGVNQK	CYNCGKPGHLARQC	RQGII	CHHCGKRGHMQKDC	RQKKQQGNNRRGPRVVPSAPPML
FIV tm2	VQTKGPRLV	CFNCKKPGHLARQC	KEAKR	CNNCGKPGHLAANC	WQGGRKTSGNEKVGRAAAPVNQVQQIVPSAPPMEEKLLDL
HIV-2 rod	AQQRKAFK	CWNCGKEGHSARQC	RAPRRQG	CWKCGKPGHIMTNC	PDRQAG
SIV pbj/bc113	VQQKGQRKIIK	CWNCGKEGHSARQC	RAPRRQG	CWKCGKAGHVMAKC	PERQAG
SIV stm	AQQQGRRTVK	CWNCGKEGHTAKQC	KAPRRQG	CWKCGKPGHQMAKC	PERQVG
SIV agm3	MQQGGQRGRPRPPVK	CYNCGKFGHMQRQC	PEPRKMR	CLKCGKPGHLAKDC	RGQVN
SIV agm40	AQGPRGRGRSRGPVR	CFRCGQIGHIQKDC	PKGGPIK	CLKCGKPGHMAKDC	RSGQAN
SIV cpz	FQKGQGAGPKRKIK	CFNCGKEGHLARNC	KAPRRKG	CWRCGQEGHQMKDC	TGRQVN
HIV-1 hxb2	MQRGNFRNQRKIVK	CFNCGKEGHTARNC	RAPRKKG	CWKCGKEGHQMKDC	TERQAN
HIV-1 mal	MQRGNFKGQKRIK	CFNCGKEGHLARNC	RAPRKKG	CWKCGKEGHQMKDC	TERQAN
HIV-1 rf	MLQKGNFRDQRKIVK	CFNCGKVGHIAKNC	RAPRKKG	CWKCGKEGHQMKDC	TNEGRQAN
HTLV-2 nra	VVQPRRPPPTQP	CFRCGKIGHWSRDC	TQPRPPPGP	CPLCQDPSHWKRDC	PQLKPPQEEGEPLLLDLSSTSGTTEEKNSLGGEI
HTLV-1 mt-2	VVQPKKPPPNQP	CFRCGKAGHWSRDC	TQPRPPPGP	CPLCQDPTHWKRDC	PRLKTIPEPEPEEDALLLDLPADIPHPKNSIGGEV
BLV japanese	VHTPGPKMPGPRQPAPKRPPPGP	CYRCLKEGHWARDC	PTKATGPPPGP	CPICKDPSHWKRDC	PTLKSKN
RSV prague c	AVVNRERDGQTGSGGRARGL	CYTCGSPGHYQAQC	PKKRKSGNSRER	CQLCNGMGHNAKQC	RKRDGNQGQRPGKGLSSGPWPGPEPPAVS
RSV sd-r	AVVNRERDGQTGSGGRARGL	CYTCGSPGHYQAQC	PKKAKSGNSRER	CQLCDGMGHNAKQC	RKRDGNQGQRPGRGLSSGPWPGPEQPAVS
ALV rsa	AVVNRERDGQTGSGGRARGL	CYTCGSPGHYQAQC	PKKAKSGNSRER	CQLCDGMGHNAKQC	RRRDGNQGQRPGKGLSSGSWPVSEQPAVS

1, the linker subdomain, box 2, and the C-terminal subdomain. Murine NCs contain an N-terminal subdomain, one box, and a C-terminal subdomain. The third characteristic of retroviral NC proteins is the presence of many basic residues (Fig. 2). These basic residues occasionally lie within the Cys-His box(es), but are more prevalent in the flanking subdomains, often forming contiguous doublets or triplets.

Because the Cys-His box is similar to the zinc finger domain found in many DNA-binding transcription factors, it was hypothesized that NC too binds zinc (Berg 1986). Early reports regarding zinc association with NC were ambiguous. A $^{65}Zn^{2+}$ blotting technique was used to demonstrate binding to AMV NC but not MA or CA (Schiff et al. 1988). However, atomic absorption spectroscopy revealed submolar levels of Zn^{2+} contained within the virion (Jentoft et al. 1988). The availability of solid-phase peptide synthesis made it possible to synthesize NC peptides including the entire protein in quantities sufficient for examination by photometric and NMR spectroscopic methods. An important outcome of these studies was the establishment that Zn^{2+}, as well as Co^{2+} and Cd^{2+}, is able to specifically associate with NC by forming a tetrahedral complex, with the metal coordinated by each cysteine and the histidine of the Cys-His box (Fitzgerald and Coleman 1991; Green and Berg 1989, 1990; Roberts et al. 1989; South et al. 1989, 1990a). Metal binding constants for either the Cys-His box or the entire protein have been established and show a strong and selective binding for zinc over other metals (Green and Berg 1990; Mely et al. 1991). Zinc binding is diminished by several orders of magnitude when any of the cysteines or the histidine of the box are mutated (Green and Berg 1990; Mely et al. 1991; Prats et al. 1991) or modified (Mely et al. 1991). During the course of working with purified NCs it was discovered that the protein is extremely sensitive to oxidation of the cysteine's sulfhydryl groups, which thereby also greatly diminishes the ability of NC to coordinate zinc (Fitzgerald and Coleman 1991; Green and Berg 1989; South et al. 1990). Cysteine oxidation may help explain why the initial study did not detect quantitative levels of zinc within virions (Jentoft et al. 1988). More recent studies of isolated virus particles clearly indicate that nearly every Cys-His box is bound to zinc in vivo (Bess et al. 1992; Chance et al. 1992; Summers et al. 1992). Virus producer cells grown in the presence of a zinc-ejecting compound

Fig. 2. Amino acid residues of the nucleocapsid protein. The NC residues of a group of 32 arbitrarily chosen retroviruses are grouped and aligned by subdomains described in the text. Viruses, identified by strain where applicable, are ordered roughly by the extent of conservation of residues in the Cys-His box(es). *Top*, viruses containing only one Cys-His box. *Bottom*, viruses containing two Cys-His boxes. Sequences and identification of mature termini were obtained from various gene banks and from Dr. Louis E. Henderson. Virus abbreviations and accession numbers are as follows: *MuLV*, murine leukemia virus; *AKV MuLV*, #P03336; *cas-br-e MuLV*, #P27460; *Moloney MuLV*, #P03332; *Rauscher MuLV*, Dr. Henderson; *Friend MuLV*, #P26805; *BaEV*, baboon endogenous virus, #P03341; *FeLV*, feline leukemia virus, #P10262; *GALV*, gibbon ape leukemia virus, #P21416; *SNV*, spleen necrosis virus, #M54993; *MPMV*, Mason-Pfizer monkey virus, #P07567; *SRV1*, simian retrovirus 1, #P04022; *MMTV*, mouse mammary tumor virus, #P11284; *BIV*, bovine immunodeficiency virus, #P19559; *EIAV*, equine infectious anemia virus, #P03351; *CAEV*, caprine arthritis encephalitis virus, #P33458; *Visna virus*, #P35955; *FIV*, feline immunodeficiency virus, #P31821; *HIV*, human immunodeficiency virus; *SIV*, simian immunodeficiency virus; *HIV-2* and *SIVs*, Dr. Henderson, #L19252, #P19504, #P31634, #P17282, #P27978; *HIV-1s*, #M17451, #P04591, #P04594; *HTLV*, human T-cell leukemia virus; *HTLV-1*, #P14077; *HTLV-2*, #L20734, Dr. Henderson; *BLV*, bovine leukemia virus, #P03344; *RSV*, Rous sarcoma virus, #P03322, #D10652; *ALV*, avian leukosis virus, #M37980

release noninfectious virus particles, indicating that zinc binding is a necessary step in the virus life cycle (RICE et al. 1993).

NMR spectroscopy of HIV-1 NC has permitted the determination of the majority of its three-dimensional conformation in solution (MORELLET et al. 1992, 1994; OMICHINSKI et al. 1991; SOUTH et al. 1990a, b, 1991; SUMMERS et al. 1990, 1992). In the absence of zinc, the Cys-His box is unstructured. Upon zinc binding, each Cys-His box adopts a compact structure consisting of three distinct, tightly constrained turns of the peptide backbone. The residue in the middle turn is spatially constrained, accounting for the conservation of glycine at this position. The compact structure appears to be further stabilized by the presence of seven hydrogen bonds that form amongst the noncoordinating conserved residues within the box. In box 1, the phenylalanine and isoleucine residues form a hydrophobic cleft on one side of the box; since nearly all Cys-His boxes contain an aromatic residue and a hydrophobic residue in the same positions as the HIV-1 box 1 Phe and Ile, this hydrophobic cleft may be a critical structural element of the box. In the M-MuLV Cys-His box, tyrosine and tryptophan occupy these positions, and zinc binding increases the extent of their interaction (CORNILLE et al. 1990; MELY et al. 1991). The other regions of NC , in contrast to the Cys-His boxes, appear to exist in an unstructured state.

Initially, some groups saw evidence that the structure of box 2 was less stable than box 1 (MORELLET et al. 1994; SOUTH et al. 1991). However, one of these groups has found that the thermolability of box 2 is not consistently observed (M.F. Summers, personal communication). In addition, one group has found evidence for an interaction between the HIV-1 boxes in a synthesized peptide, as delineated by the spatial proximity of the box 1 phenylalanine with the box 2 tryptophan (DÉMÉNÉ et al. 1994; MORELLET et al. 1992, 1994). The interaction of these two aromatic residues appears to be mediated in part by a proline residue in the linker region that separates the two boxes (MORELLET et al. 1994). Substitution of D-proline for the L-proline within the linker region, although permitting the proper folding of the individual boxes, alters the positioning of the two boxes, abrogating the interaction of the Phe with Trp. Similarly, substitution of the box 1 histidine with cysteine does not affect zinc binding but does abolish the Phe-Trp interaction (DÉMÉNÉ et al. 1994). The interaction between the boxes has not been supported by data from two other groups, either in chemically synthesized peptides (OMICHINSKI et al. 1991), particle-derived NC (SOUTH et al. 1990a; SUMMERS et al. 1992), or recombinant NC (M.F. Summers, personal communication); the reason for this discrepancy is unknown.

3.2 Effects of NC Mutations on Packaging In Vivo

Genetic analyses of NC mutants in vivo have unequivocally shown that the Cys-His box(es) and the flanking basic residues mediate the encapsidation of viral genomic RNA. Cys-His box deletions have been found to lower the level of packaged viral RNA approximately 100-fold in single-box viruses and 10- to 30-

fold in those retroviruses containing two boxes (GORELICK et al. 1990; MERIC et al. 1988). Deletion of both boxes in the double-box viruses impaired packaging 100-fold, suggesting that each box can mediate RNA packaging independently of the other box (MERIC et al. 1988). Within the box, point mutation of any of the zinc-coordinating cysteines and histidine impaired packaging significantly (ALDOVINI and YOUNG 1990; DORFMAN et al. 1993; DUPRAZ et al. 1990; GORELICK et al. 1988, 1990; REIN et al. 1994), indicating that the spatial conformation of the box, which is maintained by these residues via zinc coordination, is critical for packaging. It is also possible that these residues actually interact with the RNA. In some instances, point mutation of one of these residues has been seen to impair RNA packaging by a double-box virus to a larger extent than a deletion of the entire box, presumably by interfering with the proper functioning of the other box (ALDOVINI and YOUNG 1990; DUPRAZ et al. 1990).

Mutational analyses have shown that several other residues within the Cys-His box are involved in RNA packaging. These other residues include the conserved aromatic residue and the conserved hydrophobic residue immediately C-terminal to the histidine (DORFMAN et al. 1993; DUPRAZ et al. 1990; GORELICK et al. 1988; MERIC and GOFF 1989); as discussed in Sect. 3.3.1, these residues are thought to interact with nucleic acids. In addition, since these residues are thought to form a hydrophobic cleft which interacts with the RNA, as discussed in Sect. 3.1, the genomic RNA may interact with the hydrophobic cleft during packaging (see also Sect. 3.3.3). Substitution of the conserved glycine residue with alanine causes only slight decreases in packaging (DORFMAN et al. 1993; DUPRAZ et al. 1990), but substitution with more bulky amino acids impairs packaging more severely (DUPRAZ et al. 1990; MERIC and GOFF 1989), reflecting the requirement of this position of the box to form a spatially limited acute turn. In addition, studies characterizing NCs containing two box 1s or two box 2s indicate that the two boxes are not identical; box 1 appears to play more of a role than box 2 in HIV-1 packaging (GORELICK et al. 1993), while the converse is true for RSV (BOWLES et al. 1993). Since the two boxes differ only in the non-conserved residues, these residues must be mediating the packaging differences between the two boxes. However, since no single-point mutation of any one of these nonconserved residues has been found to significantly impair RNA packaging to date, these residues may act together, in a redundant fashion.

A 1993 paper reported that an RSV mutant lacking both Cys-His boxes was able to encapsidate its genomic RNA with high efficiency, but that the encapsidated RNA was highly prone to degradation (ARONOFF et al. 1993). However, these authors have been unable to confirm the ability of this mutant to package RNA, and more recent results coincide with the results in MuLV and HIV-1 that the Cys-His boxes are required for efficient packaging (M. Linial, personal communication).

The basic arginine or lysine residues flanking the Cys-His box(es) are required for packaging in a less constrained manner than the conserved residues of the boxes. Substitution of individual basic residues do not impair RNA packaging, but substitution of several basic residues together cause moderate reductions (Fu et al. 1988; Housset et al. 1993; Rein et al. 1994). Interestingly, since lysines can replace arginines without any effect on RNA packaging, it appears that only the basic nature of these residues, and not their identity per se, is involved in packaging (Housset et al. 1993).

The studies described above indicate that NC is required for specific RNA packaging. Does NC mediate the specific recognition of the packaging signal, or does NC instead provide essential nonspecific contacts, with the specific recognition occurring elsewhere in Gag? Several studies have used chimeric Gag proteins to probe the specificity of RNA packaging. Replacement of the RSV NC domain with the M-MuLV NC domain, generating a chimeric RSV Gag polyprotein, severely impaired its ability to package its own (RSV) RNA, but enabled the Gag polyprotein to package M-MuLV RNA (Dupraz and Spahr 1992).Replacement of most of the M-MuLV NC domain with most of the HIV-1 NC domain resulted in a fivefold decrease in packaging of an M-MuLV RNA and a fourfold increase in packaging of an HIV-1 RNA (Zhang and Barklis 1995). Replacement of the entire HIV-1 NC domain with the entire M-MuLV NC domain, generating a chimeric HIV-1 Gag, enabled Gag to package an RNA containing the M-MuLV Ψ region over an RNA lacking the Ψ region (Berkowitz et al. 1995a). A chimeric HIV-1 Gag in which the M-MuLV box was flanked by the HIV-1 N- and C-terminal subdomains was not able to package the M-MuLV Ψ RNA, indicating that the proper flanking residues are required for the ability of the M-MuLV Cys-His box to recognize the M-MuLV packaging signal. In addition, though wild-type M-MuLV was able to package HIV-1 RNAs, only a chimeric M-MuLV Gag polyprotein containing the HIV-1 NC domain was able to preferentially encapsidate the HIV-1 genomic RNA over spliced HIV-1 RNAs. These studies indicate that NC specifically recognizes the packaging signal during RNA packaging.

Interestingly, many NC mutants have been found to be noninfectious though they were not significantly impaired in genomic RNA packaging. Since these mutants were also not impaired in virion assembly, the defect was presumed to occur during infection, in the production of the DNA provirus. Several of the RSV mutants were found to contain a large percentage of the packaged viral RNA in monomeric form (Bowles et al. 1993; Dupraz et al. 1990; Meric and Spahr 1986), indicating that in RSV NC is involved in either formation or, more likely (see Sect. 4), stabilization of the 70S dimer and that dimerization of the packaged RNA was critical for infectivity. Other infectivity mutants contained dimeric RNA (Bowles et al. 1993; Gorelick et al. 1993; Meric and Goff 1989; Meric et al. 1988), suggesting that NC plays a role in a postpackaging step during viral infection, perhaps in primer tRNA annealing to the genomic RNA (Housset et al. 1993).

3.3 In Vitro Activities of the NC Protein

In vitro, retroviral NC proteins have been shown to exhibit at least six different activities that may serve the same functions in vivo: nonspecific RNA binding, sequence-specific RNA binding, promotion of RNA dimerization, RNA dimer maturation, tRNA-genomic RNA annealing, and nucleic acid strand transfer. The sequence-specific RNA binding activity of NC may reflect the specific recognition of packaging elements by the NC domain of the Gag polyprotein which mediates genomic RNA encapsidation in vivo. The nonspecific RNA-binding activity may be responsible for the nonspecific coating of the RNA genome by free NC proteins in the mature virus particle. The last four activities are most likely different manifestations of NC's ability to destabilize helices (see below) during nonspecific RNA binding.

3.3.1 Nonspecific RNA-Binding Activity

The initial studies characterizing the structure of the mature virion found the RNA genome to be tightly associated with a few thousand molecules of the NC protein (BOLOGNESI et al. 1973; DAVIS and RUECKERT 1972; FLEISSNER and TRESS 1973; QUIGLEY et al. 1972). Subsequent studies of NC protein purified from virus particles demonstrated that NC could bind to DNA and RNA in vitro, with a preference for single-stranded over duplex nucleic acids (DAVIS et al. 1976; LEIS et al. 1978; SCHULEIN et al. 1978; SMITH and BAILEY 1979; SYKORA and MOELLING 1981). The binding appears to be a slightly- or noncooperative process in vitro (JENTOFT et al. 1988; KARPEL et al. 1987; SMITH and BAILEY 1979), and is consistent with formation of three ion-pairs with the RNA per NC molecule (SECNIK et al. 1992; SMITH and BAILEY 1979). NC binds to RNA with a saturating density of 1 NC per 4–7 nucleotides, depending on the identity of the NC molecule (JENTOFT et al. 1988; KARPEL et al. 1987; KHAN and GIEDROC 1994; SMITH and BAILEY 1979; YOU and MCHENRY 1993).

The basic residues of NC play a large role in its ability to bind to nucleic acids. Binding of RSV NC to RNA was abrogated by modification of all of the basic residues, indicating that the lysine and arginine residues form necessary interactions with the RNA (LEIS and JENTOFT 1983). Point mutation of the lysine doublet in between the RSV Cys-His boxes or immediately C-terminal to the M-MuLV NC Cys-His box impaired these NCs' abilities to bind to viral RNA segments (FU et al. 1988; PRATS et al. 1991). One explanation for these results is that the amino side chain groups of the basic residues bond with the phosphate groups of the RNA backbone. Indeed, if the N-terminal subdomain of HIV-1 NC formed an α-helix during RNA binding, as proposed, the five basic residues in this subdomain would lie along the same side of the helix (SUROVOY et al. 1993). A study analyzing the interactions of NC with individual components of nucleotides found that NC bound to the base and ribose moieties, but not the phosphate group (SECNIK et al. 1992); however, the binding of NC to individual nucleotide components may not reflect the binding to whole nucleotides in an RNA.

Binding of NC to nucleic acids is mediated by nonelectrostatic interactions as well as ion pairing, implying that hydrophobic residues of NC interact with hydrophobic regions of the RNA (DELAHUNTY et al. 1992). In addition, NC was seen to bind to RNA at pH 10, when NC should have a net negative charge

(Davis et al. 1976). Fluorescence studies and magnetic resonance analysis have indicated that the side chain ring of the conserved aromatic residue in the MuLV and HIV-1 Cys-His boxes intercalates into the RNA helix and forms stacking interactions with the bases (Casas-Finet et al. 1988; Delahunty et al. 1992; Karpel et al. 1987; South et al. 1990b; Summers et al. 1992). The isoleucine residue in box 1 of HIV-1 has also been suggested to interact with a DNA substrate (Delahunty et al. 1992; South et al. 1990b).

The Cys-His box per se is involved in nonspecific RNA binding only under certain in vitro conditions. Upon zinc binding to a fragment of HIV-1 NC, box 1 underwent a decrease in its RNA site size from 2.75 to 1.75 nucleotides as well as an increase in the apparent K_m for an RNA reporter molecule by almost 2 logs (Delahunty et al. 1992). However, modification or mutation of the cysteine residues of larger NC segments did not affect the ability of these NC segments to bind to nucleic acids in vitro (Jentoft et al. 1988; Karpel et al. 1987; Prats et al. 1991; Roberts et al. 1988; Surovoy et al. 1993). In addition, NC fragments lacking the Cys-His boxes altogether retained their nonspecific RNA-binding activity (Berkowitz and Goff 1994; Surovoy et al. 1993). Since a substitution of the box 1 histidine of an HIV-1 NC peptide with cysteine significantly impaired the ability of the peptide to bind to RNA (Déméné et al. 1994), mutations in the boxes can apparently alter the conformation of the protein in such a way as to interfere with its ability to bind to RNA.

One group has reported that phosphorylation of RSV or AMV NC increases its affinity for genomic viral RNA 100-fold in vitro (Leis et al. 1984). In addition, phosphorylation of NC was found to occur in vivo, primarily on the serine residue at position 40 (Leis et al. 1984). However, mutation of this serine to alanine did not affect virus replication (Fu et al. 1988), indicating that phosphorylation of NC is not required for efficient RNA packaging in vivo.

3.3.2 Promotion of Dimerization, Dimer Maturation, tRNA Annealing, and Strand Transfer

Binding of NC to tRNA has been demonstrated to result in denaturation of the tRNA, indicating that NC contained a helix-destabilizing activity that coincided with its RNA-binding activity (Khan and Giedroc 1992). In support of this result, NC was recently found to denature a short duplex DNA (Tsuchihashi and Brown 1994). NC has also been shown to promote the annealing of complementary single-stranded nucleic acids to each other, presumably by melting out local intramolecular structures in each molecule (Dib-Hajj et al. 1993; Tsuchihashi and Brown 1994). By facilitating the denaturing and annealing of nucleic acid structures, NC lowers the kinetic barrier to structural transformations, thereby expediting the establishment of equilibrium between different nucleic acid structures (Tsuchihashi and Brown 1994; Tsuchihashi et al. 1993). This equilibrium is regulated primarily by free energy differences between the different structures.

The ability of NC to promote the formation of the most stable structure from less stable structures probably accounts for the augmentation of RNA dimerization, dimer maturation, tRNA annealing, and DNA strand transfer by NC in vitro. Efficient in vitro dimerization of retroviral RNA segments containing the appropriate sequences (see Sect. 4) was originally shown to require NC (BIETH et al. 1990; DARLIX et al. 1990, 1992; DE ROCQUIGNY et al. 1991; PRATS et al. 1990; WEISS et al. 1992a), although subsequent studies have found that efficient dimerization can occur in the absence of NC if the dimerization buffer has a high ionic strength (BERKHOUT et al. 1993; MARQUET et al. 1991, 1994; ROY et al. 1990). RNA dimers formed in vitro contain a subset of thermolabile dimers analogous to immature RNA dimers found in vivo (see Sect. 4), addition of NC to the RNA population converts the thermolabile dimers to the more thermostable forms (Y.-X. Feng, T.D. Copeland, L.E. Henderson, J.G. Levin, and A. Rein, personal communication). Primer tRNA annealing to the template RNA in vitro is normally inefficient, because of the existence of intramolecular duplex structures in both the tRNA and template RNA; addition of NC to the annealing reaction or heat denaturation of the substrate RNAs dramatically increases the extent of annealing (BARAT et al. 1989, 1991; DARLIX et al. 1992; DE ROCQUIGNY et al. 1991; PRATS et al. 1988). NC has also been found to increase the extent of DNA strand transfer during reverse transcription in vitro (ALLAIN et al. 1994; LAPADAT-TAPOLSKY et al. 1993; PELISKA et al. 1994), presumably by melting out local structures in the single-stranded donor and acceptor strands. Mutagenesis of NC indicated that its ability to facilitate each of four of these processes is dependent on the basic residues, not on the Cys-His boxes (DE ROCQUIGNY et al. 1992; LAPADAT-TAPOLSKY et al. 1993; PRATS et al. 1991; Y.-X. Feng, T.D. Copeland, L.E. Henderson, J.G. Levin, and A. Rein, personal communication). However, mutations in the Cys-His boxes can interfere with NC's ability to augment these processes in vitro (DÉMÉNÉ et al. 1994; MORELLET et al. 1994).

3.3.3 Specific RNA-Binding Activity

Almost all of the early studies which compared the binding of NC or the Gag polyprotein to different single-stranded RNAs failed to detect a preference of the protein for the viral RNA encoding that protein. Several early studies found that RSV matrix protein p19 and mammalian p12 protein could bind with specificity to viral RNA (SEN et al. 1976; SEN and TODARO 1976, 1977). However, subsequent studies indicated that the RSV matrix protein does not bind to viral RNA with specificity (STEEG and VOGT 1990) and that it was possible that these proteins were in fact NC (DICKSON et al. 1985; MERIC et al. 1984). The only other early study to detect a preference for viral sequences used a nitrocellulose filter binding assay to determine that the Friend murine leukemia virus (F-MuLV) NC protein bound to F-MuLV RNAs 5- to 20-fold better than to cellular mRNAs (NISSEN-MEYER and ABRAHAM 1980).

As mentioned in Sect. 2.4, recombinant or synthetic HIV-1 NC proteins have recently been found to bind with specificity to HIV-1 RNAs from the 5′ end of the genome in vitro (Berkowitz and Goff 1994; Berkowitz et al. 1993; Clever et al. 1995; Dannull et al. 1994; Sakaguchi et al. 1993). Binding analysis of NC mutants in these assays has demonstrated that binding requires at least one Cys-His box and its flanking basic residues, exactly like the requirements for packaging in vivo. Since the basic residues of NC appear to mediate its nonspecific RNA-binding activity and yet are required for sequence-specific RNA-binding in vitro and in vivo, it is possible that genomic RNA encapsidation involves a nonspecific binding component as well as a specific binding component. Such a packaging mechanism occurs in the hepadnaviruses, where encapsidation of the genomic RNA requires specific recognition by the P protein and nonspecific interactions mediated by multiple basic residues in the C protein (Nassal 1993). Alternatively, the basic residues may, along with the Cys-His boxes, make sequence-specific contacts with the RNA.

Comparison of NC binding to very small deoxyoligonucleotides indicated that NC bound to guanosine differently than to the other nucleotides, in a manner that is dependent on zinc (South and Summers 1993). NMR analysis of a G-containing deoxypentanucleotide bound to an HIV-1 NC fragment indicated that the guanosine inserted into a hydrophobic cleft in the Cys-His box formed by the conserved Phe, Ile, and Ala box residues; multiple molecular interactions were seen to stabilize the complex and might thus determine the specificity of binding. Whether this cleft forms part of the Ψ-binding site in vivo, and whether the molecular environment of the cleft determines which RNAs are packaged, remains to be determined.

3.4 The BLV Matrix Precursor

As discussed in Sect. 2.5.1, in vitro the BLV MA cleavage precursor p15 contains the determinants for specific recognition of the putative BLV packaging signal, while the NC protein contains only nonspecific RNA-binding activity (Katoh et al. 1991). Further processing of p15 to the mature matrix protein, p10, abrogates its ability to bind with specificity to RNAs containing the packaging signal. Evidence for the role of the p15 domain of the BLV Gag polyprotein in the specific packaging of the BLV genomic RNA in vivo has yet to be provided.

4 RNA Dimerization

In the 1970s, the 70S RNA isolated from retrovirus particles was shown by various techniques to be a noncovalent dimer of two identical 35S genomic RNAs (for review, see Coffin 1984). The two RNAs were found, by electron microscopy, to be linked near their 5′ ends (Bender et al. 1978; Bender and Davidson 1976; Kung et al. 1976). The area of linkage, termed the dimer linkage structure (DLS), was approximately 50 nucleotides in length and lay approximately 500 nucleotides from the 5′ ends of RSV and M-MuLV (MURTI et al. 1981). The dimer was thought to be held together solely by contacts between the two RNAs, since phenol extraction of the 70S RNA did not cause its dissociation into the 35S subunits; indeed, protease digestion of the dimer did not lower its melting temperature (BENDER et al. 1978). 35S monomers isolated by heat denaturation of 70S RNA in vitro did not self-assemble back into dimers in vitro (FARAS et al. 1973; KUNG et al. 1976; MANGEL et al. 1974; MURTI et al. 1981).

Several lines of evidence initially suggested that dimerization normally occurs after the RNA is packaged and the virus is released from the cell. First, researchers were unable to isolate 70S RNA from infected cells. Second, rapidly harvested RSV or visna virus particles contained monomeric RNA which dimerized minutes (RSV) or hours (visna) after particle release (BRAHIC and VIGNE 1975; CANAANI et al. 1973; CHEUNG et al. 1972; KORB et al. 1976). Third, monomeric RNA in rapidly harvested virions was able to dimerize when the virus was incubated at 40°C, but not if the virus was first solubilized with detergent (CANAANI et al. 1973). Since virus maturation also occurred only after virus release, and in one case was shown to occur at the same time as dimerization of the RNA (KORB et al. 1976), these results raised the possiblity that RNA dimerization is linked to, or even dependent on, the cleavage of the Gag and Gag-Pol polyproteins. Since the NC protein is bound to the 70S RNA in mature virions (BOLOGNESI et al. 1973; DARLIX et al. 1990; DARLIX and SPAHR 1982; DAVIS and RUECKERT 1972; FLEISSNER and TRESS 1973; MERIC et al. 1984; PRATS et al. 1990; QUIGLEY et al. 1972) and has helix-destabilizing activity (KHAN and GIEDROC 1992; LEVIN and SEIDMAN 1981; SCHULEIN et al. 1978; SYKORA and MOELLING 1981; TSUCHIHASHI and BROWN 1994), it appeared that NC could contribute to dimerization. In this model, the NC protein, after release from the Gag polyprotein, binds to each RNA and denatures local double-stranded structures near the 5′ ends, allowing the RNAs to form interstrand contacts. This model appeared to be supported by the observations that prevention of Gag polyprotein cleavage by protease inactivation (OERTLE and SPAHR 1990; STEWART et al. 1990) or mutation of NC (DUPRAZ et al. 1990) resulted in avian retrovirus particles that contained monomeric RNA.

Rapidly harvested virions of murine retroviruses were found to contain dimeric, not monomeric RNA (EAST et al. 1973; RIGGIN et al. 1974). In addition, a later study on RSV indicated that, contrary to previous studies, immature virus contained dimeric RNA which had a lower melting temperature than that of 70S RNA from mature virus and which would dissociate if extracted from the immature virions by standard techniques (STOLTZFUS and SNYDER 1975). Because of these and other studies (BOWLES et al. 1993; DUPRAZ et al. 1990; GORELICK et al. 1993; LEVIN et al. 1974; MERIC and GOFF 1989; MESSER et al. 1981), FU and REIN recently analyzed the state of M-MuLV and HIV-1 genomic RNA in mature and immature virus (FU et al. 1994; FU and REIN 1993). These authors found that protease-deficient or rapidly harvested (i.e. "immature") virions contained dimeric RNA which had a melting temperature approximately 5°C lower than the wild-type 70S RNA from mature virus. In addition, this "immature dimer" had a lower electrophoretic mobility and sedimentation coefficient than the mature dimer, indicating that the conformation of the dimer changed during its maturation. Indeed, RNA dimers sedimenting slightly slower than mature 70S dimers were detected in several of the early studies on rapidly harvested RSV (CHEUNG et al. 1972; STOLTZFUS and SNYDER 1975) and M-MuLV (EAST et al. 1973) particles. The role of NC in mediating these conformational changes in the RNA was supported by the inability of an NC mutant to mature its dimer (FU and REIN 1993) . NC has also been found to mature dimers of Harvey sarcoma virus (HaSV) RNA segments in vitro (Y.-X. Feng, T.D. Copeland, L.E. Henderson, J.G. Levin, and A. Rein, personal communication).

Thus, it has become apparent that mature NC does not mediate dimerization per se; the RNAs dimerize prior to virus maturation. Instead, NC appears to induce conformational changes in the dimer which give the dimer greater stability. Since AMV RNA dimer maturation occurs concomitantly with virus maturation (KORB et al. 1976), these conformational changes appear to occur relatively quickly for AMV. However, since dimer maturation in M-MuLV (FU and REIN 1993) and visna virus (BRAHIC and VIGNE 1975) lags behind virus maturation, as detected by western blot analysis or electron microscopy, the conformational changes are presumed to occur relatively slowly in these viruses.

If NC is not necessary for RNA dimerization, why is it difficult to anneal 35S RNA in vitro? One possibility is that early researchers did not use high enough concentrations of 35S RNA. By using high concentrations of genomic RNA segments synthesized in vitro with viral polymerases, Darlix and coworkers discovered that certain RNAs could dimerize spontaneously, and that the NC protein catalyzed the dimerization (BIETH et al. 1990; DARLIX et al. 1990, 1992; MARQUET et al. 1991; PRATS et al. 1990; ROY et al. 1990; TORRENT et al. 1994a, b). The ability of the RNAs to dimerize was diminished as their concentration was lowered (BERKHOUT et al. 1993; MARQUET et al. 1991; ROY et al. 1990). Analysis of sequences needed *in cis* for dimerization of viral RNA segments in vitro indicated that the dimerization domains coincide with the Ψ or E packaging regions in RSV (BIETH et al. 1990), M-MuLV (PRATS et al. 1990), REV-A (DARLIX et al. 1992), VL30 (TORRENT et al. 1994a, b), and HIV-1 (DARLIX et al. 1990). In some cases, small (67–207 nt) RNAs from these regions were sufficient for dimerization.

Later, evidence was provided that the linkage occurring in the in vitro-synthesized dimers was not standard Watson-Crick base-pairing, but instead might include a tetrahelical structure mediated by purine quartets (MARQUET et al. 1991). Purine quartets are regions in which four purine molecules, two from each RNA monomer, coordinate one monovalent cation. All retroviruses contain, in the 5′ portion of the viral intron, at least two polypurine tracts, sharing the 6-nucleotide consensus PuGGAPuA. In fact, polypurine tracts lie within the regions previously identified to be necessary for in vitro dimerization of M-MuLV (PRATS et al. 1990), REV-A (DARLIX et al. 1992), and HIV-1 (DARLIX et al. 1990) RNAs. A

1993 study provided biochemical evidence, i.e., a higher thermostability in the presence of potassium than other monovalent cations, that a dimer of a 98-nt HIV-1 NL4-3 RNA, spanning the *gag* start codon, was linked by guanosine quartets (G-quartets); removal of the 3′ 12 nucleotides, which contained one of the polypurine tracts, eliminated the RNA's ability to dimerize (SUNDQUIST and HEAPHY 1993). The formation of G-quartets in a similar HIV-1 RNA in vitro has also been demonstrated by another group (AWANG and SEN 1993).

However, several groups argued that polypurine tracts are not important for RNA dimerization. An RNA containing only the 5′ 255 nucleotides of the HIV-2 genome, laying entirely upstream of the PBS, could dimerize as efficiently as an HIV-2 RNA extending into the *gag* coding region and including five polypurine tracts (BERKHOUT et al. 1993). The determinants for in vitro dimerization of BLV RNA lay entirely upstream of the splice donor, in a 128-nucleotide region spanning the PBS and containing only one polypurine tract (KATOH et al. 1993). In response to these studies, the dimerization abilities of HIV-1 RNAs laying either upstream or downstream of the splice donor were re-investigated (MARQUET et al. 1994). Using RNAs from strain Mal, the authors found that, in addition to the previously identified RNA spanning nucleotides #311–415 (DARLIX et al. 1990; MARQUET et al. 1991), an RNA spanning nucleotides #1–311 could dimerize in vitro. The inability of this RNA to dimerize in previous experiments was attributed to its dissociation during electrophoresis in gels lacking Mg^{2+}. The properties of formation and maintenance of these two nonoverlapping RNAs were found to be different; RNA #1–311 dimerized relatively quickly, in a cation-dependent manner, and was relatively thermolabile. RNA #311–415, on the other hand, dimerized relatively slowly, in an extremely thermostable, cation-independent manner, presumably via G- or purine-quartets. An RNA containing both of these regions exhibited properties similar to RNA #1–311, indicating that dimerization of large RNAs containing both regions primarily involved linkages similar to those formed in the upstream region. However, since deletion of several of the downstream purine-rich regions decreased the thermostability of the large RNA in vitro (MARQUET et al. 1994; PAILLART et al. 1994), it is possible that some of the purine-rich regions may stabilize the RNA dimer in vivo via non-Watson-Crick base-pairing, perhaps involving parallel interactions. This stabilization does not appear to be mediated by G-quartets (FU et al. 1994).

What is the nature of the linkage(s) in the upstream region of the large HIV-1 RNA? Dimerization of this RNA was shown to require sequences in a small region upstream of the splice donor, termed the dimerization inititation site (DIS; SKRIPKIN et al. 1994). The DIS contains a stem-loop structure (SL1 in Fig. 1) implicated in RNA dimerization by two other groups (KATOH et al. 1993; LAUGHREA and JETTÉ 1994). In the 8-nt loop lies a 6-nt palindrome whose palindromic nature, but not its primary structure, is conserved among all HIV-1 strains. A model of dimerization (LAUGHREA and JETTÉ 1994; SKRIPKIN et al. 1994), termed the kissing loop model (LAUGHREA and JETTÉ 1994), suggests that the six palindromic loop nucleotides on one RNA bind to the same six nucleotides, in an antiparallel manner, on the other RNA (Fig. 3). During or after this binding, the stem on each RNA dissociates and then reforms with the opposing sides from the other RNA, extending the antiparallel contact between the two RNAs from 6 to 28 bp. Recently, the loop-loop antiparallel interaction has been supported by elegant in vitro compensatory mutational studies (PAILLART et al. 1994). However, studies addressing the formation of the extended duplex have yielded conflicting results (R. Marquet and C. Ehresmann; M. Laughrea and N. Jette; J. Clever and T. Parslow, personal communications).

Other retroviruses appear to have a similar palindromic stem-loop. In M-MuLV, a completely conserved, 16-nucleotide perfect palindrome precedes the GACG double hairpin motif; structural analysis of the leader RNA indicated that this palindrome formed an unstable stem-loop in the monomeric form of the RNA but was completely base-paired in the dimeric form (TOUNEKTI et al. 1992). When the leader RNA existed as a monomer, a 30-nucleotide region containing the palindrome was available for hybridization with a complementary oligonucleotide, but the oligonucleotide did not hybridize to the dimeric form of the leader RNA (PRATS et al. 1990). In BLV, a stem-loop containing a 6-nucleotide palindrome in its loop has also been predicted to lie in the leader, upstream of the splice donor (KATOH et al. 1993), although structural analysis of the BLV leader RNA has not been performed.

In summary, the studies on the in vitro dimerization of retroviral leader segments suggest that the 70S RNA found in mature virus is maintained by at least two kinds of linkages: an antiparallel linkage consisting of Watson-Crick base-pairing in the region upstream of the splice donor, and a linkage mediated by purine-rich sequences downstream from the splice donor. Given that packaged genomic RNAs initially dimerize in a relatively unstable manner, and that the dimerization of HIV-1 RNA #1–311 is a relatively quick but weak phenomenon (MARQUET et al. 1994), it is tempting to speculate that genomic RNAs initially dimerize by base-pairing upstream of the splice donor (i.e., by the kissing loop model). This idea could be addressed by characterization of the immature dimer encapsidated by protease-deficient virus. It is also tempting to speculate that, since NC catalyzes stabilizing conformational changes in the immature dimer in vivo and since the downstream linkage provides increased stability to the dimeric RNA segments in vitro, NC induces dimer maturation through the formation of the downstream linkage. Before this idea could be investigated, the exact nature of the downstream interaction would have to be identified.

5 Primer tRNA Packaging

The process of reverse transcription begins with synthesis of the minus strand of the DNA provirus. The primer for this strand is a tRNA of cellular origin, termed the primer tRNA, that is encapsidated with, but independently of, the genomic RNA (for review see COFFIN 1984; VARMUS and SWANSTROM 1984;

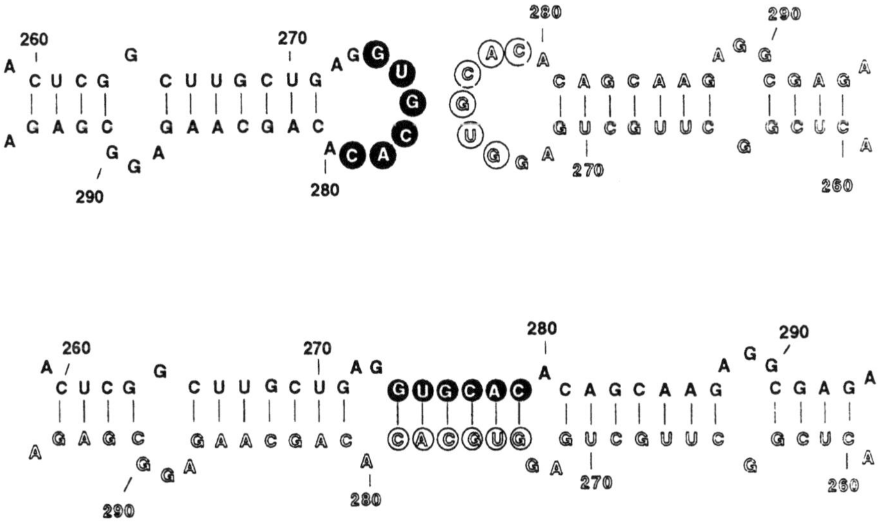

Fig. 3. Kissing loop model of the initiation of HIV-1 RNA dimerization. *Top*, stem-loop #1 (see Fig. 1) from each of two RNAs aligned in an antiparallel fashion. Sequences in the loop of one RNA (*encircled*) anneal to the same sequences of the other RNA, due to the palindromic nature of the loop. Next, the stems on each RNA dissociate, with the sides of each stem re-annealing to the complementary sides from the other RNA. *Bottom*, the structure of the duplex region after dimerization. (Courtesy of Drs. Chantal Ehresmann and Roland Marquet.) Note that the HIV-1 sequences depicted here are from strain Mal, unlike the Hxb2 sequences depicted in Fig. 1

WATERS and MULLIN 1977). Avian retroviruses use tRNA^{Trp}, REV and some mammalian retroviruses including M-MuLV use tRNA^{Pro}, and other mammalian retroviruses including HIV-1 use tRNA^{Lys3}. The 3′ 16–19 nucleotides of the primer tRNA hybridize to a perfectly complementary sequence, termed the primer-binding site (PBS), immediately downstream of the U5 region in the genomic RNA. Reverse transcriptase begins synthesizing the minus strand by extending the 3′ end of the tRNA, using the U5 region as its template (for a review of reverse transcription see TELESNITSKY and GOFF 1993).

Retrovirus particles contain a repertoire of about 30 tRNA molecules (MAK et al. 1994; SAWYER and DAHLBERG 1973), which differs from the tRNA repertoire of the infected cell, indicating that some tRNAs are encapsidated more efficiently than others. The primer tRNA has the highest efficiency of encapsidation; each virion contains six to ten molecules, representing 20%–30% of the virion-associated tRNA population (LEVIN and SEIDMAN 1979; MAK et al. 1994; PETERS et al. 1977; SAWYER and DAHLBERG 1973). In contrast, the primer tRNA comprises less than 5% of the cellular tRNA population. Approximately ten of the virion-associated tRNAs are associated with the genomic RNA dimer (ERIKSON and ERIKSON 1971; FARAS et al. 1973; SAWYER and DAHLBERG 1973); these include the two primer tRNAs which bind strongly to the two PBS elements prior to dimer maturation (CRAWFORD and GOFF 1985; STEWART et al. 1990) and eight other tRNAs which bind less strongly (FARAS et al. 1973) to other sites on the dimer during virus maturation (CANAANI et al. 1973; CHEUNG et al. 1972).

The selectivity of encapsidation of the primer tRNAs appears to be mediated by the RT domain of the Gag-Pol polyprotein. First, selective primer tRNA packaging occurs in virus that is deficient in genomic RNA (LEVIN and SEIDMAN 1979; MAK et al. 1994; PETERS and HU 1980) or that lacks the integrase domain of the Gag-Pol polyprotein (LEVIN and SEIDMAN 1981; MAK et al. 1994), but not in virus that lacks the entire Gag-Pol polyprotein (LEVIN and SEIDMAN 1981; PETERS and HU 1980; SAWYER and HANAFUSA 1979) or contains a large deletion in the RT domain (MAK et al. 1994). Point mutations in the connection domain of HIV-1 RT can also abolish specific primer $tRNA^{Lys3}$ encapsidation (J. Mak and L. Kleiman, personal communication). Second, purified mature RT protein from avian retroviruses binds to primer $tRNA^{Trp}$ with specificity in vitro in a limited number of assay systems (HASELTINE et al. 1977; PANET et al. 1975; PETERS and HU 1980). Of course, since tRNAs are encapsidated when the RT protein is contained within the Gag-Pol polyprotein and not when RT exists as a free dimer, these in vitro binding assays may reflect a nonphysiological activity.

Murine retroviruses exhibit less enrichment for the primer tRNA during tRNA packaging than avian viruses and HIV-1 (LEVIN and SEIDMAN 1981; WATERS and MULLIN 1977), due to the relevant abundance of $tRNA^{Pro}$ in the cytoplasm. Correspondingly, mature MuLV RT protein does not bind to primer $tRNA^{Pro}$ with specificity in vitro (HASELTINE et al. 1977; PANET and BERLINER1978). 70S RNA from RT-deficient MuLV virions contains annealed primer tRNA (LEVIN and SEIDMAN 1981; W. Fu and A. Rein, personal communication), while 70S RNA from RT-deficient ASV virions does not (PETERS and HU 1980; SAWYER and HANAFUSA 1979), reflecting the difference in the need for RT-mediated primer tRNA enrichment in these two viruses.

The rest of the virion-associated tRNAs appear to be packaged by non-specific interactions with either RT or NC in the context of their precursor polyproteins. Mature RTs from many retroviruses are capable of binding many, if not all, tRNAs in vitro (PANET and BERLINER 1978; PETERS and HU 1980; ROBERT et al. 1990; WEISS et al. 1992b), and NC also possesses nonspecific RNA-binding activity (see Sect. 3.3). Since RT-deficient virions encapsidate wild-type amounts of tRNA, albeit without specificity for the primer tRNA (LEVIN and SEIDMAN 1981; PETERS and HU 1980; SAWYER and HANAFUSA 1979), NC may normally mediate nonspecific tRNA encapsidation in vivo.

It is interesting to speculate on the factors that determine the amounts of the tRNA packaged by virions. For example, why are only 6–10 primer tRNA molecules packaged, when 50–100 Gag-Pol polyproteins enter the assembling virion? In addition, why are only 20–30 other tRNAs encapsidated? Perhaps the assembling virion has space for only 30 or so tRNAs, facilitating a competition between RT molecules bound to primer tRNAs and NC molecules bound to other tRNAs. Alternatively, perhaps RT mediates both primer-specific and non-specific tRNA encapsidation, with one tRNA packaged per Gag-Pol dimer. Indeed, the ratio of RT dimers to tRNA molecules in mature virions is approximately 1:1.

What part of the primer tRNA is recognized by RT during virion assembly? Cross-linking analysis of the RT-tRNA complex in vitro indicated that the anticodon loop of the tRNA is bound to RT (ARAYA et al. 1980; BARAT et al. 1989). However, $tRNA^{Lys1}$ and $tRNA^{Lys2}$ are packaged by HIV-1 as well as primer $tRNA^{Lys3}$ though they contain several different bases in the anticodon loop (MAK et al. 1994); in addition, mutation of two of the three bases of the anticodon loop did not decrease the level of tRNA incorporation (HUANG et al. 1994). Furthermore, several studies indicated that the anticodon loop does not mediate the specific binding of the primer tRNA by RT protein in vitro (HU and DAHLBERG 1983; WEISS et al. 1992b). Fragments of the RSV primer $tRNA^{Trp}$ do not bind with specificity to RSV RT in vitro (CORDELL et al. 1979; HASELTINE et al. 1977), suggesting that the entire structure of the tRNA is either directly or indirectly involved in its binding to RT. However, a tRNA containing only the 5′ 67 nucleotides could bind to RT, indicating that most of the acceptor stem is not recognized by RT (BAROUDY and CHIRIKJIAN 1980).

6 Other Issues Involved in RNA Packaging

In this review, we have tried to present a comprehensive survey of the wealth of data on RNA packaging. Although significant advances have been made towards defining the packaging signals on the genomic RNA and the packaging site on the Gag polyprotein, much remains unclear. The packaging elements in the RNA and protein must be unequivocally identified, and the exact mechanism of their interaction described. In addition, there are issues regarding the packaging process which are still unresolved at this time.

6.1 Is Dimerization Required for Packaging?

Several lines of evidence support the notion that dimerization of the genomic RNA is a prerequisite to packaging. First, the packaging elements of every retrovirus analyzed so far colocalize with the dimerization initiation or stabilization domains identified in vitro (see Sect. 2, 4). Perhaps the best example of the linkage of dimerization and packaging is the 67-nt VL30 RNA demonstrated to be sufficient for both processes (TORRENT et al. 1994b), in which a single-point mutation interfered with both processes (TORRENT et al. 1994a). Conceptually, there is no reason why the packaging and dimerization domains should overlap unless one process is dependent on the other. Second, in conditions in which virions encapsidate genomic RNA poorly, the low level of packaged RNA still consists of dimers, not monomers. When mutations were introduced into NC, the mutants which packaged very low levels of genomic RNA contained only dimeric RNA (Bowles et al. 1993; Dupraz et al. 1990; Gorelick et al. 1993; Meric and Goff 1989). In addition, when wild-type M-MuLV-infected cells were treated with actinomycin D, a drug that blocks transcription, RNA packaging was found to drop off over the course of 3 h, presumably due to the reduction in the concentration of genomic RNA in the cytoplasm (Levin et al. 1974). However, the viral RNA encapsidated during this period was dimeric, not monomeric.

It should be noted that monomeric viral genomic RNA has often been isolated from virus particles. However, because of the instablility of the immature dimer, it is likely that these monomers originated from immature dimers that did not survive the isolation procedure. As discussed in Sect. 4, any virus which has not had a chance to mature its dimer, e.g., rapidly harvested, protease-deficient, or NC mutant virus, is likely to yield monomeric RNA if the extraction procedure is performed with buffers containing strong detergents or low cation concentrations.

Several in vitro results do not support the idea that dimerization is required for packaging. First, HIV-1 NC protein has been shown to bind with specificity to monomeric HIV-1 RNA fragments as well as dimeric HIV-1 RNA fragments in vitro (Berkowitz et al. 1993). Second, retroviral RNAs lacking either of the two elements which mediate dimerization of RNAs in vitro are able to be packaged. Deletion of the 6-nt palindrome from the loop of the DIS stem-loop of HIV-1 did not affect the virus' ability to replicate (G. Harrison and A.M.L. Lever, personal communication); in addition, an HIV-1 mutant containing a deletion of the entire stem-loop was found to replicate with only moderately reduced kinetics (Kim et al. 1994). Direct examination of the status of the virion RNA in the DIS mutants has not yet been reported. In addition, a 160-nt RSV segment lacking the polypurine tracts can mediate packaging of heterologous RNAs (A. Yeo and M. Linial, personal communication), indicating that the polypurine-mediated linkage is not required for packaging of retroviral RNAs.

6.2 Is Genomic RNA Translation and Packaging Regulated?

Extremely little is known about the factors which determine whether newly synthesized viral genomic RNAs will be translated or packaged. It is possible that there is no regulation of these activities, i.e. all of the RNAs are competent for translation and packaging, and each process involves the arbitrary selection of one of these RNAs from a single pool. However, it is also possible that all of the RNAs are not equivalent; perhaps some RNAs are competent only for translation, or packaging. Indeed, it is even possible that the RNAs exist in two mutually exclusive pools – one pool which can only be translated and the other pool which can only be packaged. Experiments aimed at discriminating between these possibilities have rarely been reported, and the few published reports do not strongly favor any one of the possibilities.

An early study provided evidence for the existence of two different pools of retroviral genomic RNA (MESSER et al. 1981). MuLV-infected cells continued to produce virus for 12 h after addition of actinomycin D to the culture; however, incorporation of virion RNA stopped after the 3rd h. These results suggest that a pool of genomic RNAs competent only for translation existed in the cells and had a longer half-life than the pool of genomic RNAs competent for packaging.

In 1982 it was proposed that NC might regulate the ability of the RSV genomic RNA to be translated, since addition of NC reduced the extent of translation of genomic RNA isolated from virus in vitro (DARLIX and SPAHR 1982). However, in these assays, NC was also found to inhibit reverse transcription of the RNA. This result suggests that the conditions used for the inhibitions were not relevant to the in vivo situation, since recent papers have indicated that NC promotes reverse transcription under conditions more likely to be physiological (HOUSSET et al. 1993; LAPADAT-TAPOLSKY et al. 1993; PRATS et al. 1988; TSUCHIHASHI and BROWN 1994; WEISS et al. 1992a). Later, it was proposed that the RSV genomic RNA could exist in two different conformations, one competent only for translation and the other competent only for packaging (DARLIX 1986); however, this proposal was based mostly on computer-generated structure predictions of very large RNAs. Even later, it was demonstrated that dimerization of a 5′ segment of the RSV or HIV-1 genomic RNA inhibited translation of the RNA (BAUDIN et al. 1993; BIETH et al. 1990). Although the dimerization of the genomic RNA could, in theory, interfere with translation in vivo, no evidence for this mechanism in wild-type virus has been reported.

RNA dimerization has been suggested to be required for packaging, but a direct dependence has not been firmly established (see Sect. 6.1). Even if dimerization of the genomic RNA switched the RNA from translation to packaging, the factors which regulate dimerization would require identification. Perhaps all genomic RNAs are destined for dimerization, but only dimerize at a certain concentration or age.

All avian retroviruses contain three small open reading frames (ORFs) upstream of the *gag* gene which are conserved in position, length, and nucleotide contexts, but not in the sequence of encoded amino acids (HACKETT et al. 1992). Interestingly, translation of the 3′-most ORF, ORF3, is required for the genomic RNA to be packaged (DONZE and SPAHR 1992; MOUSTAKAS et al. 1993). Since ORF3 is adjacent to the O3 stem-loop, which serves as an important packaging element, it is possible that translation of ORF3 changes the conformation of the O3 stem-loop, perhaps creating a secondary structure that serves as the packaging element (KNIGHT et al. 1994). If this model were true, nascent avian retrovirus genomic RNAs would be competent only for translation, with competency for packaging achieved only after association with polyribosomes in the cytoplasm. Interestingly, if Gag binding to the RNA blocked subsequent initiation of translation of the RNA, polyribosome-associated RNAs would eventually be competent only for packaging.

Translation of the HIV-1 Gag polyprotein has been shown to be inhibited by the TAR stem-loop structure at the 5′ end of the genomic RNA (PARKIN et al.

1988). Binding of the La autoantigen to the TAR element, which occurs with specificity in vitro (CHANG et al. 1994), relieves the translational inhibition (SVITKIN et al. 1994). Since Gag polyproteins are found in the nucleus of productively infected cells, it is interesting to speculate, at least in the case of HIV-1, that the fate of the retroviral genomic RNA is decided in the nucleus, just after transcription. Perhaps nascent genomic RNA can bind to either the Gag polyprotein or the La autoantigen, but not both; if the RNA binds to the Gag polyprotein, it will be packaged, but if the RNA binds to the La autoantigen, it will be translated. Of course, this idea is purely speculative. To add another layer of mystery to this idea, we have recently found that the M-MuLV Gag polyprotein can bind to a nuclear RNA-binding protein, which we have termed GIP2 (K. Ålin and S.P. Goff, unpublished observations).

6.3 How is Genomic RNA Packaged over Spliced RNA?

Retroviral genomic RNA is preferentially packaged over spliced RNA. The simplest possible mechanism to insure that genomic RNA is packaged over spliced RNA involves the localization of the packaging signal to the intron of the viral genomic RNA. This mechanism may occur in REV and the murine retroviruses, where the only packaging element (the GACG double hairpin) firmly identified so far resides downstream of the splice donor (see Sect. 2.2, 2.3). However, in all of the retroviruses, there is indirect evidence for the existence of packaging elements in the spliced RNAs, probably in the 5′ region of the RNAs. First, a REV LTR-based vector containing very little sequence from the viral intron was packaged tenfold more efficiently than a heterologous vector (EMBRETSON and TEMIN 1987). Second, deletion of the viral intronic sequences, including the Ψ region, from an M-MuLV LTR-based vector reduced the transduction titer 3000-fold, but the Ψ^- vector still exhibited a titer of 10^3 cfu/ml (MANN and BALTIMORE 1985). Third, HIV-1 and RSV particles encapsidate fairly high levels of spliced RNAs, on the order of 10% of the levels of the genomic RNA (KATZ et al. 1986; LUBAN and GOFF 1994). Fourth, studies involving the cotransfection of cells with HIV-1 and nonviral vectors indicate that HIV-1-spliced RNAs are packaged much more efficiently than heterologous RNAs (BERKOWITZ et al. 1995a). Fifth, in avian retroviruses the only packaging elements identified so far reside in the spliced RNAs as well as the genomic RNA (KATZ et al. 1986; KNIGHT et al. 1994; SORGE et al. 1983). Thus, another mechanism of packaging, which takes these results into account, involves a packaging signal composed of at least two packaging elements – one in the viral intron and the other in a region of the genomic RNA shared with the spliced RNAs. Since the spliced RNAs would contain a packaging element, they would be packaged, but with less efficiency than the genomic RNA bearing two packaging elements. This model of packaging is supported by our observation that an HIV-1 mutant with a deletion in the viral intron encapsidates its spliced RNAs and its genomic RNA with the same efficiency (LUBAN and GOFF 1994).

It should be mentioned that it is hard to understand why retroviruses would contain packaging elements in their spliced RNAs. Packaging of viral *env* mRNAs would reduce the number of cells able to support viral replication, since infection of a cell by a particle containing a dimer of *env* mRNA would lead to the expression of envelope proteins in the cell, causing it to become resistant to superinfection by the normal virus particles.

Two issues related to the number of packaging elements are the number of binding sites in the Gag monomer and whether an RNA is packaged via interactions with one or multiple Gag monomers. If we assume that each Cys-His box contains one binding site for one packaging element, we can find evidence both for packaging by one monomer and packaging by multiple monomers. Evidence for the former lies in our finding that HIV-1 mutants containing only one Cys-His box encapsidate their spliced RNAs and genomic RNA with equal efficiency (Berkowitz et al. 1995a). The mutants have lost the ability to discriminate between the two RNAs presumably because a second Cys-His box is required to recognize the second packaging element. Since two mutant Gag monomers could in theory cooperate to select the genomic RNA, the RNAs are apparently being packaged by interacting with only one Gag monomer.

Other results, however, suggest either that RNAs are packaged by interacting with multiple Gag monomers or that our assumption – that one box binds one element – is wrong. First, SNV and the murine retroviruses contain only one Cys-His box, but, as mentioned previously in this section, appear to have packaging elements outside of the viral intron as well as within it. Second, in vivo results indicate that HIV-1, SNV, and the murine retroviruses may have more than one packaging element in their viral introns (see Sect. 2). Third, in vitro binding assays have demonstrated the existence of multiple binding elements in the 5′ portion of the HIV-1 and BLV viral introns (Berkowitz and Goff 1994; Clever et al. 1995; Katoh et al. 1993). The recognition of multiple elements could be achieved by one Gag monomer if each Cys-His box could bind to multiple packaging elements, or if there were other binding sites in Gag in addition to the Cys-His boxes.

Alternatively, the preferential packaging of genomic RNAs containing multiple packaging elements over spliced RNAs containing one or several packaging elements would be possible if RNAs were packaged by a simultaneous interaction with a large number of Gag polyproteins. If RNAs were in fact packaged by multiple Gag polyproteins, the efficiency of packaging of an RNA would be directly dependent on the number of packaging elements it possessed. In this regard, it is possible that the genome is peppered with packaging elements. Analysis of the primary sequence does not, in most cases (Kim et al. 1994), reveal consensus motifs, but it is possible that the elements are defined by their secondary structure instead of their primary structure.

6.4 Other Questions Not Yet Investigated

Several issues of retroviral RNA packaging are still poorly understood. What proportion of the virions actually contain genomic RNA? Why is only one dimeric RNA packaged per virion? Is the virion too small to house a second dimer? If there is little space left in the virion after genomic RNA packaging, how can so many small molecules enter the virion? How do 1500-Gag monomers bind to two RNAs? Does each Gag monomer in the cytoplasm bind to a genomic RNA, or are there "special" Gag monomers that bind to the RNAs, or is the genomic RNA only bound by a Gag multimer in the assembling capsid? If each Gag monomer binds to an RNA, and only two RNAs can fit inside of each virion, are all but two of the Gag-RNA interactions broken during virion assembly? Do different retroviruses use different and incompatible packaging signals? On this last issue, REV packages M-MuSV RNA (Yang and Temin 1994), HIV-1 packages SIV RNA (Rizvi and Panganiban 1993), and M-MuLV packages HIV-1 RNA (Berkowitz et al. 1995a; Zhang and Barklis 1995), but HIV-1 appears to be unable to package MuLV RNA (Delwart et al. 1992; Berkowitz et al. 1995a).

References

Adam MA, Miller AD (1988) Identification of a signal in a murine retrovirus that is sufficient for packaging of nonretroviral RNA into virions. J Virol 62: 3802–3806

Adkins B, Hunter T (1981) Identification of a packaged cellular mRNA in virions of Rous sarcoma virus. J Virol 39: 471–480

Aldovini A, Young R (1990) Mutations of RNA and protein sequences involved in human immunodeficiency virus type 1 packaging result in production of noninfectious virus. J Virol 64: 1920–1926

Alford RL, Honda S, Lawrence CB, Belmont JW (1991) RNA secondary structure analysis of the packaging signal for Moloney murine leukemia virus. Virology 183: 611–619

Allain B, Lapadat-Tapolsky M, Berlioz C, Darlix J-L (1994) Transactivation of the minus-strand DNA transfer by nucleocapsid protein during reverse transcription of the retroviral genome. EMBO J 13: 973–981

Anderson DJ, Lee P, Levine KL et al (1992) Molecular cloning and characterization of the RNA packaging-defective retrovirus SE21Q1b. J Virol 66: 204–216

Araya A, Keith G, Fournier M, Gandar JC, Labouesse J, Litvak S (1980) Photochemical cross-linking studies on the interactions of avian myeloblastosis virus reverse transcriptase with primer $tRNA^{Trp}$ and TTP. Arch Biochem Biophys 205: 437–448

Armentano D, Yu S, Kantoff PW, von Ruden T, Anderson WF, Gilboa E (1987) Effect of internal viral sequences on the utility of retroviral vectors. J Virol 61: 1647–1650

Aronoff R, Linial M (1991) Specificity of retroviral RNA packaging. J Virol 65: 71–80

Aronoff R, Hajjar AM, Linial ML (1993) Avian retroviral RNA encapsidation: reexamination of functional 5′ RNA sequences and the role of the nucleocapsid cys-his motifs. J Virol 67: 178–188

Awang G, Sen D (1993) Mode of dimerization of HIV-1 genomic RNA. Biochemistry 32: 11453–11457

Barat C, Lullien V, Schatz O et al (1989) HIV-1 reverse transcriptase specifically interacts with the anticodon domain of its cognate primer tRNA. EMBO J 8: 3279–3285

Barat C, Le Grice SFJ, Darlix J-L (1991) Interaction of HIV-1 reverse transcriptase with a synthetic form of its replication primer, $tRNA^{Lys,3}$. Nucleic Acid Res 19: 751–757

Baroudy BM, Chirikjian JG (1980) Structural requirements for binding of bovine $tRNA^{Trp}$ with avian myeloblastosis virus DNA polymerase. Nucleic Acid Res 8: 57–66

Bartenschlager R, Junker-Niepmann J, Schaller H (1990) The P gene product of hepatitis B virus is required as a structural component for genomic RNA encapsidation. J Virol 64: 5324-5332
Baudin F, Marquet R, Isel C, Darlix J-L, Ehresmann B, Ehresmann C (1993) Functional sites in the 5′ region of human immunodeficiency virus type 1 RNA form defined structural domains. J Mol Biol 229: 382–397
Bender MA, Palmer TD, Gelinas RE, Miller AD (1987) Evidence that the packaging signal of Moloney murine leukemia virus extends into the *gag* region. J Virol 61: 1639–1646
Bender W, Davidson N (1976) Mapping of poly(A) sequences in the electron microscope reveals unusual structure of type C oncornavirus RNA molecules. Cell 7: 595–607
Bender W, Chien Y-H, Chattopadhyay S, Vogt PK, Gardner MB, Davidson N (1978) High-molecular weight RNAs of AKR, NZB, and wild mouse viruses and avian reticuloendotheliosis virus all have similar dimer structures. J Virol 25: 888–896
Berg J (1986) Potential metal-binding domains in nucleic acid binding proteins. Science 232: 485–487
Berkhout B, Essink BBO, Schoneveld I (1993) In vitro dimerization of HIV-2 leader RNA in the absence of PuGGAPu motifs. FASEB J 7: 181–187
Berkowitz RD, Goff SP (1994) Analysis of binding elements in the human immunodeficiency virus type 1 genomic RNA and nucleocapsid protein. Virology 202: 233–246
Berkowitz RD, Luban J, Goff SP (1993) Specific binding of human immunodeficiency virus type 1 *gag* polyprotein and nucleocapsid protein to viral RNAs detected by RNA mobility shift assays. J Virol 67: 7190–7200
Berkowitz RD, Ohagen Å, Höglund S, Goff SP (1995a) Retroviral nucleocapsid domains mediate the specific recognition of genomic viral RNAs by chimeric Gag polyproteins during RNA packaging in vitro. J Virol 69: 6445–6456
Berkowitz RD, Hammarskjold M-L, Helga-Maria C, Rekosh D, Goff SP (1995b) 5′ regions of HIV-1 are not sufficient for encapsidation: implications for the HIV-1 packaging signal. Virology 212: 718–723
Bess JW Jr, Powell PJ, Issaq HJ et al (1992) Tightly bound zinc in human immunodeficiency virus type 1, human T-cell leukemia virus type 1, and other retroviruses. J Virol 66: 840–847
Bieth E, Gabus C, Darlix J-L (1990) A study of the dimer formation of Rous sarcoma virus RNA and of its effect on viral protein synthesis in vitro. Nucleic Acids Res 18: 119–127
Bishop JM, Levinson WE, Sullivan D, Fanshier L, Quintrell N, Jackson J (1970) The low molecular weight RNAs of Rous sarcoma virus . II. The 7S RNA. Virology 42: 927–937
Bolognesi DP, Luftig R, Shaper JH (1973) Localization of RNA tumor virus polypeptides I. Isolation of further virus substructures. Virology 56: 549–564
Boris-Lawrie K, Temin HM (1994) The retroviral vector: replication cycle and safety considerations for retrovirus-mediated gene therapy. Ann NY Acad Sci 716: 59–70
Bowles NE, Damay P, Spahr P (1993) Effect of rearrangements and duplications of the cys-his motifs of Rous sarcoma virus nucleocapsid protein. J Virol 67: 623–631
Brahic M, Vigne R (1975) Properties of Visna virus particles harvested at short time intervals: RNA content, infectivity, and ultrastructure. J Virol 15: 1222–1230
Buchschacher GL Jr, Panganiban AT (1992) Human immunodeficiency virus vectors for inducible expression of foreign genes. J Virol 66: 2731–2739
Canaani E, Helm KVD, Duesberg P (1973) Evidence for 30–40S RNA as precursor of the 60–70S RNA of Rous sarcoma virus. Proc Natl Acad Sci USA 72: 401–405
Casas-Finet JR, Jhon N, Maki AH (1988) p10, a low molecular weight single-stranded nucleic acid binding protein of murine leukemia retroviruses, shows stacking interactions of its single tryptophan residue with nucleotide bases. Biochemistry 27: 1172–1178
Chance MR, Sagi I, Wirt MD et al (1992) Extended x-ray absorption fine structure studies of a retrovirus: equine infectious anemia virus cysteine arrays are coordinated to zinc. Proc Natl Acad Sci USA 89: 10041–10045
Chang Y-N, Kenan DJ, Keene JD, Gatignol A, Jeang K-T (1994) Direct interactions between autoantigen La and human immunodeficiency virus leader RNA. J Virol 68: 7008–7020
Cheung K-S, Smith RE, Stone MP, Joklik WK (1972) Comparison of immature (rapid harvest) and mature Rous sarcoma virus particles. Virology 50: 851–864
Clavel F, Orenstein JM (1990) A mutant of human immunodeficiency virus with reduced RNA packaging and abnormal particle morphology. J Virol 64: 5230–5234
Clever J, Sassetti C, Parslow TG (1995) RNA secondary structure and binding sites for gag gene products in the 5′ packaging signal of human immunodeficiency virus type 1. J Virol 69: 2101–2109
Cobrinik D, Soskey L, Leis J (1988) A retroviral RNA secondary structure required for efficient initiation of reverse transcription. J Virol 62: 3622–3630

Cobrinik D, Aiyar A, Ge Z, Katzman M, Huang H, Leis J (1991) Overlapping retrovirus U5 sequence elements are required for efficient integration and initiation of reverse transcription. J Virol 65: 3864–3872
Coffin J (1984) Structure of the retroviral genome. In: Weiss R, Teich N, Varmus H, Coffin J (eds) RNA tumor viruses. Cold Spring Harbor Laboratory, Cold Spring Harbor, pp 261–368
Cordell B, Swanstrom R, Goodman HM, Bishop JM (1979) $tRNA^{Trp}$ as primer for RNA-directed DNA polymerase: structural determinants of function. J Biol Chem 254: 1866–1874
Cornille F, Mely Y, Ficheux D et al (1990) Solid phase synthesis of the retroviral nucleocapsid protein NCp10 of Moloney murine leukemia virus and related "zinc fingers" in free SH forms. Int J Pept Protein Res 36: 551–558
Crawford S, Goff SP (1985) A deletion mutation in the 5′ part of the pol gene of Moloney murine leukemia virus blocks proteolytic processing of the gag and pol polyproteins. J Virol 53: 899–907
Dannull J, Surovoy A, Jung G, Moelling K (1994) Specific binding of HIV-1 nucleocapsid protein to PSI RNA in vitro requires N-terminal zinc finger and flanking basic amino acid residues. EMBO J 13: 1525–1533
Darlix J-L (1986) Control of Rous sarcoma virus RNA translation and packaging by the 5′ and 3′ untranslated sequences. J Mol Biol 189: 421–434
Darlix J-L, Spahr P (1982) Binding sites of viral protein p19 onto Rous sarcoma virus RNA and possible controls of viral functions. J Mol Biol 160: 147–161
Darlix J-L, Gabus C, Nugeyre M, Clavel F, Barre-Sinoussi F (1990) *Cis* elements and *trans*-acting factors involved in the RNA dimerization of the human immunodeficiency virus HIV-1. J Mol Biol 216: 689–699
Darlix J-L, Gabus C, Allain B (1992) Analytical study of avian reticuloendotheliosis virus dimeric RNA generated in vivo and in vitro. J Virol 66: 7245–7252
Davis J, Scherer M, Tsai WP, Long C (1976) Low-molecular-weight Rauscher leukemia virus protein with preferential binding for single-stranded RNA and DNA. J Virol 18: 709–718
Davis NL, Rueckert RR (1972) Properties of a ribonucleoprotein particle isolated from nonidet P-40-treated Rous sarcoma virus. J Virol 10: 1010–1020
de Rocquigny H, Ficheux D, Gabus C, Fournié-Zaluski M-C, Darlix J-L, Roques BP (1991) First large scale chemical synthesis of the 72 amino acid HIV-1 nucleocapsid protein NCp7 in an active form. Biochem Biophys Res Commun 180: 1010–1018
de Rocquigny H, Gabus C, Vincent A, Fournie-Zaluski M, Roques B, Darlix J-L (1992) Viral RNA annealing activities of human immunodeficiency virus type 1 nucleocapsid protein require only peptide domains outside the zinc fingers. Proc Natl Acad Sci USA 89: 6472–6476
Delahunty MD, South TL, Summers MF, Karpel RL (1992) Nucleic acid interactive properties of a peptide corresponding to the N-terminal zinc finger domain of HIV-1 nucleocapsid protein. Biochemistry 31: 6461–6469
Delwart EL, Buchachacher GL Jr, Freed EO, Panganiban AT (1992) Analysis of HIV-1 envelope mutants and pseudotyping of replication-defective HIV-1 vectors by genetic complementation. AIDS Res Hum Retroviruses 8: 1669–1677
Déméné H, Dong CZ, Ottman M et al (1994) ^{1}H NMR structure and biological studies of the His^{23} to Cys mutant nucleocapsid protein of HIV-1 indicate that the conformation of the first zinc finger is critical for virus infectivity. Biochemistry 33: 11707–11716
Dib-Hajj F, Khan R, Giedroc DP (1993) Retroviral nucleocapsid proteins possess potent nucleic acid strand renaturation activity. Protein Sci 2: 231–243
Dickson C, Eisenman R, Fan H (1985) Protein biosynthesis and assembly. In: Weiss R, Teich N, Varmus H, Coffin J (eds) RNA tumor viruses. Cold Spring Harbor Laboratory, Cold Spring Harbor, pp 135–146
Donze O, Spahr P (1992) Role of the open reading frames of Rous sarcoma virus leader RNA in translation and genome packaging. EMBO J 11: 3747–3757
Dorfman T, Luban J, Goff SP, Haseltine WA, Gottlinger HG (1993) Mapping of functionally important residues of a cysteine-histidine Box in the human immunodeficiency virus type 1 nucleocapsid protein. J Virol 67: 6159–6169
Dupraz P, Spahr P (1992) Specificity of Rous sarcoma virus nucleocapsid protein in genomic RNA packaging. J Virol 66: 4662–4670
Dupraz P, Oertle S, Meric C, Damay P, Spahr P (1990) Point mutations in the proximal cys-his box of Rous sarcoma virus nucleocapsid protein. J Virol 64: 4978–4987

East JL, Allen PT, Knesek JE, Chan JC, Bowen JM, Dmochowski L (1973) Structural rearrangement and subunit composition of RNA from released Soehner-Dmochowski murine sarcoma virions. J Virol 11: 709–720
Embretson JE, Temin HM (1987) Lack of competition results in efficient packaging of heterologous murine retroviral RNAs and reticuloendotheliosis virus encapsidation-minus RNAs by the reticuloendotheliosis virus helper cell line. J Virol 61: 2675–2683
Erikson E, Erikson RL (1971) Association of 4S ribonucleic acid with oncornavirus ribonucleic acids. J Virol 8: 254–256
Erikson E, Erikson RL, Henry B, Pace NR (1973) Comparison of oligonucleotides produced by RNase TI digestion of 7S RNA from avian and murine oncornaviruses and from uninfected cells. Virology 53: 40–46
Faras AJ, Garapin AC, Levinson WE, Bishop JM, Goodman HM (1973) Characterization of the low-molecular-weight RNAs associated with the 70S RNA of Rous sarcoma virus. J Virol 12: 334–342
Fitzgerald DW, Coleman JE (1991) Physicochemical properties of cloned nucleocapsid protein from HIV. Interactions with metal ions. Biochemistry 30: 5195–5201
Fleissner E, Tress E (1973) Isolation of a ribonucleoprotein structure from oncornaviruses. J Virol 12: 1612–1615
Fu W, Rein A (1993) Maturation of dimeric viral RNA of Moloney murine leukemia virus. J Virol 67: 5443–5449
Fu W, Gorelick RJ, Rein A (1994) Characterization of human immunodeficiency virus type 1 dimeric RNA from wild-type and protease-deficient virions. J Virol 68: 5013–5018
Fu X, Katz RA, Skalka AM, Leis J (1988) Site-directed mutagenesis of the avian retrovirus nucleocapsid protein, pp12. J Biol Chem 263: 2140–2145
Fujimura T, Esteban R, Esteban LM, Wickner RB (1990) Portable encapsidation signal of the L-A double-stranded RNA virus of S. cerevisiae. Cell 62: 819–828
Gorelick RJ, Henderson LE, Hanser JP, Rein A (1988) Point mutants of Moloney murine leukemia virus that fail to package viral RNA: evidence for specific RNA recognition by a "zinc finger-like" protein sequence. Proc Natl Acad Sci USA 85: 8420–8424
Gorelick RJ, Nigida SM, Bess JR, Arthur LO, Henderson LE, Rein A (1990) Noninfectious human immunodeficiency virus type 1 mutants deficient in genomic RNA. J Virol 64: 3207–3211
Gorelick RJ, Chabot DJ, Rein A, Henderson LE, Arthur LO (1993) The two zinc fingers in the human immunodeficiency virus type 1 nucleocapsid protein are not functionally equivalent. J Virol 67: 4027–4036
Green LM, Berg JM (1989) A retroviral cys-xaa$_2$-cys-xaa$_4$-his-xaa$_4$-cys peptide binds metal ions: spectroscopic studies and a proposed three-dimensional structure. Proc Natl Acad Sci USA 86: 4047–4051
Green LM, Berg JM (1990) Retroviral nucleocapsid protein-metal ion interactions: folding and sequence variants. Proc Natl Acad Sci USA 87: 6403–6407
Hackett PB, Dalton MW, Johnson DP, Petersen RB (1992) Phylogenetic and physical analysis of the 5′ leader RNA sequences of avian retroviruses. Nucleic Acid Res 19: 6929–6934
Harrison GP, Lever AL (1992) The human immunodeficiency virus type 1 packaging signal and major splice donor region have a conserved stable secondary structure. J Virol 66: 4144–4153
Harrison GP, Hunter E, Lever AML (1995) Secondary structure model of the Mason-Pfizer monkey virus 5′ leader sequence: identification of a structural motif common to a variety of retroviruses. J Virol 69: 2175–2186
Haseltine WA, Panet A, Smoler D et al (1977) Interaction of tryptophan tRNA and avian myeloblastosis virus reverse transcriptase: further characterization of the binding reaction. Biochemistry 16: 3625–3632
Hayashi T, Shioda T, Iwakura Y, Shibuta H (1992) RNA packaging signal of human immunodeficiency virus type 1. Virology 188: 590–599
Hirsch RC, Lavine JE, Chang L, Varmus HE, Ganem D (1990) Polymerase gene products of hepatitis B viruses are required for genomic RNA packaging as well as for reverse transcription. Nature 344: 552–555
Housset V, De Rocquigny H, Roques B, Darlix J-L (1993) Basic amino acids flanking the zinc finger of Moloney murine leukemia virus nucleocapsid protein NCp10 are critical for virus infectivity. J Virol 67: 2537–2545
Hu JC, Dahlberg JE (1983) Structural features required for binding of $tRNA^{Trp}$ to avian myeloblastosis virus reverse transcriptase. Nucleic Acid Res 11: 4823–4833
Huang Y, Mak J, Cao Q, Li Z, Wainberg MA, Kleiman L (1994) Incorporation of excess wild-type and mutant $tRNA^{Lys}3$ into human immunodeficiency virus type 1. J Virol 68: 7676–7683

Ikawa Y, Ross J, Leder P (1974) An association between globin messenger RNA and 60S RNA derived from Friend leukemia virus. Proc Natl Acad Sci USA 71: 1154–1158
Jentoft JE, Smith LM, Fu X, Johnson M, Leis J (1988) Conserved cysteine and histidine residues of the avian myeloblastosis virus nucleocapsid protein are essential for viral replication but not "zinc-binding fingers". Proc Natl Acad Sci USA 85: 7094–7098
Karpel RL, Henderson LE, Oroszlan S (1987) Interactions of retroviral structural proteins with single-stranded nucleic acids. J Biol Chem 262: 4961–4967
Katoh I, Kyushiki H, Sakamoto Y, Ikawa Y, Yoshinaka Y (1991) Bovine leukemia virus matrix-associated protein MA(p15): further processing and formation of a specific complex with the dimer of the 5′-terminal genomic RNA fragment. J Virol 65: 6845–6855
Katoh I, Yasunaga T, Yoshinaka Y (1993) Bovine leukemia virus RNA sequences involved in dimerization and specific gag protein binding: close relation to the packaging sites of avian, murine, and human retroviruses. J Virol 67: 1830–1839
Katz RA, Terry RW, Skalka AM (1986) A conserved cis-acting sequence in the 5′ leader of avian sarcoma virus RNA is required for packaging. J Virol 59: 163–167
Kaye JF, Richardson JH, Lever AML (1995) cis-acting sequences involved in human immunodeficiency virus type 1 RNA packing. J Virol 69: 6588–6592
Kaye JF, Lever AML (1996) *trans*-acting proteins involved in RNA encapsidation and viral assembly in human immuno deficiency virus type 1. J Virol 70: 880–886
Khan R, Giedroc DP (1992) Recombinant human immunodeficiency virus type 1 nucleocapsid (NCp7) protein unwinds tRNA. J Biol Chem 267: 6689–6695
Khan R, Giedroc DP (1994) Nucleic acid binding properties of recombinant Zn_2 HIV-1 nucleocapsid protein are modulated by COOH-terminal processing. J Biol Chem 269: 22538–22546
Kim H-J, Lee K, O'Rear JJ (1994) A short sequence upstream of the 5′ major splice site is important for encapsidation of HIV-1 genomic RNA. Virology 198: 336–340
Knight JB, Si ZH, Stoltzfus CM (1994) A base-paired structure in the avian sarcoma virus 5′ leader is required for efficient encapsidation of RNA. J Virol 68: 4493–4502
Konings DAM, Nash MA, Maizel JV, Arlinghaus RB (1992) Novel GACG-hairpin pair motif in the 5′ untranslated region of type C retroviruses related to murine leukemia virus. J Virol 66: 632–640
Korb J, Trávnícek M, Riman J (1976) The oncornavirus maturation process: quantitative correlation between morphological changes and conversion of genomic virion RNA. Intervirology 7: 211–224
Koyama T, Harada F, Kawai S (1984) Characterization of a Rous sarcoma virus mutant defective in packaging its own genomic RNA: biochemical properties of mutant TK15 and mutant-induced transformants. J Virol 51: 154–162
Kung H-J, Hu S, Bender W et al (1976) RD-114, baboon, and woolly monkey viral RNAs compared in size and structure. Cell 7: 609–620
Lapadat-Tapolsky M, De Rocquigny H, Van Gent D, Roques B, Plasterk R, Darlix J-L (1993) Interactions between HIV-1 nucleocapsid protein and viral DNA may have important functions in the viral life cycle. Nucleic Acid Res. 21: 831–839
Laughrea M, Jetté L (1994) A 19-nucleotide sequence upstream of the 5′ major splice donor is part of the dimerization domain of human immunodeficiency virus 1 genomic RNA. Biochemistry 33: 13464–13474
Leis J, Jentoft J (1983) Characteristics and regulation of interaction of avian retrovirus pp12 protein with viral RNA. J Virol 48: 361–369
Leis J, Johnson S, Collins LS, Traugh JA (1984) Effects of phosphorylation of avian retrovirus nucleocapsid protein pp12 on binding of viral RNA. J Biol Chem 259: 7726–7732
Leis JP, McGinnis J, Green RW (1978) Rous sarcoma virus p19 binds to specific double-stranded regions of viral RNA: effect of p19 on cleavage of viral RNA by RNase III. Virology 84: 87–98
Lever A, Gottlinger H, Haseltine W, Sodroski J (1989) Identification of a sequence required for efficient packaging of human immunodeficiency virus type 1 RNA into virions. J Virol 63: 4085–4087
Levin JG, Seidman JG (1979) Selective packaging of host tRNA's by murine leukemia virus particles does not require genomic RNA. J Virol 29: 328–335
Levin JG, Seidman JG (1981) Effect of polymerase mutations on packaging of primer $tRNA^{Pro}$ during murine leukemia virus assembly. J Virol 38: 403–408
Levin JG, Grimley PM, Ramseur JM, Berezesky IK (1974) Deficiency of 60 to 70S RNA in murine leukemia virus particles assembled in cells treated with actinomycin D. J Virol 14: 152-161
Linial M, Medeiros E, Hayward WS (1978) An avian oncovirus mutant (SE21Q1b) deficient in genomic RNA: biological and biochemical characterization. Cell 15: 1371–1381

Lipsick JS, Ibanez CE, Baluda MA (1986) Expression of molecular clones of v-myb in avian and mammalian cells independently of transformation. J Virol 59: 267–275
Luban J, Goff SP (1991) Binding of human immunodeficiency virus type 1 (HIV-1) RNA to recombinant HIV-1 *gag* polyprotein. J Virol 65: 3203–3212
Luban J, Goff SP (1994) Mutational analysis of cis-acting packaging signals in human immunodeficiency virus type 1 RNA. J Virol 68: 3784–3793
Mak J, Jiang M, Wainberg MA, Hammarskjold M-L, Rekosh D, Kleiman L (1994) Role of Pr160$^{gag\text{-}pol}$ in mediating the selective incorporation of tRNA^{lys} into human immunodeficiency virus type 1 particles. J Virol 68: 2065–2072
Mangel WF, Delius H, Duesberg PH (1974) Structure and molecular weight of the 60–70S RNA and the 30–40S RNA of the Rous sarcoma virus. Proc Natl Acad Sci USA 71: 4541–4545
Mann R, Baltimore D (1985) Varying the position of a retrovirus packaging sequence results in the encapsidation of both unspliced and spliced RNAs. J Virol 54: 401–407
Mann R, Mulligan RC, Baltimore D (1983) Construction of a retrovirus packaging mutant and its use to produce helper-free defective retrovirus. Cell 33: 153–159
Mansky LM, Krueger AE, Temin HM (1995) The bovine leukemia virus encapsidation signal is discontinuous and extends into the 5′ end of the *gag* gene. J Virol 69: 3282–3289
Marquet R, Baudin F, Gabus C et al (1991) Dimerization of human immunodeficiency virus (type 1) RNA: stimulation by cations and possible mechanism. Nucleic Acids Res 19: 2349-2357
Marquet R, Paillart J-C, Skripkin E, Ehresmann C, Ehresmann B (1994) Dimerization of human immunodeficiency virus type 1 RNA involves sequences located upstream of the splice donor site. Nucleic Acid Res 22: 145–151
Martin P, Henry C, Ferre F et al (1986) Characterization of a myc-containing retrovirus generated by propagation of an MH2 viral subgenomic RNA. J Virol 57: 1191–1194
Mely Y, Cornille F, Fournié-Zaluski M-C, Darlix J-L, Roques BP, Gérard D (1991) Investigation of zinc-binding affinities of Moloney murine leukemia virus nucleocapsid protein and its related zinc finger and modified peptides. Biopolymers 31: 899–906
Meric C, Spahr P (1986) Rous sarcoma virus nucleic acid-binding protein p12 is necessary for viral 70S RNA dimer formation and packaging. J Virol 60: 450–459
Meric C, Goff SP (1989) Characterization of Moloney murine leukemia virus mutants with single-amino-acid substitutions in the cys-his box of the nucleocapsid protein. J Virol 63: 1558-1568
Meric C, Darlix J-L, Spahr P (1984) It is Rous sarcoma virus protein p12 and not p19 that binds tightly to Rous sarcoma virus RNA. J Mol Biol 173: 531–538
Meric C, Gouilloud E, Spahr P (1988) Mutations in Rous sarcoma virus nucleocapsid protein p12 (NC): deletions of cys-his boxes. J Virol 62: 3328–3333
Messer LI, Levin JG, Chattopadhyay SK (1981) Metabolism of viral RNA in murine leukemia virus-infected cells: evidence for differential stability of viral message and virion precursor RNA. J Virol 40: 683–690
Morellet N, Jullian N, de Rocquigny H, Maigret B, Darlix J-L, Roques BP (1992) Determination of the structure of the nucleocapsid protein NCp7 from the human immunodeficiency virus type 1 by ^{1}H NMR. EMBO J 11: 3059–3065
Morellet N, de Rocquigny H, Mély Y et al (1994) Conformational behaviour of the active and inactive forms of the nucleocapsid NCp7 of HIV-1 studied by ^{1}H NMR. J Mol Biol 235: 287–301
Moustakas A, Sonstegard TS, Hackett PB (1993) Alterations of the three short open reading frames in the Rous sarcoma virus leader RNA modulate viral replication and gene expression. J Virol 67: 4337–4349
Murphy JE, Goff SP (1989) Construction and analysis of deletion mutations in the U5 region of Moloney murine leukemia virus: effects on RNA packaging and reverse transcription. J Virol 63: 319–327
Murti KG, Bondurant M, Tereba A (1981) Secondary structural features in the 70S RNAs of Moloney murine leukemia and Rous sarcoma viruses as observed by electron microscopy. J Virol 37: 411–419
Nassal M (1992) The arginine-rich domain of the hepatitis B core protein is required for pregenome encapsidation and productive viral positive-strand DNA synthesis but not for virus assembly. J Virol 66: 4107–4116
Nishizawa M, Koyama T, Kawai S (1985) Unusual features of the leader sequence of Rous sarcoma virus packaging mutant TK15. J Virol 55: 881–885
Nissen-Meyer J, Abraham AK (1980) Specificity of RNA binding by the structural protein (p10) of Friend murine leukemia virus. J Mol Biol 142: 19–28

Norton PA, Coffin JM (1985) Bacterial β-galactosidase as a marker of Rous sarcoma virus gene expression and replication. Mol Cell Biol 5: 281–290
Oertle S, Spahr P (1990) Role of the *gag* polyprotein precursor in packaging and maturation of Rous sarcoma virus genomic RNA. J Virol 64: 5757–5763
Omichinski JG, Clore GM, Sakaguchi K, Appella E, Gronenborn AM (1991) Structural characterization of a 39-residue synthetic peptide containing the two zinc binding domains from the HIV-1 p7 nucleocapsid protein by CD and NMR spectroscopy. FEBS Lett 292: 25–30
Paillart J-C, Marquet R, Skripkin E, Ehresmann B, Ehresmann C (1994) Mutational analysis of the bipartite dimer linkage structure of human immunodeficiency virus type 1 genomic RNA. J Biol Chem 269: 27486–27493
Panet A, Berliner H (1978) Binding of tRNA to reverse transcriptase of RNA tumor viruses. J Virol 26: 214–220
Panet A, Haseltine WA, Baltimore D, Peters G, Harada F, Dahlberg JE (1975) Specific binding of tryptophan transfer RNA to avian myeloblastosis virus RNA-dependent DNA polymerase (reverse transcriptase). Proc Natl Acad Sci USA 72: 2535–2539
Parkin NT, Cohen EA, Darveau A, Rosen C, Haseltine W, Sonenberg N (1988) Mutational analysis of the 5′ non-coding region of human immunodeficiency virus type 1: effects of secondary structure on translation. EMBO J 7: 2831–2837
Parolin C, Dorfman T, Palu G, Gottlinger H, Sodroski J (1994) Analysis in human immunodeficiency virus type 1 vectors of *cis*-acting sequences that affect gene transfer into human lymphocytes. J Virol 68: 3888–3895
Peliska JA, Balasubramanian S, Giedroc DP, Benkovic SJ (1994) Recombinant HIV-1 nucleocapsid protein accelerates HIV-1 reverse transcriptase catalyzed DNA strand transfer reactions and modulates RNase H activity. Biochemistry 33: 13817–13823
Peters G, Harada F, Dahlberg JE, Panet A, Haseltine WA, Baltimore D (1977) Low-molecular-weight RNAs of Moloney murine leukemia virus: identification of the primer for RNA-directed DNA synthesis. J Virol 21: 1031–1041
Peters GG, Hu J (1980) Reverse transcriptase as the major determinant for selective packaging of tRNA's into avian sarcoma virus particles. J Virol 36: 692–700
Prats A, Sarih L, Gabus C, Litvak S, Keith G, Darlix J-L (1988) Small finger protein of avian and murine retroviruses has nucleic acid annealing activity and positions the replication primer tRNA onto genomic RNA. EMBO J 7: 1777–1783
Prats A, Roy C, Wang P et al (1990) Cis elements and trans-acting factors involved in dimer formation of murine leukemia virus RNA. J Virol 64: 774–783
Prats A, Housset V, de Billy G et al (1991) Viral RNA annealing activities of the nucleocapsid protein of Moloney murine leukemia virus are zinc independent. Nucleic Acids Res 19: 3533–3541
Pugatsch T, Stacey DW (1983) Identification of a sequence likely to be required for avian retroviral packaging. Virology 128: 505–511
Quigley JP, Rifkin DB, Compans RW (1972) Isolation and characterization of ribonucleoprotein substructures from Rous sarcoma virus. Virology 50: 65–75
Rein A, Harvin DP, Mirro J, Ernst SM, Gorelick RJ (1994) Evidence that a central domain of nucleocapsid protein is required for RNA packaging in murine leukemia virus. J Virol 68: 6124–6129
Rice WG, Schaeffer CA, Harten B et al (1993) Inhibition of HIV-1 infectivity by zinc-ejecting aromatic C-nitroso compounds. Nature 361: 473–475
Richardson JH, Child LA, Lever AML (1993) Packaging of human immunodeficiency virus type 1 RNA requires *cis*-acting sequences outside the 5′ leader region. J Virol 67: 3997-4005
Riggin CH, Bondurant MC, Mitchell WM (1974) Differences between murine leukemia virus and murine sarcoma virus: effects of virion age and multiplicity of infection on viral RNA. Intervirology 2: 209–221
Rizvi TA, Panganiban AT (1993) Simian immunodeficiency virus RNA is efficiently encapsidated by human immunodeficiency virus type 1 particles. J Virol 67: 2681–2688
Robert D, Sallanfranque-Andreola M-L, Bordier B et al (1990) Interactions with $tRNA^{Lys}$ induce important structural changes in human immunodeficiency virus reverse transcriptase. FEBS Lett 277: 239–242
Roberts WJ, Elliott JI, McMurray WJ, Williams KR (1988) Synthesis of the p10 single-stranded nucleic acid binding protein from murine leukemia virus. Pept Res 1: 74–80
Roberts WJ, Pan T, Elliott JI, Coleman JE, Williams KR (1989) p10 single-stranded nucleic acid binding protein from murine leukemia virus binds metal ions via the peptide sequence Cys^{26}-X_2-Cys^{29}-X_4-His^{34}-X_4-Cys^{39}. Biochemistry 28: 10043–10047

Roy C, Tounekti N, Mougel M et al (1990) An analytical study of the dimerization of in vitro generated RNA of Moloney murine leukemia virus MoMuLV. Nucleic Acids Res 18: 7287-7292
Sakaguchi K, Zambrano N, Baldwin ET et al (1993) Identification of a binding site for the human immunodeficiency virus type 1 nucleocapsid protein. Proc Natl Acad Sci USA 90: 5219–5223
Sakalian M, Wills JW, Vogt VM (1994) Efficiency and selectivity of RNA packaging by Rous sarcoma virus gag deletion mutants. J Virol 68: 5969–5981
Sawyer RC, Dahlberg JE (1973) Small RNAs of Rous sarcoma virus: characterization by two-dimensional polyacrylamide gel electrophoresis and fingerprint analysis. J Virol 12: 1226-1237
Sawyer RC, Hanafusa H (1979) Comparison of the small RNAs of polymerase-deficient and polymerase-positive Rous sarcoma virus and another species of avian retrovirus. J Virol 29: 863–871
Schiff LA, Nibert ML, Fields BN (1988) Characterization of a zinc blotting technique: evidence that a retroviral *gag* protein binds zinc. Proc Natl Acad Sci USA 85: 4195–4199
Schlesinger S, Makino S, Linial M (1994) cis-acting genomic elements and trans-acting proteins involved in the assembly of RNA viruses. Semin Virol 5: 39–49
Schulein M, Burnette WN, August JT (1978) Stoichiometry and specificity of binding of Rauscher oncovirus 10,000-dalton (p10) structural protein to nucleic acids. J Virol 26: 54–60
Schwartzberg P, Colicelli J, Goff SP (1983) Deletion mutants of Moloney murine leukemia virus which lack glycosylated *gag* protein are replication competent. J Virol 46: 538–546
Secnik J, Gelfand CA, Jentoft JE (1992) Retroviral nucleocapsid protein specifically recognizes the base and the ribose of mononucleotides and mononucleotide components. Biochemistry 31: 2982–2988
Sen A, Todaro GJ (1976) Specificity of in vitro binding of primate type C viral RNA and the homologous viral p12 core protein. Science 193: 326–328
Sen A, Todaro GJ (1977) The genome-associated, specific RNA binding proteins of avian and mammalian type C viruses. Cell 10: 91–99
Sen A, Sherr CJ, Todaro GJ (1976) Specific binding of the type C viral core protein p12 with purified viral RNA. Cell 7: 21–32
Shields A, Witte WN, Rothenberg E, Baltimore D (1978) High frequency of aberrant expression of Moloney murine leukemia virus in clonal infections. Cell 14: 601–609
Skripkin E, Paillart J-C, Marquet R, Ehresmann B, Ehresmann C (1994) Identification of the primary site of the human immunodeficiency virus type 1 RNA dimerization in vitro. Proc Natl Acad Sci USA 91: 4945–4949
Smith BJ, Bailey JM (1979) The binding of an avian myeloblastosis virus basic 12,000 dalton protein to nucleic acids. Nucleic Acids Res 7: 2055–2072
Sorge J, Ricci W, Hughes SH (1983) Cis-acting RNA packaging locus in the 115-nucleotide direct repeat of Rous sarcoma virus. J Virol 48: 667–675
Sorge J, Wright D, Erdman VD, Cutting AE (1984) Amphotropic retrovirus vector system for human cell gene transfer. Mol Cell Biol 4: 1730–1737
South TL, Summers MF (1993) Zinc- and sequence-dependent binding to nucleic acids by the N-terminal zinc finger of the HIV-1 nucleocapsid protein: NMR structure of the complex with the Psi-site analog, dACGCC. Protein Sci 2: 3–19
South TL, Kim B, Summers MF (1989) ^{113}Cd NMR studies of a 1: 1Cd adduct with an 18-residue finger peptide from HIV-1 nucleic acid binding protein p7. J Am Chem Soc 111: 395–396
South TL, Blake PR, Sowder III RC, Arthur LO, Henderson LE, Summers MF (1990a) The nucleocapsid protein isolated from HIV-1 particles binds zinc and forms retroviral-type zinc fingers. Biochemistry 29: 7786–7789
South TL, Kim B, Hare DR, Summers MF (1990b) Zinc fingers and molecular recognition. Structure and nucleic acid binding studies of an HIV zinc finger-like domain. Biochem Pharmacol 40: 123–129
South TL, Blake PR, Hare DR, Summers MF (1991) C-terminal retroviral-type zinc finger domain from the HIV-1 nucleocapsid protein is structurally similar to the N-terminal zinc finger domain. Biochemistry 30: 6342–6349
Stacey DW (1979) Messenger activity of virion RNA for avian leukosis viral envelope glycoprotein. J Virol 29: 949–956
Steeg CM, Vogt VM (1990) RNA-binding properties of the matrix protein (p19^{gag}) of avian sarcoma and leukemia viruses. J Virol 64: 847–855
Stewart L, Schatz G, Vogt VM (1990) Properties of avian retrovirus particles defective in viral protease. J Virol 64: 5076–5092
Stoker AW, Bissell MJ (1988) Development of avian sarcoma and leukosis virus-based vector-packaging cell lines. J Virol 62: 1008–1015

Stoltzfus CM, Snyder PN (1975) Structure of B77 sarcoma virus RNA: stabilization of RNA after packaging. J Virol 16: 1161–1170
Summers MF, South TL, Kim B, Hare DR (1990) High-resolution structure of an HIV zinc fingerlike domain via a new NMR-based distance geometry approach. Biochemistry 29: 329–340
Summers MF, Henderson LE, Chance MR et al (1992) Nucleocapsid zinc finger detected in retroviruses: EXAFS studies of intact viruses and the solution-state structure of the nucleocapsid protein from HIV-1. Protein Sci 1: 563–574
Sundquist WI, Heaphy S (1993) Evidence for interstrand quadruplex formation in the dimerization of human immunodeficiency virus 1 genomic RNA. Proc Natl Acad Sci USA 90: 3393–3397
Surovoy A, Dannull J, Moelling K, Jung G (1993) Conformational and nucleic acid binding studies on the synthetic nucleocapsid protein of HIV-1. J Mol Biol 94–104
Svitkin YV, Pause A, Sonenberg N (1994) La autoantigen alleviates translational repression by the 5′ leader sequence of the human immunodeficiency virus type 1 mRNA. J Virol 68: 7001–7007
Svoboda J, Dvorák M, Guntaka R, Geryk J (1986) Transmission of (LTR, v-src, LTR) without recombination with a helper virus. Virology 153: 314–317
Sykora KW, Moelling K (1981) Properties of the avian viral protein p12. J Gen Virol 55: 379–391
Taylor JM, Cordell-Stewart B, Rohde W, Goodman HM, Bishop M (1975) Reassociation of 4S and 5S RNA's with the genome of avian sarcoma virus. Virology 65: 248–259
Telesnitsky A, Goff SP (1993) Strong stop strand transfer during reverse transcription. In: Skalka AM, Goff SP (eds) Reverse transcriptase. Cold Spring Harbor Laboratory, Cold Spring Harbor, pp 49–84
Torrent C, Bordet T, Darlix J-L (1994a) Analytical study of rat retrotransposon VL30 RNA dimerization in vitro and packaging in murine leukemia virus. J Mol Biol 240: 434–444
Torrent C, Gabus C, Darlix J-L (1994b) A small and efficient dimerization/packaging signal of rat VL30 RNA and its use in murine leukemia virus-VL30-derived vectors for gene transfer. J Virol 68: 661–667
Tounekti N, Mougel M, Roy C et al (1992) Effect of dimerization on the conformation of the encapsidation Psi domain of Moloney murine leukemia virus RNA. J Mol Biol 223: 205–220
Tsuchihashi Z, Brown PO (1994) DNA strand exchange and selective DNA annealing promoted by the human immunodeficiency virus type 1 nucleocapsid protein. J Virol 68: 5863-5870
Tsuchihashi Z, Khosla M, Herschlag D (1993) Protein enhancement of hammerhead ribozyme catalysis. Science 262: 99–102
Ullu E, Murphy S, Melli M (1982) Human 7S L RNA consists of a 140 nucleotide middle-repetitive sequence inserted in an Alu sequence. Cell 29: 195–202
Varmus H, Swanstrom R (1984) Replication of retroviruses. In: Weiss R, Teich N, Varmus H, Coffin J (eds) RNA tumor viruses. Cold Spring Harbor Laboratory, Cold Spring Harbor, pp 369–512
Vicenzi E, Dimitrov DS, Engelman A et al (1994) An integration-defective U5 deletion mutant of human immunodeficiency virus type 1 reverts by eliminating additional long terminal repeat sequences. J Virol 68: 7879–7890
Vile RG, Ali M, Hunter E, McClure MO (1992) Identification of a generalised packaging sequence for D-type retroviruses and generation of a D-type retroviral vector. Virology 189: 786–791
Walker TA, Pace NR, Erikson RL, Erikson E, Behr F (1974) The 7S RNA common to oncornaviruses and normal cells is associated with polyribosomes. Proc Natl Acad Sci USA 71: 3390–3394
Watanabe S, Temin HM (1982) Encapsidation sequences for spleen necrosis virus, an avian retrovirus, are between the 5′ long terminal repeat and the start of the *gag* gene. Proc Natl Acad Sci USA 79: 5986–5990
Waters LC, Mullin BC (1977) Transfer RNA in RNA tumor viruses. Prog Nucleic Acid Res Mol Biol 20: 131–160
Weiss S, König B, Morikawa Y, Jones I (1992a) Recombinant HIV-1 nucleocapsid protein p15 produced as a fusion protein with glutathione *S*-transferase in *Escherichia coli* mediates dimerization and enhances reverse transcription of retroviral RNA. Gene 121: 203–212
Weiss S, König B, Müller H-J, Seidel H, Goody RS (1992b) Synthetic human $tRNA^{Lys3}$ and natural bovine $tRNA^{Lys3}$ interact with HIV-1 reverse transcriptase and serve as specific primers for retroviral cDNA synthesis. Gene 111: 183–197
Yang S, Temin HM (1994) A double hairpin structure is necessary for the efficient encapsidation of spleen necrosis virus retroviral RNA. EMBO 13: 713–726
You JC, McHenry CS (1993) HIV nucleocapsid protein: expression in *Escherichia coli*, purification, and characterization. J Biol Chem 268: 16519–16527
Zhang Y, Barklis E (1995) Nucleocapsid protein effects on the specificity of retrovirus RNA encapsidation. J Virol 69: 5716–5722

Role of Auxiliary Proteins in Retroviral Morphogenesis

É.A. Cohen[1], R.A. Subbramanian[1], and H.G. Göttlinger[2]

1 Introduction

Though both the oncoviruses and the lentiviruses belong to the Retroviridae family, the lentiviral subfamily is comparatively complex both from the point of view of the number of viral proteins encoded by these viruses and in the regulation of their expression. In addition to the *gag, pol,* and *env* open reading frames (ORFs) present in all retroviruses, lentiviral genomes also contain novel ORFs generally not found in prototypic retroviruses. These ORFs code for a variety of auxiliary proteins which account for the tight and often intricate regulation of gene expression observed in lentiviruses. Auxiliary protein involvement in viral replication has been particularly well characterized in the most extensively studied member of the lentiviral subfamily, the human immunodeficiency virus type 1 (HIV-1), the causative agent of the acquired immune deficiency syndrome (AIDS).

HIV-1 codes for at least six auxiliary proteins, namely Vif, Vpr, Tat, Rev, Vpu, and Nef. These proteins can be classified into two groups based on the temporal regulation of their expression. Tat, Rev, and Nef are Rev-independent proteins coded early in the viral life cycle, while Vif, Vpr, and Vpu are Rev-dependent late

[1]Laboratoire de Rétrovirologie Humaine, Département de Microbiologie et Immunologie, Université de Montréal, CP 6128, Succursale Centre Ville, Montréal, H3C 3J7, Canada,
[2]Division of Human Retrovirology, Dana-Farber Cancer Institute, and Department of Pathology, Harvard Medical School, 44 Binney Street, Boston, MA 02115, USA

proteins that are expressed during active viral assembly. HIV-2 and the simian immunodeficiency viruses (SIVs) generally do not code for a Vpu protein, but code for another late protein, Vpx, not found in HIV-1. The functional roles of the early proteins, Tat, Rev, and Nef, have been extensively reviewed elsewhere and are not dealt with in this work (CULLEN 1992; SUBBRAMANIAN and COHEN 1994). The focus of this review is on the late auxiliary proteins that are coexpressed with the viral structural proteins Gag, Pol and Env and hence likely to be functional during active virion assembly.

Recent investigations into the function of the late auxiliary proteins suggest that the expression of these proteins ensures efficient virion release as well as optimal infectivity of virions generated in their presence. Most auxiliary proteins are not packaged into progeny virions and have to be synthesized de novo in infected cells. Notable exceptions are the auxiliary proteins Vpr and its homologue Vpx, both of which are efficiently packaged into virions (COHEN et al. 1990a; YUAN et al. 1990a,b). Virion incorporation of viral products may allow effective regulation of functions that immediately follow infection, before de novo protein synthesis can be initiated. In fact, virion-incorporated proteins may confer some advantageous qualities to lentiviruses not at the disposal of their oncoviral relatives. HIV-1 Vpr has been demonstrated to be one of the nucleophilic viral determinants that may aid proviral DNA transport to the nucleus in nondividing target cells such as macrophages (HEINZINGER et al. 1994). In this regard, it is interesting to note that oncoviruses, which lack a Vpr-like protein, do not efficiently infect nondividing cells (LEWIS and EMERMAN 1994). In addition to the virally encoded Vpr and Vpx proteins, some members of the lentiviral family appear to recruit and incorporate cellular factors such as cyclophilins into the progeny virions to benefit their own life cycle, as discussed later in this review. In fact, it is likely that all lentiviral auxiliary proteins function, at least in part, by targeting specific cellular factors and thus manipulating existing cellular processes to optimize viral propagation. Coding for a panoply of such auxiliary viral products, which may exploit different cellular processes to advance the viral life cycle, may actually distinguish the lentiviruses from other prototypic retroviruses.

2 Vif

A Vif (viral infectivity factor) protein is encoded by all primate immunodeficiency viruses and by most other lentiviruses, including feline immunodeficiency virus (FIV), bovine immunodeficiency virus (BIV), and visna virus. The lentiviral *vif* genes encode proteins with electrophoretic mobilities in SDS-PAGE gels of between 23K and 29K. Vif has a crucial role in the replication cycle of HIV-1, HIV-2, and FIV, and its evolutionary conservation in this family suggests that Vif may be of similar importance in other lentiviruses. A coherent picture of its functional importance and mechanism of action is beginning to emerge, primarily from the intensive analysis of HIV-1.

Early studies of the HIV-1 *vif* gene showed that it is a bona fide viral coding sequence and that its product is immunogenic in vivo, indicating that Vif is expressed during natural infections (ARYA and GALLO 1986; KAN et al. 1986; LEE et al. 1986). Disruption of the HIV-1 *vif* gene revealed that Vif is not absolutely required for virus replication in vitro, although the transmission of *vif*-deficient mutants was inefficient in some $CD4^+$ cell lines (SODROSKI et al. 1986; FISHER et al. 1987).

Interestingly, the *vif*-deficient mutants were severely impaired in establishing a cell-free infection mediated by virions alone, while they retained the ability to spread efficiently by cell-to-cell transmission, suggestive of a defect in the progeny virions generated in the absence of Vif. Cells infected with *vif*-deficient HIV-1 mutants produced virions with comparable efficiency to the wild-type strain, but the mutant virions were as much as 1000 times less infectious than wild-type virions (STREBEL et al. 1987). The defect associated with the mutant virions could be complemented, at least in part, by providing Vif *in trans*. These results pointed to a role of Vif during virus morphogenesis, consistent with the Rev-dependent expression of Vif late in the viral life cycle (GARRETT et al. 1991; SCHWARTZ et al. 1991). However, other studies indicate that in some cell types Vif is also crucial for cell-to-cell transmission, though cell-free transmission is generally more severely affected (FISHER et al. 1987; SAKAI et al. 1991; FAN and PEDEN 1992; GABUZDA et al. 1992; VON SCHWEDLER et al. 1993).

It has become clear that the requirement for Vif differs dramatically in different cell types. The *vif* genes of HIV-1, HIV-2, and SIV_{mac} are crucial for viral infectivity in natural target cells such as peripheral blood T lymphocytes and monocyte/macrophages. These primary cells show highly restricted permissivity to Vif-deficient viruses suggestive of a requirement for Vif in these cells in vivo (MICHAELS et al. 1993; GABUZDA et al. 1994; PARK et al. 1994). While some established cell lines are also nonpermissive for *vif*-deficient strains, most cell lines appear to be either permissive or semipermissive (FAN and PEDEN 1992; GABUZDA et al. 1992; BLANC et al. 1993; SAKAI et al. 1993; SOVA and VOLSKY 1993; VON SCHWEDLER et al. 1993). In permissive cells such as the T-lymphoid SupT1 line, Vif is dispensable even for cell-free transmission (GABUZDA et al. 1992; VON SCHWEDLER et al. 1993).

Important insights into the function of Vif were gained in particular from the study of virus derived from semipermissive cells such as some derivatives of the T-lymphoid CEM cell line. Virions produced by *vif*-deficient mutants in semipermissive cells are poorly infectious even for permissive cells (GABUZDA et al. 1992; VON SCHWEDLER et al. 1993). However, after one round of replication in permissive cells, *vif*-deficient mutants continue to spread with wild type kinetics (VON SCHWEDLER et al. 1993). Conversely, *vif*-deficient viruses produced in permissive cells can efficiently infect semipermissive as well as nonpermissive cells and proceed through one replication cycle with close to wild-type kinetics (GABUZDA et al. 1992; VON SCHWEDLER et al. 1993). However, further transmission of the virus, which requires additional replication cycles, is impaired (von Schwedler et al. 1993). Evidently, the consequences of a defect in the *vif* gene depend on the cell type in which the virus is produced rather than on the target cells. This conclusion is consistent with the observation that *vif*-deficient HIV-1 mutants can be complemented in producer but not in target cells by providing Vif *in trans* (Gabuzda et al. 1992; von Schwedler et al. 1993). Since vif is not incorporated into virions in readily detectable amounts (Von Schwedler et al. 1993), it is currently unclear whether this protein acts immediately before or during virion morphogenesis or in the next cycle immediately before virion entry.

While accumulating evidence points to a role of Vif in virion morphogenesis, the exact nature of the defect exhibited by virus particles produced in the absence of Vif remains to be determined. The amount and processing of viral proteins in wild type and *vif*-deficient virions appears similar. Also, no difference in the viral RNA content of virions produced in the presence and absence of Vif is detectable (Gabuzda et al., 1992; von Schwedler et al. 1993). It was reported that Vif acts as a protease and cleaves the transmembrane glycoprotein gp41 near its C-terminus (Guy et al. 1991). The results of that particular study seemed to imply that the Vif-mediated C-terminal processing of gp41 is crucial for virion infectivity. However, others have reported that Vif does not cleave the C-terminal end of gp41 (von Schwedler et al. 1993; Ma et al. 1994). Also, C-terminal truncations of gp41 do not alter the effect of Vif on virion infectivity (Gabuzda et al. 1992). Furthermore, the *vif*-defective phenotype could not be overcome by pseudotyping HIV-1 particles with the envelope glycoproteins of a murine retrovirus, which does not encode a Vif-like product (von Schwedler et al. 1993). The latter result suggested that the Vif defect, though effected earlier during morphogenesis of the virions, only becomes apparent in the next cycle of replication. Indeed, a striking impairment of viral DNA synthesis was noted in cells that were exposed to Vif-deficient virus, unless the mutant virions were produced in permissive cells (Sova and Volsky 1993; von Schwedler et al. 1993). Since the efficiency of viral entry was normal even when the mutant virions were produced in semipermissive cells (von Schwedler et al. 1993), it appears that the absence of a functional Vif during virus production leads to a defect in uncoating and/or reverse transcription of the genomic viral RNA in newly infected target cells.

It can currently be surmised that vif-effect early in the replication cycle results from a function at the level of virion morphogenesis, a late event in the preceding viral life cycle. While the morphology of Vif-deficient virus particles is not grossly altered, subtle effects of Vif on the morphology of the mature virion can be detected (Höglund et al. 1994). The characteristic cone-shaped core of mature wild-type virions contains electron-dense material which appears homogeneously distributed throughout the core structure. The electron-dense material is thought to represent the viral genomic RNA complexed with viral proteins. By contrast, virions produced in semipermissive cells in the absence of Vif frequently have a core structure that exhibits nonhomogeneous packing of the electron-dense material (HÖGLUND et al. 1994). In these mutant viruses, the viral ribonucleoprotein complex is concentrated at the broad end of the cone-shaped core, while the narrow end of the core appears transparent. These morphological changes are likely to be functionally relevant, since they are only apparent in cells in which Vif is required (HÖGLUND et al. 1994).

It is likely that the effect of Vif on viral DNA synthesis following entry is a consequence of its effect on the packing of the viral ribonucleoprotein complex. It has become clear that Vif exists both in a soluble cytosolic and in a membrane-associated form in infected cells (MICHAELS et al. 1993; GONCALVES et al. 1994). It is conceivable that the membrane-bound Vif could affect the coordinated assembly of the virion components at the plasma membrane. Consistent with this hypothesis, the membrane-association of Vif appears to be important for its effect on virion infectivity (GONCALVES et al. 1994).

The dispensability of Vif in certain cell types suggests that in some circumstances cellular factors may be able to compensate for Vif. Alternatively, Vif might counteract the activity of a cell-type specific factor that suppresses viral replication. As this auxiliary protein appears to play an essential role in natural target cells such as primary lymphocytes and macrophages, further analysis of its function at the molecular level may open new gene therapeutic avenues aimed at late steps in the viral replication cycle.

3 Vpr

The open reading frame coding for viral protein R (Vpr) is common to all primate immunodeficiency viruses. Vpr sequences frequently overlap *vif* sequences on the 5′ end and *tat* sequences on the 3′ end. The full-length HIV-1 Vpr is a 96 amino acid long protein which is expressed primarily from a singly spliced, Rev-dependent mRNA (ARRIGO and CHEN 1991; GARRETT et al. 1991). The HIV-2 and SIV_{mac} Vpr gene products are slightly longer proteins of 105 and 101 amino acids, respectively. Vpr is one of the retroviral auxiliary proteins that is packaged into virions (COHEN et al. 1990a; YUAN et al. 1990a,b). Consistent with its incorporation into progeny virions, Vpr is a late protein that is produced together with the structural proteins of the virion. Indeed, Vpr is incorporated into virions in molar amounts comparable to Gag.

Important progress in our understanding of the mechanisms that govern Vpr incorporation has been made in recent years. It is now clear that neither the viral envelope glycoproteins nor the genomic viral RNA play a role in Vpr incorporation (LU et al. 1993; LAVALLÉE et al. 1994; PAXTON et al. 1993). As Vpr is not synthesized as part of a polyprotein precursor, it must independently associate with the assembling capsid structure. Several recent studies have clearly shown that Vpr uses a virion-association motif in the C-terminal p6 domain of the viral Gag polyprotein. Removal or truncation of the p6 domain abolished the incorporation of Vpr into HIV-1 virions (PAXTON et al. 1994; CHECROUNE et al. 1995). More importantly, when the HIV-1 p6 domain was fused to the C-terminus of the Gag polyprotein of Moloney murine leukemia virus (Mo-MLV), a distantly related retrovirus that lacks a p6 domain, efficient incorporation of Vpr into the heterologous capsids was achieved (KONDO et al. 1995). Further mutational analysis showed that the Vpr virion-association motif is located within residues 1–46 of the HIV-1 p6 domain (KONDO et al. 1995). Consistent with its role in Vpr incorporation, a p6 domain is present in the Gag polyproteins of all primate immunodeficiency viruses, but is not found in other retroviruses.

While the role of the p6 domain in the virion association of Vpr is well established, it is not known whether p6 and Vpr interact directly. In contrast to the Gag p6 domain, which is highly variable among different HIV-1 genotypes, the Vpr protein is well conserved. A region within the N-terminal half of Vpr has been predicted to assume an alpha-helical structure with well-demarcated hydrophobic and hydrophilic faces. Mutations which are predicted to disrupt the helix or which affect either the hydrophobic or hydrophilic interface residues within the helix impair Vpr incorporation, suggesting that this amphipathic helix constitutes an interface involved in the incorporation process (MAHALINGAM et al. 1995; YAO et al. 1995. The C-terminal basic domain of the protein appears to be of lesser importance for the virion association of Vpr (MAHALINGAM et al. 1995).

The trafficking of Vpr within the infected cell can readily be traced by immunofluorescence. Vpr expressed in the absence of Gag proteins localizes to the nucleus. However, Vpr becomes associated with membrane compartments late in the replication cycle, when virion assembly and budding occur (LU et al. 1993). Within the virion, Vpr and its SIV homologue Vpx appear to be located between the viral envelope and the viral core (YU et al. 1993; LISKA et al. 1994; WANG et al. 1994).

The virion association of Vpr is suggestive of a role early after infection, before de novo synthesis of viral proteins can occur. Several studies indicate that Vpr confers rapid growth kinetics to viruses expressing the protein both in vitro and in vivo (OGAWA et al. 1989; COHEN et al. 1990b; LANG et al. 1993; CONNOR et al. 1995). In vitro, Vpr-positive strains grow faster and release moderately higher levels of virus than their Vpr-negative counterparts, an effect that is particularly pronounced in primary macrophages (HATTORI et al. 1990; WESTERVELT et al. 1992; CONNOR et al. 1995). Vpr expressing molecular clones maintained high viral loads in vivo in SIV-infected rhesus monkeys, while Vpr mutants were associated with lower viral burden in the same system (LANG et al. 1993). It appears that Vpr plays an important role in pathogenesis in vivo, as a Vpr-defective SIV_{mac} either reverted to a Vpr-positive phenotype in infected rhesus monkeys or caused only a low viral burden and no clinical disease (LANG et al. 1993).

An initial study of the mode of action of Vpr revealed a moderate ability to transactivate gene expression driven by the HIV-1 LTR as well as by heterologous promoters (COHEN et al. 1990b). This transactivating ability of Vpr could be advantageous early in the infectious cycle, before the strong viral transactivator Tat becomes available. It is conceivable that Vpr has a role in augmenting the basal promoter activity of the long terminal repeat during a period immediately following integration, in which Tat is not yet available. Such activation of the long terminal repeat may occur via changes in the cellular environment induced by the incoming Vpr or Vpr synthesized following integration. Results from a recent study suggest that the augmented replication evident in macrophages may primarily result from Vpr synthesized in situ and not by the virion-associated Vpr (CONNOR et al. 1995). It is likely that in the macrophage lineage Vpr may act upon transcriptional regulation of cellular genes and indirectly galvanize a cellular milieu that will support optimal viral replication. This concept is supported by the interesting finding that Vpr, in the absence of any other viral proteins, induces differentiation and growth arrest in a rhabdomyosarcoma cell line, demonstrating that Vpr can modify basic cellular pathways (LEVY et al. 1993). Also, it has recently become clear that Vpr inhibits the proliferation of chronically infected $CD4^+$ T cells, which again can be related to the cell growth arrest first identified in the muscle cell model (LEVY et al. 1993; ROGEL et al. 1995). On the basis of these observations, it has been suggested that one of the functions of Vpr may be to prevent the expansion of $CD4^+$ T cells specific for viral antigens. Vpr may thus contribute to the functional energy of helper cells documented in AIDS patients (ROGEL et al. 1995).

Recent studies show that extracellular Vpr is a powerful activator of HIV-1 expression. Exogeneously added Vpr reactivated virus expression in latently infected cell lines and induced virus replication in newly infected resting peripheral blood mononuclear cells (LEVY et al. 1994, 1995). Purified recombinant Vpr in minute concentrations was also capable of inducing a long-lasting state of increased cellular permissiveness for HIV-1 replication when added prior to infection. These effects provide further evidence that Vpr acts indirectly by modifying the cellular milieu. Such pervasive alteration of the cellular environment most likely involves specific associations with cellular proteins, and one such candidate of relative molecular weight 200K has recently been identified. However, the functional relevance of this association is not known (ZHAO et al. 1994).

It was also reported that Vpr facilitates the nuclear localization of the viral DNA in the absence of mitosis (HEINZINGER et al. 1994). In contrast to oncoretroviruses, HIV-1 does not require mitosis for entry of the viral genome into the nucleus, which is a necessary step of the retroviral life cycle (ROE et al. 1993; LEWIS and EMERMAN 1994). It appears that HIV-1 has two redundant determinants that confer nucleophilic properties to the viral preintegration complex. One of these nucleophilic determinants is located in the matrix protein, a fraction of which remains associated with the viral nucleic acid after entry (BUKRINSKY et al. 1993). The nuclear localization signal in the matrix protein is required in the absence of Vpr for viral replication in growth-arrested cells and terminally differentiated macrophages. Interestingly, in the presence of a functional Vpr, the karyophilic determinant in the matrix protein was dispensable for replication in nondividing host cells, suggesting that Vpr functions as a second determinant that facilitates the transport of the preintegration complex into the nucleus (HEINZINGER et al. 1994). It remains to be seen whether this function of Vpr results from its effects on cellular proliferation and differentiation. Such a correlation may help unify the pleiotropic effects associated with this auxiliary protein.

4 Vpx

The *vpx* gene is unique to the HIV-2/SIV_{mac} group of primate immunodeficiency viruses and is not present in HIV-1. Vpx shows homology with Vpr, and is similarly well conserved (TRISTEM et al. 1992; WESTERVELT et al. 1992). The members of the HIV-2/SIV_{mac} group also possess a *vpr* gene, which is located immediately downstream of the *vpx* gene. The genome organization as well as the sequence similarities between Vpx and Vpr suggest that the *vpx* gene most likely arose from the duplication of the *vpr* gene in a common ancestor of the HIV-2/SIV_{mac} group of primate immunodeficiency viruses (TRISTEM et al. 1990, 1992). Like Vpr, Vpx is packaged into virions and confers a growth advantage in certain cells (YUAN et al. 1990a,b; KAPPES et al. 1991; YU et al. 1991). It appears that within the virion, Vpx, like the homologous Vpr, is located between the outer protein shell formed by the matrix protein and the viral core structure formed by the capsid protein (YU et al. 1993; LISKA et al. 1994). At the cellular level, Vpx seems to associate with the capsid protein p27 (HORTON et al. 1994). Early work by HENDERSON and coworkers (1988) indicated that Vpx can interact with single-stranded RNA. This observation suggests that Vpx may remain associated with the viral nucleic acid following entry. Vpx, like Vpr, is packaged into virions via a direct or indirect interaction with the C-terminal p6 domain of the viral Gag polyprotein (WU et al. 1994). While the same Gag domain mediates the virion-association of both Vpx and Vpr, the precise requirements for the incorporation of these proteins differ. C-terminal sequences within the p6 domain, which are required for the incorporation of Vpr, are dispensable for the incorporation of Vpx. Also, Vpx is not efficiently incorporated into HIV-1 virions, indicating that the HIV-1 p6 domain lacks a virion-association motif for Vpx (KAPPES et al. 1993; WU et al. 1994).

5 Vpu

The viral protein U (Vpu) is unique to the HIV-1 group of primate immunodeficiency viruses (COHEN et al. 1988; STREBEL et al. 1988; HUET et al. 1990). The *vpu* gene overlaps the 5′ end of the *env* gene and is translated from the same bicistronic mRNA (SCHWARTZ et al. 1990). The expression of Vpu is Rev dependent; hence Vpu, like Vif and Vpr, is a late protein (ARRIGO and CHEN 1991; GARRETT et al. 1991). However, unlike Vpr, Vpu is not incorporated into virions. Biochemical studies show that Vpu is a type-1 integral membrane protein with a hydrophobic N-terminal portion which provides a membrane anchor and a hydrophilic C-terminal portion which constitutes the cytoplasmic domain of this protein. It was also shown that Vpu is capable of homo-oligomerization (MALDARELLI et al. 1993).

Vpu is not essential for productive viral replication in T cells in vitro. However, Vpu increases the levels of virions released from infected cells (STREBEL et al. 1988). Vpu does not augment viral protein synthesis or the assembly of structural proteins into viral particles. Rather, Vpu facilitates the release of assembled virions from the cell surface (STREBEL et al. 1989; TERWILLIGER et al. 1989). This view is supported by electron microscopic studies which revealed an accumulation of budding virions at the cell surface in the absence of Vpu. In addition, Vpu may be involved in the proper assembly and maturation of virions, as aberrant budding structures are frequently seen in the absence of Vpu (KLIMKAIT et al. 1990, YAO et al. 1993). Furthermore, a lack of Vpu leads to an increased frequency of budding at intracellular membranes, rather than at the plasma membrane (KLIMKAIT et al. 1990). However it is not clear if this phenomenon may result from the endocytosis of unreleased viral particles tethered at the membrane back into the cell in the absence of Vpu. It is clear that Vpu need not be expressed from the viral genome, and can equally function *in trans* (TERWILLIGER et al. 1989; YAO et al. 1992). The effect of Vpu on virion release is not dependent on virion maturation, since Vpu facilitates release both in the presence and absence of a viral protease (GÖTTLINGER et al. 1993). Also, Vpu can significantly enhance particle production by Gag proteins from retroviruses distantly related to HIV-1, suggesting that the effect of Vpu is unlikely to require highly specific interactions with Gag proteins (GÖTTLINGER et al. 1993). Rather, it appears that Vpu enhances retroviral budding indirectly through modification of the cellular environment.

Recently, a second function associated with Vpu expression has been identified: efficient and selective degradation of the CD4 molecule in the endoplasmic reticulum (ER). In cells infected with HIV-1, the viral envelope precursor gp160 and CD4 can form complexes which are retained in the ER, thus affecting the maturation and transport of this viral protein (CRISE et al. 1990; JABBAR and NAYAK 1990; BOUR et al. 1991). Vpu can release gp160 from these intracellular complexes, thus allowing the maturation of gp160 to gp120 and gp41 during transport through the secretory pathway (WILLEY et al. 1992a). The release of gp160 from intracellular complexes appears to be a consequence of Vpu-induced degradation of CD4 retained in the ER (WILLEY et al. 1992b). The facilitation of virion release by Vpu is not directly related to the Vpu-mediated degradation of CD4, since augmented capsid release is evident even in the absence of the envelope glycoproteins and CD4 (YAO et al. 1992; GERAGHTY and PANGANIBAN 1993).

Specific sequences in the cytoplasmic domain of CD4 appear to be essential for its susceptibility to Vpu-induced degradation (LENBURG and LANDAU 1993; VINCENT et al. 1993; WILLEY et al. 1994). Recently, it has become evident that this CD4 region is likely to constitute an alpha-helical structural motif which may be crucial for Vpu–mediated degradation (YAO et al. 1995). Two independent studies have also implicated the transmembrane domain of CD4 to be necessary for the Vpu mediated degradation of the protein (BUONOCORE et al. 1994; RAJA et al. 1994). Within Vpu, sequences critical for this function are located in its hydrophilic C-terminal domain (SCHUBERT and STREBEL 1994). However, the Vpu membrane anchor does appear to play an important role as Vpu failed to degrade CD4 unless it was associated with the same membrane compartment as CD4 (CHEN et al. 1993). A recent study has demonstrated that Vpu can physically form a complex with CD4 (BOUR et al. 1995). However, the involvement of other proteins in this complex cannot be convincingly ruled out at this time .

It was recently shown that phosphorylation of Vpu affects both known functions of Vpu, though clearly to varying extents. While the virion release function was only partially affected by Vpu phosphorylation mutants, CD4 degradation was adversely affected by the same mutants (SCHUBERT and STREBEL 1994; FRIBORG et al. 1995). The phosphorylated residues in Vpu are located within a central, highly conserved dodecapeptide motif. Two seryl residues within this conserved motif (amino acids 52 and 56) are phosphorylated by a casein kinase II related protein (FRIBORG et al. 1995; SCHUBERT et al. 1992).

Vpu shares both structural and biochemical similarities with the influenza M2 ion channel protein, which modifies luminal conditions in membrane compartments. Based on these similarities, it has been speculated that Vpu may wield similar ion channel activities, though this notion has not been addressed experimentally to date (STREBEL et al. 1988). An ion channel model appears attractive, since it could accommodate the apparently separate functions associated with Vpu. Ion channels are known to regulate cytoskeletal elements which are likely to participate in the terminal events during viral budding. Simultaneously, an ion channel activity of Vpu may affect local conditions in the ER and thereby induce degradation of the retained CD4.

6 Cyclophilin A

As a consequence of a relatively small genome size and limited coding capacity, RNA viruses need to rely heavily on proteins encoded by the host cell to complete their replication cycle. Although retroviral Gag proteins are capable of forming virus-like particles in the absence of other viral proteins (GHEYSEN et al. 1989; WILLS and CRAVEN 1991; HUNTER 1994), recent evidence suggests that interactions between Gag and host cell proteins can play a crucial role for virion infectivity. It has been known for some time that the capsid proteins of certain murine leukemia viruses harbor host range determinants, which appear to be recognized by the product of a single cellular gene, Fv-1 (HOPKINS et al. 1977; DESGROSEILLERS and JOLICOEUR 1983). In the case of HIV-1, host cell proteins which specifically interact with the capsid domain of the Gag polyprotein were recently identified using the yeast GAL4 two-hybrid system (LUBAN et al. 1993). All of the human cDNA clones which scored positive in this assay encoded either cyclophilin A or B, two members of a family of proteins that bind the immunosuppressant cyclosporin A (CsA), possess peptidyl-prolyl *cis-trans* isomerase activity, and assist in the folding of other proteins (SCHREIBER 1991; WALSH et al. 1992). The cyclophilins also bound to HIV-1 Gag protein in vitro, confirming the specificity of the interaction (LUBAN et al. 1993). While cyclophilin B, but not cyclophilin A, also bound to the Gag polyprotein of a closely related simian immunodeficiency virus, binding to the Gag proteins of more distantly related retroviruses was not observed (LUBAN et al. 1993).

Interestingly, the in vitro interaction between cyclophilins and HIV-1 Gag could be inhibited by CsA, suggesting that the binding sites for Gag and for the immunosuppressive ligand overlap (LUBAN et al. 1993). It is thought that the cyclophilin-CsA complex binds to and inhibits the phosphatase activity of calcineurin, which is required for T-cell receptor-mediated signaling (WALSH et al. 1992). Therefore, the formation of a complex between HIV-1 Gag and cyclophilins raised the intriguing possibility that HIV-1 Gag protein might mediate immunosuppression in a manner analogous to CsA. However, this seems unlikely, since the Gag-cyclophilin complex does not bind to calcineurin in vitro (LUBAN et al. 1993). Nevertheless, it remains possible that the HIV-1 Gag-cyclophilin interaction has adverse effects on the host cell by competing with interactions between cyclophilins and their putative physiological ligands.

In view of the potentially harmful consequences of the Gag-cyclophilin interaction for the host, the question arises whether the interaction confers any advantage to the virus. This question was recently answered by showing that cyclophilin A is required for the formation of fully infectious HIV-1 virions (THALI et al. 1994). It was demonstrated that purified HIV-1 virions contain a protein of relative molecular mass 18K, which reacts with anti-cyclophilin A antibody (FRANKE et al. 1994; THALI et al. 1994). Remarkably, cyclophilin A was found in HIV-1 virions at levels comparable to those of the *pol*-encoded proteins (FRANKE et al. 1994; THALI et al. 1994). Cyclophilin B, which resides in the endoplasmic reticulum and also binds to HIV-1 Gag protein in vitro, was not present in HIV-1 virion preparations, consistent with the cytosolic location of the newly synthesized Gag polyprotein. Virions of other retroviruses, even of primate immunodeficiency viruses closely related to HIV-1, do not contain cyclophilin A (FRANKE et al. 1994; THALI et al. 1994), demonstrating that the presence of cyclophilin A in HIV-1 virions is not simply a consequence of its abundance in the host cell. Moreover, another abundant cytoplasmic immunophilin, FKBP12, is not incorporated into HIV-1 virions (FRANKE et al. 1994), providing further evidence that cyclophilin A is incorporated in a specific manner.

Several lines of evidence indicate that the incorporation of cyclophilin A is mediated by a specific interaction with the capsid domain of the HIV-1 Gag polyprotein. First, the capsid protein by itself is capable of binding to cyclophilin A in vitro (Luban et al. 1993). Second, small deletions throughout the N-terminal half of the capsid domain prevented the virion association of cyclophilin A (Thali et al. 1994). The deletions abolished viral replication, but did not affect viral particle formation, which demonstrates that the Gag-cyclophilin interaction is dispensable for HIV-1 Gag polyprotein assembly and virion release. Third, the replacement of the CA domain of the Mo-MLV Gag polyprotein by the CA domain of HIV-1 resulted in the formation of viral particles which, unlike particles formed by the parental Mo-MLV Gag polyprotein, contained cyclophilin A at levels comparable to those found in HIV-1 virions (Thali et al. 1994). It can therefore be concluded that sequences within the HIV-1 capsid domain are sufficient to mediate the virion association of cyclophilin A. Important determinants for cyclophilin A incorporation appear to reside in a proline-rich region of the HIV-1 capsid domain, since the transfer of this region into the Gag polyprotein of SIV resulted in binding to cyclophilin A (Franke et al. 1994). Conversely, the ability to incorporate cyclophilin A was lost when the analogous region from SIV was transferred into HIV-1 (Franke et al. 1994). Mutagenesis revealed that one of the four conserved prolines in this region is essential for the virion association of the prolyl-isomerase cyclophilin A (Franke et al. 1994). Interestingly, this proline residue is also critical for HIV-1 replication (Franke et al. 1994), indicating that the Gag-cyclophilin A interaction is functionally relevant.

The importance of the interaction between Gag and cyclophilin A for HIV-1 replication was confirmed by treatment of virus-producing cells with drugs that compete with Gag for binding to cyclophilin A. When producer cells were incubated with CsA, the incorporation of cyclophilin A into HIV-1 virions was inhibited in a dose-dependent manner (Thali et al. 1994). Importantly, the infectivity of the virions was inhibited with the same dose dependence. The antiviral effect of the drug did not depend on its immunosuppressive activity, since SDZ NIM 811, a nonimmunosuppressive CsA analogue that retains the capacity to bind to cyclophilin A, was equally effective (Thali et al. 1994). SDZ NIM 811 exhibited antiviral activity against laboratory strains and against primary isolates from geographically distinct regions, and was active both in established cell lines and in primary lymphocytes and monocytes (Rosenwirth et al. 1994). As one might have predicted, the drug was inactive against SIV_{mac}, a primate immunodeficiency virus closely related to HIV-1, which does not incorporate cyclophilin A (Thali et al. 1994).

Although the Gag-cyclophilin A interaction is dispensable for HIV-1 virion assembly and release, and virion morphology appears unaffected by either CsA or SDZ NIM 811, cyclophilin A may yet have a subtle role in virion morphogenesis. Since cyclophilin A binds to the capsid protein, it is tempting to speculate that the requirement for the Gag-cyclophilin A interaction reflects a crucial role of cyclophilin A in the function of the HIV-1 virion core. Although the function of the core structure remains largely unknown, possible hints come from the study of murine leukemia viruses. The capsid protein of Mo-MLV remains a component of the viral nucleoprotein complex following viral entry (Bowerman et al. 1989), suggesting a role during the early stages of virus replication. Furthermore, a role early after infection is suggested by the presence of tropism determinants in the capsid proteins of certain murine leukemia viruses which affect the integration of the viral DNA into the host genome. It is interesting in this respect that SDZ NIM 811 has been reported to induce a defect both at the level of reverse transcription of the HIV-1 genome and at the level of transport of the HIV-1 preintegration complex into the nucleus (Steinkasserer et al. 1995). It is possible that cyclophilin A affects the fine architecture of the HIV-1 virion core structure, perhaps by facilitating the correct folding of the capsid protein, which may be critical for efficient reverse transcription, nuclear transport, and/or integration of the viral genome. In any case, it is likely that further investigation of the role of cyclophilin A in the HIV-1 life cycle will yield important insights into the function of retroviral capsid proteins.

Acknowledgments. E.A.C. is a recipient of a National Health Research and Development Program (NHRDP) of Canada Career award. This work was supported by grants from the Medical Research Council and NHRDP to E.A.C., and by National Institutes of Health grants AI29873 and AI34267 to H.G.G.

References

Arrigo SJ, Chen IS (1991) Rev is necessary for translation but not the cytoplasmic accumulation of HIV-1 vif, vpr, and env/vpu 2 RNAs. Genes Dev 5: 808–819

Arya SK, Gallo RC (1986) Three novel genes of human T-lymphotropic virus type III: immune reactivity of their products with sera from acquired immune deficiency syndrome patients. Proc Natl Acad Sci USA 83: 2209–2213

Blanc D, Patience C, Schulz TF, Weiss R, Spire B (1993) Transcomplementation of VIF$^-$ HIV-1 mutants in CEM cells suggests that Vif affects late steps of the viral life cycle. Virology 193: 186–192

Bour S, Boulerice F, Wainberg MA (1991) Inhibition of gp160 and CD4 maturation in U937 cells after both defective and productive infections by human immunodeficiency virus type 1. J Virol 65: 6387–6396

Bour S, Schubert U, Strebel K (1995) The human immunodeficiency virus type 1 Vpu protein specifically binds to the cytoplasmic domain of CD4: implications for the mechanism of degradation. J Virol 69: 1510–1520

Bowerman B, Brown PO, Bishop JM, Varmus HE (1989) A nucleoprotein complex mediates the integration of retroviral DNA. Genes Dev 3: 469–478

Bukrinsky MI, Haggerty S, Dempsey MP, Sharova N, Adzhubel A, Spitz L, Lewis P, Goldfarb D, Emerman M, Stevenson M (1993) A nuclear localization signal within HIV-1 matrix protein that governs infection of non-dividing cells. Nature 365: 666–669

Buonocore L, Turi TG, Crise B, Rose JK (1994) Stimulation of heterologous protein degradation by the Vpu protein of HIV-1 requires the transmembrane and cytoplasmic domains of CD4. Virology 204: 482–486

Checroune F, Yao XJ, Göttlinger HG, Bergeron D, Cohen EA (1995) Incorporation of Vpr into human immunodeficiency virus type 1: role of conserved regions within the p6 domain of Pr55^{gag}. J AIDS Hum Retrovirol 10: 1–7

Chen MY, Maldarelli F, Karczewski MK, Willey RL, Strebel K (1993) Human immunodeficiency type 1 Vpu protein induces degradation of CD4 in vitro: the cytoplasmic domain of CD4 contributes to Vpu sensitivity. J Virol 67: 3877–3884
Cohen EA, Terwilliger EF, Sodroski JG, Haseltine WA (1988) Identification of a protein encoded by the vpu gene of HIV-1. Nature 334: 532–534
Cohen EA, Dehni G, Sodroski JG, Haseltine WA (1990a)Human immunodeficiency virus Vpr product is a virion-associated regulatory protein. J Virol 64: 3097–3099
Cohen EA, Terwilliger EF, Jalinoos Y, Proulx J, Sodroski JG, Haseltine WA (1990b) Identification of HIV-1 Vpr product and function. J Acquir Immune Defic Syndr 3: 11–18
Connor RI, Chen BK, Choe S, Landau NR (1995) Vpr is required for efficient replication of human immunodeficiency virus type 1 in mononuclear phagocytes. Virology 206: 935–944
Crise B, Buonocore L, Rose JK (1990) CD4 is retained in the endoplasmic reticulum by the human immunodeficiency virus envelope glycoprotein precursor. J Virol 64: 5585–5593
Cullen BR (1992) Mechanism of action of regulatory proteins encoded by complex retroviruses. Microbiol Rev 56: 375–394
DesGroseillers L, Jolicoeur P (1983) Physical mapping of the Fv-1 tropism host range determinant of Balb/c murine leukemia viruses. J Virol 48: 685–696
Fan L, Peden K (1992) Cell-free transmission of Vif mutants of HIV-1. Virology 190: 19–29
Fisher AG, Ensoli B, Ivanoff L, Chamberlain M, Petteway S, Ratner L, Gallo RC, Wong-Staal F (1987) The sor gene of HIV-1 is required for efficient virus transmission in vitro Science 237: 888–893
Franke EK, Yuan HEH, Luban J (1994) Specific incorporation of cyclophilin A into HIV-1 virions Nature 372: 359–362
Friborg J, Ladha A, Göttlinger H, Haseltine WA, Cohen EA (1995) Functional analysis of the phosphorylation sites on the human immunodeficiency virus type 1 Vpu protein. J AIDS and Hum Retrovirol 8: 10–22
Gabuzda DH, Lawrence K, Langhoff E, Terwilliger E, Dorfman T, Haseltine WA, Sodroski J (1992) Role of vif in replication of human immunodeficiency virus type 1 in $CD4^+$ T lymphocytes. J Virol 66: 6489–6495
Gabuzda DH, Li H, Lawrence K, Vasir BS, Crawford K, Langhoff E (1994) Essential role of vif in establishing productive HIV-1 infection in peripheral blood T lymphocytes and monocytes/ macrophages. J Acquir Immune Defic Syndr 7: 908–915
Garrett ED, Tiley LS, Cullen BR (1991) Rev activates expression of the human immunodeficiency virus type 1 vif and vpr gene products. J Virol 65: 1653–1657
Geraghty RJ, Panganiban AT (1993) Human immunodeficiency type 1 Vpu has a $CD4^-$ and an envelope glycoprotein-independent function. J Virol 67: 4190–4194
Gheysen D, Jacobs E, de Foresta F, Thiriart C, Francotte M, Thines D, and De Wilde M (1989) Assembly and release of HIV-1 precursor $Pr55^{gag}$ virus-like particles from recombinant baculovirus-infected insect cells. Cell 59: 103–112
Goncalves J, Jallepalli P, Gabuzda DH (1994) Subcellular localization of the Vif protein of human immunodeficiency virus type 1. J Virol 68: 704–712
Göttlinger HG, Dorfman T, Cohen EA, Haseltine WA (1993) Human immunodeficiency virus type 1 Vpu enhances the production of capsids from widely divergent retroviruses. Proc Natl Acad Sci USA 90: 7381–7385
Guy B, Geist M, Dott K, Spehner D, Kieny MP, Lecocq JP (1991) A specific inhibitor of cysteine proteases impairs a Vif-dependent modification of human immunodeficiency virus type 1 Env protein. J Virol 65: 1325–1331
Hattori N, Michaels F, Fargnoli K, Marcon L, Gallo RC, Franchini G (1990) The human immunodeficiency virus type 2 vpr gene is essential for productive infection of human macrophages. Proc Natl Acad Sci USA 87: 8080–8084
Heinzinger NK, Bukrinsky MI, Haggerty SA, Ragland AM, Kewalramani V, Lee MA, Gendelman HE, Ratner L, Stevenson M, Emerman M (1994) The Vpr protein of human immunodeficiency virus type 1 influences nuclear localization of viral nucleic acids in nondividing host cells. Proc Natl Acad Sci USA 91: 7311–7315
Henderson LE, Sowder RC, Copeland TD, Benveniste RE, Oroszlan S (1988) Isolation and characterization of a novel protein (X-orf product) from SIV and HIV-2. Science 241: 199–201
Höglund S, Öhagen A, Lawrence K, Gabuzda D (1994) Role of vif during packing of the core of HIV-1. Virology 201: 349–355
Hopkins N, Schindler J, Hynes R (1977) Six NB-tropic murine leukemia viruses derived from a B-tropic virus of Balb/c have altered P30. J Virol 21: 309–318

Horton R, Spearman P, Ratner L (1994) HIV-2 viral protein X associates with the Gag p27 capsid protein. Virology 199: 453–457
Huet T, Cheynier R, Meyerhans A (1990) Genetic organization of a chimpanzee lentivirus related to HIV-1. Nature 345: 356–359
Hunter E (1994) Macromolecular interactions in the assembly of HIV and other retroviruses. Semin Virol 5: 71–83
Jabbar MA, Nayak DP (1990) Intracellular interaction of human immunodeficiency virus type 1 (ARV-2) envelope glycoprotein gp160 with CD4 blocks the movement and maturation of CD4 to the plasma membrane. J Virol 64: 6297–6304
Kan NC, Franchini G, Wong-Staal F, DuBois GC, Robey WG, Lautenberger JA, Papas TS (1986) Identification of HTLV-III/LAV sor gene product and detection of antibodies in human sera. Science 231: 1553–1555
Kappes JC, Conway JA, Lee SW, Shaw GM, Hahn BH (1991) Human immunodeficiency virus type 2 Vpx protein augments viral infectivity. Virology 184: 197–209
Kappes JC, Parkin JS, Conway JA, Kim J, Brouillette CG, Shaw GM, Hahn BH (1993) Intracellular transport and virion incorporation of Vpx requires interaction with other virus type-specific components. Virology 193: 222–233
Klimkait T, Strebel K, Hoggan MD, Martin MA, Orenstein JM (1990) The human immunodeficiency virus type 1 specific protein Vpu is required for efficient virus maturation and release. J Virol 64: 621–629
Kondo E, Mammano F, Cohen EA, Göttlinger HG (1995) The $p6^{gag}$ domain of human immunodeficiency virus type 1 is sufficient for the incorporation of Vpr into heterologous viral particles. J Virol 69: 2759–2764
Lang SM, Weeger M, Stahl-Hennig C, Coulibaly C, Hunsmann G, Müller J, Müller-Hermelink H, Fuchs D, Wachter H, Daniel MM, Desrosiers RC, Fleckenstein B (1993) Importance of Vpr for infection of rhesus monkeys with simian immunodeficiency virus. J Virol 67: 902–912
Lavallée C, Yao XY, Ladha A, Göttlinger H, Haseltine WA, Cohen EA (1994) Requirement of the Pr55 Gag precursor for incorporation of Vpr product into human immunodeficiency virus type 1 viral particles. J Virol 68: 1926–1934
Lee TH, Coligan JE, Allan JS, McLane MF, Groopman JE, Essex M (1986) A new HTLV-III/LAV protein encoded by a gene found in cytopathic retroviruses. Science 231: 1546–1549
Lenburg ME, Landau NR (1993) Vpu induced degradation of CD4: requirement for specific amino acid residues in the cytoplasmic domain of CD4. J Virol 67: 7238–7245
Levy DN, Fernandes LS, Williams WV, Weiner DB (1993) Induction of cell differentiation by human immunodeficiency virus 1 Vpr. Cell 72: 541–550
Levy DN, Refaeli Y, MacGregor RR, Weiner DB (1994) Serum Vpr regulates production, infection and latency of human immunodeficiency virus type 1. Proc Natl Acad Sci USA 91: 10873–10877
Levy ND, Refaeli Y, Weiner DB (1995) Extracellular Vpr protein increases cellular permissiveness to human immunodeficiency virus replication and reactivates virus from latency. J Virol 69: 1243–1252
Lewis PL, Emerman M (1994) Passage through mitosis is required for oncoretroviruses but not for the human immunodeficiency virus. J Virol 68: 510–516
Liska V, Spehner D, Mehtali M, Schmitt D, Kirn A, Aubertin AM (1994) Localization of viral protein X in simian immunodeficiency virus macaque strain and analysis of its packaging requirement. J Gen Virol 75: 2955–2962
Lu YL, Spearman P, Ratner L (1993) Human immunodeficiency virus type 1 viral protein R localization in infected cells and virions. J Virol 67: 6542–6550
Luban J, Bossolt KL, Franke EK, Kalpana GV, Goff SP (1993) Human immunodeficiency virus type 1 Gag protein binds to cyclophilins A and B. Cell 73: 1067–1078
Ma XY, Sova P, Chao W, Volsky DJ (1994) Cysteine residues in the Vif protein of human immunodeficiency virus type 1 are essential for viral infectivity. J Virol 68: 1714–1720
Mahalingam S, Khan SA, Jabbar MA, Monken CE, Collman RG, Srinivasan A (1995) Identification of residues in the N-terminal acidic domain of HIV-1 Vpr essential for virion incorporation. Virology 207: 297–302
Maldarelli F, Chen MY, Willey RL, Strebel K (1993) Human immunodeficiency virus type 1 Vpu protein is an oligomeric type I integral membrane protein. J Virol 67: 5056-5061
Michaels FH, Hattori N, Gallo RC, Franchini G (1993) The human immunodeficiency virus type 1 (HIV-1) Vif protein is located in the cytoplasm of infected cells and its effect on viral replication is equivalent in HIV-2. AIDS Res Hum Retroviruses 9: 1025–1030

Ogawa K, Shibata R, Kiyomasu T, Higuchi I, Kishida Y, Ishimoto A, Adachi A (1989) Mutational analysis of the human immunodeficiency virus vpr open reading frame. J Virol 63: 4110–4114
Park IW, Myrick K, Sodroski J (1994) Effects of vif mutations on cell-free infectivity and replication of simian immunodeficiency virus. J Acquir Immune Defic Syndr 7: 1228–1236
Paxton W, Connor RI, Landau NR (1993) Incorporation of Vpr into human immunodeficiency virus type 1 virions: requirement for the p6 region of gag and mutational analysis. J Virol 67: 7229–7237
Raja NU, Vincent MJ, Jabbar MA (1994) Vpu-mediated proteolysis of gp160/CD4 chimeric envelope glycoproteins in the endoplasmic reticulum: requirement of both the anchor and cytoplasmic domains of CD4. Virology 204: 357–366
Roe T, Reynolds TC, Yu G, Brown PO (1993) Integration of murine leukemia virus DNA depends on mitosis. EMBO J 12: 2099–2108
Rogel ME, Wu LI, Emerman M (1995) The human immunodeficiency virus type 1 vpr gene prevents cell proliferation during chronic infection. J Virol 69: 882–888
Rosenwirth B, Billich A, Datema R, Donatsch P, Hammerschmid F, Harrison R, Hiestand P, Jaksche H, Mayer P, Peichl P, Quesniaux V, Schatz F, Schuurman HJ, Traber R, Wenger R, Wolff B, Zenke G, Zurini M (1994) Inhibition of human immunodeficiency virus type 1 replication by SDZ NIM 811, a nonimmunosuppressive cyclosporine analog. Antimicrob Agents Chemother 38: 1763–1772
Sakai K, Ma X, Gordienko I, Volsky D (1991) Recombinational analysis of a natural noncytopathic human immunodeficiency virus type 1 (HIV-1) isolate: role of the vif gene in HIV-1 infection kinetics and cytopathicity. J Virol 65: 5765–5773
Sakai H, Shibata R, Sakuragi JI, Sakuragi S, Kawamura M, Adachi A (1993) Cell-dependent requirement of human immunodeficiency virus type 1 Vif protein for maturation of virus particles. J Virol 67: 1663–1666
Schreiber SL (1991) Chemistry and biology of the immunophilins and their immunosuppressive ligands. Science 251: 283–287
Schubert U, Schneider T, Henklein P, Hoffmann K, Berthold E, Hauser H, Pauli G, Porstmann T (1992) Human immunodeficiency virus type 1 encoded Vpu protein is phosphorylated by casein kinase II. Eur J Biochem 204: 875–883
Schubert U, Strebel K (1994) Differential activities of the human immunodeficiency virus type 1 encoded Vpu protein are regulated by phosphorylation and occur in different cellular compartments. J Virol 68: 2260–2271
Schwartz S, Felber BK, Fenyö EM, Pavlakis GN (1990) Env and Vpu proteins of human immunodeficiency virus type 1 are produced from multiple bicistronic mRNAs. J Virol 64: 5448–5456
Schwartz S, Felber BK, Pavlakis GN (1991) Expression of human immunodeficiency virus type 1 vif and vpr mRNAs is Rev-dependent and regulated by splicing. Virology 183: 677–686
Sodroski J, Goh WC, Rosen C, Tartar A, Portetelle D, Burny A, Haseltine W (1986) Replicative and cytopathic potential of HTLV-III/LAV with sor gene deletions. Science 231: 1549–1553
Sova P, Volsky DJ (1993) Efficiency of viral DNA synthesis during infection of permissive and nonpermissive cells with vif-negative human immunodeficiency virus type 1. J Virol 67: 6322–6326
Steinkasserer A, Harrison R, Billich A, Hammerschmid F, Werner G, Wolff B, Peichl P, Palfi G, Schnitzel W, Mlynar E, Rosenwirth B (1995) Mode of action of SDZ NIM 811, a nonimmunosuppressive cyclosporin A analog with activity against human immunodeficiency virus type 1 (HIV-1): interference with early and late events in HIV-1 replication. J Virol 69: 814–824
Strebel K, Daugherty D, Clouse K, Cohen D, Folks T, Martin MA (1987) The HIV 'A' (sor) gene product is essential for virus infectivity. Nature 328: 728–730
Strebel K, Klimkait T, Martin MA (1988) A novel gene of HIV-1, vpu, and its 16-kilodalton product. Science 241: 1221–1223
Strebel K, Klimkait T, Maldarelli F, Martin MA (1989) Molecular and biochemical analyses of human immunodeficiency virus type 1 Vpu protein. J Virol 63: 3784–3791
Subbramanian RA, Cohen EA (1994) Molecular biology of the human immunodeficiency virus accessory proteins. J Virol 68: 6831–6835
Terwilliger EF, Cohen EA, Lu YC, Sodroski JG, Haseltine WA (1989) Functional role of human immunodeficiency virus type 1 vpu. Proc Natl Acad Sci USA 86: 5163–5167
Thali M, Bukovsky A, Kondo E, Rosenwirth B, Walsh CT, Sodroski J, Göttlinger HG (1994) Functional association of cyclophilin A with HIV-1 virions. Nature 372: 363–365
Tristem M, Marshall C, Karpas A, Petrik J, Hill F (1990) Origin of vpx in lentiviruses. Nature 347: 341–342

Tristem M, Marshall C, Karpas A, Hill F (1992) Evolution of the primate lentiviruses: evidence from vpx and vpr. EMBO J 11: 3405–3412

Vincent MJ, Raja NU, Jabbar MA (1993) Human immunodeficiency virus type 1 Vpu protein induces degradation of chimeric envelope glycoproteins bearing the cytoplasmic and anchor domain of CD4: role of the cytoplasmic domain in vpu-induced degradation in the endoplasmic reticulum. J Virol 67: 5538–5549

Von Schwedler U, Song J, Aiken C, Trono D (1993) vif is crucial for human immunodeficiency virus type 1 proviral DNA synthesis in infected cells. J Virol 67: 4945–4955

Walsh CT, Zydowsky LD, McKeon FD (1992) Cyclosporin A, the cyclophilin class of peptidylprolyl isomerases, and blockade of T cell signal transduction. J Biol Chem 267: 13115–13118

Wang JJ, Lu YL, Ratner L (1994) Particle assembly and Vpr expression in human immunodeficiency virus type 1 infected cells demonstrated by immunoelectron microscopy. J Gen Virol 75: 2607–2614

Westervelt P, Henkel T, Trowbridge DB, Orenstein J, Heuser J, Gendelman HE, Ratner L (1992) Dual regulation of silent and productive infection in monocytes by distinct human immunodeficiency virus type 1 determinants. J Virol 66: 3925–3931

Willey RL, Maldarelli F, Martin MA, Strebel K (1992a) Human immunodeficiency virus type 1 Vpu protein regulates the formation of intracellular gp160-CD4 complexes. J Virol 66: 226–234

Willey RL, Maldarelli F, Martin MA, Strebel K (1992b) Human immunodeficiency virus type 1 Vpu protein induces rapid degradation of CD4. J Virol 66: 7193–7200

Willey RL, Buckler-White A, Strebel K (1994) Sequences present in the cytoplasmic domain of CD4 are necessary and sufficient to confer sensitivity to the human immunodeficiency virus type 1 Vpu protein. J Virol 68: 1207–1212

Wills JW, Craven RC (1991) Form, function, and use of retroviral Gag proteins. AIDS 5: 639–654

Wu X, Conway JA, Kim J, Kappes JC (1994) Localization of Vpx packaging signal within the C terminus of the human immunodeficiency virus type 2 Gag precursor protein. J Virol 68: 6161–6169

Yao XJ, Göttlinger H, Haseltine WA, Cohen EA (1992) Envelope glycoprotein and CD4 independence of Vpu facilitated HIV-1 capsid export. J Virol 66: 5119–5126

Yao XJ, Garzon S, Boisvert F, Haseltine WA, Cohen EA (1993) The effect of Vpu on HIV-1-induced syncytia formation. J Acquir Immune Defic Syndr 6: 135–141

Yao XJ, Friborg J, Checroune F, Gratton S, Boisvert F, Sékaly RP, Cohen EA (1995) Degradation of CD4 induced by human immunodeficiency virus type 1 Vpu protein: a predicted alpha-helix structure in the proximal cytoplasmic region of CD4 contributes to Vpu sensitivity. Virology 209: 615–623

Yao XJ, Subbramanian R, Rougeau N, Boisvert F, Bergeron D, Cohen EA (1995) Mutagenic analysis of Hiv-1Vpr: role of a predicted N-terminal alpha helical structure on Vpr nuclear localization and virion incorporation. J virol 69: 7032–7044

Yu XF, Yu QC, Essex M, Lee TH (1991) The vpx gene of simian immunodeficiency virus facilitates efficient viral replication in fresh lymphocytes and macrophages. J Virol 65: 5088–5091

Yu XF, Matsuda Z, Yu QC, Lee TH, Essex M (1993) Vpx of simian immunodeficiency virus is localized primarily outside the virus core in mature virions. J Virol 67: 4386–4390

Yuan X, Matsuda Z, Matsuda M, Essex M, Lee TH (1990a) Human immunodeficiency virus vpr gene encodes a virion-associated protein. AIDS Res Hum Retroviruses 6: 1265–1271

Yuan X, Matsuda M, Essex M, Lee TH (1990b) Open reading frame vpr of simian immunodeficiency virus encodes a virion-associated protein. J Virol 64: 5688–5693

Zhao LJ, Mukherjee S, Narayan O (1994) Biochemical mechanism of HIV-1 Vpr function: specific interaction with a cellular protein. J Biol Chem 289: 15827–15832

Use of Heterologous Expression Systems to Study Retroviral Morphogenesis

P. BOULANGER[1] and I. JONES[2]

1 Introduction

Cloning and expression of viral proteins in heterologous systems is often the method of choice for obtaining the large quantities of purified proteins necessary for structural and functional analyses. The recombinant expression technologies involved were developed initially to satisfy the needs of the medical and manufacturing industries for large-scale production of viral components at low cost for use in vaccine or diagnostic applications (ADA 1988; OSTERHAUS et al. 1985; PORTETELLE et al. 1991; SHIDA et al. 1987). It soon became apparent, however, that these methodologies could also be used on the analytical scale to provide much information on the more fundamental aspects of virus physiology and pathogenesis. There are few more impressive examples of this than the pro-

[1]Laboratoire de Virologie Moléculaire, Centre National de la Recherche Scientifique (URA-1487) and Faculté de Médecine, 2, Boulevard Henri IV, 34060 Montpellier, France
[2]Institute of Virology and Environmental Microbiology, Natural Environment Research Council, Mansfield Road, Oxford OX1 3SR, United Kingdom

gress made through the use of heterologous expression systems in understanding the morphogenesis of retrovirus particles.

One of the key factors governing the dissemination of any virus is the efficiency of the late steps of its life cycle, viz. (a) the virion assembly, (b) the maturation process, which transforms a non-infectious particle into an infectious virion, and (c) the release from the host cell and the subsequent infection of neighbouring cells. The relatively modest coding capacity of many virus genomes, including retroviruses, means that their capsids are composed of a restricted number of structural components, and in most cases of only a single major capsid protein. The specific and regular association of these capsid proteins often leads to assembly intermediates sometimes referred to as previrions. In subsequent steps, additional protein components and nucleic acid are incorporated into the assembling structure, resulting, finally, in virions that are stable, mature and infectious.

It follows that the assembly of a virus capsid consists, at least at discrete stages of the process, of the homopolymerisation of a unique protein species. When this process is mimicked by the expression of the single viral gene encoding the structural protein concerned, it provides a unique opportunity to analyse the mutual interactions of the protein monomers, their multimerisation and their assembly into capsids or precapsids (in vivo and in vitro). At the outset, it might be thought that the expression of a single viral gene in its homologous host-cell system would have the advantage of placing the gene product in its natural environment. As such, virus capsid morphogenesis might be expected to proceed as it would in vivo. In practice, however, the expression of the normal cell-specific regulation pathways often results in the level of viral gene product being lower than could be achieved in a heterologous expression system. This can constitute a block in analysis if the signal-to-background ratio of an assembly phenomenon is low or the quantification of a suppressing effect is required. In addition, homologous systems contain, by definition, the full complement of the normal cellular ligands for the viral components under study, e.g. cytoskeletal scaffold, transport pathways and receptor or chaperon proteins. These factors can make it dificult to assess the intrinsic properties of a viral capsid protein as opposed to the functions that depend on partner protein interaction.

By contrast, the expression of virus genes in heterologous systems offers a number of advantages. (a) The deregulated or inducible gene expression of heterologous systems usually results in high-level synthesis of the cloned gene product. (b) Expression of single capsid proteins which encode all the information necessary for oligomerisation and self-assembly within the coding region itself, assembly into pre-capsids or capsids can be analysed in vivo, or in vitro (after cell lysis and isolation). (c) The effect of mutations can be studied in *cis*, and also in *trans*, in experiments where mutants of the same protein are co-expressed in the same cell. (d) Co-expression of two or more different viral genes coding for physiological partners within the host cell or within the virus particle can also be envisaged, allowing an analysis of the interactions between these two gene products in vivo.

2 Expression Systems Available for Studying Retrovirus Assembly

2.1 Metabolic Requirements

There are a number of prerequisites for the successful use of heterologous expression systems in general, and for the study of the retrovirus morphogenetic pathway in particular. (a) As protein conformation is critical for the formation of legitimate molecular contacts between capsomers, the correct folding of the protein(s) under study is a key factor for the initial steps of the process. (b) If the acquisition of the functional three-dimensional structure requires the intervention of a chaperon protein, this chaperon has to be ubiquitous and present in sufficient quantity in the heterologous cell used for its expression so it would not be the limiting factor. (c) Post-translational modifications, such as glycosylation, phosphorylation, N-acylation, ADP-ribosylation or ubiquitination can modify the conformation of a viral protein profoundly so allowing access to, or masking, critical domains involved in the interactions that lead to assembly. They can also influence protein transport to, or/and compartmentalization in, the site of capsid assembly. (d) Finally, the presence or absence of the general regulatory mechanisms of gene expression, such as mRNA splicing or ribosomal frame-shifting, must be considered before the choice of a heterologous system is made. The possible interference by minor frameshifted products or from proteins resulting from translation of alternately spliced mRNA species must be also considered on a case by case basis. For instance, ribosomal frameshifting occurs efficiently in yeast (Kramer et al. 1986) and in insect cells (Hughes et al. 1993; Sommerfelt et al. 1993) to such a level that can lead to a ratio of unprocessed to processed Gag antigen that is insufficient for particle formation to be observed (Gheysen et al. 1989; Overton et al. 1989). Similarly, the relative levels of protease and Gag expression in mammalian cells can control the level of discernible Gag particle formation (Karacostas et al. 1993; Nam et al. 1988). All heterologous expression systems provide some of the prerequisites necessary for functional study of Gag precursor polyprotein assembly, but few systems provide them all, a fact that is reflected in the level of progress achieved in the study of capsid assembly in different systems to date. Nevertheless, the considerable body of information on the morphogenetic pathway of retroviruses accumulated during the last decade by using this kind of approach has largely compensated for the minor drawbacks imposed by its limitations.

2.2 Bacteria

In general, it is true to say that assembly of discernible virus-related structures from expressed capsid antigens has been restricted to higher eucaryotic expression systems, although exceptions are emerging (Campbell and Vogt 1995; Ehrlich et al. 1992, 1994; Klikova et al. 1995). Intermediates and the analysis of subdomains of retrovirus Gag precursor have, however, benefited from a number of expression studies in bacteria. Full-length HIV-1 Gag precursor polyprotein ($Pr55^{gag}$) can be expressed in bacteria but suffers from considerable degradation (Luban and Goff 1991; Luban et al. 1993b; Wagner et al. 1992). There is little evidence for assembly of the polyprotein into particle intermediates, although other properties of the molecule, such as RNA binding and the ability to act as an efficient substrate for the protease, appear faithfully represented (Berkowitz et al. 1993; Partin et al. 1990). Genetic evidence suggests that HIV-1 $Pr55^{gag}$ expressed in *Escherichia coli* has limited ability to interact, and this can be used to map the sequences necessary for the interaction (Luban and Goff 1991), but the relationship between these findings and the ability to assemble into pre-core-like structures has yet to emerge. By contrast, Gag precursor polyprotein of the primate retrovirus Mason-Pfizer monkey virus (M-PMV) expressed at high levels in bacteria assembles into capsid-like structures, 80–90 nm in diameter, morphologically indistinguishable from capsids observed in HeLa cells expressing the same recombinant protein from a similar T7-based vector (Klikova et al. 1995). The occurrence of spiral-like structures in *E. coli* cells suggests that assembly of the M-PMV capsid may involve sequential addition of capsomers to a nucleation complex (Klikova et al. 1995).

Each of the subdomains of HIV-1 Gag has been stably expressed at high level in *E. coli* (Burnette et al. 1992; Ehrlich et al. 1990; Mills et al. 1992), facilitating a number of structure and function studies. In many cases the purified antigen has shown properties of limited assembly, consistent with the presence of sequences required for protein interaction (Campbell and Vogt 1995; Ehrlich et al. 1992, 1994). However, these studies need to be correlated with those based on particle proficient expression systems, before a direct role for the identified sequences in the assembly process can be assessed. In HIV-1, there is a very clear morphological change in the appearance of the virion following maturation. This is assumed to be linked to the cleavage of the Gag precursor polyprotein and the re-alignment of individual Gag domains (Wills and Craven 1991), although recent findings suggest this interpretation may be naive (Kaplan et al. 1994). For heterologous expression systems that represent only part of the assembly process, it is essential to distinguish which of the Gag interactions is being mapped: those leading to the ordered alignment of $Pr55^{gag}$, or those that guide re-alignment following proteolytic cleavage. Notwithstanding these difficulties, *E. coli* expression of individual domains of HIV-1 $Pr55^{gag}$ has led to the derivation of the three-dimensional structure of p17 (Massiah et al. 1994; Matthews et al. 1994), the first high-resolution structure of any complete Gag-derived protein. In addition, crystals of p24 complexed with an Fab have been reported (Prongay et al. 1990) that should lead, in due course, to the three-dimensional structure of the p24 protein. As these two proteins together (as $Pr41^{gag}$) contain, in essence, most of the information necessary for HIV-1 particle assembly (see below), their three-dimensional structure, even in isolation, ought to benefit an understanding of the way in which Gag monomers interact.

2.3 Yeast

Expression in yeast cells has offered little advance over *E. coli* expression studies in general. Even though N-myristylation of HIV-1 Gag precursor and plasma membrane targeting occur, no assembly and budding of Gag particles has been observed (BIEMANS et al. 1992; JACOBS et al. 1989). Similar observations have been made for other retroviral Gag proteins, e.g. the $Pr76^{gag}$ of RSV (BONNET and SPAHR 1990). One set of studies that positively relies on the expression of functional Gag proteins in yeast is the two-hybrid genetic analysis of Gag domains that interact in *trans* (FRANKE et al. 1994a; LUBAN et al. 1992). This system has also identified cellular factors interacting with HIV-1 Gag and involved in virus infectivity (FRANKE et al. 1994b; LUBAN et al. 1993a; THALI et al. 1994). However, the failure to assemble in most cases a structure that resembles the retrovirus particle in bacteria and lower eucaryotes has led to a predominance of higher eucaryotic heterologous expression systems for the study of Gag assembly. The preferred systems for the expression of retrovirus Gag have been therefore vertebrate and invertebrate recombinant viruses, although unvaluable data have been obtained from provirus-transfected cell lines (GÖTTLINGER et al. 1991).

2.4 Animal Viruses

Recombinant animal viruses have been extensively used for the expression of retrovirus capsid proteins, viz. human adenovirus (Ad), Semliki forest virus (SFV; BERGLUND et al. 1993), vaccinia virus (VV) and baculovirus. Recombinant vaccinia viruses (SMITH and MACKETT 1992) are grown in mammalian cells of human, simian and rodent origins and provide, in many cases, the host-cell environment of the original retrovirus gene product. Furthermore, synthetic or heterologous poxvirus promoters have been designed which result in a higher level of expression than that driven by the natural late promoters of VV genes (DAVISON and MOSS, 1989a,b; 1990; PEARSON et al. 1991). The VV vector technology has also been greatly improved by the introduction of new vectors which allow foreign genes to be expressed in an inducible manner in VV-infected cells (DUNCAN and SMITH 1992; RODRIGUEZ and SMITH 1990; ZHANG and MOSS 1991).

In the case of the baculovirus expression system, the polyhedrin or p10 protein promoters are exceptionally active and can lead to the production of large amounts of foreign gene products (DOERFLER 1988; LUCKOW and SUMMERS 1989; STEWART and POSSEE 1993). The most widely used expression vehicle is the *Autographa californica* nuclear polyhedrosis virus that produces multiple nucleocapsids per occlusion body (or polyhedron), abbreviated AcMNPV (O'REILLY et al. 1992).

Post-translational modifications of foreign gene proteins occur in baculovirus-infected insect cells as they would in natural mammalian cells. These include glycosylation, phosphorylation, specific proteolytic processing and signal peptide cleavage (JARVIS and SUMMERS 1989; KURODA et al. 1986, 1990; MATSUURA et al. 1987; NYUNOYA et al. 1988; OVERTON et al. 1989; POSSEE 1986). Recombinant proteins can also be efficiently secreted, via the introduction of a heterologous secretory signal sequence to the 5′ extremity of the gene constructs (STEWART et al. 1991), or via natural secretion sequence(s) present within proteins, as observed with retrovirus HIV-1 Gag precursor (ROYER et al. 1992; CHAZAL et al. 1995). A major distinction between baculovirus-expressed proteins synthesized in insect cells and their mammalian-expressed counterparts is that they undergo a glycosylation process different from that in mammalian cells. Insect cell-derived glycoproteins have simple, unbranched oligo-saccharides with a high mannose content (FENOUILLET et al. 1994; KURODA et al. 1990). In spite of this difference, the correct three-dimensional conformation is clearly achieved by many recombinant proteins, as shown by their capacity to generate specific complexes and/or assemblies following expression in insect cells, faithfully reproducing that observed in mammalian cells. This is typified by the complex assembly of bluetongue virus core-like particles and tubules (FRENCH et al. 1990; FRENCH and ROY 1990; URAKAWA and ROY 1988), and for the SV40 T antigen-p53 complexes (O'REILLY and MILLER 1988). As a result of their suitability for this type of study, heterologous expression systems using recombinant vaccinia virus-infected mammalian cells or recombinant baculovirus-infected insect cells have largely contributed to the current model for the mechanisms of retrovirus capsid assembly.

3 Results

3.1 Morphogenetic Information for Retrovirus Capsid Assembly

Most retrovirus capsid protein precursors contain all their morphogenetic information within their reading frame. In all retroviruses isolated to date, the major virion structural proteins are encoded by the *gag* and *env* genes. Results from many expression systems have demonstrated that the expression of only the retrovirus *gag* gene, encoding the Gag precursor polyprotein (Pr^{gag}), is sufficient to obtain intracellular assembly and extracellular release of membrane-enveloped particles typical of the immature virion from expressing cells. These particulate structures have been variously termed "core-like particles", "Gag particles", "retrovirus-like" particles or "virus-like particles" (VLP), as their morphological characteristics mimic those of immature particles containing an uncondensed core. This finding iconographically illustrates the carriage of all the genetic information for the Pr^{gag} assembly within the *gag* gene sequence, and that self-assembly of the Pr^{gag} can be studied in a context where the other viral late genes coding for major structural proteins, Env, or accessory proteins, Vpu or Vpx, are not present. The only exception to date seems to be the human spumaretrovirus HSRV, which has been found to be incapable of self-assembly when expressed in insect cells using a recombinant baculovirus (Morin, Gay and Boulanger, unpublished data).

However, it has to be noted that HIV-1 envelope glycoprotein (gp) could influence the site of budding and release of the Gag particles. Co-expression of HIV-1 Gag and gp160 in polarized epithelial cells using two recombinant vaccinia viruses results in the selective release of Gag particles into the basolateral medium, in contrast to single Gag expression (OWENS et al. 1991). The polarized budding has been found to be mediated by the intracytoplasmic domain of the gp41 (LODGE et al. 1994).

Recombinant HIV-1 $Pr55^{gag}$ has been been found to assemble and be released as membrane-enveloped retrovirus-like Gag particles from HeLa cells infected by an adenovirus-HIV recombinant (VERNON et al. 1991), and from simian cells harbouring an SV40 virus vector (SMITH et al. 1990). Likewise, recombinant Pr^{gag} from HIV-1 or Friend leukaemia virus have been shown to assemble in recombinant vaccinia virus-infected cells (FLEXNER et al. 1988; GOWDA et al. 1989; HAFFAR et al. 1990; HOSHIKAWA et al. 1991; HU et al. 1990; KARACOSTAS et al. 1989; MIYAZAWA et al. 1990; SHIODA and SHIBUTA, 1990; VON POBLOTZKI et al. 1993; WAGNER et al. 1992, 1994). Pr^{gag} from HIV-1, HIV-2, HTLV-1, M-PMV, SIV, FIV and BIV also assemble Gag particles in baculovirus-infected insect cells (DELCHAMBRE et al. 1989; GHEYSEN et al. 1989; HUGHES et al. 1993; LUO et al. 1990, 1994; MADISEN et al. 1987; MORIKAWA et al. 1991; OVERTON et al. 1989; RASSMUSSEN et al. 1990; ROYER et al. 1991; SOMMERFELT et al. 1993; ZHAO et al. 1994). In all cases, particle assembly only occurred in the absence of functional retroviral protease, with the sole constituent of the particles formed being unprocessed Gag precursor molecules. The presence of RNAs, heterogeneous in size and of both viral and cellular origins, within these particles has been reported (GHEYSEN et al. 1989; SHIODA and SHIBUTA 1990), but their significance and role in Gag assembly process is not clear.

The high level of expression of recombinant Pr^{gag} in adenovirus-, vaccinia virus- or baculovirus-infected cells has facilitated the isolation of membrane-enveloped Gag particles sedimenting at the density of retrovirions (1.15–1.18 g/cm^3) during sucrose gradient analysis of infected cell culture fluid samples. Abundant budding particles typical of the size and shape of the authentic immature particles were also visible at the cell surface under the EM (Fig. 1). The high efficiency of retrovirus-like particle assembly in the baculovirus system has made possible a statistical analysis of antibody labelling by quantitative immunoelectron microscopy (QIEM) of the population of retroviral Gag particles released by budding from the plasma membrane of Sf9 cells (CARRIÉRE et al. 1995). HIV-1 Gag particles that have been probed with a panel of characterised monoclonal antibodies have been scored for the accessibility of epitopes localised in different Gag domains determined by QIEM. This approach is based on

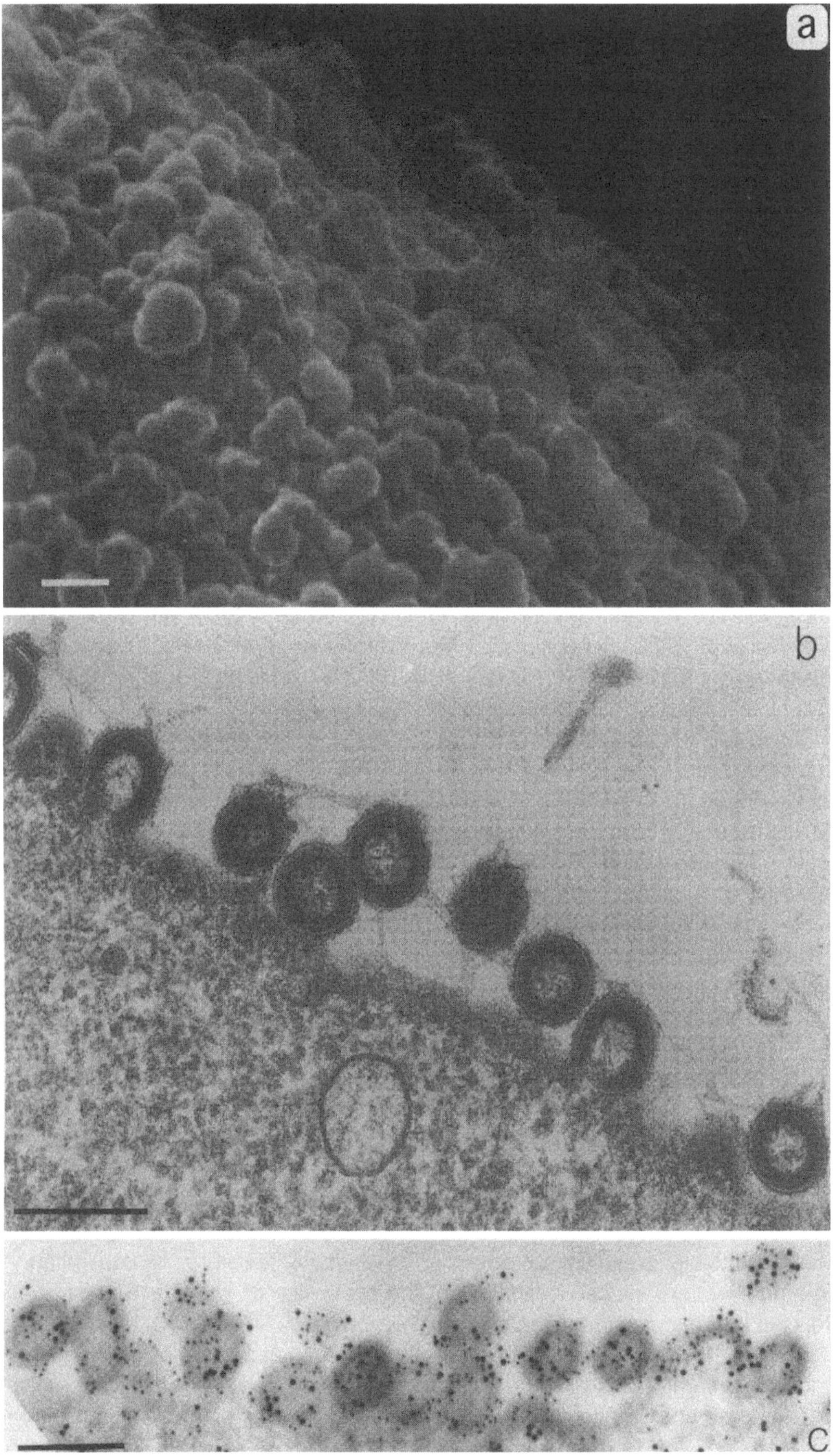
a
b
c

the hypothesis that highly reactive Gag epitopes represent exposed regions which are not involved in Gag interactions, whereas poorly or non-reactive epitopes would be good candidates for interacting domains.

Using this approach, five regions have been found to be immunoreactive by QIEM analysis of recombinant HIV-1 Gag particles, viz. residues 11–25 near the N-terminus of the MA, 113–132 near the MA-CA junction, 201–233 and 285–341 in the CA, and the C-terminal moiety of the p6 domain (CARRIÉRE et al. 1995). By corollary, these observations delineate four antigenically silent windows within the $Pr55^{gag}$ sequence: (a) between residues 26 and 112 in MA, (b) residues 133–200 and (c) 234–284 in CA, and (d) within residues 357–462, overlapping the sp2, NCp7 and sp1 domains (Fig. 2). These inaccessible domains represent potential sites of interaction between Gag precursor molecules within the particle. In particular, the first domain includes MA residues 54–65, which has been found by mutagenetic analysis to be critical for particle assembly (CHAZAL et al. 1995; FREED et al. 1994; YU et al. 1992), and which constitutes a distinct helix (helix 2) in the three-dimensional structure of p17 (MASSIAH et al. 1994; MATTHEWS et al. 1994). The fourth region coincides with a portion of the HIV-1 Pr^{gag} sequence, which has been found to have a major influence in vivo in the *trans*-rescue process of carboxytruncated Gag mutants (residues 358–374), and in vitro in Gag ligand affinity blotting assays (residues 375–426), as described below. The domains highlighted by this analysis are also consistent with the recent report that the minimal domain required for HIV-1 Gag polyprotein multimerization in the GAL4 two-hybrid system is the region 240–434 (FRANKE et al. 1994a).

3.2 Cellular Assembly Pathway of Recombinant Retroviral Capsids

Retroviruses are subdivided into types B, C and D on the basis of their morphology. In the type C retroviruses, the most represented group, the assembly and budding of particles occur concurrently at the plasma membrane. By contrast, types B and D retroviruses pre-assemble immature A-type particles within the cytoplasm, and then translocate to the cell surface, where they are released by budding. M-PMV is the prototypic representative of type D retroviruses. The assembly pathway of recombinant retrovirus capsid proteins in heterologous cell expression systems, and particularly in non-mammalian cell types, seems to follow faithfully that observed in the natural host cell. For N-myristylated $Pr55^{gag}$ of HIV-1, as well as for the equivalent molecule of other type C retroviruses,

◄

Fig. 1a–c. Retrovirus-like particles budding from the plasma membrane of recombinant baculovirus-infected insect cells (Sf9) expressing HIV-1 WT $Pr55^{gag}$. **a** Scanning electron microscopy (EM). **b** Transmission EM. **c** Immunoelectron microscopic analysis by double immunogold labelling of composite Gag particles constituted of two Gag species, HIV-1 WT $Pr55^{gag}$ and HIV-HSRV Gag chimera (CARRIÉRE et al. 1995). The two Gag precursors are distinguished by their respective MA domains: HIV-1 MA antibody is labelled with 5-nm colloidal gold grains, whilst the human spumaretrovirus MA antibody is labelled by 10-nm gold grains. *Bar*, 200 nm

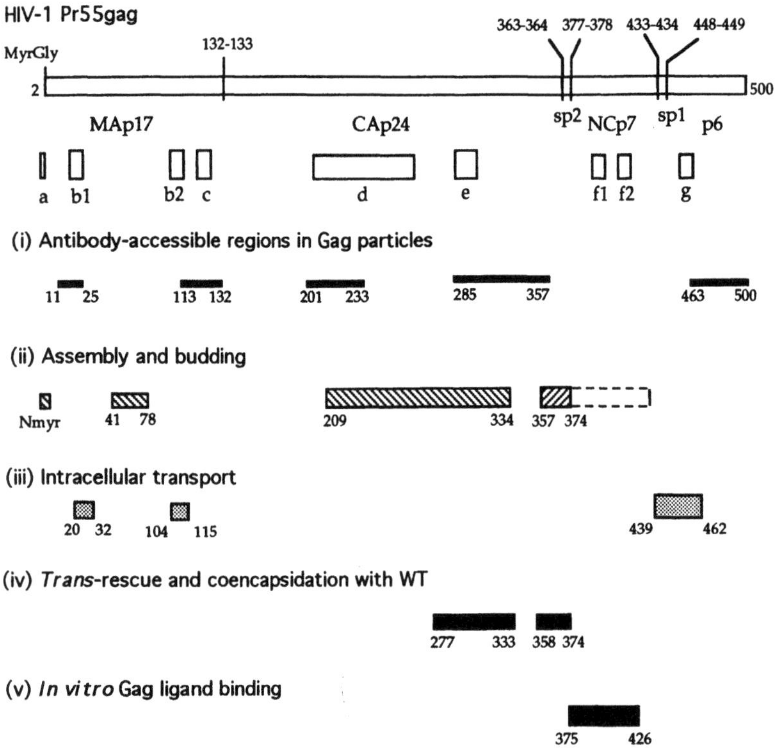

Fig. 2. Diagrammatic illustration of the sequence landmarks and functional domains in the HIV-1 $Pr55^{gag}$, as deduced from data obtained with recombinant *gag*-polyproteins expressed in various heterologous systems. The major domains (MA, CA, NCp7 and p6) and spacer peptides sp2 and sp1, defined by the HIV-1 protease cleavage sites (*vertical bars*), are shown on a linear representation of the Gag precursor, and numbered from the initiator Met^1 of the Gag_{LAI} sequence. *Open boxes* mark the position of the following sequence elements: *a*, N-myristylation signal; *b1, b2*, polybasic motifs at residues 26–32 and 110–114 in the MA; *c*, decapeptide ^{119}AADTGHSSQV^{128}, conserved in structural proteins of non-related RNA viruses (Blomberg and Medstrand 1990); *d*, cyclophilin-binding region (177–277; Luban et al. 1993a); *e*, major homology region (*MHR*), within residues 285–304 (Mammano et al. 1994; Wills and Craven 1991); *f1, f2*, His-Cys boxes at positions 392–405 and 413–426, respectively; *g*, conserved PTAPP sequence, at position 455–459 in p6. (*i*) Regions of encapsidated Pr^{gag} accessible to monoclonal antibodies are indicated by *horizontal bars* and their positions indicated by the amino acid numbers in the Pr^{gag} sequence (Carrière et al. 1995). (*ii*) The regions in the MA (residues 41–78), the CA (209–334) and at the CA-sp2 junction (374–358) of which integrity is required for Gag particle assembly, budding and release are shown by *hatched boxes*. The *dotted area* represents the NCp7 domain, which is required for assembly and budding of HIV Gag particles from recombinant vaccinia virus-infected cells (Hoshikawa et al. 1991) but not from baculovirus-infected insect cells (Hockley et al. 1994; Jowett et al. 1992; Royer et al. 1991). (*iii*) *Stippled boxes* represent the regions involved in intracellular transport, viz. within residues 20–32 and 104–115 in the MA (Chazal et al. 1995), and 439–462 at the sp1-p6 junction. (*iv*) The regions essential for *trans*-rescue and co-encapsidation of carboxy-truncated Gag mutants with WT Pr^{gag} in vivo (277–333 and 358–374) and (*v*) the region essential for Gag–Gag interactions in vitro in Gag ligand affinity blotting assays (375–426) are indicated by *filled boxes*

recombinant Gag particles are found to assemble at, and bud from, the cell surface. By contrast, when M-PMV Pr^{gag} is expressed using the baculovirus system, most of the capsid shells are found to assemble in the insect cell cytoplasm, from where they are later targeted to the plasma membrane (SOMMERFELT et al. 1993).

Expression of the wild-type M-PMV *gag-pro-pol* genes, however, results in a significant proportion of particles assembling at the plasma membrane, as in type C morphogenesis (SOMMERFELT et al. 1993). On the other hand, when the three Gag-Pol polyprotein precursors are expressed containing a non-functional protease (D26N substitution) the morphogenesis is exclusively D type (SOMMERFELT et al. 1993). Since the only difference between the wild-type and the mutant recombinant products is the activity of the viral protease, a role for this protease in redirecting the assembly pathway of a D-type recombinant Gag polyprotein to the cell surface is suggested, although this might be effected via a control of the local levels of Gag polyprotein present. The different pathways of morphogenesis typified by type C and types B and D retroviruses has been suggested to be less the consequence of intrinsic differences in their assembly mechanisms, than the result of a sufficient concentration of precursor molecules at local cellular sites where they can aggregate and self-assemble (RHEE and HUNTER 1990, 1991).

It is possible that the data obtained with the M-PMV protease mutant indicate a differential protease regulation mechanism between M-PMV on the one hand, and other retroviruses on the other (RHEE and HUNTER 1991). In contrast to the results obtained with M-PMV, there has been no observation of HIV and FIV capsid assembly and release in the presence of an active protease, either expressed in *cis* or in *trans* in the baculovirus expression system (HUGHES et al. 1993; MORIKAWA et al. 1991; OVERTON et al. 1989). In the vaccinia system, when HIV-1 (or HTLV-1) protease was co-expressed in *cis*, using a genetic construction which maintained the frameshift signal, the Gag particles found in the extracellular medium were formed of both uncleaved and cleaved Gag precursors (NAM et al. 1988; SHIODA and SHIBUTA 1990). Overexpression of HIV-1 protease by in-phase fusion of the *pro-pol* gene to the *gag* gene sequence has been found to result in the loss of the cell capacity to assemble and release Gag particles (KARACOSTAS et al. 1993; SHIODA and SHIBUTA 1990), suggesting the relative levels of precursor Gag and protease are a key determinant of Gag particle assembly.

3.3 Interactions and Assembly of Recombinant Structural Proteins of Retroviruses In Vitro

As it became clear that all the morphogenetic information for capsid assembly was present within the retrovirus Gag precursor protein sequence, it was a logical progression to develop in vitro assays to analyse, stepwise, the mechanisms of the assembly process and the polypeptide domains involved in capsid protein interactions. As noted above, progress has been achieved following the purification of recombinant HIV-1 CAp24 expressed in *E. coli* (EHRLICH et al. 1992, 1994). Purified CAp24 protein spontaneously forms dimers (48 kDa) and higher oligomeric forms (presumably dodecamers of about 300 kDa) in vitro, and mainly dimers and tetramers in the presence of homobifunctional cross-linking reagents (EHRLICH et al. 1992). When examined by EM, recombinant CAp24 oligomers appear as flexible elongated fibres of about 200 nm or longer, or rigid rod-like structures, depending upon the buffer used. The elongated structures can dissociate into grossly spherical subunits of 10–20 nm in diameter, and these particles could be the dodecamer subunits, and represent the true intermediates in the assembly pathway (EHRLICH et al. 1992). Limited digestion with endoproteases, used as structural probes to map protease-sensitive sites in the oligomeric forms, has provided insight into the sequences involved in their stabilization. It has also generated large protein fragments of functional and structural interest. The suggestion from these experiments is that HIV-1 CA is composed of two structural domains separated by a surface-exposed, immunodominant nonstructured central region around the arginine residue 100 in the CA sequence (232 in the Gag precursor sequence; EHRLICH et al. 1992, 1994). The carboxy-terminal domain contains the retroviral Gag major homology region, which lies between residues 152–171 in the HIV-1 CA protein sequence, or 285–304 in the Gag precursor sequence (MAMMANO et al. 1994; WILLS and CRAVEN 1991).

Experiments with purified retrovirus structural proteins in vitro have also provided a valuable method for the manipulation of conditions for oligomerisation and possible cofactors implicated in capsid assembly. As an approach to the *cis*-regulation mechanisms of protease activation and polyprotein processing in relation to the virus particle assembly, WT and mutants of in vitro cell-free system synthesised or baculovirus-expressed Gag-Pol polyproteins of ALV have been analysed with respect to their proteolytic cleavage at the Gag-Pol junction in vitro (STEWART and VOGT 1994). The results are consistent with a model in which the Gag-Pol junction sequence within isolated polyproteins adopts a protease-resistant coiled coil conformation reminiscent of a leucine zipper. By contrast, this junction would be held in an extended conformation within immature particles, as a result of interactions between neighbouring Gag and Gag-Pol precursors. The alternate extended or collapsed structural conformation of this region in vivo and in vitro, respectively, would explain why the first protease cleavage site which occurs in vivo is confined to this junction, leading to the activation of the viral reverse transcriptase in ALV virions (STEWART and VOGT 1994).

Reconstruction of virus-like particles from purified, recombinant retrovirus structural proteins has also been successfully achieved. Efficient in vitro assembly of capsid-like structures has been obtained with bacterially expressed M-PMV Gag polyprotein following complete urea-solubilisation and subsequent renaturation in low salt-alkaline conditions (KLIKOVA et al. 1995). Likewise, using bacterially expressed ALV and HIV-1 CA-NC polyprotein, roughly spherical particles, 60–100 nm in diameter, have been obtained (CAMPBELL and VOGT 1995). Addition of RNA results in a better efficiency in particle assembly, and cylindrical structures, sometimes showing a tail of protein-covered RNA, were also detected. The diameter of the particles re-assembled in vitro from CA-NC and RNA appeared similar to the diameter reported for mature retrovirus cores as seen in EM sections (CAMPBELL and VOGT 1995). These data confirm the validity of the in vitro approach to study retrovirus assembly and demonstrate the value of combining several experimental methods to the dissection of Gag functions.

Thus, in a semi in vitro study, it has been shown that HIV-1 Pr^{gag} isolated as soluble polyprotein from baculovirus-infected cells has the capacity to bind to homologous Pr^{gag} immobilised onto nitrocellulose membrane in ligand affinity blotting assay (CARRIÉRE et al. 1995). Using this assay and a panel of carboxy-truncated mutants of HIV-1 Pr^{gag}, it has been found that the Gag sequence between residues 375 and 426, which overlaps the spacer peptide sp2 at the CA-NC junction and the two zinc fingers in the NC domain, is important for stable Gag interactions and Gag particle assembly (CARRIÉRE et al. 1995), consistent with the results from various analyses using recombinant viral protein methodologies.

3.4 Expression of *gag* Mutants in Heterologous Systems: Mutational Effects in *cis* and Identification of Functional Domains in Gag Assembly

The peptide sequences involved in assembly of retrovirus capsids have been investigated by extensive mutagenesis of recombinant *gag* gene products expressed in various systems. This includes deletions at the N- and C-termini of Pr^{gag} (DELCHAMBRE et al. 1989; GHEYSEN et al. 1989; HUGHES et al. 1993; JOWETT et al. 1992; OVERTON et al. 1989; ROYER et al. 1991, 1992; WAGNER et al. 1992), internal deletions (CHAZAL et al. 1995; WAGNER et al. 1994), insertions (CHAZAL et al. 1994) and substitutions (HONG and BOULANGER 1993; VON POBLOTZKI et al. 1993) in its different domains, MA, CA, and NC. The results were consistent with the data obtained with mutants made in the provirus, and confirmed the existence of three major discrete domains in the retrovirus Pr^{gag} molecule which are essential for Gag particle assembly (WILLS and CRAVEN 1991). The first critical domain corresponds to the MA N-terminus and its myristylation (or more generally acylation) acceptor site. The second domain lies in the N-terminal moiety of Pr^{gag}, within residues 41–78 in HIV-1 MA (CHAZAL et al. 1995; FREED et al. 1994; YU et al. 1992), or within the p2 peptide located between the MA and the p10-coding sequences of RSV Gag (WELDON and WILLS 1993; WILLS et al. 1994). The third domain comprises the carboxy-terminal moiety of the CA and extends into the NC, spanning its zinc finger(s) and polybasic sequence (BENNETT et al. 1993; WELDON and WILLS 1993). All the reported mutations within the region 209–374 of recombinant HIV-1 Pr^{gag} have been found to be deleterious to Gag assembly (CHAZAL et al. 1994; HONG and BOULANGER 1993; VON POBLOTZKI et al. 1993; ZHAO et al. 1994). This region overlaps the major homology region (MHR; WILLS and CRAVEN 1991), between residues 286 and 304 in HIV-1. It also contains a motif (^{317}MTETLxxQNA^{326} in HIV-1), which is conserved in structural proteins of many RNA viruses, and the residue Cys^{330}, which is common to the CA proteins of HIV, SIV, EIV, HTLV-I and maedi-visna virus (ARGOS 1989).

The role of the carboxy-terminal domains, including p6, NC and the spacer peptide sp2 at the CA-NC junction, in the assembly of Gag particles has been specifically analysed using baculovirus-expressed, N-myristylation-competent HIV-1 Gag precursor proteins carrying carboxy-terminal truncations of increasing lengths. The downstream boundary for the capacity of assembly and budding of HIV-1 retrovirus-like particles in Sf9 cells appeared to be located within the spacer peptide sp2 at the CA-NC junction (GHEYSEN et al. 1989; HOCKLEY et al. 1994; JOWETT et al. 1992; ROYER et al. 1991).

The intracellular transport of Gag precursor molecules is indirectly linked to the retrovirus morphogenetic process, since capsid structural proteins must be transported to the sites of assembly. The N-terminus of many retrovirus Pr^{gag} is modified by addition of a myristyl radical at the glycine acceptor residue at position 2 in the Pr^{gag} sequence, and this signal has been found to be indispensable to the plasma membrane-targeting and extracellular budding of membrane-enveloped Gag particles (WILLS and CRAVEN 1991). N-Myristylation-defective mutants of HIV-1 and SIV Pr^{gag} expressed in baculovirus-infected cells fail to be released by budding into the extracellular medium, but self-assemble intracellularly, in majority within the nucleus (DELCHAMBRE et al. 1989; GHEYSEN et al. 1989; HUGHES et al. 1993; OVERTON et al.1989; ROYER et al. 1991, 1992). In the natural targeting of HIV and SIV Gag precursors therefore, the myristylated N-terminus is dominant as a membrane-targeting signal over other putative transport signals present in other domains of both molecules, e.g. karyophilic signals in CA or MA (BIEMANS et al. 1992; BUKRINSKY et al. 1993). Two polybasic motifs in HIV-1 MA have been found to be involved in the intracellular transport of N-myristylated $Pr55^{gag}$ expressed in Sf9 cells (CHAZAL et al. 1995), and a nuclear localisation function has been mapped by deletion analysis to the carboxy-terminal domain of unmyristylated HIV-1 Pr^{gag} (ROYER et al. 1991). The latter lies within residues 439–462, overlapping the spacer peptide sp1 at the NC-p6 junction (Carriére, Gay and Boulanger, unpublished data). The relevance of the nuclear assembly of recombinant Gag particles to the HIV-1 life cycle remains to be elucidated.

A current summation of the important functional regions and peptide sequence elements of HIV-1 $Pr55^{gag}$ is illustrated in Fig. 2.

3.5 Expression of *gag* Assembly Mutants in Heterologous Systems: Effects in *trans*

A further advantage of recombinant virus technology when applied to retrovirus morphogenesis is that assembly-defective mutants of Pr^{gag} can also be analysed in complementation assays with the wild-type (WT) allele of Pr^{gag}, by co-infecting the same cell with two recombinants, one containing the mutant and the other one the WT *gag* gene. In the case of a recessive mutation, the WT domain would compensate for the morphogenetic defect, and the Pr^{gag} mutant would be rescued in *trans* and co-encapsidated with the WT Pr^{gag}. Mutant and WT Pr^{gag} can be advantageously constructed to become immunologically and biochemically distinguishable for this analysis, e.g. by making use of natural antigenic variation, by epitope tagging or by chimeric constructions. IEM using double immunogold labelling of budding Gag particles can provide the visual confirmation of the occurrence of mutant Pr^{gag} co-encapsidation (Fig. 1c), and biochemical analysis of extracellular particles isolated from the culture medium allows the degree of efficacy of the *trans*-rescue process to be assessed. Such an analysis applied to HIV-1 Pr^{gag} expressed in recombinant baculovirus-infected cells has shown that various assembly-defective mutants were efficiently complemented in *trans* by the recombinant WT Pr^{gag}, and could be copackaged into budding Gag particles, whereas other mutants failed to be rescued by the WT provided in *trans* (Carrière et al. 1995; Chazal et al. 1990, 1994, 1995; Hong and Boulanger 1993; Zhao et al. 1994). The results suggest that two discrete regions, within residues 277–333 (overlapping the MHR) and within residues 358–374 (overlapping the sp2 peptide), are essential for the *trans*-rescue phenomenon (Fig. 2).

On the other hand, a phenomenon of negative *trans*-dominance has been observed with mutations localised in certain domains of HIV-1 Pr^{gag}. The first observation of a *trans*-dominant negative mutation within a structural gene (*gag*) which negatively affects the cellular capacity to support HIV-1 replication and release of the WT virus was reported by Trono et al. (1989). An insertion mutant has been constructed at the same position in the recombinant HIV-1 Pr^{gag} (Gag*in*209) and expressed in baculovirus-infected cells. Gag*in*209 showed a significant deleterious effect on the assembly of other Pr^{gag} mutants and of WT Pr^{gag} in co-expression assays (Chazal et al. 1994). This suggests that the overall dominant negative effect on HIV-1 replication and release (Trono et al. 1989) was, in part, due to the involvement of mutant Gag polyprotein in some steps of capsid morphogenesis and budding.

Likewise, two assembly-defective mutants of unmyristylated recombinant HIV-1 Pr^{gag} expressed in insect cells, L268P and L322S, of which mutations are located on each side of the MHR, have been found to negatively affect the intracellular assembly pathway of the full-length HIV-1 Pr^{gag} when co-expressed in *trans* (Hong and Boulanger 1993). Biochemical and EM analyses suggest that this negative effect was due to illegitimate interactions between mutant and

WT recombinant Gag proteins (Hong and Boulanger 1993). Two other genetic constructs of baculovirus-expressed HIV-1 Pr^{gag} corresponding to a carboxy-truncated MA domain (*amb*120), and to the MA with a short additional sequence from the CA (*och*180), have also been found to have a *trans*-dominant inhibitory phenotype on assembly and budding, following co-expression with WT Pr^{gag} (Chazal et al. 1995). EM analysis of co-infected cells has shown that this effect was due to a massive redirection of the Gag particle assembly pathway to intracellular vesicles in the case of *amb*120, and to the nucleus for *och*180 (Chazal et al. 1995).

These results serve to illustrate the complexity of phenotypes observable for Gag mutants. They highlight the involvement of host cell functions in the cytoplasmic anchorage of retrovirus Pr^{gag}, Pr^{gag} intracellular transport to the sites of assembly and, finally, successful extracellular export.

3.6 Role of Host-Cell Factors in Retrovirus Assembly

In some heterologous expression systems, no retrovirus-like Gag particle assembly has been detected, despite the fact that active viral protease was not co-expressed. This is the case for the HIV-1 Pr^{gag} expressed in bacteria (Luban and Goff 1991; Luban et al. 1993b; Wagner et al. 1992) and yeast (Biemans et al. 1992; Jacobs et al. 1989), and for the human spumavirus HSRV in insect cells (Morin, Gay and Boulanger, unpublished data). These results suggest that certain factors, which are necessary for Gag particle assembly, are present in only some cell types, or, alternatively, that only certain retrovirus Gag precursor polyproteins are capable of legitimate interactions with ubiquitous cell factors. This is typified by the difference observed in assembly capacity of HIV-1 and M-PMV Gag precursors expressed in bacteria (refer to Sect. 2.2). If the phenomenon of retrovirus Gag assembly requires bacterial chaperonin molecules, the efficiency of M-PMV Gag precursor to assemble into capsid-like structures in *E.coli* cells (Klikova et al. 1995), unlike HIV-1 Gag, would imply that these chaperonins are competent for the folding and self-assembly of M-PMV Gag, but not of HIV-1 Gag precursor.

Using the GAL4 double-hybrid screening method to identify cellular components interacting with retrovirus capsid proteins, it has beeen found recently that cyclophilins A, B and C, which have peptidy-prolyl *cis-trans* isomerase activity, and bind to HIV-1 Pr^{gag}, are potential cellular partners of the Pr^{gag}. In particular, cyclophilin A, which is virion incorporated, could serve as a chaperon for the folding of Pr^{gag} and help it gain its functional conformation (Luban et al. 1993a). In this case, one could hypothesise that some cellular factors would be missing or present at too low a level in yeast and insect cells to provide the required assembly function for HIV-1 and HSRV assembly, respectively. However, more recent data suggest that the Gag-cyclophilin A complex is required for replication and for the infectivity of HIV-1 virions, but not for Pr^{gag}-directed viral particle formation (Franke et al. 1994b; Thali et al. 1994).

Likewise, the difference in the budding behaviour between p6-deleted mutant proviral HIV-1 expressed in a human T-cell line and a p6-deleted mutant of recombinant Gag precursor expressed in baculovirus-infected cells was first attributed to hypothetical membrane factor(s) absent from human cells but provided in sufficient quantities by insect cells. The budding Gag particles of the p6-mutant provirus fail to be released into the culture medium of Jurkat cells (GÖTTLINGER et al. 1991) and other mammalian cell types, whereas they are released at WT levels from the baculovirus-infected insect cell surface (JOWETT et al., 1992; ROYER et al. 1991, 1992). It has since been suggested that functions essential for the virus release are associated with the Vpu accessory protein. Thus, Vpu could have a role in the cell dependence of the virus release-defective phenotype of p6 mutants (GÖTTLINGER et al. 1993).

The role of host-cell factors in retrovirus Pr^{gag} assembly, therefore, remains largely a matter of conjecture, although it is clear that the use of recombinant Pr^{gag} and a more systematic screening of cell protein expression libraries will provide valuable information in the near future on the virus-cell interactions that occur at the stage of viral capsid assembly. The role of cell factors might explain some of the subtle differences observed in the assembly patterns of recombinant Gag proteins from the same retrovirus in different expression systems, or from closely related retroviruses in similar expression systems. For example, the absence of any detectable budding particles in human cells infected by a recombinant vaccinia virus expressing the HIV-1 $Pr41^{gag}$ (HOSHIKAWA et al. 1991), compared to baculovirus-infected insect cells expressing the same construct which still produce a few Gag particles at their surface (JOWETT et al. 1992; ROYER et al. 1991, 1992), could reflect some cellular participation in the assembly and budding process. Similarly, no viral assemblies have been found to occur with recombinant HIV-1 MA alone expressed in baculovirus- (CHAZAL et al. 1995; Morikawa and Jones, unpublished data), or in vaccinia virus-infected cells (HOSHIKAWA et al. 1991), whereas recombinant SIV MA has been reported to be capable of self-assembly following expression in insect cells (GONZALEZ et al. 1993).

4 Conclusion and Open Questions

The morphogenesis of viruses in general and of retroviruses in particular raise a number of issues. (a) The first and most important question is the determination of whether the viral capsid protein encodes all its own morphopoietic information. In other words, can capsid protein assembly occur in a system (homologous or heterologous) in which the capsid gene is solely expressed? (b) If this is the case, what are the domains involved in this phenomenon, either directly by providing the intermolecular contact sites, or indirectly, via their influence on overall conformation? (c) Have post-translational modifications, accessory signals (e.g. N-myristylation) and/or cellular location any role in initiating or influencing the efficiency and final yield of capsid assembly? (d) What are the scaffold or chaperon protein(s), virus-coded proteins or host-cell components which are necessary for assembly? (e) As a corollary, is the capacity of assembly of virus capsomers restricted to some permissive cell lines, or can a variety of cell types provide the required morphopoietic factors?

The expression of retrovirus capsid protein genes in various heterologous systems has largely contributed to the elucidation of a number of these points, at least for most of the known retroviruses. In general, the extensive analysis of recombinant Pr^{gag} and their various mutants has confirmed the data obtained with mutants made in proviruses, suggesting that most systems are faithful mimics of the authentic host cell. In addition, the expression of a retroviral structural protein in a heterologous context, by 'isolating' this single assembly step from the overall retrovirus life cycle and replication pathway, has provided a substantial body of information on its homopolymerisation and higher order self-assembly reaction, as well as an early glimpse of the possible interactions with host cell components.

Although a general model for retrovirus morphogenetic pathway has not emerged yet from these studies, a certain number of points are now well established. Due to the diversity in the gene sequence of all retroviruses, with the exception of the major homology region (WILLS and CRAVEN 1991), each retrovirus Pr^{gag} has its specific assembly domains. However, discrete regions implicated, directly or indirectly, in the retrovirus particle assembly process have been delineated for many retrovirus Pr^{gag} molecules. They have generally shared locations, viz. within the MA domain, at the C-terminal third of the CA, and in the NC domain (WILLS and CRAVEN 1991; and refer to Fig. 2). A role for dimerisation of retroviral Pr^{gag} has been suggested on the basis of the in vitro cross-linking of bacterially expressed HIV-1 Pr^{gag} (EHRLICH et al. 1990, 1994), in vitro cross-linking of baculovirus expressed HIV-1 MA domain (MORIKAWA et al. 1995), and by the NMR-derived structure of the isolated MA domain (MATTHEWS et al. 1994). However, the dimer formation has been a matter of controversy (MASSIAH et al. 1994), and SIV MA has been found to crystallize as a trimer (RAO et al. 1995). Furthermore, the higher order structure of the Gag particle has been suggested by high-resolution EM studies of recombinant WT Gag particles (NERMUT et al. 1994). These data have suggested that the Pr^{gag} molecules are rod like in shape and are packed together in hexameric arrays that lead, in a cell background that is competent for assembly, to the formation of a spherical particle with icosahedral symmetry. The possible role of nucleic acid molecules from the cellular pool in the initial step of retrovirus Pr^{gag} assembly and how nucleic acid is incorporated in the correct stoichiometric amounts remains an open question (CAMPBELL and VOGT 1995; DARLIX et al. 1990; SAKALIAN et al. 1994).

There is no ideal heterologous system of gene expression for studying assembly of retroviruses at the moment, but it is possible to imagine the paradigm for this in the near future, and what features currently need to be improved in order to meet it. (a) A more efficient and more physiological system that would provide the same post-translational modifications as those occurring in the natural environment of the virus, so that their proper folding and native three-dimensional conformation are assured. (b) Instead of a permanent over-expression of the cloned gene, as is the case for most of the systems used at present, a temporally regulated expression incorporating controllable gene promoters should be invented. IPTG-inducible vaccinia virus vectors which allow regulated gene expression have already been designed, but this type of vector could be improved and generalised for other systems. (c) After the current era of single (or double) retrovirus structural protein expression vectors, one can envisage the complete reconstitution of viral or pre-viral particles from multiple structural proteins co-expressed in *cis* or in *trans* within the same cell using several recombinant viruses harbouring one, or several, retroviral genes (French et al. 1990; Thomsen et al. 1994). This will make possible a dissection, step by step, of the chronology of events and of the interactions between the different processed and/or unprocessed virus components, as well as between virus and host cell proteins, that lead eventually to a stable, mature, virus particle.

References

Ada GL (1988) Prospectives for HIV vaccines. J Acquir Immune Defic Syndr 1: 295–303

Argos P (1989) A possible homology between immunodeficiency virus p24 core protein and picornaviral VP2 coat protein: prediction of HIV p24 antigenic sites. EMBO J 8: 779–785

Bennett RP, Nelle TD, Wills JW (1993) Functional chimeras of the Rous sarcoma virus and human immunodeficiency virus Gag proteins. J Virol 67: 6487–6498

Berglund P, Sjöberg M, Garoff H, Atkins G, Sheahan B, Liljeström P (1993) Semliki Forest virus expression system: production of conditionally infectious recombinant particles. Biotechnology 11: 916–920

Berkowitz RD, Luban J, Goff SP (1993) Specific binding of human immunodeficiency virus type 1 Gag polyprotein and nucleocapsid protein to viral RNAs detected by RNA mobility shift assays. J Virol 67: 7190–7200

Biemans R, Thines D, Gheysen D, Rutgers T, Cabezon T (1992) Subcellular localization of recombinant truncated Gag precursor proteins of HIV expressed in *Saccharomyces cerevisiae*. AIDS 6: 541–546

Blomberg J, Medstrand P (1990) A sequence in the carboxylic terminus of the HIV-1 matrix protein is highly similar to sequences in membrane-associated proteins of other RNA viruses: possible functional implications. New Biol 2: 1044–1046

Bonnet D, Spahr PF (1990) Rous sarcoma virus expression in *Saccharomyces cerevisiae*: processing and membrane targeting of the *gag* gene product. J Virol 64: 5628–5632

Bukrinsky M, Haggerty S, Dempsey MP, Sharova N, Adzhubel A, Spitz L, Lewis P, Goldfarb D, Emerman M, Stevenson M (1993) A nuclear localization signal within HIV-1 matrix protein that governs infection of non-dividing cells. Nature 365: 666–669

Burnette R, Kahn R, Glover CJ, Felsted RL (1992) Bacterial expression, purification and in vitro N-myristoylation of HIV-1 p17 *gag*. Protein Exp Purif 3: 395–402

Campbell S, Vogt VM (1995) Self-assembly in vitro of purified CA-NC proteins from Rous sarcoma virus and human immunodeficiency virus type 1. J Virol 69: 6487–6497

Carrière C, Gay B, Chazal N, Morin N, Boulanger P (1995) Sequence requirements for encapsidation of deletion mutants and chimeras of human immunodeficiency virus type 1 Gag precursor into retrovirus-like particles. J Virol 69: 2366–2377

Chazal N, Carrière C, Gay B, Boulanger P (1994) Phenotypic characterization of insertion mutants of the human immunodeficiency virus type 1 Gag precursor expressed in recombinant baculovirus-infected cells. J Virol 68: 111–122
Chazal N, Gay B, Carrière C, Tournier J, Boulanger P (1995) Human immunodeficiency virus type 1 MA deletion mutants expressed in baculovirus-infected cells: *cis* and *trans* effects on the Gag precursor assembly pathway. J Virol 69: 365–375
Darlix J, Gabus C, Ngeyre M, Clavel F, Barré-Sinoussi F (1990) *Cis* element and *trans*-acting factors involved in the RNA dimerization of the human immunodeficiency virus HIV-1. J Mol Biol 216: 689–699
Davison AJ, Moss B (1989a) Structure of vaccinia virus early promoters. J Mol Biol 210: 749–769
Davison AJ, Moss B (1989b) Structure of vaccinia virus late promoters. J Mol Biol 210: 771–784
Davison AJ, Moss B (1990) New vaccinia virus recombination plasmids incorporating a synthetic late promoter for high level expression of foreign proteins. Nucl Acids Res 18: 4285–4286
Delchambre M, Gheysen D, Thines D, Thiriart C, Jacobs E, Verdin E, Horth M, Burny A, Bex F (1989) The GAG precursor of simian immunodeficiency virus assembles into virus-like particles. EMBO J 8: 2653–2660
Doerfler W (1988) Expression of the *Autographa californica* nuclear polyhedrosis virus genome in insect cells: homologous viral and heterologous vertebrate genes – the baculovirus vector system. In: Muzyczka N (ed) Viral expression vectors. Springer, Berlin Heidelberg New York, pp 131–172 (Current topics in microbiology and immunology, vol 158)
Duncan SA, Smith GL (1992) Identification and characterization of an extracellular envelope glycoprotein affecting vaccinia virus egress. J Virol 66: 1610–1621
Ehrlich LS, Krausslich HG, Wimmer E, Carter CA (1990) Expression in *Escherichia coli* and purification of human immunodeficiency virus type 1 capsid protein (p24). AIDS Res Hum Retroviruses 6: 1169–1175
Ehrlich LS, Agresta BE, Carter CA (1992) Assembly of recombinant human immunodeficiency virus type 1 capsid protein in vitro. J Virol 66: 4874–4883
Ehrlich LS, Agresta BE, Gelfand CA, Jentoft J, Carter CA (1994) Spectral analysis and tryptic susceptibility as probes of HIV-1 capsid protein structure. Virology 204: 515–525
Fenouillet E, Gluckman JC, Jones IM (1994) Functions of HIV envelope glycans. Trends Biochem Sci 19: 65–70
Flexner C, Broyles SS, Earl P, Chakrabarti S, Moss B (1988) Characterization of human immunodeficiency virus *gag/pol* gene products expressed by recombinant vaccinia viruses. Virology 166: 339–349
Franke EK, Yuan HEH, Bossolt KL, Goff SP, Luban J (1994a) Specificity and sequence requirements for interactions between various retroviral Gag proteins. J Virol 68: 5300–5305
Franke EK, Yuan HEH, Luban J (1994b) Specific incorporation of cyclophilin A into HIV-1 virions. Nature 372: 359–362
Freed EO, Orenstein JM, Buckler-White AJ, Martin MA (1994) Single amino acid changes in the human immunodeficiency virus type 1 matrix protein block virus particle production. J Virol 68: 5311–5320
French TJ, Roy P (1990) Synthesis of bluetongue virus (BTV) corelike particles by a recombinant baculovirus expressing the two major structural core proteins of BTV. J Virol 64: 1530–1536
French TJ, Marshall JJ, Roy P (1990) Assembly of double-shelled viruslike particles of bluetongue virus by the simultaneous expression of four structural proteins. J Virol 64: 5695–5700
Gheysen D, Jacobs E, de Foresta F, Thiriart C, Francotte M, Thines D, De Wilde M (1989) Assembly and release of HIV-1 precursor $Pr55^{gag}$ virus-like particles from recombinant baculovirus-infected insect cells. Cell 59: 103–112
Gonzalez SA, Affranchino JL, Gelderblom HR, Burny A (1993) Assembly of the matrix protein of simian immunodeficiency virus into virus-like particles. Virology 194: 548–556
Göttlinger HG, Dorfman T, Sodroski JG, Haseltine WA (1991) Effect of mutations affecting the p6 *gag* protein on human immunodeficiency virus particle release. Proc Natl Acad Sci USA 88: 3195–3199
Göttlinger HG, Dorfman T, Cohen EA, Haseltine WA (1993) Vpu protein of human immunodeficiency virus type 1 enhances the release of capsids produced by *gag* gene constructs of widely divergent retroviruses. Proc Natl Acad Sci USA 90: 7381–7385
Gowda SD, Stein BS, Steimer KS, Engleman EG (1989) Expression and processing of human immunodeficiency virus type 1 *gag* and *pol* genes by cells infected with a recombinant vaccinia virus. J Virol 63: 1451–1454

Haffar O, Garrigues J, Travis B, Moran P, Zarling J, Hu SL (1990) Human immunodeficiency virus-like nonreplicating particles assemble in a recombinant vaccinia virus expression system. J Virol 64: 2653–2659

Hockley DJ, Nermut MV, Grief C, Jowett JBM, Jones IM (1994) Comparative morphology of Gag protein structures produced by mutants of the *gag* gene of immunodeficiency virus type 1. J Gen Virol 75: 2985–2997

Hong SS, Boulanger P (1993) Assembly-defective point mutants of the immunodeficiency virus type I Gag precursor phenotypically expressed in recombinant baculovirus-infected cells. J Virol 67: 2787–2798

Hoshikawa N, Kojima A, Yasuda A, Takayashiki E, Masuko S, Chiba J, Sata T, Kurata T (1991) Role of the *gag* and *pol* genes of human immunodeficiency virus in the morphogenesis and maturation of retrovirus-like particles expressed by recombinant vaccinia virus: an ultrastructural study. J Gen Virol 72: 2509–2517

Hu S-L, Travis BM, Garrigues J, Zarling JM, Sridhar P, Dykers T, Eichberg JW, Alpers C (1990) Processing, assembly and immunogenicity of human immunodeficiency virus core antigens expressed by recombinant vaccinia virus. Virology 179: 321–329

Hughes BP, Booth TF, Belyaev AS, McIlroy D, Jowett J, Roy P (1993) Morphogenic capabilities of human immunodeficiency virus type 1 *gag* and *gag-pol* proteins in insect cells. Virology 193: 242–255

Jacobs E, Gheysen D, Thines D, Francotte M, De Wilde M (1989) The HIV-1 Gag precursor $Pr55^{gag}$ synthesized in yeast is myristoylated and targeted to the plasma membrane. Gene 79: 71–81

Jarvis DL, Summers MD (1989) Glycosylation and secretion of human tissue plasminogen activator in recombinant baculovirus-infected insect cells. Mol Cell Biol 9: 214–223

Jowett JBM, Hockley DJ, Nermut MV, Jones IM (1992) Distinct signals in HIV-1 Pr55 necessary for RNA binding and particle formation. J Gen Virol 73: 3079–3086

Kaplan AH, Manchester M, Swanstrom R (1994) The activity of the protease of human immunodeficiency virus type 1 is initiated at the membrane of infected cells before the release of viral proteins and is required for release to occur with maximum efficiency. J Virol 68: 6782–6788

Karacostas V, Nagashima K, Gonda MA, Moss B (1989) Human immunodeficiency virus-like particles produced by a vaccinia virus expression vector. Proc Natl Acad Sci USA 86: 8964–8967

Karacostas V, Wolffe EJ, Nagashima K, Gonda MA, Moss B (1993) Overexpression of the HIV-1 *gag-pol* polyprotein results in intracellular activation of HIV-1 protease and inhibition of assembly and budding of virus-like particles. Virology 193: 661–671

Klikova M, Rhee SS, Hunter E, Ruml T (1995) Efficient in vivo and in vitro assembly of retroviral capsids from Gag precursor proteins expressed in bacteria. J Virol 69: 1093–1098

Kramer RA, Schaber MD, Skalka AM, Ganguly K, Wong-Staal F, Reddy EP (1986) HTLV-III *gag* protein is processed in yeast cells by the virus *pol*-protease. Science 231: 1580–1584

Kuroda K, Gröner A, Frese K, Drenckhahn D, Hauser C, Rott R, Doerfler W, Klenk H-D (1986) Expression of the influenza virus hemagglutinin in insect cells by a baculovirus vector. EMBO J 5: 1359–1365

Kuroda K, Geyer H, Geyer R, Doerfler W, Klenk H-D (1990) The oligosaccharides of influenza virus hemagglutinin expressed in insect cells by a baculovirus vector. Virology 174: 418–429

Lodge R, Goettlinger H, Gabuzda D, Cohen EA, Lemay G (1994) The intracytoplasmic domain of gp41 mediates polarized budding of human immunodeficiency virus type 1 in MDCK cells. J Virol 68: 4857–4861

Luban J, Goff SP (1991) Binding of human immunodeficiency virus type 1 (HIV-1) RNA to recombinant HIV-1 *gag* polyprotein. J Virol 65: 3203–3212

Luban J, Alin KB, Bossolt KL, Humaran T, Goff SP (1992) Genetic assay for multimerisation of retroviral *gag* polyproteins. J Virol 66: 5157–5160

Luban J, Bossolt KL, Franke EK, Kalpana GV, Goff SP (1993a) Human immunodeficiency virus type 1 *gag* protein binds to cyclophilins A and B. Cell 73: 1067–1078

Luban J, Lee C, Goff SP (1993b) Effect of linker insertion mutations in the human immunodeficiency virus type 1 *gag* gene on activation of viral protease expressed in bacteria. J Virol 67: 3630–3634

Luckow VA, Summers MD (1989) High level expression of nonfused foreign genes with *Autographa californica* nuclear polyhedrosis virus expression vectors. Virology 170: 31–39

Luo L, Li Y, Kang CY (1990) Expression of *gag* precursor protein and secretion of virus-like Gag particles of HIV-2 from recombinant baculovirus infected insect cells. Virology 179: 874–880

Luo L, Li Y, Dales S, Kang CY (1994) Mapping of functional domains for HIV-2 Gag assembly into virus-like particles. Virology 205: 496–502

Madisen L, Travis B, Hu S-L, Purchio AF (1987) Expression of the human immunodeficiency virus *gag* gene in insect cells. Virology 158: 248–250

Mammano F, Hagen A, Höglund S, Göttlinger HG (1994) Role of the major homology region of human immunodeficiency virus type 1 in virion morphogenesis. J Virol 68: 4927–4936

Massiah MA, Starich MR, Paschall CM, Summers MF, Christensen AM, Sundquist WI (1994) Three-dimensional structure of the human immunodeficiency virus type 1 matrix protein. J Mol Biol 244: 194–223

Matsuura Y, Possee RD, Overton HA, Bishop DHL (1987) Baculovirus expression vectors: the requirements for high level expression of proteins including glycoproteins. J Gen Virol 67: 1515–1529

Matthews S, Barlow P, Boyd J, Barton G, Russell R, Mills H, Cunningham M, Meyers N, Burns N, Clark N, Kingsman S, Kingsman A, Campbell I (1994) Structural similarity between the p17 matrix protein of HIV-1 and interferon-γ. Nature 370: 666–668

Mills HR, Berry N, Burns NR, Jones IM (1992) Simple and efficient production of the core antigens of HIV-1 HIV-2 and simian immunodeficiency virus using pGEX expression vectors in *Escherichia coli*. AIDS 6: 437–439

Miyazawa M, Nishio J, Chesebro B (1990) Partial protection of susceptible mice against Friend retrovirus-induced leukemia with vaccinia-Friend *gag* recombinant vaccines. In: Brown F, Chanock RM, Ginsberg HS, Lerner RA (eds) Vaccine 90. Cold Spring Harbor Laboratory, Cold Spring Harbor, pp 407–412

Morikawa S, Booth TF, Bishop DL (1991) Analysis of the requirements for the synthesis of virus-like particles by feline immunodeficiency virus *gag* using baculovirus vector. Virology 183: 288–297

Morikawa Y, Kishi T, Zhang WH, Nermut MV, Hockley DJ, Jones IM (1995) A molecular determinant of human immunodeficiency virus particle assembly located in matrix antigen p17. J Virol 69: 4519–4523

Nam SH, Kidokoro M, Shida H, Hatanaka M (1988) Processing of *gag* precursor polyprotein of human T-cell leukemia virus type 1 by virus-encoded protease. J Virol 62: 3718–3728

Nermut MV, Hockley DJ, Jowett JBM, Jones IA, Garreau M, Thomas D (1994) Fullerene-like organization of HIV *gag*-protein shell in virus-like particles produced by recombinant baculovirus. Virology 198: 288–296

Nyunoya H, Akagi T, Ogura T, Maeda S, Shimojo K (1988) Evidence for phosphorylation of trans-activator p40X of human T-cell leukemia virus type 1 produced in insect cells with a baculovirus expression vector. Virology 167: 538–544

O'Reilly DR, Miller LK (1988) Expression and complex formation of simian virus 40 large T antigen and mouse p53 in insect cells. J Virol 62: 3109–3119

O'Reilly DR, Miller LK, Luckow VA (1992) Baculovirus expression vectors: a laboratory manual. Freeman , New York, pp 1–347

Osterhaus A, Weijer K, Uytdehaag F, Jarrett O, Sundquist B, Morein BJN (1985) Induction of protective immune response in cats with feline leukemia virus. J Immunol 135: 591–596

Overton HA, Fuji Y, Price IR, Jones IM (1989) The protease and *gag* gene products of the human immunodeficiency virus: authentic cleavage and post-translational modification in an insect cell expression system. Virology 170: 107–116

Owens RJ, Dubay JW, Hunter E, Compans RW (1991) Human immunodeficiency virus envelope protein determines the site of virus release in polarized cells. Proc Natl Acad Sci USA 88: 3987–3991

Partin K, Krausslich HG, Ehrlich LS, Wimmer E, Carter CA (1990) Mutational analysis of a native substrate of the human immunodeficiency proteinase. J Virol 64: 3938–3947

Pearson A, Richardson C, Yuen L (1991) The 5' noncoding region sequence of the *Choristoneura biennis* entomopoxvirus spheroidin gene functions as an efficient promoter in mammalian vaccinia expression system. Virology 180: 561–566

Portetelle D, Limbach K, Burny A, Mammerickx M, Desmettre P, Riviere M, Zavada J, Paoletti E (1991) Recombinant vaccinia virus expression of the bovine leukaemia virus envelope gene and protection of immunized sheep against infection. Vaccine 9: 194–200

Possee RD (1986) Cell-surface expression of influenza virus haemagglutinin in insect cells using a baculovirus vector. Virus Res 5: 43–59

Prongay AJ, Smith TJ, Rossmann MG, Ehrlich LS, Carter CA, McLure J (1990) Preparation and crystallisation of a human immunodeficiency virus p24-Fab complex. Proc Natl Acad Sci USA 87: 9980–9984
Rao Z, Belyaev AS, Fry E, Roy P, Jones IM, Stuart DI (1995) Crystal structure of SIV matrix antigen and implications for virus assembly. Nature 378: 743–747
Rasmussen L, Battles JK, Ennis WH, Nagashima K, Gonda MA (1990) Characterization of virus-like particles produced by a recombinant baculovirus containing the *gag* gene of the bovine immunodeficiency-like virus. Virology 178: 435–451
Rhee SS, Hunter E (1990) A single amino acid substitution within the matrix protein of a type D retrovirus converts its morphogenesis to that of a type C retrovirus. Cell 63: 77–86
Rhee SS, Hunter E (1991) Amino acid substitutions within the matrix protein of type D retroviruses affects assembly, transport and membrane association of a capsid. EMBO J 10: 535–546
Rodriguez JF, Smith GL (1990) Inducible gene expression from vaccinia virus vectors. Virology 177: 239–250
Royer M, Cerutti M, Gay B, Hong SS, Devauchelle G, Boulanger P (1991) Functional domains of HIV-1 *gag*-polyprotein expressed in baculovirus-infected cells. Virology 184: 417–422
Royer M, Hong SS, Gay B, Cerutti M, Boulanger P (1992) Expression and extracellular release of human immunodeficiency virus type 1 Gag precursors by recombinant baculovirus-infected cells. J Virol 66: 3230–3235
Sakalian M, Wills JW, Vogt VM (1994) Efficiency and selectivity of RNA packaging by Rous sarcoma virus Gag deletion mutants. J Virol 68: 5969–5981
Shida H, Tochikura T, Sato T, Konno T, Hirayoshi K, Seki M et al (1987) Effects of the recombinant vaccinia virus that express HTLV-1 envelope gene on HTLV-1 infection. EMBO J 6: 3379–3384
Shioda T, Shibuta H (1990) Production of human immunodeficiency virus (HIV)-like particles from cells infected with recombinant vaccinia virus carrying the *gag* gene of HIV. Virology 175: 139–148
Smith GL, Mackett M (1992). In: Binns MM, Smith GL (eds) Recombinant poxviruses. CRC, Boca Raton, pp 81–122
Smith AJ, Hammarskjöld M-L, Rekosh D (1990) Human immunodeficiency virus type 1 $Pr55^{gag}$ and $Pr160^{gag\text{-}pol}$ expressed from a simian virus 40 late replacement vector are efficiently processed and assembled into viruslike particles. J Virol 64: 2743–2750
Sommerfelt MA, Roberts CR, Hunter E (1993) Expression of simian type D retroviral (Mason-Pfizer monkey virus) capsids in insect cells using recombinant baculovirus. Virology 192: 298–306
Stewart L, Vogt VM (1994) Proteolytic cleavage at the Gag-Pol junction in avian leukosis virus: differences in vitro and in vivo. Virology 204: 45–59
Stewart LMD, Possee RD (1993) Baculovirus expression vectors. In: Davison AJ, Elliott RM (eds) Molecular virology: a practical approach. IRL Press, Oxford, pp 229–256
Stewart LMD, Hirst M, Lopez-Ferber M, Merrywheather AT, Cayley PJ, Possee RD (1991) Construction of an improved baculovirus insecticide containing an insect-specific toxin gene. Nature 352: 85–88
Thali M, Bukovsky A, Kondo E, Rosenwirth B, Walsh C, Sodroski J, Göttlinger HG (1994) Functional association of cyclophilin A with HIV-1 virions. Nature 372: 363–365
Thomsen DR, Roof LL, Homa FL (1994) Assembly of herpes simplex virus (HSV) intermediate capsids in insect cells infected with recombinant baculoviruses expressing HSV capsid proteins. J Virol 68 2442–2457
Trono D, Feinberg MB, Baltimore D (1989) HIV-1 Gag mutants can dominantly interfere with the replication of the wild-type virus. Cell 59: 113–120
Urakawa T, Roy P (1988) Bluetongue virus tubules made in insect cells by recombinant baculoviruses: expression of the NS1 gene of bluetongue virus serotype 10. J Virol 62: 3919–3927
Vernon SK, Murthy S, Wilhelm J, Chanda PK, Kalyan N, Lee SG, Hung PP (1991) Ultrastructural characterization of human immunodeficiency virus type 1 Gag-containing particles assembled in a recombinant adenovirus vector system. J Gen Virol 72: 1243–1251
Von Poblotzki A, Wagner R, Niedrig M, Wanner G, Wolf H, Modrow S (1993) Identification of a region in the $Pr55^{gag}$-polyprotein essential for HIV-1 particle formation. Virology 193: 981–985
Wagner R, Fliessbach H, Wanner G, Motz M, Niedrig M, Deby G, von Brunn A, Wolf H (1992) Studies on processing particle formation and immunogenicity of the HIV-1 *gag* gene product: a possible component of a HIV vaccine. Arch Virol 127: 117–137
Wagner R, Deml L, Fliessbach H, Wanner G, Wolf H (1994) Assembly and extracellular release of chimeric HIV-1 $Pr55^{gag}$ retrovirus-like particles. Virology 200: 162–175

Weldon RA Jr, Wills JW (1993) Characterization of a small (25-kilodalton) derivative of the Rous sarcoma virus Gag protein competent for particle release. J Virol 67: 5550–5561
Wills JW, Craven RC (1991) Form, function and use of retroviral *gag* proteins. AIDS 5: 639–654
Wills JW, Cameron CE, Wilson CB, Xiang Y, Bennett RP, Leis J (1994) An assembly domain of the Rous sarcoma virus Gag protein required late in budding. J Virol 68: 6605–6618
Yu XF, Yuan X, Matsuda Z, Lee TH, Essex M (1992) The matrix protein of human immunodeficiency virus type 1 is required for incorporation of viral envelope protein into mature virions. J Virol 66: 4966–4971
Zhang Y, Moss B (1991) Inducer-dependent conditional-lethal mutant animal viruses. Proc Natl Acad Sci USA 88: 1511–1515
Zhao Y, Jones IM, Hockley DJ, Nermut MV, Roy P (1994) Complementation of human immunodeficiency virus (HIV-1) Gag particle formation. Virology 199: 403–408

Morphogenesis at the Retrotransposon-Retrovirus Interface: Gypsy and Copia Families in Yeast and *Drosophila*

S.B. SANDMEYER and T.M. MENEES

1 Introduction

Long terminal repeat (LTR) retrotransposons have been identified in a wide variety of organisms including fungi (BOEKE and SANDMEYER 1991), plants (VOYTAS and AUSUBEL 1988; FLAVELL et al. 1995; SMYTH et al. 1989; GRANDBASTIEN et al.

Department of Microbiology and Molecular Genetics, College of Medicine, University of California, Irvine, CA 92717, USA

1989), invertebrates (SPRINGER et al. 1991; BRITTEN 1995; BINGHAM and ZACHAR 1989), and vertebrates (FLAVELL and SMITH 1992). Many of these LTR retrotransposons can be sorted into two superfamilies named after the prototypic *Drosophila* elements copia and gypsy (mdg4). The early history of the characterization of these elements has been reviewed elsewhere (BOEKE and CORCES 1995; BINGHAM and ZACHAR 1989). Expression of representative elements results in the formation of intracellular particles analogous to the retroviral core particle. This review will focus on morphogenesis of that particle — assembly, processing, and reverse transcription — in members of these families for which the DNA sequence has been determined in the fruit fly *Drosophila* and in the yeasts *Saccharomyces cerevisiae* and *Schizosaccharomyces Pombe*. Collectively, these systems reveal tremendous diversity as well as striking similarities among the mechanisms members of these two LTR retrotransposon families and retroviruses have used to solve common morphogenetic problems.

Copialike and gypsylike elements are structurally similar to retroviruses. LTRs of several hundred bp flank an internal domain which contains coding information for proteins required for replication. Retrotransposons have been distinguished from retroviruses by the apparently exclusively intracellular cycles of retrotransposon replication and integration. However, this distinction has been eroded in recent years by the discoveries that retroviruses can undergo limited cycles of intracellular replication and integration (TCHENIO and HEIDMANN 1991) and that some members of the gypsylike family, and gypsy itself, encode envelope proteins (SONG et al. 1994; TANDA et al. 1994; PELISSON et al. 1994) and are transmitted extracellularly (KIM et al. 1994; PELISSON et al. 1994). Whether, for gypsy, replication is a function of extracellular transmission, as for vertebrate retroviruses, or also occurs effciently intracellularly, as for related retrotransposons, is not yet known.

Copialike and gypsylike families are grossly distinguished by protein sequence differences in the conserved polymerase domain of reverse transcriptase (RT) (XIONG and EICKBUSH 1990; BRITTEN 1995) and in the order of the integrase (IN) and RT domains. Phylogenetic relationships of LTR retrotransposons have been described elsewhere (XIONG and EICKBUSH 1990) as have relationships of elements within the copialike (FLAVELL 1995) and gypsylike (SPRINGER and BRITTEN 1993) families. In the copialike family, the IN-coding domain precedes the RT-coding domain, but in the gypsylike family, the order is reversed. In sequence and organization, gypsylike elements are more similar to retroviruses than are the copialike elements. Given the pervasive presence of both types of elements in the plant and animal kingdoms, however, one question posed by these comparisons is why there are no copialike retroviruses. Other properties which, given their constancy in retroviruses, a priori might be expected to be consistent within either the copialike or gypsylike family, such as the mechanism by which stoichiometry of major structural proteins to catalytic proteins is maintained, the nature of the minus-and plus-strand primers, and the presence of a nucleocapsid metal finger motif are conserved between closely related elements, but not within gypsylike or copialike classes per se. Salient sequence-derived features of the genomes of representative members of the copialike and gypsylike families from *Drosophila* and yeast for which the sequence is known are shown in Table 1.

2 Properties of Particles

Where it has been examined, reverse transcription of retrotransposons occurs associated with a particulate nucleoprotein complex (FUETTERER and HOHN 1987). The term viruslike particle (VLP) will be used to refer to the intracellular nucleoprotein particle comprising retrotransposon RNA and/or DNA and retrotransposon-encoded proteins. The term major structural protein will be used to refer to noncatalytic VLP proteins found in stoichiometric excess relative to the catalytic proteins RT and IN. The sequences of the major structural proteins of the LTR retrotransposons are not conserved, other than between very closely-related members of the copialike and gypsylike families.

Copia VLPs were characterized starting with observations in the early 1980s (FALKENTHAL and LENGYEL 1980; SHIBA and SAIGO 1983; FLAVELL 1984; EMORI et al. 1985; MIYAKE et al. 1987). Since that time, the VLPs for Ty1, a copialike element in *Saccharomyces cerevisiae* (GARFINKEL et al 1985), and Ty3 (HANSEN et al. 1992) and Tf1 (LEVIN et al. 1990) gypsylike elements in *S. cerevisiae* and *Schizosaccharomyces pombe*, respectively, and virus particles for gypsy (PELISSON et al. 1994) have been investigated. The basic picture which emerges is that VLPs are the functional and structural equivalent of the viral core particle. Electron microscopic analysis of cells producing VLPs or of concentrated VLPs has shown heterogeneous, ovoid to spherical particles of 50 to 60 nm in size. Although observers have generally considered the core to be somewhat electron dense, populations visualized thus far have not been as homogeneous as retroviral cores, probably because they reflect multiple stages of maturation. VLPs are associated with genomic RNA as well as low molecular weight heterologous RNA species. The particles typically have a major capsid protein of 26–33 kilodaltons (kDa) and lower amounts of other proteins including: nucleocapsid, in some cases; an aspartyl protease (PR); RT/RNase H; and IN. For the purposes of this review, the abbreviations used for these proteins in retroviruses (CA, NC, PR, RT, and IN, respectively) (LEIS et al. 1988) have been adopted. Studies of virtually all of these elements have relied upon synthetic constructs or genetic systems which are conducive to high levels of expression in order to achieve preparative or readily observable amounts of VLPs. No specific subcellular site of particle nucleation has been identified for any of these retroviruslike elements. However, whether there is a transient membrane-associated phase, such as occurs in particle formation of some retroviral types, would not necessarily be known from the observations which have been made, because, as suggested above, the majority of VLPs observed in steady-state preparations are likely to be

Table 1. Properties of selected retrotransposons

Organism element (class)[a]	Element/LTR length (bp)	Inverted repeat[b]	Duplication of target site (bp)	– Strand primer binding site/primer[c]	+ Strand primer[d]	ORFs/Gag metal finger motif?[e]	*gag-pol* ORFs overlap (nt)[f]	Frameshift/site[g]	References[h]
Saccharomyces cerevisiae									
Ty1 (copia)	5918/334–338	TG...CA	5	]TGGTAGCGCC iMet-tRNA	GGGUGGUA[	*TYA, TYB*/no (Gag) (Pol)	38	+1 CUU AGG C	1
Ty2 (copia)	5961/332	TG...CA	5	]TGGTAGCGCC iMet-tRNA	GGGUGGUA[	*TYA, TYB*/no (Gag) (Pol)	44	+1 CUU AGG C	2
Ty3 (gypsy)	5351/340	TGTTGTAT...ATACAACA	5	]ccTGGTAGCG iMet-tRNA	GAGAGAGAGGAAGA[	*GAG3, POL3*/yes (Gag) (Pol)	38	+1 GCG AGU U	3
Ty4 (copia)	6226/371	TGTTG...CAACA	5	]TGGCGACCCCAGTGAGGG Asn-tRNA	AAGGGAGCA[	*TYA4, TYB4*/yes (Gag) (Pol)	227	+1 CUU AGG C	4
Ty5 (copia)	5375/251	TGTTGA...TCAACA	5	]GGTTATGAGCCCT iMet-tRNA fragment	GGGGGGA[	1 ORF/ yes (Gag, Pol)	NA	NA	5
S.Pombe									
Tf1 (gypsy)	4941/358	TGTtAGC...GCTaACA	5	]ATAACTGAACT 5′end of Tf1 mRNA	GGGGAGGGCAA[	1 ORF/no (Gag, Pol)	NA	NA	6
Drosophila									
gypsy	7469/482	AgTTA...TAAtT	4	T]GGCGCCCAAC Lys-tRNA	GAGGGGGGAGU[	ORF1, ORF2, ORF3/no (Gag) (Pol) (Env)	58	–1 AAU UUU UUA GGG	7
17.6 (gypsy)	7439/512	AgTgaCA...TGcaAtT	4	T]GGCGCAGTCGATGTGAT Ser-tRNA	AAGGGAAGGGA[	ORF1, ORF2, ORF3/no (Gag) (Pol) (Env)	46	–1 GAA AAU UUU CAG	8
297 (gypsy)	6995/414	AGTgA...TtACT	4	T]GGCGCAGTCGGTAGGAT Ser-tRNA	AAGGGAAGGGG[	ORF1, ORF2, ORF3/no (Gag) (Pol) (Env)	46	–1 GAA AAU UUU CGG	9
tom (gypsy)	7060/474	AGTgA...TtACT	4 or 5	T]GGCGCAGTCGGTAGGAT Ser-tRNA	AAGGGGGGAGG[	ORF1, ORF2, ORF3/no (Gag) (Pol) (Env)	40	–1 GAA AAU UUU CAG	10
412 (gypsy)	7440/481	TGTAgT...AtTACA	4	]TGGCGACCGTGACAGTCG Arg-tRNA	AAAAGGAGGGAGA[	ORF3, ORF4/no (Gag) (Pol)	88	–1	11

ulysses (gypsy)	10653/2136	TGTT...AACA	4	]ataTGGCGCCCAA Lys-tRNA	GGUGA[	ORF1, ORF2/no (Gag) (Pol)	32	+1	12
micropia (gypsy)	5487/505	TGTCG...CGACA	4	]GAAGTGGGAT Leu-tRNA	GAAATGTCAGAATGGCCG[	1 ORF/yes (Gag, Pol)	NA	NA	13
copia	5146/276	TGTTG...CAACA	5	]GGTTATGGGCCCAG iMet-tRNA fragment	GAGGGGGCG[	1 ORF/ yes (Gag, Pol)	NA	NA	14

[a] Class indicates order of *pol* gene products: copia, IN-RT/RH; gypsy, RT/RH-IN, where IN is integrase and RT/RH is reverse transcriptase/RNAse H.

[b] Refers to sequence at ends of LTRs. Uppercase denotes part of repeat structure; lower case denotes not part of repeat structure.

[c] Uppercase denotes complementarity with primer, lowercase denotes no complementarity with primer.] indicates border with 5′ LTR sequence. iMet-tRNA, initiator methionine tRNA.

[d] Presumed plus strand primer = polypurine tract adjacent to 3′ LTR in element's genomic RNA. [indicates border with 3′ LTR sequence.

[e] ORF, open reading frame. Identities of gene products encoded by ORF's are indicated in parentheses below each ORF. By analogy to retroviruses: Gag, structural proteins; Pol, catalytic proteins; Env, envelope proteins. Gag metal finger motif is a cys/his sequence of the type found in the retroviral nucleocapsid proteins, which are required for binding/packaging of retroviral genomic RNA.

[f] NA, not applicable because the element has a single ORF.

[g] +1 and −1 indicate the number of nucleotides that ORF2 is shifted relative to ORF1. Site is the sequence of the mRNA where the frameshift occurs. Codons indicated represent o frame. NA, not applicable because the element has a single ORF.

[h] 1, Clare and Farabaugh (1985), Belcourt and Farabaugh (1990), Boeke and Sandmeyer (1991)(and references therein); 2, Warmington et al. (1985), Boeke and Sandmeyer (1991)(and references therein); 3, Hansen and Sandmeyer (1990), Boeke and Sandmeyer (1991)(and references therein); 4, Stucka et al. (1989), Janetzky and Lehle (1992), Boeke and Sandmeyer (1991) (and references therein); 5, Voytas and Boeke (1992, 1993), Zou et al. (1996); 6, Levin et al. (1990, 1993), H. Levin (personal communication); 7, Marlor et al. (1986), Bingham and Zachar (1989) (and references therein); 8, Saigo et al. (1984), Bingham and Zachar (1989) (and references therein); 9, Inouye et al. (1986), Bingham and Zachar (1989) (and references therein); 10, Tanda et al. (1988, 1994); 11, Yuki et al. (1986), Bingham and Zachar (1989) (and references therein); 12, Scheinker et al. (1990), Evgen'ev et al. (1992); 13, Lankenau et al. (1988); 14. Emori et al. (1985), Mount and Rubin (1985), Bingham and Zachar (1989) (and references therein).

representative of the more prolonged stages in particle formation. Details of the foregoing observations have been presented in previous reviews (Boeke and Corces 1989; Finnegan 1989; Boeke and Sandmeyer 1991). This discussion will primarily focus on features which distinguish particular systems or relate to recent discoveries.

2.1 Copia Particles

The copia element contains a single long open reading frame (ORF) of 4227 nucleotides (nts) (Mount and Rubin 1985). The major structural proteins are encoded in a 2-kilobase (kb) RNA derived from a splicing event which joins a splice donor at position 1604 to an acceptor close to the end of the ORF (Yoshioka et al 1990; Brierley and Flavell 1990; Miller et al. 1995), creating a smaller ORF capable of encoding a protein of 48 kDa, consistent with the approximate mass of the protein previously observed to be expressed from the 2-kb transcript (Flavell et al. 1980). Analysis of the predicted sequence of this protein revealed a single copy of the metal finger motif found in NC in retroviruses, beginning at amino acid number 232 in the protein and an aspartyl PR-active site motif, DSG, beginning at amino acid position 292 (Mount and Rubin 1985).

Transformation of clones for a complete copia element and cDNA for the 2-kb RNA into *D. hydei,* which does not contain endogenous copia elements, allowed analysis in vivo of the proteins encoded by the complete and spliced transcripts (Yoshioka et al. 1990). Both constructs sponsored accumulation of copia particles. The major protein species detected by antisera generated against VLPs is 33 kDa. In vitro translation of the 2-kb species showed an additional polypeptide of 23 kDa, which was inferred to be the protease based on the observation that it did not react with the antiserum which recognized the amino-terminal 33-kDa species. Based on the presumption that a mutation introduced into the putative PR-active site aspartyl residue would allow identification of the predicted higher molecular weight precursor species, a mutation was introduced, changing it to an alanyl residue. This construct was used to generate RNA, which was translated in vitro, and a precursor species of about 50 kDa was identified.

Copia particle formation is sensitive to the proper ratio of the major structural and putative catalytic proteins. While in vivo expression of the 2-kb RNA species described above resulted in accumulation of nuclear VLPs grossly similar to wild-type (wt) particles, expression of a 5-kb form of the RNA in which the 3′ splice site was eliminated, but the sequence of the encoded protein was unchanged, did not result in a detectable level of particle proteins. Elimination of precursor processing also destabilized particle protein in vivo, as the PR-active site mutant did not generate detectable particles in the context of the 2-kb spliced construct or the full-length clone.

A particularly interesting feature of copia VLPs is that they accumulate in the nucleus. Miyake et al. (1987) estimated only about 2 of 500 VLPs to be cytoplasmic. A nuclear localization signal has not been identified, but nuclear localization of VLPs also occurs with overexpression of the 2-kb RNA in yeast where the nuclear membrane does not break down during the cell cycle (YOSHIOKA et al. 1992). This finding suggests the presence of a conserved signal for nuclear localization. Circular copia DNAs produced from reverse transcribed species represent a significant fraction of copia extrachromosomal DNA (FLAVELL and ISH-HOROWICZ 1981; FLAVELL 1984) and some of these resemble the circular copies of retroviral genomes (FLAVELL and ISH-HOROWICZ 1983). Nuclear-localized extrachromosomal retroviral DNA also accumulates as circular forms (COFFIN 1990), and the high percentage of circular copia DNA could reflect efficient nuclear localization of the particles. Circular forms of extrachromosomal copia DNA have not yet been shown to recombine into the genome at a high frequency, but they could play a role in copia genomic rearrangements by virtue of their representation in the nucleus. In yeast, recombination of reverse transcribed cDNAs of Ty1 has been observed for Ty1 IN mutants (SHARON et al. 1994).

2.2 Copialike Ty Particles

The *S. cerevisiae* elements Ty1 (CLARE and FARABAUGH 1985; HAUBER et al. 1985), Ty2 (WARMINGTON et al. 1985; FULTON et al. 1985), Ty4 (STUCKA et al. 1989; JANETZKY and LEHLE 1992), and Ty5 (OLIVER et al. 1992; VOYTAS and BOEKE 1992) are copialike in sequence and organization. Ty1 and Ty2 elements are about 5.9 kilobase pairs (kbp) in length, including 334–338-bp LTRs. These LTRs are also found as isolated insertions in the genome and are referred to as delta elements. Ty1 and Ty2 are transcribed into polyadenylated 5.7-kb RNAs beginning and ending at overlapping positions in the LTRs and containing two ORFs. The first ORF, *TYA*, is 440 codons in length and the second, *TYB*, is 1328 codons in length in a representative element, Ty1-912. The nt sequences of the internal domains of the Ty1 and Ty2 elements contain two regions of dissimilarity, the first covering much of *TYA* and the second, covering a smaller region in *TYB*. The difference in protein sequence is less, suggesting that function has been conserved.

Ty1 particle formation has been studied by fusion of sequences upstream of the transcription initiation site to strong promoters such as those for the phosphoglycerate kinase (*PGK*) (MELLOR et al. 1985b), the *GALI* (GARFINKEL et al. 1985), and the *ADHI* genes (MÜLLER et al. 1987). Overexpression of two distinct Ty1 elements has shown accumulation of 60-nm particles (MELLOR et al. 1985b; GARFINKEL et al. 1985). More quantitative examination of large numbers of individual particles has shown that they are heterogeneous; in one study, negatively stained particles ranged from 15 to 39 nm, but unstained cryological preparations showed particles about 30% larger in diameter (BURNS et al. 1992). Electron micrographs showed spheroid to ovoid shapes and some central density, although not to the extent observed for retroviruses. Examination of several strains overexpressing Ty1 (GARFINKEL etl al. 1985) showed that particles occurred cytoplasmically, but in some strains hundreds of particles were collected into single clusters, while in other strains particles were isolated.

In order to identify VLP components, extracts of cells expressing high levels of Ty1 RNA and proteins have been fractionated by velocity sedimentation over sucrose or glycerol gradients. Translation of the *TYA* reading frame is predicted to produce a protein of 49 kDa. The mobility of this protein is apparently particularly sensitive to electrophoretic conditions and thus empirical estimates of the mass of this protein based on experiments in different laboratories have not been in good agreement. A protein of apparent mobility of a protein between 56 and 62 kDa (p1) present in cells expressing *TYA* alone and in cells expressing a PR mutant (Mellor et al. 1985b; Adams et al. 1987a; Müller et al. 1987; Youngren et al. 1988) appears to be unprocessed *TYA* protein. The estimated size of the mature capsid protein processed from this precursor ranges from 47 to 55 kDa (p2) (Mellor et al. 1985b; Adams et al. 1987a; Youngren et al. 1988; Müller et al. 1987). A smaller processing product between 5 and 7 kDa may also be present, but thus far has eluded identification. Both the *TYA* protein precursor and the processed form have been reported to have nucleic acid binding activity (Mellor et al. 1985a,b; Müller et al. 1987). Particles produced from PR mutant Ty1 elements or from *TYA* alone are more irregular in shape than wt particles. Expression of a trancated *TYA* protein designed to simulate the major processed form resulted in production of particles which were more uniform in size than wt particles and corresponded in range of sizes to the low end of that observed in wild-type (wt) Ty1 populations (Burns et al. 1992). Incubation of immature particles formed from *TYA*-encoded p1 and artificially "matured" particles formed by expressing the truncated version of *TYA*-encoded p2 with antisera against peptides representing amino- and carboxyl-terminal epitopes, showed that amino-terminal, but not carboxyl-terminal epitopes were accessible to antibodies (Brookman et al.1995). These data are consistent with exposure of the aminoterminus of p1 on the surface of Ty1 particles, although it is possible that there are significant differences between these particles and ones formed from expressing *TYA* and *TYB*.

Wt Ty1 particles are associated with 5.7-kb Ty1 RNA, reverse transcriptase (Mellor et al. 1985b; Garfinkel et al. 1985), and integrase activity (Eichinger and Boeke 1988). The reverse transcriptase activity has been measured by both endogenous and exogenous template primer assays. Ty1 reverse transcription and transposition are temperature sensitive. In vitro reverse transcriptase activity is optimal at 20°C (Garfinkel et al. 1985) and transposition measured in induced cultures is optimal at 18°–23°C.

Availability of the Ty1 sequence permitted directed mutagenesis of Ty1 elements in order to test the possible functions of domains of the predicted *TYB*-encoded protein. Antibodies against the putative RT domain of Ty1 showed that it was present in the particulate fraction described above and that mutations introduced into this domain abolished particle-associated RT activity (Garfinkel et al. 1985). Small insertions and deletions (Adams et al. 1987a; Müller et al. 1987; Youngren et al 1988) have been introduced to the region encoding PR. In cells expressing these mutants, particles were present, but mutants differed morphologically, apparently depending on the particular element and on the mutation introduced. Generally PR mutants did not have the condensed core found in wt particles. VLPs produced from Ty1 PR mutant elements contained precursor *TYA* (56–62 kDa) and *TYA-TYB* protein (190 kDa) (p3). Interestingly, in two PR mutants, RT activity measured by exogenous assay was substantial (MULLER et al. 1987; YOUNGREN et al. 1988).

A Ty1 PR-dependent processing pathway for polyproteins into mature species has been defined using antibodies directed against specific domains encoded by *TYA* and *TYB* to analyze extracts from cells pulse-labeled with radioactive amino acid and chased with an excess of nonradioactive amino acid during expression of Ty1 (CURCIO and GARFINKEL 1992; YOUNGREN et al 1988). *TYA* protein is processed to some extent even as early as 15 min after initiation of labeling, but processing is not complete for about 8h. A 30-min pulse followed by a 1-h chase showed *TYA-TYB* fusion protein (190 kDa), PR-IN-RT (160 kDa), and IN-RT (140 kDa). After 24 h, mature IN (90 kDa) and RT (60 kDa) were apparent. A 23-kDa PR species was present, but was not recovered reproducibly. Mature species were quite stable and persisted in cells longer than 36 h.

Whether the heterogeneity of nonsynchronous populations of Ty1 particles can be completely explained by the presence of particles in different stages of maturation represented is not known. The finding that particles which are unprocessed or composed exclusively of *TYA* protein appear more variable, distorted, or larger than particles composed of an artificially truncated species is consistent with this interpretation. Nevertheless, the persistence of some precursor species in pulse-chase experiments suggests that even in synchronous populations, particles contain a substantial amount of unprocessed species. This is apparently particularly true for particles in cells which are not undergoing artificially elevated levels of Ty1 expression.

The system of overexpression of individual Ty1 elements has surfaced a puzzling aspect of the relationship between Ty1 expression and transposition (CURCIO and GARFINKEL 1991a). In standard laboratory haploid yeast strains, Ty1 transcripts can constitute from 0.1% to 0.8% of total RNA and, correspondingly, a substantial percentage of the poly(A) RNA. Spontaneous transposition of a marked, chromosomal element is estimated to occur at 1-3 $\times$ 10^{-7} per Ty1 element per generation (CURCIO and GARFINKEL 1991b). In the case of a genetically marked, plasmid-borne, *GALl*-promoter driven element under inducing conditions, levels of the 5.7-kb RNA can be increased 220- to 225-fold. However, the transposition rate of the marked element in these cells increases by a significantly greater factor — up to 10 to 28 $\times$ 10^{3}-fold the spontaneous level. Thus, high-level expression of a single Ty1 has a 45- to 125-fold greater effect on the transposition rate than on RNA levels. This could be explained to some extent if a significant fraction of endogenous elements are defective. However, this model was not supported by a study of a limited set of cloned chromosomal elements and a set of 39 gap-repaired Ty1 elements performed by Curcio and Garfinkel (CURCIO et al. 1988; CURCIO and GARFINKEL 1992, 1994). In the study of gap-repaired elements, 74% were transpositionally competent. No evidence was obtained which suggested that the inactive elements had significant repressive effects on the activity of other elements. CURCIO and GARFINKEL (1992) examined steps in Ty1 particle formation and processing in order to determine at what level quantitative changes coincident with overexpression might correlate with increased transposition. Processing of the *TYA* protein precursor occurred much more effciently in cells undergoing induction than in noninduced cells. This finding suggested that endogenous PR activity could be rate limiting in populations of cells not undergoing artificial induction of Ty1 expression. In order to test for availability of Ty1-dependent catalytic activity in populations of cells not undergoing induction of an active element, Curcio and Garfinkel tested the ability of uninduced populations to complement defects in induced, defective elements. As a control for the level of background transposition activity, they used an isogenic *spt3* mutant strain which does not express Ty1 from the native promoter and so can have no complementing activity. This study showed that a PR mutant and an IN mutant were not complemented by endogenous expression but that two RT mutants and one IN mutant were complemented and transposed at a frequency 20- to 30-fold higher if endogenous elements were expressed. Thus, they argue that overexpression of a specific, active element results in higher levels of transposition because of complementation of endogenous elements with limiting, possibly cis-acting processing activity. In the later study, however, CURCIO and GARFINKEL (1994) demonstrated that expression of elements results in *trans* as well as *cis* mobilization. Therefore it is unlikely that the effect of overexpression of a single element is exclusively due to saturation of a *cis* requirement for PR. Three observations may shed additional light on the nature of the limitation on processing activity. First, PR activity may be required in higher levels than is reverse transcriptase or integrase. This is suggested by the observation that some retroviruses, avain leukosis and sarcoma virus and mouse mammary tumor virus (COFFIN 1990), as well as at least one retrotransposon, copia (BINGHAM and ZACHAR 1989), express much higher levels of PR than either RT or IN. Second, in retroviruses PR must dimerize to be active, and this is likely to be the case for the Ty1 PR as well. Thus, if elements with PR defects exists, even low-level formation of heterologous PR dimers could exert a *trans*-dominant negative effect on processing. Third, electron micrographs of cells expressing Ty1 PR mutants show that particles have aberrant morphologies and appear cracked. It is possible that processing must occur within a certain kinetic window in order for particle formation to occur productively. Thus processing could be more sensitive than other steps to limiting catalytic activity. In summary, probably several factors contribute to the discrepancy between RNA levels and transposition activity. Among these are the presence of some inactive elements and a threshhold requirement for proteolytic processing activity.

The Ty1 5.7-kb RNA is found associated with wt Ty1 particles (MELLOR et al. 1985b; GARFINKEL et al. 1985) and with the artificial particles produced by a truncated *TYA* protein (BURNS et al. 1992). In the latter form, it is at least partially protected against benzonase, an endonuclease of 30 kDa, but not against the smaller 13.7-kDa RNaseA (BURNS et al. 1992). This prompted Burns et al. to propose that the VLP contains the RNA within an open, cage-like structure. However, the extent to which these uniform particles lacking *TYB* proteins simulate naturally mature particles has not yet been established. Genetic assays in which one Ty1 contributes a *TRP* "donor" transcript with the *cis*-acting determinants to be reverse transcribed into DNA and integrated and a second Ty1 provides element-encoded proteins, but is not competent for reverse transcription, have been used to define the Ty1 *cis*-acting determinants required for transposition (XU and BOEKE 1990b). The smallest donor Ty1 elements found to transpose efficiently were expressed as RNAs having R-U5 and 285 nt of 5′ internal sequence on the 5′ end and 23 bp of internal sequence and U3-R on the 3′ end. A construct with 120 nt of 5′ internal sequence displayed an order of magnitude less transposition. Biochemical analysis of the composition of different helper-donor particles belied the straightforward nature of these genetic results. Although RNAs with less than 285 nt of internal sequence were not associated with Ty1 particles, analysis of *TRPI* and *HIS3* showed that these RNAs expressed under the *GALI* promoter with no Ty1 sequences were VLP associated as was the RNA for the ribosomal protein RP51 expressed from its native promoter. These RNAs were not present in the particulate fraction when no Ty1 particles were produced, and so they were likely to have been associated with the particles. Because VLPs are not ultimately segregated in virions, however, it has been difficult to discriminate effectively between specific packaging of replication template and fortuitous particle association. Other transcripts present at high level, such as the *GALI* transcript and transcripts encoding actin and pyruvate kinase, did not occur in association with Ty1 VLPs. Thus, the Ty1 sequence may include both positive and negative determinants of packaging, and it is possible that the *TRP1* RNA actually contributed sequences required for packaging of the Ty1 donor element. Because the number of genomic Ty1 RNAs included per particle is unknown, whether RNAs associated with the deficient donor RNA could also facilitate packaging is unclear. If dimerization occurs and is completed after particle formation, as is the case for retroviruses (COFFIN 1990), this would be unlikely.

Much less is known about two other types of copialike elements, Ty4 and Ty5. The Ty4 element has sequence and organizational similarity to Ty1 and Ty2 (JANETZKY and LEHLE 1992). The LTRs of this element are 371 bp in length and as isolated insertions are referred to as tau (STUCKA et al. 1989) elements. The Ty4 *TYA4* frame could encode a protein of 414 aa, but this protein has not yet been identified. *TYB4* overlaps *TYA4* in the +1 frame and is 1465 codons in length. One copy of the NC metal finger motif is represented in the carboxyl-terminal region of the predicted *TYA4* protein.

Ty5 was identified in the region of the type X subtelomeric repeat in the sequence of chromosome III (OLIVER et al. 1992; VOYTAS and BOEKE 1992). The element originally identified contained recognizable NC, PR, and RT-RNaseH motifs which placed it in the copialike class. A transpositionally active element has now been identified (ZOU et al. 1996), so information concerning the encoded protein species should be forthcoming.

2.3 *Drosophila* Gypsy and Gypsylike Elements

The *Drosophila* gypsy and gypsylike elements for which the sequence is known include 17.6 (SAIGO et al. 1984), 297 (INOUYE et al. 1986), tom (TANDA et al. 1994), 412 (YUKI et al. 1986), and gypsy (MARLOR et al. 1986). The 17.6, 297, tom, and gypsy elements contain three ORFs, as does the related element transposable element D (TED) from the moth *Trichoplusia ni* (FIESEN and NISSEN 1990). The first ORF of each of these elements encodes a putative Gag-like protein. Proteins of 15, 23, 30 and 34 kDa have been observed binding to nucleic acids and occurring in the particle preparations from cells producing gypsy (FALKENTHAL and LENGYEL 1980), but the derivation of these species has not been determined. The sequences of the proteins predicted from the first ORFs of 17.6, 297, and tom show that they form a gypsylike subfamily (INOUYE et al. 1986; TANDA et al. 1988; SPRINGER and BRITTEN 1993). The second ORF of the *Dropsophila* gypsy and gypsylike elements encodes proteins with PR, RT/RNaseH, and IN sequence motifs and overlaps the first ORF in the -1 frame. Less is known about these proteins than is known about the proteins encoded by Ty1 and Ty3, although RT activity is associated with gypsy virions and particles (TANDA et al. 1994). The predicted proteins encoded in the third ORFs contain hydrophobic, potentially membrane-spanning regions and appear to constitute envelope proteins.

A distinguishing feature of retroviruses when compared to retrotransposons was believed to be the obligatory retroviral cycle of extracellular infection. An important physical feature which correlates with this capacity for infection is the presence of an envelope protein. In retroviruses this ORF is expressed from a spliced mRNA (COFFIN 1990). If envelope proteins are expressed from gypsy, that implies an extracellular phase in the life cycles of these elements. SYOMIN et al. (1993) reported particles larger than typical VLPs in the culture supernatants of *D. melanogaster* and *D. virilis* cells and the presence of gypsy RNA and DNA. Other workers have gathered genetic evidence for extracellular infection by gypsy. The *ovo* gene is a hotspot for gypsy insertion, and gypsy insertion into the antimorph allele, *ovo*^{D1} (which causes sterility in a heterozygote), can revert the mutant phenotype. KIM et al. (1994) exploited this assay to demonstrate that injection of egg plasm from embryos of a strain with actively transposing gypsy elements into embryos of a strain not containing gypsy elements "contaminated" recipient embryos at a low level so that they tested positive by the reversion assay for the presence of active elements. In a second set of experiments, developing flies were fed homogenates of pupae from strains which were active or inactive for gypsy transposition. Flies fed on active strain extracts acquired genomic elements.

The data described above supporting extracellular infection by the gypsy element together with the existence of the third ORF suggested that an envelope function might be associated with gypsy particles. PELISSON et al. (1994) showed that expression of a spliced 2059-nt RNA not containing nts 569–5550 encoding the putative 482-aa (54-kDa) envelope protein correlated in follicle cells of young, homozygous *flam* females with gypsy germline transposition activity. Monoclonal antibodies against the putative envelope protein were used to show that this species was present in tissues expressing the 2.1-kb RNA. The predicted protein has the features of an envelope protein, including an amino-terminal 13-aa hydrophobic signal sequence, a potential dibasic cleavage site, RSRR (which is the same sequence as the human T-cell leukemia virus envelope cleavage site producing surface and transmembrane proteins), and a 23-aa hydrophobic putative transmembrane domain in the carboxyl-terminal region. SONG et al. (1994) used monoclonal antibodies against the predicted protein to demonstrate the presence of a glycoprotein of approximately the expected size.

The nature of the gypsy particle was also investigated in extracts from transposition permissive flies (SONG et al. 1994). Sucrose gradient fractionation showed several peaks of reverse transcriptase activity. One of these occurred at a position of lower density in the gradient than the Ty1 VLP control. This fraction showed incorporation of radioactive dNTPs into gypsy sequence in an endogenous RT assay. Examination of this fraction by immuno-electron microscopy showed irregular particles of about 100 nm which reacted with antibodies to a recombinant envelope protein. Using the ovo^{D1} reversion assay for gypsy mobilization described above, these investigators showed that flies not containing active gypsy elements and exposed as larvae to the appropriate sucrose gradient fractions gave rise to ovo^{D1} revertant progeny. At least some of these events were caused by gypsy insertions into ovo^{D1}, although a significant number of reversion events in one experiment were linked to copia elements. These experiments and those described above showed that gypsy particles could be transmitted extracellularly, that the presence of an envelope protein correlated with the ability of the element to mobilize in vivo, and that the features of the envelope protein are in good agreement with those expected for a retroviral envelope protein. Thus, gypsy is a retrovirus. Whether gypsy also transposes efficiently intracellularly in tissues which do not produce envelope protein and thus is also a "facultative" retrotransposon has not yet been determined.

The tom transposable element of *D. ananassae* also contains a third ORF. In this case, the genomic 6.8-kb RNA is spliced to form a 2.8-kb RNA lacking nts from position 1276 to position 5159,16 bp downstream from the beginning of ORF3. Tom is mobilized at high frequency in the germ lines of females from the *ca; px* strain. A comparison of the expression patterns of tom RNA in the germarium of this strain and in the related nonpermissive *ca* strain showed that the active strain accumulates only about threefold higher levels of both spliced and unspliced RNAs, but the permissive strain has high levels of tom RNA in stem cells, cystoblasts, and oocyte nuclei where it is not found in the nonpermissive strain. The predicted protein has a mass of 70.7 kDa and, if processed to remove a leader of 157 aa encoded in the first ORF, would yield a protein of 52.4 kDa. Cleavage at a basic sequence within this protein would produce a protein of 32.5 KDa and a protein of 19.9 kDa containing a hydrophobic transmembrane domain. Immunoblot analysis using antibodies against a recombinant protein based on the third ORF demonstrated reactive proteins of 59 and 42 kDa in extracts of permissive flies. Thus, tom represents a second member of the gypsy family which has a putative envelope protein and viral properties.

2.4 Gypsylike Ty Particles

The gypsylike elements are represented in *S. cerevisiae* by Ty3 (Hansen et al. 1988) and in *Schizosaccharomyces pombe* by Tf1 and Tf2 (Levin et al. 1990). Ty3 contains two overlapping ORFs. The first ORF, designated *GAG3*, in order to indicate its organizational and sequence differences with the other Ty elements and similarity to retroviral *gag*, encodes a predicted protein of 290 aa. There is a partial copy of the major homology region (MHR) motif ($QGX_2EX_5FX_3L$) and one copy of the NC metal finger motif $CX_2CX_4HX_4C$. The second ORF, *POL3*, overlaps *GAG3* and encodes the catalytic proteins PR, RT-RNaseH, and IN.

Ty3 is naturally induced in haploid cells by mating pheromones and is consequently mobilized during mating (Kinsey and Sandmeyer 1995). In order to study cells in which prolonged expression of Ty3 resulted in accumulation of VLPs, the U3 region of a Ty3 was replaced with the *GALl* upstream activating sequence and this Ty3 was cloned into a high-copy yeast vector (Hansen et al. 1988). Cells containing the galactose-inducible Ty3 construct make high levels of Ty3 proteins and undergo retrotransposition estimated at about 6×10^{-3} events/generation. Inspection of cells expressing high levels of Ty3 showed large clusters of 50-nm particles (Hansen et al. 1992). Antibodies raised against peptides designed from the predicted *GAG3* and *POL3* protein sequences were used to identify the Ty3 proteins. Extracts of cells expressing Ty3, fractionated on sucrose step gradients, showed the presence of Ty3 proteins, genomic RNA, and extrachromosomal full-length DNA in the 70%-30% interface. These particles are about 156S. Examination of the proteins by immunoblot analysis showed a *GAG3* precursor species of 38 kDa and a major 26-kDa species and minor 31-kDa and 39-kDa species which react with the antibody directed toward a determinant at the amino-terminal region of the predicted Gag3 protein. Analysis with an antibody against the carboxyl-terminal region of the predicted Gag3 protein showed the 38-kDa and 39-kDa species as well as a major 9-kDa and a minor 11-kDa species. Based on reactivity to different antibodies and amino-terminal sequence analysis, putative PR species of 16 kDa, RT of 55 kDa, RT-IN fusion of 115 kDa, and IN species of 61 and 58 kDa were identified.

The 26-kDa CA protein plays a central role in particle formation. Overexpression by galactose induction of a Ty3 element truncated after the CA-coding region showed that particles were formed (Kirchner et al. 1992) which could be concentrated by velocity sedimentation The two-hybrid system of Fields and Song (1989) was used to examine interactions within this domain (Orlinsky et al. 1995). Expression of CA fused to the Gal4p DNA binding domain and Gal4p activation domain in cells with a copy of *lacZ* under control of the transcriptional activator Gal4p resulted in activation of lacZ expression, demonstrating that CA proteins interact. The MHR domain of retroviruses plays a role in replication and in particle formation (STRAMBIO-DE-CASTILLIA and HUNTER 1992; MAMMANO et al. 1994). Mutations introduced into the Ty3 MHR motif ($QGX_2EX_5FX_3L$) changing the conserved glycyl residue to alanyl or valyl residues resulted in complete loss of particle formation (ORLINSKY et al. 1996). Mutations changing the conserved glutamyl residue to an aspartyl, asparaginyl, or lysyl residue blocked formation of extrachromosomal DNA and in the case of the lysyl residue mutation blocked Gag3 processing. These results are similar to the loss of particle formation or loss of infectivity and replication which results from mutations in the MHR of retroviruses.

Ty3 NC contains one copy of the metal finger motif found in one or two copies in retrovirus NC proteins. Mutations in the metal finger domain of Ty3 NC resulted in low levels of particle formation and disrupted protein processing (ORLINSKY and SANDMEYER 1994). Whether loss of processing is a secondary effect of lack of genomic RNA was difficult to assess given the low level of particles present in cells expressing these mutants. These mutants could be complemented in *trans*, showing that there was no primary defect in ability of the precursor Gag3 to act as a substrate for processing. Although NC is encoded in *GAG3* and therefore expressed together with CA at high levels, there is evidence that NC is not required at equimolar levels with CA. A *GAG3-POL3* fusion mutant which transposes only at extremely low levels could be fully complemented for transposition by expression of CA in *trans* (KIRCHNER et al. 1992). In addition to its proposed scaffolding function for genomic RNA, retroviral NC has been demonstrated in vitro to have activities (melting and tRNA annealing to genomic RNA) compatible with other functions in replication. It is therefore possible that NC supplied in *cis* with RT has a specific function in reverse transcription. A *GAG3-POL3* fusion mutant was complemented in *trans* with a truncated Ty3 expressing only *GAG3*. Although an NC metal finger contributed from *GAG3-POL3* or *GAG3* was essential for transposition, the domain could be supplied from either source, demonstrating that the metal finger of NC supplied within the same polyprotein precursor as RT is not essential for priming reverse transcription (ORLINSKY and SANDMEYER 1994).

The Ty3 PR plays a central role in Ty3 particle morphogenesis (KIRCHNER and SANDMEYER 1993). The putative PR domain contains the sequence DSG, which is found in some retrovirus PR-active sites and is related to the sequence DTG, which is found in others. In order to test whether this motif was required for production of the Ty3 protein species found in the particles, the aspartyl residue of the active site was changed to an isoleucyl residue. This mutation blocked processing and resulted in accumulation of a low level of the predicted 38-kDa and 173-kDa Gag3 and Gag3-Pol3 precursor proteins. Mutating the active site seryl to threonyl residue to mimic the commonly occurring retroviral active site motif did not have a major effect on processing or on transposition. Ty3 elements with the PR-active site mutation had low levels of RT activity and did not transpose at a detectable frequency. Sequence analysis of the amino termini of the 9–, 16–, 55–, 61–, and 58-kDa species allowed them to be mapped precisely to the polyprotein precursor. Inference of the Ty3 PR processing sites from these data showed that, as in the case of retroviral PR, the cleavage context is hydrophobic (P3 through P2′ uncharged) and the P1 positions do not contain a branched beta carbon.

Analysis of cells expressing a Gag3-Pol3 fusion mutant produced by deletion of a single nt in the overlap region together with a CA protein allowed assignment of the 39-kDa and 11-kDa proteins as Gag3 and NC species, respectively, derived from processing of the Gag3-Pol3 precursor at a position close to or coincident with the PR amino-terminus (Kirchner et al. 1992; Orlinsky and Sandmeyer 1994). Identification of the size and position of PR and the amino terminus of RT showed that there is an intervening region which potentially encodes a protein of approximately 10 kDa. Deletion of this region and substitution with a chimeric processing site permitted processing of PR and RT and did not produce gross anomalies in transposition of a *GAL1*-driven Ty3 element (J. Claypool and S.B. Sandmeyer, unpublished results). The 115-kDa species contains both RT and IN determinants. Whether the active RT species is a heterodimer with RT and RT-IN subunits is unknown. Truncation of the carboxyl-terminal region of IN by as little as 27 residues reduces RT activity to background (Kirchner and Sandmeyer 1995), which suggests that, at the very least, the RT tertiary structure is dependent on intact IN for correct folding.

Tf1 and Tf2 gypsylike elements of *S. pombe* (Levin et al. 1990) have several intriguing structural differences (Levin et al. 1990) with other gypsylike family members. The Tf1 element has LTRs of 358 nt and is transcribed into a 4.4-kb mRNA. A transpositionally active Tf1 element contains a single ORF of 1340 codons (Levin et al. 1990; Levin and Boeke 1992). Upstream of the region which can be inferred to encode PR based on the presence of the DTG motif, there is coding capability for a protein of about 260 aa, and a protein of 27 kDa has been identified which is encoded in this region (Levin et al. 1993). Antibodies directed against proteins encoded by different portions of the ORF demonstrate that initially a 140-kDa polyprotein is produced and that this is processed to the 27-kDa CA and other species, including a 56-kDa species containing IN motifs. In Tf1 elements with a mutation introduced into the PR-active site, the full-length polyprotein accumulates, indicating that at least some aspects of processing are dependent on the element-encoded PR.

3 Control of Relative Amounts of Structural and Catalytic Proteins

The mechanisms through which the LTR retrotransposons maintain the appropriate ratio of major structural proteins to catalytic proteins are striking in their diversity. In addition, classification based on these mechanisms is not congruent with copialike and gypsylike families except among closely-related elements.

Accordingly, the following discussion will be organized by mechanism rather than by family affiliation.

The correct balance of Gag and Pol proteins is established in retroviruses by ribosomal readthrough mechanisms — either nonsense codon suppression or -1 frameshifting from *gag* into *pol* (JACKS 1989). In the case of -1 frameshifting, a 7-nt sequence is present so that tRNAs positioned at the A and P sites of the ribosome can simultaneously slip into the -1 position while maintaining pairing at least at nonwobble positions. The motifs U UUA, U UUU, or A AAC are found where the first complete triplet indicates the position of the first 0 frame codon. A pseudoknot structure just downstream of the shift site enhances the frequency of frameshifting.

3.1 The -1 Frameshift Elements

Portions of the sequence of the overlap of the first and second ORFs of the *Drosophila* elements gypsy (AAU UUU UUA GGG) (Marlor et al. 1986), 17.6 (GAA AAU UUU CAG) (SAIGO et al. 1984), and 297 (GAA AAU UUU CGG) (INOUYE et al. 1986) readily accommodate the retrovirus simultaneous slippage model and are related to known slippery sites. The sequence AAU UUU UUA GGG found in gypsy is the same as the sequence found at the frameshifting site of HIV-1 (JACKS et al. 1988). The sequence A AAU UUU C found in 17.6 and 297 contains shifting codons as the sequence of the frameshifting context for Mason-Pfizer monkey virus (JACKS et al. 1987). None of these sites has been demonstrated to function in the *Drosophila* element context, nor has a junction protein been sequenced, but frameshifting activity seems reasonable to suppose based on the retrovirus precedents. This mechanism is also used in the yeast L-A dsRNA virus, where 2% -1 ribosomal frameshifting from the first ORF (*gag*), which encodes an 80-kDa major coat protein, into the second ORF (*pol*), which encodes an RNA-dependent RNA polymerase, results in production of a Gag-Pol fusion protein. Frameshifting has been shown to occur on the sequence G GGU UUA and a pseudoknot which begins 4 nt downstream of the end of that sequence contributes to the frameshifting context (DINMAN et al. 1991).

3.2 Plus 1 Frameshifting of Ty1

In consideration of the retroviruslike frameshifting of the yeast dsRNA virus, it is ironic that none of the LTR retrotransposons in yeast appear to frameshift via simultaneous slippage of tRNAs in the ribosomal P and A sites. In the case of the copialike elements Ty1, 2, and 4, transition from *TYA* to *TYB* is into the +1 frame. Although, as in the case of retroviruses, a 7-nt sequence has been shown to be sufficient for at least a low level of shifting, the Ty1 frameshift differs mechanistically (BELCOURT and FARABAUGH 1990). The Ty1 sequence, CUU AGG C, found in the 38-nt overlap between *TYA* and *TYB* mediates from 20% to 40% frameshifting, depending on the context. The salient features of this sequence were defined by monitoring the frameshifting activity of a cassette fused between coding sequences for the *HIS4* and *lacZ* proteins. They are as follows: (1) CUU is decoded by a $tRNA^{Leu}{}_{UAG}$, which has an unmodified U at the wobble position and which can decode all six leucine codons; (2) the +1 overlapping triplet is a leucine codon (UUA), which can be decoded by the same tRNA; and (3) AGG, the next 0 frame codon, is decoded by $tRNA^{Arg}{}_{CCU}$, which is a minor tRNA species encoded by a single gene. These features suggested that pausing prior to occupancy of the A site by the $tRNA^{Arg}{}_{CCU}$ could permit +1 shifting by the peptidyl $tRNA^{Leu}$ in the P site. That this was the case rather than retroviruslike simultaneous slippage was demonstrated by sequence analysis of the junction region in the frameshifted product which showed the sequence Leu-Gly rather than Leu-Arg, as would have been produced if slipping had occurred with tRNAs in both the P and A sites. In fact, expression of the rare $tRNA^{Arg}{}_{CCU}$ species is inversely proportional to frameshifting (Xu and Boeke 1990a; Kawakami et al. 1993). Nevertheless, a codon specifying a low-abudance tRNA is not sufficient to ensure frameshifting, as substitution of codons for other either naturally or artificially low abundance tRNAs for AGG did not sustain frameshifting. The 7-nt sequence described as the Ty1 frameshifting context also occurs in the 44-nt overlap of the first and second ORFs of Ty2 and in the 227-bp overlap of the first and second ORFs of Ty4 (Janetzky and Lehle 1992).

3.3 Plus 1 Frameshift by the Incoming $tRNA^{Val}$ in Ty3

Ty3 synthesis of Gag3-Pol3 also depends upon a +1 ribosomal frameshift from *GAG3* into *POL3* (Kirchner et al. 1992). However, this frameshift differs in recoding site and in context from the Ty1 frameshift described above. Deletion and point mutagenesis of the 38-bp overlap between *GAG3* and *POL3* defined a 21-nt region which is essential for efficient frameshifting (Farabaugh et al. 1993). The sequence at the beginning of the 21-bp region, GCG AGU U, was identified as the recoding site by sequence analysis of the frameshifted junction protein. In addition to identifying the site of frameshifting, this analysis showed that alanine (GCG) is followed by valine (GUU) in the protein product of frameshifting, thus excluding a simultaneous slippage model. Overexpression of the $tRNA^{Ser}{}_{GCU}$, which decodes the second 0 frame triplet in the recoding site, resulted in about a ten-fold reduction in frameshifting. This implies that a ribosome with the P site occupied by $tRNA^{Ala}$ pauses prior to decoding in the 0 frame. In order to better understand the mechanism of shifting, two additional types of experiments were performed. Sixty-one other codons were substituted for GCG to determine whether anticodons which allowed weak pairing in the 0 frame were sufficient to support +1 shifting (Vimaladithan and Farabaugh 1994). Codons CUU, GCG, and CCG allowed +1 shifting, but not other codons, even ones which would be predicted to be slippery based on stability of anticodon pairing in the 0 and +1 frames. These findings argue against slipping and imply a role for sequences in the $tRNA^{Ala}$ structure outside of the anticodon. A prediction of this model is that another tRNA with the same anticodon would not produce the shift. Overexpression of $tRNA^{Ala}{}_{UGC}$ mutated to CGC in the anticodon resulted in a twofold reduction of frameshifting, suggesting that this isoacceptor, although it decoded GCG, did not promote frameshifting and competed with the correct isoacceptor species. Farabaugh proposes that $tRNA^{Ala}{}_{CGC}$ could, by interaction with an incoming noncognate species during ribosomal pausing, change the kinetics of EF-1α proofreading in favour of that noncognate species, and that the structure of $tRNA^{Val}{}_{IAC}$ in the +1 frame might be susceptible to this stabilization. Retroviral frameshifting contexts include a predicted RNA pseudoknot which contributes to ribosomal pausing. The nature of the 7.5-fold enhancement of frameshifting contributed by the 14 nt following the 7-nt recoding sequence in Ty3 is intriguing but is not known.

3.4 Regulated Translation of Major Structural Proteins from LTR Retrotransposons with Single ORFs

Surprisingly, copia (MOUNT and RUBIN 1985), and representatives of copialike (Ty5) (Zou et al. 1996) and gypsylike (Tf1) (LEVIN et al. 1990) families in *Drosophila* and yeast, contain single ORFs encoding both the major structural, non-envelope proteins and the catalytic proteins. In each case, elements with single ORFs have been shown to have transposition activity. The copia element is transcribed into a major 5-kb RNA species which is spliced to a 2-kb RNA species (YOSHIOKA et al. 1990; BRIERLEY and FLAVELL 1990; MILLER et al. 1995). As described above, particle formation by copia is dependent upon expression of the 2-kb RNA and processing by the protease encoded in that RNA. The surprising observation in this system is that levels of CA, which is encoded in the 2-kb transcript, are apparently not controlled at the level of splicing. In order to investigate potential translational control of the relative expression from the 5-kb and 2-kb RNAs, BRIERLEY and FLAVELL (1990) tested the expression of a beta-galactosidase-coding region fused to different positions in the copia sequence. They found that fusions of the beta-galactosidase-coding region to the 3′ end of a synthetically spliced construct were expressed efficiently, but that a fusion at the same position in a construct in which the 3′ acceptor site was disrupted by site-directed mutagenesis (i.e., the 5-kb transcript) did not produce equivalent amounts of protein. These investigators argued that differential expression from the 2- and 5-kb RNAs is consistent with translational regulation of the major structural and catalytic proteins.

Ty5 provides the first example of a single ORF retroelement in *Saccharomyces*. It is a copialike element originally identified in the sequence of *S. cerevisiae* chromosome III, adjacent to the type X subtelomeric repeat (OLIVER et al. 1992; VOYTAS and BOEKE 1992). However, identification of more complete elements found in *S. paradoxus* has allowed determination of the sequence of an active element (Zou et al. 1995, 1996). The mechanism, if any, by which an excess of certain structural proteins is produced is not yet known.

Tf1 and Tf2 are two related elements from *Schizosaccharomyces pombe*. The DNA sequence of an active Tf1 element was determined and was shown to contain a single ORF of 1340 codons (Levin et al. 1990; Levin and Boeke 1992). Tf1 may not be the only gypsylike element to contain a single ORF as the SURL element from sea urchins also contains a single ORF (Springer et al. 1991), although transposition of that element has not yet been demonstrated. In the case of Tf1, early in the growth of induced cultures equivalent amounts of CA and a representative catalytic protein (IN) and low levels of DNA reverse transcription intermediates are detected. In stationary phase cultures, however, the ratio of CA to IN is about 26 : 1 and this transition correlates with the appearance of a eightfold higher level of approximately full-length DNA (Atwood et al.1996).

4 Priming and Reverse Transcription of LTR Retrotransposons

4.1 Plus and Minus Strand Primers for Reverse Transcription

Genetic experiments with Ty1 showed transposition through an RNA intermediate and reiteration of regions present uniquely at each end of the genomic RNA at the other end in the DNA copy. Thus, retrotransposition displays the hallmarks of replication by retroviruslike reverse transcription (Boeke et al. 1985).

Inspection of the putative plus-strand priming polypurine tract (ppt) sequences and minus-strand primer-binding site (pbs) of LTR retrotranposons, however, reveals unexpected differences with retroviruses (Table 1). The plus-strand primer is a short polypurine tract in the LTR retrotransposons. It ranges in length from 6 purines out of the 8 positions in Ty1 to 14 consecutive purines in Ty3. The pbs of copia (Kikuchi et al. 1986) has complementarity to an internal region in a fragment of initiator $tRNA^{Met}$, which has been shown to function as the minus-strand primer. The putative pbs sequence of the copialike element Ty5 is consistent with priming from a similar position in $tRNA^{Met}$ (Voytas and Boeke 1993). The Ty1 (copialike) and Ty3 (gypsylike) elements prime from the 3′ end of initiator $tRNA^{Met}$ (Chapman et al. 1992; Keeney et al. 1995). The pbs for these elements, which is 10 nt in length for Ty1 and 8 nt in length for Ty3, is relatively short compared to those of retroviruses. Other retrotransposons have more substantial tracts of complementarity (see Table 1). The related gypsy elements 17.6, 297, and tom have putative pbs sequences which are complementary to $tRNA^{Ser}$. Gypsy and ulysses pbs sequences are complementary to $tRNA^{Lys}$, but the 412 and micropia members of this class are inferred to be primed by $tRNA^{Arg}$ and $tRNA^{Leu}$, respectively.

The Tf1 element defines a new class of elements based on its mechanism of minus-strand priming. Inspection of the Tf1 and Tf2 sequences failed to reveal significant complementarity to a known tRNA sequence but uncovered an 11 nt inverted repeat between the region typically occupied by the pbs at the U5-internal domain junction and the 5′ end of the genomic RNA. Changes in either the pbs region or the 5′ terminal region blocked production of reverse transcripts but the combination of these changes designed to restore the predicted foldback structure was permissive for reverse transcription. These results raise some interesting questions about the priming molecule itself, among them whether priming is on a 3′ OH created by processing close to the 5′ end of one genomic RNA and is then templated by a second molecule or occurs on the 2′ OH as observed for msDNA (FURUICHI et al. 1987). Inspection of a closely-related element, CfT-I, from the fungal tomato pathogen *Cladosporium fulvum* (MCHALE et al. 1992), which has weak complementarity to known tRNA sequences, revealed that the sequence would be consistent with a 9-bp foldback structure between the pbs region and a region potentially occurring at the 5′ end of the RNA (LEVIN 1995). The existence of this foldback-priming mechanism in retroelements is interesting because it offers concrete evidence that tRNAs are not the unique solution to the problem of priming retroviruslike reverse transcriptase. In addition, the existence of a mechanism for autologous priming in retroelements could have conferred independence of host tRNA populations and thus promoted horizontal transmission between distantly related species.

In addition to functioning as a scaffolding protein, NC has been implicated in tRNA annealing and priming (WILLS and CRAVEN 1991). Given the almost complete conservation of the metal finger motif in NC proteins of retroviruses, the broad inconstancy of this feature within either the copialike or gypsylike families is striking. One might speculate that some other feature of the priming apparatus might be conserved along with NC in the representative LTR retrotransposons (Ty3, Ty4, possibly Ty5, and copia). Yet the presence of the NC metal finger motif does not correlate with relatedness of the RT polymerase domain or the identity of the tRNA primer or even the position of the end of the primer in the tRNA. Additional alignments of RT may eventually reveal a motif conserved or function maintained which is common among elements and viruses containing a metal finger NC. Whether small NC-type proteins exist for the nonmetal finger *gag*-like proteins is also not known. Processing of a small, carboxyl-terminal domain occurs for at least one of these elements (Ty1), but nucleic acid-binding activity has also been reported for the mature form of this protein, which does not contain this domain.

The position of the internal pbs relative to the adjacent, upstream LTR is a variant feature of the LTR retrotransposons. It appears to be a primary determinant of whether trimming activity is required by IN prior to integration. In the case of Ty3, the position of the 3′ end of the tRNA primer is 2 nt displaced from the internal 3′ end of the upstream LTR as is the case for retroviruses. Species with 2 extra nt at the outside ends are predicted by the position of pbs. These extra nt have been shown to be present in extrachromosomal Ty3 DNA and the 3′ ends have been shown to be processed in wt, but not mutant IN VLPs (KIRCHNER and SANDMEYER 1995). The position of the Ty1 pbs, which anneals to the same primer species, initiator $tRNA^{Met}$, is exactly juxtaposed to the internal 3′ end of the LTR. Based on the retroviral model, Ty1 minus-strand strong stop begins with the 3′ nt of the U5 region of the RNA (CHAPMAN et al. 1992). Data from in vitro integration also indirectly support this positioning of the primer. Ty1 DNA with blunt ends and no extra nt acts efficiently in integration reactions, whereas DNA with a 4-nt 5′ overhang does not act as an efficient donor (EICHINGER and BOEKE 1990). The copialike Ty4, gypsylike 412, and copia also juxtapose the pbs to the 3′ end of U5. The putative pbs sequences of the gypsy, 17.6, and 297 elements overlap the 3′ end of U5 by 1 nt. Although admittedly novel, the position of the pbs sequences of this family, which overlap the 3′ nt of U5 with RNA, can be accomodated by the generic retroviral model of replication. In the case of the gypsy, 17.6, and 297 elements, the 5′ end of the minus-strand strong-stop DNA species is, in theory, incomplete relative to the integrated form by one deoxyribonucleotide. Studies with HIV have shown, however, that RNaseH preferentially cleaves the tRNA primer so that the final ribonucleotide remains associated with the DNA. If this is also the case for gypsy, then this ribonucleotide could template synthesis of the 3′ end of the plus strand. The 3′ end is directly involved in strand transfer upon integration, while a 5′ ribonucleotide at the end of the minus strand could be degraded during the subsequent gap repair process. The deduced LTR ends of the gypsy and gypsylike elements are also unusual in that they do not have the conserved TG/CA dinucleotide found at the termini of other LTR retrotransposons and retroviruses.

4.2 Gypsy, Gypsylike and Ty Reverse Transcription Intermediates

DNA species associated with VLPs have been examined for mdg1, mdg3, gypsy (ARKHIPOVA et al. 1986), Ty1 (EICHINGER and BOEKE 1988; MÜLLER et al 1991; POCHART et al. 1993a; CHAPMAN et al. 1992), and Ty3 (HANSEN et al. 1992; KIRCHNER and SANDMEYER 1996). Strong-stop species consistent with a retroviruslike mode of replication have been observed in each case. For gypsy expressed at steady-state levels, the level of plus-strand strong stop was estimated to be five to ten times higher than the level of minus-strand strong stop. For mdg1, mdg3, gypsy, and Ty1, treatment of a minus-strand strong-stop species with RNase results in a reduction in size consistent with removal of the putative tRNA primer. For mdg1, 3, and 4 and Ty1 and 3, the plus-strand strong-stop species is observed as a doublet. In the case of the mdg-gypsy elements the larger species correspond to the length of the LTR extended by the length of the region of complementarity with the putative tRNA primer. In the case of the yeast elements Ty1 (LAUERMANN and BOEKE 1994) and Ty3 (KIRCHNER and SANDMEYER 1996) primed by initiator $tRNA^{Met}$, the plus-strand strong stop is extended for 12 nts, up to the first modified base of initiator $tRNA^{Met}$.

4.3 Ty Priming by Initiator $tRNA^{Met}$

The function of initiator $tRNA^{Met}$ in Ty1 and Ty3 replication has been investigated using a combination of biochemical and genetic strategies (KEENEY et al. 1995; CHAPMAN et al. 1992). Incubation of Ty1 VLPs with [α-^{32}P]dTTP resulted in labeling of a 75-nt RNA species, consistent with the size of the predicted initiator $tRNA^{Met}$ primer plus one radioactive nucleotide. In the presence of all four dNTPs, a species of 171 nt is radiolabeled which is reduced to 96 nt by treatment with RNase. In order to demonstrate that the presence of this species is specifically dependent upon the complementarity to initiator $tRNA^{Met}$, a genetic strategy was used. Five nt changes were introduced into the pbs of Ty1 and shown to virtually abolish Ty1 transposition. Transposition of wt Ty1 was also greatly reduced in a strain in which the sole initiator $tRNA^{Met}$ was supplied from a plasmid-borne copy of the $tRNA^{Met}$ gene with the changes complementary to those introduced into the pbs and compensatory changes in the 5′ stem (*imt4-9*). However, when the mutant Ty1 and tRNA gene were expressed in the same cells, Ty1 DNA production and transposition were restored. Northern blot analysis using an initiator $tRNA^{Met}$-specific probe showed that disruption of complementarity between the pbs and the priming tRNA did not affect inclusion of initiator $tRNA^{Met}$ in the particle. In a separate study, POCHART et al. (1993b) showed that initiator $tRNA^{Met}$ is concentrated about 11-fold in Ty1 particles relative to its concentration in total tRNA. Interestingly, a minor species, $tRNA\text{-}^{Ser}{}_{GCU}$, was similarly represented in VLPs compared to $tRNA^{Met}$. This represented about a 150-fold concentration relative to its cytoplasmic representation.

Genetic manipulation of the primer tRNA gene has had two other interesting applications. Although it is straightforward to deduce the nt in the aminoacyl stem which pair with the pbs, the residues which contribute to inclusion of the tRNA in the particle and which may be involved in interactions with NC or RT are not self-evident. A strain containing deletions of the chromosomal *IMT1-4* genes and transformed with two plasmids, one carrying Ty1 and a copy of the mutated initiator $tRNA^{Met}$ which is active in translation but does not prime wt Ty1 reverse transcription, *imt4-9,* and one carrying a mutated, tester initiator $tRNA^{Met}$, was used to probe these primer functions. Mutations which affected the level of this tester tRNA or which were located at essential positions for priming disrupted transposition. In the first analysis, mutations were introduced to 11 positions of the anticodon stem and loop region, which has been shown to be important for HIV RT-$tRNA^{Lys}$ interaction (BARAT et al. 1989), and 4 mutations were introduced in the TΨ arm, which has been shown to be important for avian RT interaction (HU and DAHLBERG 1983). Single mismatches in two independent positions in the region which pairs directly with the pbs blocked transposition, but the effects of the mutations were reversed when compensatory mutations were introduced into the pbs of the Ty1 element tested, demonstrating that the effect was at the level of primer annealing. Interestingly, although single base mismatches were not tolerated, a G:U base pair between the pbs and the tRNA did not disrupt transposition. Mutations in the TΨ arm at positions 50, 54, 60, and 64 reduced

Ty1 transposition. The extent of the effect depended on the nature of the individual mutation or combination of mutations, with the greatest effect being observed with changes of A54 to G or C, mutations which extended the arm structure, or changes which disrupted structure important for initiator function at positions U50:A64. Experiments in which chimeric initiator $tRNA^{Met}$ genes between *S. cerevisiae* and *Arabidopsis thaliana* and *S. pombe* gene were tested showed that chimeras with differences from the natural primer in the D loop stem did not produce active primers. Together these experiments suggested that the domain created by interaction of the TΨC and D loops in the L-shaped tertiary structure could play a role in primer recognition or function.

Ty3 contains a pbs region with 8 nt of complementarity to initiator $tRNA^{Met}$. Ty3 does not transpose in the strain carrying the initiator $tRNA^{Met}$ gene with the five compensatory mutations in the aminoacyl stem (*imt4-9*) (KEENEY et al. 1995), showing that $tRNA^{Met}$ is likely to be the Ty3 primer. The effect of point mutations in the initiator tRNA on Ty3 transposition were also tested. Despite the differences noted above between Ty3 and Ty1 (weak similarity in RT, presence of NC in Ty3 but not in Ty1), Ty3 transposition showed a qualitatively similar sensitivity to mutations in the primer to that displayed by Ty1 (KEENEY et al. 1995). In view of the apparent functionality of G:U pairing in Ty1 priming described in the preceding section, the interactions between the initiator $tRNA^{Met}$ and the Ty3 pbs probably extend past the 8-nt complementary tract through the next two bases which could form G:U pairs, giving a pbs 10 nt in length. However, this is still significantly shorter than the 18 nts observed for retroviral pbs sequences. The shorter tract of complementarity may be tolerated because of the lower growth temperature of yeast compared to animal retroviruses. Ty1 and Ty3 RTs are temperature sensitive even in exogenous assays, so the role of temperature in annealing has not been tested directly.

4.4 Genetic Evidence that the tRNA Stem Is Not Copied into the Productive Plus-Strand Strong-Stop Species

As noted above, where a plus-strand strong-stop species has been observed for LTR retrotransposons (mdg1, mdg3, and gypsy) (ARKHIPOVA et al. 1986) and Ty1 (MÜLLER et al. 1991; POCHART et al. 1993a) and Ty3 (KIRCHNER and SANDMEYER 1995), there are two species. The longer and predominant species found for the mdg1, mdg3, and gypsy are lengths consistent with copying of the primer tRNA through the region of complementarity to the pbs as occurs for retroviruses (TELESNITSKY and GOFF 1993). However, in the case of Ty1 and Ty3, which have pbs sequences of 10 and 8 nt, respectively, the length of the plus-strand strong stop appears to be consistent with copying of the tRNA primer up to the first modified base 12 nt into the initiator $tRNA^{Met}$ (LAUERMANN and BOEKE 1994; KIRCHNER and SANDMEYER 1995). In the case of Ty1, this species has been shown to hybridize to a pbs-specific probe (MÜLLER et al. 1991). Indeed, RNA-templated products have been shown to correct pbs mutations in retroviruses, suggesting that this species is extended to form the plus strand (TELESNITSKY and GOFF 1993). However, a similar pbs templating on the tRNA primer appears not to occur for the Ty elements. If the Ty plus-strand strong-stop species is transferred to the 5′ end of the nascent minus-strand DNA and extended, the pbs of Ty elements would have been converted into a tract of 12 nt of complementarity with the primer $tRNA^{Met}$. LAUERMANN and BOEKE (1994) specifically tested for the transfer of information from the tRNA primer into the Ty1 pbs. A single-bp mutation was introduced at position 7 of the Ty1 pbs and its correction was monitored in preintegrative DNA by sequence analysis of amplified single strands and within integrated elements by selective restriction digestion. The pbs mutation was not corrected by tRNA templating in the extrachromosomal DNA or in the transposed elements. A system where information is transferred from the tRNA into the pbs in theory would allow a single RNA to template genomic replication even assuming that RNaseH activity degraded the 5′ end of the RNA template for minus-strand strong-stop DNA. In the case of the Ty system if the 5′ end of the RNA template for the minus-strand-strong-stop DNA is degraded by RNaseH, then the data suggest that the target of transfer of the plus-strand strong stop is a minus-strand DNA which has been copied past the pbs into the R-U5 region of overlap, necessarily on a separate RNA template. Since the Ty plus-strand strong-stop species extended into the tRNA would be unpaired for at least two bases at its 3′ end, it may fail completely to function as a primer and accumulate while the form not containing 12 nt from the tRNA is incorporated into full-length plus strand. Alternatively, the unpaired 3′ end of the plus strand could target it for degradation and repair templated by the pbs, rather than the tRNA-derived sequence. Ultimately, assuming degradation of the RNA template for the minus-strand strong stop, these observations would seem to argue for functional Ty genomic diploidy and corresponding increases in genome fluidity.

4.5 Ty1 Replication Intermediates

In addition to the strong-stop species described above, full-length Ty1 and Ty3 DNA is found associated with their respective VLPs (EICHINGER and Boeke 1988; HANSEN et al 1992). In the case of Ty1, significant amounts of intermediates occur as well (MÜLLER et al. 1991; POCHART et al. 1993a). Studies with two different Ty1 elements have shown that in steady-state VLP populations the major species are comprised of full-length minus strands and a 2.1-kb plus strand with its 3′ end at the 3′ end of the mature DNA. Mapping of the 5′ end of this intermediate shows that it lies near the internal sequence TGGGTGGTA, which is also the sequence at the internal domain-U3 junction identified as the plus-strand primer (POCHART et al. 1993a). As observed for other elements and mammalian retroviruses, there are circular forms of Ty1. Circular Ty1 DNAs recovered as plasmids (Eichinger and Boeke 1988) were 1-LTR circles. However, electron micrographic analysis of Ty1 species suggested that 2-LTR species might also be represented. Both studies (Müller et al. 1991; Pochart et al. 1993a) identified plus-strand species extending from the upstream end of the RNA, about 0.4 kb in one case and over 3 kb in the other. These studies suggested that full-length, completely replicated, linear Ty1 DNA is a relatively minor species. Whether this is a strain-dependent phenomenon is not known. It will be interesting to determine whether an additional discrete step exists at which completion of reverse transcription is regulated.

4.6 Regulation of Ty3 Replication

Retrovirus reverse transcription is not completed until infection of new host cells. In addition, viral replication is sensitive to the state of replication of the host cell. Rous sarcoma virus, spleen necrosis virus, and Moloney murine leukemia virus do not complete reverse transcription in quiescent host cells. Similarly, although HIV can replicate in nondividing macrophage, it does not complete reverse transcription in quiescent human T lymphocytes (reviewed in Menees and Sandmeyer 1994). In growing cells induced under a regulated promoter to high levels of Ty3 transcription, a major Ty3 DNA species is observed, which migrates as a heterogeneous species of about 5.4 kb (Hansen et al. 1992). Because, it can be digested in central and terminal regions with restriction enzymes, it presumably represents a full-length duplex DNA species, rather than the DNA-RNA heteroduplex inferred for Ty1 based on incomplete plus-strand synthesis. However, under physiological conditions, Ty3 expression is induced by the pheromone signal transduction initiated when haploid cells are exposed to cells of the opposite mating type (Bilanchone et al. 1993). Cells in this condition are not dividing but are arrested in G_1 prior to mating, at which time the synchronized populations fuse and S phase ensues. This context may be analogous in some respects to retroviruses in the original host cell or in nondividing cells.

It could be speculated that Ty3 would have greater potential for proliferation and consequently a selective advantage, if it transposed after, rather than before, cell-cell fusion. *MAT***a** cells were arrested in G_1 by treatment with α-factor (Menees and Sandmeyer 1994). After complete arrest, cells carrying no endogenous Ty3 elements, but a high-copy plasmid with Ty3 under control of the *GAL1* UAS, were induced for Ty3 expression and transposition was measured using a polymerase chain reaction (PCR) assay. In log cells transposition occurred within 3 h of induction of Ty3 transcription. Examination of arrested cells showed that although induction kinetics were normal and particles were present, full-length DNA was not detectable. No block occurred in *far 1* mutant cells, which respond to α-factor signal transduction, but do not arrest. If cells were switched to a repressing carbon source for Ty3 expression and α-factor was removed, DNA synthesis and transposition occurred within 1–2 h. Whether the mechanism of regulation depends upon reduced levels of dNTPs in G_1 cells or is through some direct effect on RT is not known. These findings predict that in natural populations of mating cells Ty3 particles form prior to mating but reverse transcription and integration occur in the new diploid cell. This was directly tested with a Ty3 element on a high-copy plasmid in one mating type and a selectable target plasmid in the opposite mating type. After mating, diploid cells which had undergone transposition of this type were identified by selection for cells where the Ty3 element from one parent had activated a suppressor tRNA on the plasmid target of the other parent (KINSEY and SANDMEYER 1995). Thus Ty3 undergoes what might be considered an inefficient infection cycle by virtue of sexual fusion of cells containing unreplicated particles with target cells in which reverse transcription is completed

5 Regulation of Turnover of Ty Particles

Ty1 and Ty3 particle proteins appear to be relatively stable. For example, Ty1 proteins labeled in a 30-min pulse under steady-state conditions of Ty1 expression were processed to mature 90-kDa and 60-kDa forms which were still present after a 36-h chase (GARFINKEL et al. 1991). Ty3 proteins produced under inducing conditions are relatively unchanged in levels after 8 h of growth in glucose, a repressing carbon source (Menees and Sandmeyer, unpublished data). Despite this stability, as described below, the turnover of Ty particles is subject to modulation arising from the state of the cells.

5.1 Turnover of Ty1 Particles in Pheromone Arrested Cells

XU and BOEKE (1991) tested the effect of α-factor arrest on Ty1 transposition in *MAT***a** cells. As is observed for Ty3, transposition is blocked by α-factor treatment. Cells blocked in α-factor were assayed genetically for evidence of transposition. Ty1 expression was induced in arrested cells. After a period of 9 h, cells were plated and monitored for transposition. Transposition was decreased by more than tenfold in cells arrested by pheromone treatment during induction. In order to determine the level at which transposition was affected, Ty1 VLPs were isolated and analyzed by Western, Northern, and Southern blotting. *TYA* proteins migrated with a mobility intermediate between the precursor form of 58 kDa and the processed 54-kDA form. Labeling with ^{32}P showed that *TYA* proteins in α-factor-treated cells were phosphorylated. In addition, loss of reactivity to an antibody with an amino-terminal epitope suggested that this protein was modified, perhaps by amino-terminal proteolysis. *TYB*-encoded 160-kDa and 90-kDa proteins were shown to be unstable, although this did not appear to be the case for the 60-kDa RT species. The level of DNA present in the particle was also greatly reduced.

5.2 Turnover of Ty3 Particles by Components of the Stress Response

Ty1 and Ty3 transposition is temperature sensitive. Ty1 transposition estimated from insertions of Ty at *ADH2* and *ADH4* selected by antimycin A resistance was 100-fold more frequent at 15°C than at 30°C (PAQUINN and WILLIAMSON 1988). Ty1 RT activity monitored in VLPs in vitro is maximal at about 15°C, but is barely detectable at 37°C (GARFINKEL et al. 1985), suggesting an explanation for the observed temperature dependence of transposition. Transposition of Ty3 was not grossly different between 23°C and at 30°C in a qualitative colony hybridization assay (HANSEN et al 1988), but was greatly reduced at 37°C. Ty3 RT is also temperature sensitive with an optimum of 25°C (HANSEN et al. 1992).

Recent investigation of the state of Ty3 VLP proteins and DNA in extracts of cells grown under inducing conditions at 30°C and 37°C showed that cells grown at 30°C contained similar amounts of the 38-kDa *GAG3* protein and 26-kDa mature CA protein, but that cells grown at 37°C contained about one-fifth that amount of Gag3 precursor and no detectable CA protein. DNA which was readily detectable in 30°C grown cells was not detectable in the 37°C grown cells. A temperature shift experiment showed that particle turnover occurred with a half-life of only about 40 min in cells at 37°C and greater than 3 h at 30°C. These observations prompted an investigation into the nature of the temperature restriction of Ty3 transposition. A mild state of heat shock is produced by growth of yeast cells at 37°C (CRAIG et al. 1993; PARSELL and LINDQUIST 1993). In order to determine whether the stress response was responsible for the low level of Ty3 particle formation at 37°C, cells were stressed by treatment with 1.55 *M* ethanol (PLESSET et al. 1982) or genetically in the *ssa1 ssa2* mutant background (STONE and CRAIG 1990). These cells also exhibited low levels of Gag3 precursor proteins and no mature CA, suggesting that the low level of Ty3 particle formation in cells grown at 37°C reflects the elevated levels of stress response proteins, rather than physical denaturation of the particle (MENEES and SANDMEYER 1996). Growth of *ssa1 ssa2* cells at 23°C, a condition which causes increased expression of the *SSB1* and *SSB2* genes for the cytosolic hsp70s (CRAIG and JACOBSEN 1985), rescued the Ty3 particle formation defect. It has been suggested that state of the *ssa1 ssa2* mutant represents a relative deficit of chaperone assembly function relative to stress response degradation functions (CRAIG et al. 1993). The formation of Ty3 particles at 23°C is consistent with this interpretation, although other explanations (such as increased stability of Ty3 VLPs at 23°C) are also reasonable. The observation that Ty3 is a substrate for turnover by the stress response suggests that it may also act an inducer of stress response proteins and that it is a substrate for ubiquitination. The effect of Ty3 on stress response protein production was monitored using antibodies against Hsp104p (a gift from S. Lindquist, University of Chicago). *HSP104* is readily induced by heat shock and has been shown by in vitro assay to have ATP-dependent disaggregating activity (PARSELL et al. 1994). Ty3 expression under galactose induction increased the level of Hsp104p relative to control cells grown in galactose (Menees and Sandmeyer, unpublished data). Ubiquitin-removing proteases can redirect proteins from degradative pathways in the cell (Baker et al. 1992). In order to test whether Ty3 in *ssa1 ssa2* cells is targeted for degradation by ubiquitination, the effect of overexpressing *UBP3*, the gene for a ubiquitin-removing protease (a gift from E. Craig, University of Wisconsin), on Ty3 particle formation was determined. Expression of *UBP3* elevated levels of Ty3 Gag3 and CA, and promoted reverse transcription in the *ssa1 ssa2* mutants expressing Ty3. These observations showed that Ty3 particles are both an inducer and a substrate for the stress response system. In addition, these experiments suggest that retrotransposons provide access to genetic systems with which to study the positive and negative roles of chaperones and ubiquitin in retroviruslike replication.

6 Application of Particles Based on Heterologous Ty Fusions

The facility with which high levels of Ty proteins can be expressed with the concomitant formation of stable intracellular particles has had several novel applications. As early as 1987 Adams et al (1987b) and Clare et al. (1988) showed that the reading frames of Ty1 could be fused to heterologous coding sequences and stable fusion proteins expressed. Adams et al. showed that when these fusion proteins included the Ty1 capsid domain, particles resulted which were effective immunogens. The utility of these polyvalent particles in eliciting cellular and humoral immune responses has been reviewed recently (Adams et al. 1995). In addition, fusion of proteins to Ty1 capsid have been used ingeniously to enhance detection and perhaps activity of reverse transcriptases including those of the LINE-like element CREI from the trypanosomatid *Crithidia fasciculata* (Gabriel and Boeke 1991), the human LINE element L1.2A (Mathias et al. 1991), the *coxI* intron i1 from the fungus *Podospora anserina* (Fassbender et al. 1994), and duck hepatitis virus B (Tavis et al. 1994). In each case, conventional techniques had failed to demonstrate activity. It is clear that expression in yeast allowed high levels of protein to be produced and the multimerization mediated by Ty1 capsid facilitated concentration and purification of the protein for testing. Fortuitous features of this system are that substrates can diffuse readily into the particles and the proteins are relatively stable and therefore may be protected from proteolysis. It may also be that this context facilitates folding and multimerization by concentrating proteins within the cell. Whether these advantages are greater for RT, which naturally occurs in this context, or whether they could also apply to other proteins, the enzymatic assay of which has been problematic, remains to be tested. That non-RT species can be active within particles is suggested by the activity of the nuclease barnase in the Ty1 particle context

(Natsoulis and Boeke 1991). Fusion of barnase to Ty1 capsid was shown to have a dominant negative effect on activity of a Ty1 element expressed in *trans*. The long-range application of targeting heterologous enzymatic activities to retroviral particles is clear. These experiments have led directly to the demonstration of model retroviral capsid-nuclease fusions as antivirals (Natsoulis et al. 1995).

7 Prospects

Genomic sequencing has only recently begun to suggest the extent of the representation of LTR retrotransposons in eukaryotic organisms. The study of these elements is in its infancy compared to investigations of retroviruses. Information about LTR retrotransposons has already posed interesting questions about the structure-function relationships of some retroviral proteins. In addition, it may provide insights into the evolution of retroviral mechanisms of replication which complement findings in retroviral systems. The coexistence of retrotransposons and retroviruses within single organisms like fruit flies and fish and the close relatedness of the gypsylike elements and viruses implies a potentially more direct interaction between these evolving types of retroelements than previously appreciated. Finally, the molecular characterization of particle morphogenesis of some of the elements described here will facilitate the application of genetics in fruit flies and yeast to the problem of the role of host proteins in particle morphogenesis.

In addition to common objectives with retrovirology, retrotransposon biology has separate objectives. The widespread nature of retrotransposons suggests that they may have significant cellular and genomic impact independent of their relationship to retroviruses, for example, the contribution of reverse transcriptase at different times in cellular development, the latent sculpting of target genomes by elements with insertion specificity, and the production of intracellular particles which may affect maintenance of the host cell stress response. Retrotransposons as naturally minimalist systems may have particular utility for applications, such as the design of retrovirus vectors with specificity or antiviral properties, the generation of effective antiviral immune responsiveness, or the study of catalytic proteins from pathogenic organisms. These developing avenues of investigation will continue to require investigation of the central problem of morphogenesis of the VLP structure.

References

Adams SE, Mellor J, Gull K, Sim RB, Tuite MF, Kingsman SM, Kingsman AJ (1987a) The functions and relationships of Ty-VLP proteins in yeast reflect those of mammalian retroviral proteins. Cell 49: 111–119

Adams SE, Dawson KM, Gull K, Kingsman SM, Kingsman AJ (1987b) The expression of hybrid HIV: Ty virus-like particles in yeast. Nature 329: 68–70

Adams SE, Burns NR, Layton GT, Kingsman AJ (1995) Hybrid Ty virus-like particles. Int Rev Immunol 11: 133–141

Arkhipova IR, Mazo AM, Cherkasova VA, Gorelova TV, Schuppe NG, Ilyin YV (1986) The steps of reverse transcription of *Drosophila* mobile dispersed genetic elements and U3-R-U5 structure of their LTRs. Cell 44: 555–563

Atwood A, Lin J-H, Levin HL (1996) The retrotransposon Tf1 assembles virus-like particles that contain excess Gag relative to integrase because of a regulated degradation process. Mol cell Biol 16: 338 346

Baker RT, Tobias JW, Varshavsky A (1992) Ubiquitin-specific proteases of *Saccharomyces cerevisiae*. J Biol Chem 267: 23364–23375

Barat C, Lullien V, Schatz O, Keith G, Nugeyre MT, Gruninger-Leitch F, Barre-Sinoussi F, LeGrice SF, Darlix JL (1989) HIV-1 reverse transcriptase specifically interacts with the anticodon domain of its cognate primer tRNA. EMBO J 8: 3279–3285

Belcourt MF, Farabaugh PJ (1990) Ribosomal frameshifting in the yeast retrotransposon Ty: tRNAs induce slippage on a 7 nucleotide minimal site. Cell 62: 339–352

Bilanchone VW, Claypool JA, Kinsey PT, Sandmeyer SB (1993) Positive and negative regulatory elements control expression of the yeast retrotransposon Ty3. Genetics 134: 685–700

Bingham PM, Zachar Z (1989) Retrotransposons and the FB transposon from *Drosophila melanogaster*. In: Berg DE, Howe MM (eds) Mobile DNA. American Society for Microbiology, Washington DC, pp 485–502

Boeke JD, Corces VG (1989) Transcription and reverse transcription in retrotransposons. Annu Rev Microbiol 43: 402–433

Boeke JD, Sandmeyer SB (1991) Yeast transposable elements. In: Broach JR, Jones EW, Pringle J (eds) The molecular and cellular biology of the yeast *Saccharomyces cerevisiae*. Cold Spring Harbor Laboratory, Cold Spring Harbor, pp 193–261

Boeke JD, Garfinkel DJ, Styles CA, Fink GR (1985) Ty elements transpose through an RNA intermediate. Cell 40: 491–500

Brierley C, Flavell AJ (1990) The retrotransposon copia controls the relative levels of its gene products post-transcriptionally by differential expression from its two major mRNAs. Nucleic Acid Res 18: 2947–2951

Britten RJ (1995) Active *gypsy/Ty3* retrotransposons or retroviruses in *Caenorhabditis elegans*. Proc Natl Acad Sci USA 92: 599–601

Brookman JL, Stott AJ, Cheeseman PJ, Burns NR, Adams SE, Kingsman AJ, Gull K (1995) An immunological analysis of Ty1 virus-like particle structure. Virology 207: 59–67

Burns NR, Saibil HR, White NS, Pardon JF, Timmins PA, Mark S, Richardson H, Richards BM, Adams SE, Kingsman SM et al (1992) Symmetry, flexibility and permeability in the structure of yeast retrotransposon virus-like particles. EMBO J 11: 1155–1164

Chapman KB, Byström AS, Boeke JD (1992) Initiator methionine tRNA is essential for Ty1 transposition in yeast. Proc Natl Acad Sci USA 89: 3236–3240

Clare J, Farabaugh P (1985) Nucleotide sequence of a yeast Ty element: evidence for an unusual mechanism of gene expression. Proc Natl Acad Sci USA 82: 2829–2833

Clare JJ, Belcourt M, Farabaugh PJ, (1988) Efficient translational frameshifting occurs within a conserved sequence of the overlap between the two genes of a yeast Ty1 transposon. Proc Natl Acad Sci USA 85: 6816–6820

Coffin JM (1990) *Retroviridae* and their replication. In: Fields BN, Knipe DM (eds) Virology, 2nd edn. Raven, New York, pp 1437–1500

Craig EA, Jacobsen K (1985) Mutations in cognate genes of *Saccharomyces cerevisiae hsp 70* result in reduced growth rates at low temperatures. Mol Cell Biol 5: 3517–3524

Craig EA, Gambill BD, Nelson RJ (1993) Heat shock proteins: molecular chaperones of protein biogenesis. Microbiol Rev 57: 402–414

Curcio MJ, Garfinkel DJ (1991a) Regulation of retrotransposition in *Saccharomyces cerevisiae*. Mol Microbiol 5: 1823–1829
Curcio MJ, Garfinkel DJ (1991b) Single-step selection for Ty1 element retrotransposition. Proc Natl Acad Sci USA 88: 936–940
Curcio MJ, Garfinkel DJ (1992) Posttranslational control of Ty1 retrotransposition occurs at the level of protein processing. Mol Cell Biol 12: 2813–2825
Curcio MJ, Garfinkel DJ (1994) Heterogeneous functional Ty1 elements are abundant in the *Saccharomyces cerevisiae* genome. Genetics 136: 1245–1259
Curcio MJ, Sanders NJ, Garfinkel DJ (1988) Transpositional competence and transcription of endogenous Ty elements in *Saccharomyces cerevisiae*: implications for regulation of transposition. Mol Cell Biol 8: 3571–3581
Dinman JD, Icho T, Wickner RB (1991) A –1 ribosomal frameshift in a double-stranded RNA virus of yeast forms a *gag-pol* fusion protein. Proc Natl Acad Sci USA 88: 174–178
Eichinger DJ, Boeke JD (1988) The DNA intermediate in yeast Ty1 element transposition copurifies with virus-like particles: cell-free Ty1 transposition. Cell 54: 955–966
Eichinger DJ, Boeke JD (1990) A specific terminal structure is required for Ty1 transposition. Genes Dev 4: 324–330
Emori Y, Shiba S, Inouye I, Yuki S, Saigo K (1985) The nucleotide sequences of copia and copia-related RNA in *Drosophila* virus-like particles. Nature 315: 773–776
Evgen'ev MB, Corces VG, Lankenau D-H (1992) Ulysses transposable element of *Drosophila* shows high structural similarities to functional domains of retroviruses. J Mol Biol 225: 917–924
Falkenthal S, Lengyel JA (1980) Structure, translation, and metabolism of the cytoplasmic copia ribonucleic acid of *Drosophila melanogaster*. Biochemistry 19: 5842–5850
Farabaugh PJ, Zhao H, Pande S, Vimaladithan A (1993) Translational frameshifting expresses the *POL3* gene of retrotransposon Ty3 of yeast. Cell 74: 93–103
Fassbender S, Bruhl K-H, Ciriacy M, Kuck U (1994) Reverse transcriptase activity of an intron encoded polypeptide. EMBO J 13: 2075–2083
Fields S, Song O (1989) A novel genetic system to detect protein-protein interactions. Nature 340: 245–246
Finnegan DJ (1989) F and related elements in *Drosophila melanogaster*. In: Berg DE, Howe MM (eds) Mobile DNA. American Society for Microbiology, Washington DC, pp 519–529
Flavell AJ (1984) Role of reverse transcription in the generation of extrachromosomal copia mobile genetic elements. Nature 310: 514–516
Flavell AJ (1992) Ty1-copia group retrotransposons and the evolution of retroelements in the eukaryotes. Genetica 86: 203–214
Flavell AJ, Ish-Horowicz D (1981) Extrachromosomal circular copies of the eukaryotic transposable element *copia* in cultured *Drosophila cells*. Nature 292: 591–595
Flavell AJ, Ish-Horowicz D (1983) The origin of extrachromosomal circular *copia* elements. Cell 34: 415–419
Flavell AJ, Smith DB (1992) A Ty1-copia group retrotransposon sequence in a vertebrate. Mol Gen Genet 233: 322–326
Flavell AJ, Ruby SW, Toole JJ, Roberts BE, Rubin GM (1980) Translation and developmental regulation of RNA encoded by the eukaryotic transposable element copia. Proc Natl Acad Sci USA 77: 7107–7111
Flavell AJ, Dunbar E, Anderson R, Pearce SR, Hartly R, Kumar A (1995) Ty1-copia group retrotransposons are ubiquitous and heterogeneous in higher plants. Nucleic Acids Res 94: 3639–3644
Friesen PD, Nissen MS (1990) Gene organization and transcription of TED, a lepidopteran retrotransposon integrated within the baculovirus genome. Mol Cell Biol 10: 3067–3077
Fuetterer J, Hohn T (1987) Involvement of nucleocapsids in reverse transcription: a general phenomenon? Trends Biochem Res 12: 92–95
Fulton AM, Mellor J, Dobson MJ, Chester J, Warmington JR, Indge KJ, Oliver SG, de la PP, Wilson W, Kingsman AJ et al (1985) Variants within the yeast Ty sequence family encode a class of structurally conserved proteins. Nucleic Acids Res 13: 4097–4112
Furuichi T, Inouye S, Inouye M (1987) Biosynthesis and structure of stable branched RNA covalently linked to the 5′ end of multicopy single-stranded DNA of *Stigmatella auranticaca*. Cell 48: 55–62
Gabriel A, Boeke JD (1991) Reverse transcriptase encoded by a retrotransposon from the trypanosomatid *Crithidia fasciculata*. Proc Natl Acad Sci USA 88: 9794–9798

Garfinkel DJ, Boeke JD, Fink GR (1985) Ty element transposition: reverse transcriptase and virus-like particles. Cell 42: 507–517
Garfinkel DJ, Hedge AM, Youngren SD, Copeland TD (1991) Proteolytic processing of pol-TYB proteins from the yeast retrotransposon Ty1. J Virol 65: 4573–4581
Grandbastien M-A, Spielmann A, Caboche M (1989) Tnt1, a mobile retroviral-like transposable element of tobacco isolated by plant cell genetics. Nature 337: 376–380
Hansen LJ, Sandmeyer SB (1990) Characterization of a transpositionally active Ty3 element and identification of the Ty3 integrase protein. J Virol 64: 2599–2607
Hansen LJ, Chalker DL, Sandmeyer SB (1988) Ty3, a yeast retrotransposon associated with tRNA genes, has homology to animal retroviruses. Mol Cell Biol 8: 5245–5256
Hansen LJ, Chalker DL, Orlinsky KJ, Sandmeyer SB (1992) Ty3 *GAG3* and *POL3* genes encode the components of intracellular particles. J Virol 66: 1414–1424
Hauber J, Nelböck-Hochstetter P, Feldmann H (1985) Nucleotide sequence and characteristics of a Ty element from yeast. Nucleic Acids Res 13: 2745–2758
Hu JC, Dahlberg JE (1983) Structural features required for the binding of $tRNA^{Trp}$ to avian myeloblastosis virus reverse transcriptase. Nucleic Acids Res 11: 4823–4833
Inouye S, Yuki S, Saigo K (1986) Complete nucleotide sequence and genome organization of a *Drosophila* transposable genetic element, 297. Eur J Biochem 154: 417–425
Jacks T (1989) Translational suppression in gene expression in retroviruses and retrotransposons. In: Swanstrom R, Vogt PK (eds) Retroviruses. Strategies of replication. Springer, Berlin Heidelberg New York, pp 93–124 (Current topics in microbiology and immunology, vol 157)
Jacks T, Townsley K, Varmus HE, Majors J (1987) Two efficient ribosomal frameshifting events are required for synthesis of mouse mammary tumor virus *gag*-related polyproteins. Proc Natl Acad Sci USA 84: 4298–4302
Jacks T, Power MD, Masiarz FR, Luciw PA, Barr PJ, Varmus HE (1988) Characterization of ribosomal frameshifting in HIV-1 *gag-pol* expression. Nature 331: 280–283
Janetzky B, Lehle L (1992) Ty4, a new retrotransposon from *Saccharomyces cerevisiae*, flanked by tau-elements. J Biol Chem 267: 19798–19805
Kawakami K, Pande S, Faiola B, Moore DP, Boeke JD, Farabaugh PJ, Strathern JN, Nakamura Y, Garfinkel DJ (1993) A rare tRNA-Arg(CCU) that regulates Ty1 element ribosomal frameshifting is essential for Ty1 retrotranspositon in *Saccharomyces cerevisiae*. Genetics 135: 309–320
Keeney JB, Chapman KB, Lauermann V, Voytas DF, Åstrom SU, von Pawel-Rammingen U, Byström A, Boeke JD (1995) Multiple molecular determinants for retrotransposition in a primer tRNA. Mol Cell Biol 15: 217–226
Kikuchi Y, Ando Y, Shiba T (1986) Unusual priming mechanism of RNA-directed DNA synthesis in copia retrovirus-like particles of *Drosophila*. Nature 323: 824–826
Kim A, Terzian C, Santamaria P, Pélisson A, Prud'homme N, Bucheton A (1994) Retroviruses in invertebrates: the gypsy retrotransposon is apparently an infectious retrovirus of *Drosophila melanogaster*. Proc Natl Acad Sci USA 91: 1285–1289
Kinsey P, Sandmeyer S (1995) Ty3 transposes in mating populations of yeast: a novel transposition assay for Ty3. Genetics 139: 81–94
Kirchner J, Sandmeyer S (1993) Proteolytic processing of Ty3 proteins is required for transposition. J Virol 67: 19–28
Kirchner J, Sandmeyer SB (1996) Ty3 integrase cleaves a dinucleotide from the 3′ termini of Ty3 DNA and may have a role in reverse transcription: analysis of integrase mutations. J Virol (in press)
Kirchner J, Sandmeyer SB, Forrest DB (1992) Transposition of Ty3 *GAG3-POL3* fusion mutant is limited by availability of capsid protein. J Virol 66: 6081–6092
Lankenau DH, Huijser P, Jansen E, Miedema K, Hennig W (1988) Micropia: a retrotransposon of *Drosophila* combining structural features of DNA viruses, retroviruses and non-viral transposable elements. J Mol Biol 204: 233–246
Lauermann V, Boeke JD (1994) The primer tRNA sequence is not inherited during Ty1 transposition. Proc Natl Acad Sci USA 91: 9847–9851
Leis J, Baltimore D, Bishop JM, Coffin J, Fleissner E, Goff SP, Oroszlan S, Robinson H, Skalka AM, Temin HM et al. (1988) Standardized and simplified nomenclature for proteins common to all retroviruses. J Virol 62: 1808–1809
Levin HL (1995) A novel mechanism of self-primed reverse transcription defines a new family of retroelements. Mol Cell Biol 15: 3310–3317
Levin HL, Boeke JD (1992) Demonstration of retrotransposition of the Tf1 element in fission yeast. EMBO J 11: 1145–1153

Levin HL, Weaver DC, Boeke JD (1990) Two related families of retrotransposons from *Schizosaccharomyces pombe*. Mol Cell Biol 10: 6791–6798

Levin HL, Weaver DC, Boeke JD (1993) Novel gene expression mechanism in a fission yeast retroelement: Tf1 proteins are derived from a single primary translation product. EMBO J 12: 4885–4895

Mammano F, hagen Å, Hoglund S, Göttlinger HG (1994) Role of the major homology region of human immunodeficiency virus type 1 in virion morphogenesis. J Virol 68: 4927–4936

Marlor RL, Parkhurst SM, Corces VG (1986) The *Drosophila melanogaster* gypsy transposable element encodes putative gene products homologous to retroviral proteins. Mol Cell Biol 6: 1129–1134

Mathias SL, Scott AF, Kazazian Jr, HH, Boeke JD, Gabriel A (1991) Reverse transcriptase encoded by a human transposable element. Science 254: 1808–1810

McHale MT, Roberts IN, Noble SM, Beaumont C, Whitehead MP, Seth D, Oliver RP (1992) CFT-I: an LTR-retrotransposon in *Cladosporium fulvum*, a fungal pathogen of tomato. Mol Gen Genet 233: 337–347

Mellor J, Fulton AM, Dobson MJ, Roberts NA, Wilson W, Kingsman AJ, Kingsman SM (1985a) The Ty transposon of *Saccharomyces cerevisiae* determines the synthesis of at least three proteins. Nucleic Acids Res 13: 6249–6263

Mellor J, Malim MH, Gull K, Tuite MF, McCready S, Dibbayawan T, Kingsman SM, Kingsman AJ (1985b) Reverse transcriptase activity and Ty RNA are associated with virus-like particles in yeast. Nature 318: 583–586

Menees TM, Sandmeyer SB (1994) Transposition of the yeast retroviruslike element Ty3 is dependent on the cell cycle. Mol Cell Biol 14: 8229–8240

Menees TM, Sandmeyer SB (1996) Ubiquitination and inhibition of yeast retroviruslike element Ty3 particle formation. Proc Natl Acad Sci USA (in press)

Miller K, Rosenbaum J, Zbrzezna V, Pogo AO (1995) The nucleotide sequence of *Drosophila melanogaster* copia-specific 2.1 kb mRNA. Nucleic Acids Res 17: 2134

Miyake T, Mae N, Shiba T, Kondo S (1987) Production of virus-like particles by the transposable genetic element, *copia*, of *Drosophila melanogaster*. Mol Gen Genet 207: 29–37

Mount SM, Rubin GM (1985) Complete nucleotide sequence of the *Drosophila* transposable element copia: homology between copia and retroviral proteins. Mol Cell Biol 5: 1630–1638

Müller F, Bruhl KH, Freidel K, Kowallik KV, Ciriacy M (1987) Processing of Ty1 proteins and formation of Ty1 virus-like particles in *Saccharomyces cerevisiae*. Mol Gen Genet 207: 421–429

Müller F, Laufer W, Pott U, Ciriacy M (1991) Characterization of products of Ty1-mediated reverse transcription in *Saccharomyces cerevisiae*. Mol Gen Genet 226: 145–153

Natsoulis G, Boeke JD (1991) New antiviral strategy using capsid-nuclease fusion proteins. Nature 352: 632–635

Natsoulis G, Seshaiah P, Federspiel MJ, Rein A, Hughes SH, Boeke JD (1995) Targeting of a nuclease to murine leukemia virus capsids inhibits viral multiplication. Proc Natl Acad Sci USA 92: 364–368

Oliver SG et al (1992) The complete DNA sequence of yeast chromosome III. Nature 357: 38–46

Orlinsky KJ, Sandmeyer SB (1994) The Cys-His motif of Ty3 NC can be contributed by Gag3 or Gag3–Pol3 polyproteins. J Virol 68: 4152–4166

Orlinsky KJ, Gu J, Hoyt M, Sandmeyer SB, Menees TM (1996) Mutations in the Ty3 major homology region affect multiple steps in Ty3 retrotransposition. J Virol (in press)

Paquin CE, Williamson VM (1988) Effect of temperature on Ty transposition. Ban Rpt 30: 235–243

Parsell DA, Lindquist S (1993) The function of heat-shock proteins in stress tolerance: degradation and reactivation of damaged proteins. Annu Rev Genet 27: 437–496

Parsell DA, Kowal AS, Singer MA, Lindquist S (1994) Protein disaggregation mediated by heat–shock protein Hsp 104. Nature 372: 475–478

Pélisson A, Song SU, Prud'homme N, Smith PA, Bucheton A, Corces VG (1994) Gypsy transposition correlates with the production of a retroviral envelope-like protein under the tissue-specific control of the *Drosophila flamenco* gene. EMBO J 13: 4401–4411

Plesset J, Palm C, McLaughlin CS (1982) Induction of heat shock proteins and thermotolerance by ethanol in *Saccharomyces cerevisiae*. Bioc Biophy Res Comm 108: 1340–1345

Pochart P, Agoutin B, Rousset S, Chanet R, Doroszkiewicz V, Heyman T (1993a) Biochemical and electron microscope analyses of the DNA reverse transcripts present in the virus-like particles of the yeast transposon Ty1. Identification of a second origin of Ty1 DNA plus strand synthesis. Nucleic Acids Res 21: 3513–3520

Pochart P, Agoutin B, Fix C, Keith G, Heyman T (1993b) A very poorly expressed $tRNA^{Ser}$ is highly concentrated together with replication primer initiator $tRNA^{Met}$ in the yeast Ty1 virus-like particles. Nucleic Acids Res 21: 1517–1521

Saigo K, Kugimiya W, Matsuo Y, Inouye S, Yoshioka K, Yuki S (1984) Identification of the coding sequence for a reverse transcriptase-like enzyme in a transposable genetic element in *Drosophila melanogaster*. Nature 312: 659–661

Scheinker VS, Lozovskaya ER, Bishop JG, Corces VG, Evgen'ev MB (1990) A long terminal repeat-containing retrotransposon is mobilized during hybrid dysgenesis in *Drosophila virilis*. Proc Natl Acad Sci USA 87: 9615–9619

Sharon G, Burkett TJ, Garfinkel DJ, (1994) Efficient homologous recombination of Ty1 element cDNA when integration is blocked. Mol Cell Biol 14: 6540–6541

Shiba T, Saigo K (1983) Retrovirus-like particles containing RNA homologous to the transposable element copia in *Drosophila melanogaster*. Nature 302: 119–124

Smyth DR, Kalitsis P, Joseph JL, Sentry JW (1989) Plant retrotansposon from *Lilium henryi* is related to Ty3 of yeast and the gypsy group of *Drosophila*. Proc Natl Acad Sci USA 86: 5015–5019

Song Su, Gerasimova T, Kurkulos M, Boeke JD, Corces V (1994) An Env-like protein encoded by a *Drosophila* retroelement: evidence that *gypsy* is an infectious retrovirus. Genes Dev 8: 2046–2057

Springer MS, Britten RJ (1993) Phylogenetic relationships of reverse transcriptase and RNase H sequences and aspects of genome structure in the gypsy group of retrotransposons. Mol Biol Evol 10: 1370–1379

Springer MS, Davidson EH, Britten RJ (1991) Retroviral-like element in a marine invertebrate. Proc Natl Acad Sci USA 88: 8401–8404

Stone DE, Craig EA (1990) Self-regulation of 70–kilodalton heat shock proteins in *Saccharomyces cerevisiae*. Mol Cell Biol 10: 1622–1632

Strambio-de-Castillia C, Hunter E (1992) Mutational analysis of the major homology region of Mason-Pfizer monkey virus by use of saturation mutagenesis. J Virol 66: 7021–7032

Stucka R, Lochmüller H, Feldmann H (1989) Ty4, a novel low-copy number element in *Saccharomyces cerevisiae*: one copy is located in a cluster of Ty elements and tRNA genes. Nucleic Acids Res 17: 4993–5001

Syomin BV, Kandror KV, Semakin AB, Tsuprun VL, Stepanov AS (1993) Presence of the gypsy (MDG4) retrotransposon in extracellular virus-like particles. FEBS Lett 323: 285–288

Tanda S, Shrimpton AE, Chueh LL, Itayama H, Matsubayashi H, Saigo K, Tobari YN, Langley CH (1988) Retrovirus-like features and site specific insertions of a transposable element, tom, in *Drosophila ananassae*. Mol Gen Genet 214: 405–411

Tanda S, Mullor J, Corces VG (1994) The *Drosophila* tom retrotransposon encodes an envelope protein. Mol Cell Biol 14: 5392–5401

Tavis JE, Perri S, Ganem D (1994) Hepadnavirus reverse transcription initiates within the stem-loop of the RNA packaging signal and employs a novel strand transfer. J Virol 68: 3536–3543

Tchenio T, Heidmann T (1991) Defective retroviruses can disperse in the human genome by intracellular transposition. J Virol 65: 2113–2118

Telesnitsky A, Goff SP (1993) Strong-stop strand transfer during reverse transcription. In: Skalka AM, Goff SP (eds) Reverse transcriptase. Cold Spring Harbor Laboratory Press, Cold Spring Harbor, pp 49–83

Vimaladithan A, Farabaugh PJ (1994) Special peptidyl-tRNA molecules can promote transitional frameshifting without slippage. Mol Cell Biol 14: 8107–8116

Voytas DF, Ausubel FM (1988) A copia-like transposable element family in *Arabidopsis thaliana*. Nature 336: 242–244

Voytas DF, Boeke JD (1992) Yeast retrotransposon revealed. Nature 358: 717

Voytas DF, Boeke JD (1993) Yeast retrotransposons and tRNAs. TIGS 9: 421–427

Warmington JR, Waring RB, Newlon CS, Indge KJ, Oliver SG (1985) Nucleotide sequence characterization of Ty1–17, a class II transposon from yeast. Nucleic Acids Res 13: 6679–6693

Wills JW, Craven RC (1991) Form, function, and use of retroviral gag proteins (editorial). AIDS 5: 639–654

Xiong Y, Eickbush TH (1990) Origin and evolution of retroelements based upon their reverse transcriptase sequences. EMBO J 9: 3353–3362

Xu H, Boeke JD (1990a) Host genes that influence transposition in yeast: the abundance of a rare tRNA regulates Ty1 transposition frequency. Proc Natl Acad Sci USA 87: 8360–8364

Xu H, Boeke JD (1990b) Localization of sequences required in cis for yeast Ty1 element transposition near the long terminal repeats: analysis of mini-Ty1 elements. Mol Cell Biol 10: 2695–2702

Xu H, Boeke JD (1991) Inhibition of Ty1 transposition by mating pheromones in *Saccharomyces cerevisiae*. Mol Cell Biol 11: 2736–2743

Yoshioka K, Honma H, Zushi M, Kondo S, Togashi S, Miyake T, Shiba T (1990) Virus-like particle formation of *Drosophila copia* through autocatalytic processing. EMBO J 9: 535–541

Yoshioka K, Fujita A, Kondo S, Miyake T, Sakaki Y, Shiba T (1992) Production of a unique multi-lamella structure in the nuclei of yeast expressing *Drosophila copia gag* precursor. FEBS Lett 302: 5–7

Youngren SD, Boeke JD, Sanders NJ, Garfinkel DJ (1988) Functional organization of the retrotransposon Ty from *Saccharomyces cerevisiae:* Ty protease is required for transposition. Mol Cell Biol 8: 1421–1431

Yuki S, Inouye S, Ishimaru S, Saigo K (1986) Nucleotide sequence characterization of a *Drosophila* retrotransposon, 412. Eur J Biochem 158: 403–410

Zou S, Wright DA, Voytas DF (1995) The *Saccharomyces* Ty5 retrotransposon family is associated with origins of DNA replication at the telomeres and the silent mating locus *HMR*. Proc Natl Acad Sci USA 92: 920–924

Zou S, Ke N, Kim JM, Voytas DF (1996) The *Saccharomyces* retrotransposon Ty5 integrates preferentially into regions of silent chromatin at the telomeres and mating loci. Genes Dev 10: 634–645

Hepatitis B Virus Morphogenesis

M. NASSAL

1 Introduction

Hepatitis B viruses, or hepadnaviruses (*hepa*totropic *DNA* viruses), comprise a family of small enveloped DNA viruses that replicate through reverse transcription of an RNA intermediate; their replication cycle is hence a cyclic permutation of that of retroviruses which are RNA viruses replicating through a DNA intermediate. Hepadnaviruses are characterized by narrow host range and pronounced liver tropism. The type member, and causative agent of B-type hepatitis

Center for Molecular Biology, University of Heidelberg, Im Neuenheimer Feld 282, 69120 Heidelberg, Germany

in man, hepatitis B virus (HBV), infects only humans and higher primates; related viruses have been found in other mammals like the woodchuck (WHV), or the ground squirrel (GSHV), and also in birds like the Pekin duck (DHBV) or the grey heron (HHBV).

In addition to their unique replication properties hepadnaviruses are of considerable medical importance, as HBV is still one of the major human pathogens (Lok 1994): in adults some 5%–10%, in young children up to 90% of acute infections progress into chronic hepatitis, with a highly increased risk for liver cirrhosis and eventually primary liver carcinoma. The number of chronic virus carriers is estimated to exceed 300 million worldwide. Since currently no generally applicable therapy is available, understanding the HBV life-cycle in detail is a vital prerequisite for the development of new antiviral strategies, including inhibitors of virus morphogenesis.

One of the main experimental obstacles for such an analysis is the lack of infectable cell lines. Apart from the much restricted possibility of employing in vitro infection of primary human hepatocytes (Gripon et al. 1993; Galle et al. 1994), the major tools available are the separate expression of those viral gene products that can be obtained in biologically active form from heterologous systems, transfection of cloned HBV DNA into suitable liver cell lines (yielding complete virions), and the use of the animal hepadnaviruses as models. Each of these systems has yielded valuable information regarding the individual viral components and some of their interactions in the process of hepatitis B virus morphogenesis; however, we are still far from a complete molecular understanding of how the infectious hepatitis B virion is formed.

Below I will focus on the progress that has been made during the last few years concerning the supramolecular structures of viral capsids, envelopes and virions. Before that I will briefly survey the major differences between hepadna- and retroviruses, and give a short update on the basic replication cycle of hepatitis B viruses; this appears adequate as hepadnaviral replication and assembly are extremely tightly coupled. The reader interested in the general molecular biology of hepatitis B viruses is referred to previously published reviews by Ganem and Varmus (1987) and monographs edited by McLachlan (1991) and, in this series, Mason and Seeger (1991), which also contain chapters covering the topic of this review; more recent articles concerning replication (Nassal and Schaller 1993a), capsid assembly (Nassal and Schaller 1993b) and the core protein (Seifer and Standring 1995) are also available.

2 Hepatitis B Viruses Versus Retroviruses – A Short Overview

Reverse transcription as replication principle, and the presence and to some extent order of the genes for the principal components Gag-Pol-Env (termed Core-P-preS/S in hepadnaviruses) are common characteristics shared by hepatitis B and retroviruses. However, many aspects are distinctly different in both virus families (NASSAL and SCHALLER 1993a; ROTHNIE et al. 1994), partly due to the extremely small size of the hepadnaviral genome (3.0–3.2 kb), and the need to efficiently exploit this restricted genetic space as evidenced, for instance, by the largely overlapping arrangement of both coding regions and regulatory elements (Fig. 1A). In the hepadnaviruses, extracellular virions contain DNA rather than RNA; integration is not an obligatory step in replication; functional mRNAs are produced from several internal promoters on the circular DNA genome rather than one LTR-based promoter, and RNA splicing does not appear to play a critical role in the basic replication cycle of at least the mammalian viruses. Although another distinction, that hepadnaviruses could synthesize their first DNA strand in a continuous fashion from a 3′-proximal origin on the RNA template, had to be given up in favour of a mechanism more closely resembling retroviral reverse transcription (see below), the differences are not confined to the genetic level; rather, they extend to the strategies employed to bring together the viral gene products that make up the infectious enveloped virion, i.e. the capsid, P and envelope proteins, and the genomic RNA. First, there is no evidence that any of the protein products requires processing; second, envelopment appears to be necessary for export of viral capsids; third, this process occurs at an internal rather than the plasma membrane; fourth, P protein, not core protein, mediates specific RNA pregenome packaging; fifth, this protein/RNA interaction rather than a protein/protein interaction is also responsible for P encapsidation. Finally, all hepadnaviruses secrete empty envelopes, termed subviral or S particles.

3 Genetic Organization of Hepadnaviruses

Figure 1A shows the genome organisation of HBV and DHBV. The HBV genome contains four major open reading frames (ORFs): preC/C, P, preS1/S2/S and X. The X gene product encodes a pleiotropically acting transcriptional activator which in the WHV model has been shown to be required for the establishment of infection (CHEN et al. 1993). Despite intensive efforts, its mode of action has not been finally established.

The preC/C ORF codes for two distinct products: one is the core protein forming the protein shell of the nucleocapsid, the other, made by translation of the joint preC/C ORF, is the precore protein which is targeted into the cell's secretory pathway, processed at both ends and eventually found in the serum of infected individuals as HBeAg. Both products are translated from genomic, terminally redundant 3.5-kb transcripts with slightly different 5′-ends. The longer precore mRNAs contain the preC initiation codon, the shorter core mRNA lacks it. The P-ORF covers some 80% of the genome and encodes the viral replication enzyme P, which is also an indispensable component in the assembly process (see below). P protein is translated from the same genomic RNA that directs

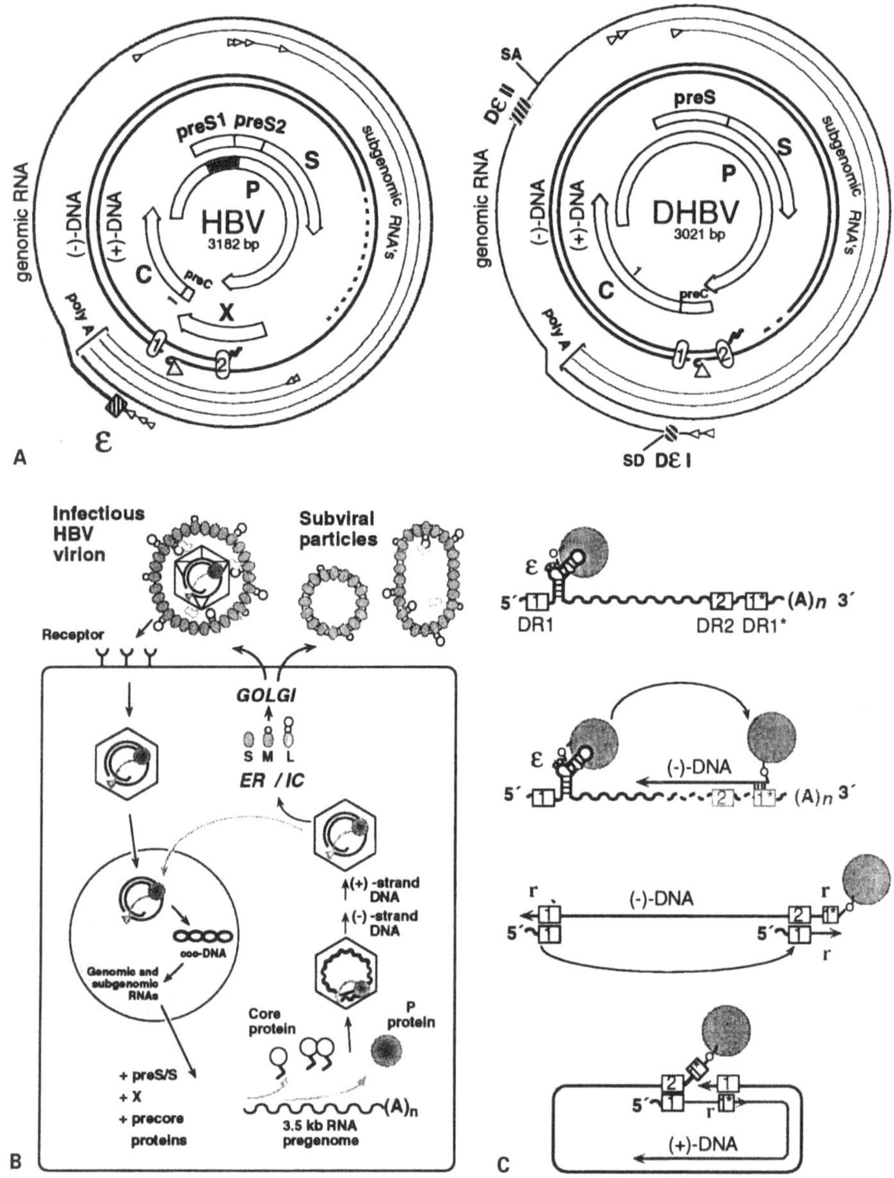

Fig. 1A–C. Fundamental aspects of hepadnavirus biology. **A** Genetic organization of HBV and DHBV. *Inner circles* represent the partially double-stranded circular DNA genomes found in extracellular virions. *Boxes marked DR1 and DR2* represent the direct repeats, the *triangle* the TP domain of P protein which is covalently attached to the (–)-DNA, and the *zigzag line* the RNA primer at the 5′-end of (+)-DNA. Open reading frames are indicated by *arrows*. *Hatched region* in the HBV P-ORF represents a non-essential spacer. *Numbers* are nt positions; for HBV, the numbering system of Pasek et al. (1979) is used, which starts with the initiator codon of the C gene; in DHBV, nt position 1 is defined by a unique *Eco*RI site inside the C gene. *Outer lines* represent the viral transcripts, *arrowheads* the approximate start sites. ε, HBV encapsidation signal; *D*εI, *D*εII, regions I and II of DHBV encapsidation signal; *SD, SA,* splice donor and acceptor sites. **B** Basic infectious cycle of hepatitis B virus. The HBV

core protein synthesis and later serves as RNA pregenome. The *env* gene consists of three in-phase ORFs, termed in 5′- to 3′-direction preS1, preS2 and S. S can be separately expressed to give the small S, or S protein; cotranslation of preS2/S yields the middle S, or M protein, that of the entire preS1/preS2/S gene the large S, or L protein. Thus the S domain is common to all three forms of Env protein. As for the preC/C ORF, this is achieved by the generation of mRNAs with staggered 5′-ends in which the initiator codons of the preS1, the preS2 or the S region are the first to be encountered by translating ribosomes. L protein is translated from a 2.4-kb mRNA, M and S from a set of 2.1-kb transcripts. Due to the more favourable context of the S-AUG, S protein may also be translated from mRNAs containing the preS2 initiator codon (SHEU and LO 1992). All viral transcripts are 3′-terminally colinear, ending after a unique polyadenylation signal located in the C gene.

The slightly smaller DHBV genome is similarly organized, but has some unique features: the X-ORF is absent; the preC/C gene is substantially larger than in HBV, coding for a core protein of 262 aa rather than 183 or 185 aa as in HBV; the *env* gene consists of only two regions, preS and S, and consequently only two products, the large and the small surface protein, are made from corresponding mRNAs of 2.35 and 2.13 kb (BŸSCHER et al. 1985); finally, recent data indicate that a second preS mRNA is produced from the pregenome by splicing (OBERT et al. 1996); hence the avian hepadnaviruses may be more closely related to retroviruses than their mammalian counterparts.

4 Basic Replication Cycle of Hepadnaviruses

The fundamental aspects of the hepadnaviral life-cycle are outlined in Fig. 1B (adapted from NASSAL and SCHALLER 1993a). The infectious DNA-containing virion binds to its target cell via interaction of the L protein (NEURATH et al. 1986,

◀

virion attaches to its target cell via interactions between the L protein and (a) still unidentified cellular receptor(s). After delivery into the nucleus, the partially dsDNA genome is repaired into cccDNA from which genomic and subgenomic RNAs are transcribed. The RNA pregenome interacts with both its translation products, core and P protein, to form a replication-competent nucleocapsid in which the RNA is reverse transcribed into DNA. At the ER, or a subsequent compartment, capsids acquire their envelope of surface proteins and are exported; alternatively, they may recycle the viral genomes to the nucleus. The surface proteins alone form spherical and filamentous S particles. *Irregularly shaped symbols* shown in filaments and virions represent the cytosolic chaperone *Hsc70*, which is associated with DHBV S particles, and probably complete virions. See text for details. **C** Simplified model of hepadnaviral replication. (–)-DNA synthesis initiates by P-protein-catalysed production of a short DNA primer with part of ε serving as template. The covalent P/primer complex is translocated to DR1* and extended to the 5′-end of the pregenome. The RNA template is degraded, except for a short oligonucleotide from the 5′-end containing the DR1 sequence, which is then transferred to DR2 and extended to the 5′-end of (–)-DNA. Due to a short terminal redundancy ("r"), the 3′-end of the (+)-DNA can use the 3′-end of (–)-DNA to continue (+)-strand synthesis

1992) with (a) still unidentified cellular receptor(s); for DHBV, the importance of the preS domain was directly demonstrated by infectivity competition (Klingmüller and Schaller 1993). Probably by a pH-independent mechanism (Rigg and Schaller 1992), the nucleocapsid is delivered into the cytoplasm of the host cell; the viral genome gains access to the nucleus and is transformed into covalently closed circular (ccc) DNA by cellular enzymes (Köck and Schlicht 1993). Subgenomic and genomic RNAs are transcribed from the cccDNA, exported into the cytoplasm and used as templates for translation of the various gene products. The *env* gene products are targeted to the endoplasmic reticulum (ER), and most of them are exported in the form of empty S particles via the constitutive secretory pathway. Core and P protein remain in the cytoplasm, where they interact specifically with the one species of genomic transcript from which they were translated, forming the nucleocapsid with its core protein shell and incorporated P protein and RNA pregenome. Inside the capsid, P protein reverse transcribes the RNA, yielding the characteristic circular, partially double-stranded DNA molecule with P protein covalently attached to the 5′-end of the first DNA strand (Fig. 1C). These replication-competent nucleocapsids either leave the cell by interaction with the *env* gene products still present in internal membranes, or they may cycle their nucleic acid back into the nucleus, replenishing the pool of cccDNA (Tuttleman et al. 1986; Wu et al. 1990). It is unlikely that major revisions to this replication scheme will be required in the future; however, our current knowledge of many details and even some fundamental cell biological aspects of the assembly of nucleocapsids, S particles and especially complete virions is still limited.

5 Supramolecular Hepatitis B Virus Structures

During hepatitis B virus infection three kinds of particulate, supramolecular structures are formed: infectious enveloped virions ("Dane particles" of about 42 nm diameter) containing an inner core, empty subviral particles (spherical and filamentous forms about 20–22 nm in diameter), and naked isometric capsids, or core particles (about 28 nm in diameter). Virions and S particles are found in serum; core particles can be liberated from virions by detergent treatment, or can be found as such inside infected hepatocytes. Electron micrographs of these particles as seen by negative staining are shown in Fig. 2A,B.

5.1 The Nucleocapsid (Core Particle)

The protein shell of the hepadnaviral nucleocapsid consists, as far as we know, exclusively of multiple subunits of a single species of core protein. Inside it harbours the viral genome, initially in the form of probably one copy of the RNA

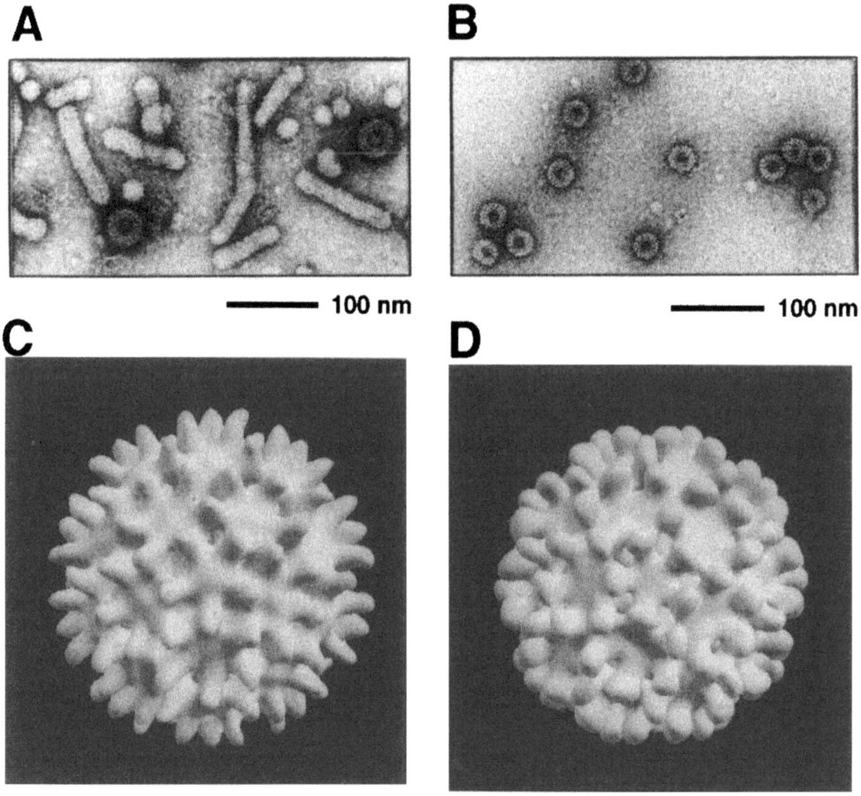

Fig. 2A–D. Ultrastructure of hepatitis B virus particles. Electron micrographs of negatively stained hepatitis B virions and S particles from patient serum (**A**) and recombinant HBV core particles produced in *E. coli* (**B**) *Larger round objects* in **A** represent complete virions (42- nm Dane particles), *smaller structures* are spherical and filamentous forms of S particles (Courtesy of Dr. H. Zentgraf, Applied Tumor Virology, German Cancer Research Center, Heidelberg, Germany.) Image reconstructions of HBV (**C**) and DHBV (**D**) core particles produced by *E. coli*. Reconstructions are based on electron micrographs from frozen-hydrated samples. The particles shown, viewed along a fivefold axis, exhibit T = 4 symmetry; smaller particles with T = 3 symmetry are also present in the preparations (Adapted from KENNEY et al. 1995)

pregenome, which is then reverse transcribed, by the co-encapsidated P protein, into the DNA genome found in extracellular virions. Most of the structural results referred to below have been obtained using the intrinsic capability of the core protein to self-assemble even in the absence of other viral components. However, P protein and RNA pregenome have a profound influence on the assembly pathway of core particles. This is not unexpected as virus propagation requires efficient co-encapsidation of these two essential replication components. Data pertaining to the assembly of such replication-competent cores are discussed separately (Sect. 5.1.5).

5.1.1 The Core Protein

The primary sequence length of the core proteins (see Fig. 3A) from the mammalian hepadnaviruses varies between 183 (HBV subtype ayw) and 188 aa (WHV); the avian hepadnavirus core proteins are substantially larger (262 aa for DHBV). The sizes of the proteins isolated from core particles correspond to

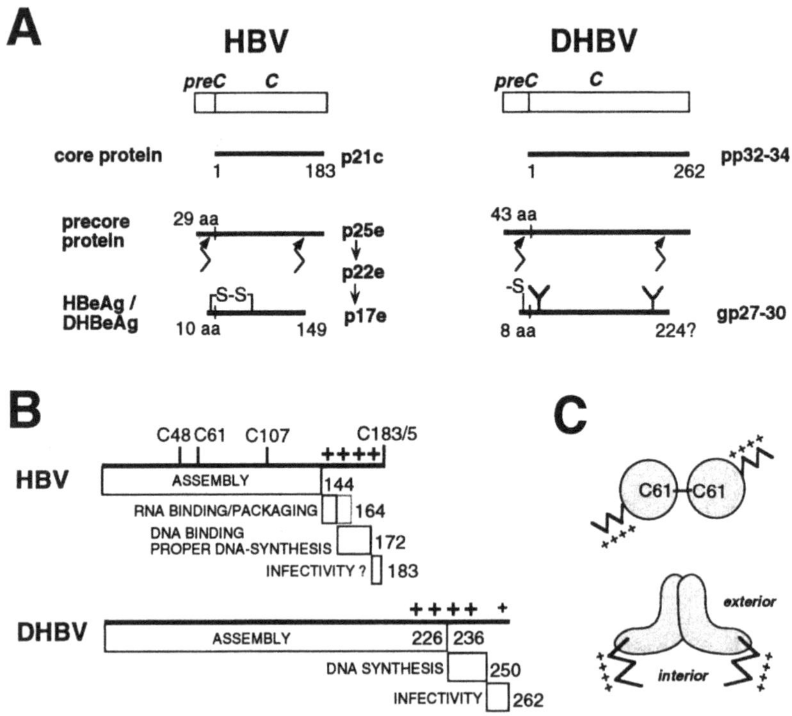

Fig. 3A–C. Hepadnaviral core proteins. **A** Correlation of preC/C genes and their products. *Open bars* represent genes, *filled bars* proteins. The capsid proteins are the primary translation products of the C genes. The secretory precore proteins arise by translation of the complete preC/C ORFs, and hence contain N-terminal extensions. During export, they are proteolytically processed at both ends (*arrows*); the cleavage sites for the DHBV protein have not been experimentally determined. In HBeAg, an intramolecular S-S-bridge blocks the assembly competence of the resulting p17e; in DHBeAg a similarly located Cys residue forms an intermolecular S-S-bridge (M. Nassal and A. Rieger, unpublished data). *Numbers* are aa positions; *designations to the right* show the apparent molecular masses in kilodaltons; *p*, protein; *pp*, phosphoprotein; *gp*, glycoprotein. *Y-shaped symbols represent* glycan residues; *filled symbols* indicate complete, *stippled symbols* partial glycosylation. **B** Functional domains in hepadnaviral core proteins. *Bars* represent the assembly domains, the strongly basic C-terminal regions are indicated by +. Cys residues are indicated by their aa position. Diagram is a composite of data obtained in heterologous expression systems and by transfection of cloned viral genomes. **C** Topological models for the HBV core protein dimer. *Upper model* is derived from cross-linking data. *Spheres* represent the assembly, *zigzag lines* the nucleic acid binding domain. *Lower model* is compatible with the electron densities observed in cryo-EM studies, which suggest a hammerhead-like structure in which the protruding domains from two tightly associated monomers form one of the visible spikes. See text for details

those of the primary translation products; hence there is apparently no processing. The only known chemical modification is phosphorylation, which might be involved in regulating the different functions of the core proteins in the viral lifecycle (see Sect. 5.1.6). All core proteins can be relatively efficiently expressed in various heterologous systems, and they retain the ability to self-associate into regular particulate structures. However, attempts to produce crystals suitable for an X-ray analysis have so far failed. Hence neither the overall architecture of the core particle nor the three-dimensional structure of the core protein itself is known at atomic resolution. Theoretical predictions have suffered from the lack in the structural data base of capsid proteins with strong sequence similarity to the hepadnaviral core proteins. The Mengovirus VP3 based model for the HBV core protein proposed by ARGOS and FULLER (1988) may be correct, but it has not been further substantiated by additional experimental evidence. Thus the major tools for the study of core protein structure and function are site-directed mutagenesis and analysis of the phenotypic consequences by expressing the variant core proteins either alone or in the context of a complete viral genome; accessibility of certain regions to antibodies and proteases; and, for complete particles, electron microscopy.

Mutational studies have provided convincing evidence for the two-domain structure of the HBV core protein, which is already reflected in the marked accumulation of basic residues in the 34 C-terminal aa residues, most of them clustered in four blocks (indicated by +'s in Fig. 3B), of which three contain the motif SPRRR(R). The self-assembly capability resides in the 144 N-terminal aa residues of the HBV core protein; mutants comprising this primary sequence are assembly competent (GALLINA et al. 1989; BIRNBAUM and NASSAL 1990); those lacking the sequence following residues 138 and 139 are not. N-Terminal deletions of more than a few aa are also deleterious, in both the HBV (e.g. CHANG et al. 1994) and DHBV (YANG et al. 1994) core protein. The basic C-terminal region constitutes an apparently non-sequence-specific nucleic acid binding domain (HATTON et al. 1992) and mediates RNA encapsidation even in *E. coli*, apparently with some preference for the mRNA the protein was translated from (BIRNBAUM and NASSAL 1990). This preference might be explained by a local concentration phenomenon; certainly it is mechanistically different from the specific pregenome encapsidation process mediated by P protein and the RNA encapsidation signal ε described below. Cross-linking studies (NASSAL et al. 1992; ZHENG et al. 1992; ZHOU and STANDRING, submitted) confirmed the two-domain topology, since except for those involving the C-terminal Cys residue C183 identical crosslinks between the internal Cys residues 48, 61 and 107 (see Fig. 3B) were found in a C-terminally truncated and the full-length core protein. The most prominent covalent link observed was a homologous S-S-bridge between two Cys 61 residues, suggesting that these two residues are in close spatial proximity, most likely in a symmetrical dimeric arrangement (Fig. 3C). While a complete dissociation and reassociation of particles from the full-length core protein has not yet been reported, this is possible with the C-terminally truncated variants, indicating a stabilizing influence of interactions between the basic core protein tail and RNA. In such experiments, the dissociated subunits were always found to migrate as dimers during gel filtration under native conditions, regardless of the presence or absence of Cys 61 (NASSAL et al. 1992). These data strongly support the view that the core protein has a high intrinsic propensity to form dimers. The C-terminal Cys residues are also easily oxidized; in particles produced in *E. coli*, the resulting S-S-bridges cross-link neighbouring dimers into a polymeric network (NASSAL et al. 1992; ZHENG et al. 1992); in *X. laevis* oocytes homologous cross-links within one dimer appear to be preferred (ZHOU and STANDRING, submitted). The basis for this difference is currently not clear; it may be related to the time course of oxidation relative to the progress of assembly which itself is dependent on the core protein concentration (see below), or the use of core proteins from different HBV subtypes with slightly differing primary sequences.

Further confirmation for the two-domain model comes from the observation that engineered fusion proteins with the foreign sequence attached to the extreme N-terminus, or after aa 144 of the HBV core protein, are still assembly competent (e.g. SCHÖDEL et al. 1992; BORISOVA et al. 1993). Such hybrid core particles hold promise as potent subunit vaccines.

An alignment of the DHBV core protein (see Fig. 3A) with the mammalian hepadnavirus core proteins shows a markedly higher homology in the N- and C-terminal parts than in the centre (SPRENGEL et al. 1985; ARGOS and FULLER 1988; STANDRING 1991). Possibly, the homologous sequences form the assembly domain while the extra sequences present in the DHBV protein are looped out, but this prediction has not yet been directly tested. In general, even small sequence alterations in the N-terminal and central regions of the DHBV core protein appear to severely compromise its self-assembly competence (YANG et al. 1994). However, the two-domain structure (Fig. 3B) seems to be conserved since deletions of the C-terminal region do not inhibit self-assembly, either in *E. coli* (YANG et al. 1994) or in transfected mammalian cells (SCHLICHT et al. 1989a; YU and SUMMERS 1991). This region is also highly basic and involved in nucleic acid binding although the positively charged residues are not all present in similarly ordered clusters as in the mammalian virus proteins.

5.1.2 The Secretory Core Gene Product HBeAg

A characteristic feature of all hepadnaviruses is that they use the core gene, in conjunction with the short in-phase preC-ORF, to produce a second, non-particulate and antigenically distinct gene product known as HBeAg. The function of the antigen which is used as a serological marker for ongoing viral replication is still obscure. The discovery in patients of mutant HBeAg^{-} viruses and their apparent selection under conditions of immune surveillance or interferon treatment (for review: BLUM 1993; MISKA and WILL 1993) has roused considerable interest in the medical community. Obviously, HBeAg is not essential for the basic viral life-cycle, but it may be a modulator of the host response to viral infection. HBeAg biosynthesis (see Fig. 3A) involves translation of the joint preC/C ORF to produce the precore precursor protein, its targeting into and processing in the secretory pathway of the cell, and finally export as a soluble, non-particulate protein which appears to be tightly associated with serum proteins. In HBV, the final product differs from the core protein by the presence of a 10-aa N-terminal extension and the absence of the basic C-terminus. Hence, the entire assembly domain of the core protein is present in HBeAg; yet its properties are fundamentally different. The solution to this structural puzzle is the formation of a specific intramolecular disulphide bridge connecting a Cys residue in its unique N-terminal extension with Cys61 (Fig. 3A), which in core protein resides at the dimer interface. The net result is the prevention of dimerization and subsequent assembly (WASENAUER et al. 1993; NASSAL and RIEGER 1993; SCHÖDEL et al. 1993).

5.1.3 Architecture and Structure of Core Particles

Various reports describing the electron microscopic appearance of negatively stained, heterologously produced HBV core particles have been cited in previous reviews; these data showed that HBV cores have an approximate diameter of 28 nm (Fig. 2B); in addition, slightly smaller particles can be seen on a number of the published electron micrographs (COHEN and RICHMOND 1982; ONODERA et al. 1982), but were usually not explicitly discussed. A powerful method bridging the gap between conventional electron microscopy and X-ray diffraction is the combination of cryo-electron microscopy and computer-assisted image reconstruction techniques (DUBOCHÉT et al. 1988). The principal advantage of cryo-electron microscopy is that the biological specimen is not dehydrated but rather embedded in vitreous (non-crystalline) ice; hence the authentic structures should be preserved much better than by conventional techniques. Applying this method to the full-length and a variant HBV core protein lacking the Arg-rich C-terminus expressed in *E. coli*, CROWTHER et al. (1994) have recently obtained strong evidence that the majority of particles (diameter approximately 34 nm) from both proteins have icosahedral symmetry (triangulation number T = 4) and are built from 240 subunits of the core protein (cf. Fig. 2C). A fraction of the particles have a smaller diameter (approximately 30 nm) and consist of only 180 subunits (T = 3). Possibly it is this heterogeneity that has so far prevented the formation of sufficiently ordered crystals for an X-ray analysis, in particular since the overall packing of subunits is extremely similar in both types of particles. In accord with the biochemical and genetic data outlined above, the reconstructions revealed a tight association between pairs of subunits. Thus, the particles are more accurately described as consisting of 120, or 90, respectively, core protein dimers. The electron density maps can be interpreted to indicate a hammerhead-like structure of the dimer (Fig. 3C), in which part of the proteins form a surface protrusion, similar to the "protruding" P domain in many RNA virus capsids. Also in accord with the biochemical data, most of the particles from the truncated variant had no detectable inner contents, while those of the full-length protein showed an internal density, probably corresponding to packaged RNA.

We have recently confirmed these data and extended the analysis to HBV cores isolated from infected human liver, and DHBV cores expressed in *E. coli* (KENNEY et al. 1995). The T = 4 protein shell of the liver-derived HBV cores appears very similar to that of the *E. coli* material shown in Fig. 2C, corroborating earlier inferences from their antigenic similarity; also this preparation contained smaller particles with apparent T = 3 symmetry. Despite the much larger size of its core protein, the basic architecture of the DHBV capsid is very similar to that of HBV and hence an evolutionarily conserved feature: the majority of particles also consist of 240 subunits, which are arranged in hammerhead-like dimers (Fig. 2D). Some of the extra mass is apparently located in the spikes, which are more oblong-shaped than in HBV. Smaller capsids with T = 3 symmetry are also found.

HBV and DHBV cores appear to have holes, in accord with the permeability of the core shell for small molecules like nucleotides. This is obviously important for reverse transcription inside the particle and is experimentally utilized in the "endogenous polymerase reaction", which provides a sensitive assay for replication-competent core particles: if naked cores are incubated with dNTPs, the endogenous (i.e. encapsidated) P protein will elongate the initiated DNA strands and convert them into more or less completely double-stranded molecules. Another structural feature may also relate to the fact that the core particle is not only a protective container for the hepadnaviral genome but a transcription machine: RNA-containing cores from *E. coli*, and liver-derived particles which should contain the complete RNA pregenome and/or various amounts of partially double-stranded DNA (and P protein), show an inner electron density, well separated from the peak density of the protein shell, except for apparently icosahedrally ordered contacts. Thus it appears that the basic C-terminal regions of the core protein keep most of the nucleic acid apart from the outer shell to allow sufficient flexibility for the P protein catalysed reverse transcription process inside the particle.

Two major unresolved issues are the nature of the molecular switch that determines formation of the complex 240 subunit T = 4 structure rather than the simple T = 1 arrangement of 60 identical subunits, and the significance of the T = 3 and T = 4 particles. In many T = 3 plant RNA viruses, T = 1 structures form in vitro if the basic nucleic acid binding regions of the capsid proteins are proteolytically removed. It is believed that interaction with the RNA mediates establishment of the T = 3 arrangement, which requires the capsid protein to be able to adopt three similar but not identical ("quasi-equivalent") conformations (for a review: ROSSMANN and JOHNSON 1989). In hepadnaviruses, the information to properly fold into distinct, only quasi-equivalent conformations possibly resides in the primary sequence of the assembly domain, as even C-terminally truncated core protein variants form predominantly particles with T = 4 symmetry. However, a potential influence of RNA on the arrangement of subunits in the particle still cannot be excluded, as even the particles from the truncated core protein contain some RNA (about 1/10th to 1/20 the amount found in particles from the full-length protein; BIRNBAUM and NASSAL 1990; HATTON et al. 1992); an analysis of core particles reconstituted in vitro after complete digestion of any nucleic acid should reveal such a potential influence. A possibly interesting finding is that the shortest of a series of C-terminally truncated DHBV core proteins that is still assembly competent appears to form smaller particles of about 20 nm (YANG et al. 1994); if confirmed, this might indicate that at least part of the information to build the complex T = 3 and T = 4 structures resides in the very C-terminal residues of the assembly domain.

Whether the T = 4 or T = 3 particles represent the biologically relevant form remains to be determined. The higher abundance of the T = 4 particles in *E. coli* and, more importantly, in cores from infected human liver suggests that these larger, more complex particles are the relevant species. Applying the above-described techniques to complete virions should reveal whether the two particle classes are discriminated between during envelopment.

The resolution of the EM data outlined above is still at the level of protein subunits. However, additional topological information can be derived from studies analysing the accessibility of certain regions of the core protein to either antibodies or proteases. The major HBcAg epitope ("c") is located around aa 80 (SALFELD et al. 1989; SÄLLBERG et al. 1991), and hence surface exposed; in addition, the region encompassing residues 127–133 was found to react with site-specific monoclonals while regions 9–20 and 133–145 were only exposed after denaturation (BICHKO et al. 1993; PUSHKO et al. 1994). Protease sensitivity studies have shown that cores produced in *E. coli* (e.g. DALSEG 1990) or in *Xenopus* oocytes (SEIFER and STANDRING 1994) have only one major sensitive site located between aa 145 and 150, i.e. the region connecting the assembly and the basic nucleic acid binding domain; interestingly, only one-third to one-half of the subunits are cleaved by trypsin or Asp-N, possibly a reflection of the only quasi-equivalent arrangement of subunits in a T = 3 or T = 4 particle. The entire assembly domain, by contrast, is highly resistant to protease and hence likely to be compactly folded.

The accessibility of the region around aa 150 to protease, and the ability of an antipeptide antiserum directed against the very terminal aa residues of DHBV core protein (SCHLICHT et al. 1989a), have been taken as arguments for a surface exposure of the entire basic region. However, a monoclonal antibody against (phosphorylated) aa 165–175 fails to react with intact HBV cores (MACHIDA et al. 1991); HBV cores from *E. coli* with at least part of the basic region contain RNA which is protected from RNAse (BIRNBAUM and NASSAL 1990; HATTON et al. 1992), and the only major difference visible in the cryo-EM studies between particles from full-length and the variant core protein lacking the basic region is the internal electron density presumably representing RNA; finally, in cores produced in animal cells proper reverse transcription of the viral RNA pregenome requires the presence of the basic region (NASSAL 1992). Hence the nucleic acid binding domain is, at least in its greater part, buried inside the particle. Combining the large panel of monoclonal antibodies with known linear epitopes and well-designed core protein mutants with the cryo-EM techniques should help to further refine the structural model of the hepadnaviral core particle.

In summary, HBV core particles consist of an outer protein shell made from the tightly folded assembly domains of 120, or 90, respectively, dimeric core protein subunits, in which regions around aa 80, 130 and 150 are surface exposed; at least the majority of the basic nucleic acid binding regions face the interior of the capsid, where it is available for binding, and possibly shaping, the viral RNA pregenome for the complex transformation into partially dsDNA catalysed by P protein (see below).

5.1.4 Core Particle Assembly

Two systems are currently used to study the dynamics of core particle assembly: direct injection of core mRNA into *Xenopus laevis* oocytes, and in vitro transcription/translation; informative data on the reconstitution of particles from purified, dissociated core protein subunits are not yet available. The advantage of both the *Xenopus* and in vitro translation systems is that the concentration of core protein produced can be controlled by the amount of input mRNA. The major results with the oocyte system were first that dimers, but not higher aggregates, are the main intermediates in assembly (Zhou and Standring 1992). Thus, unlike many other viral systems, there is no evidence for the accumulation of, for instance, pentameric or hexameric intermediates (which might more easily explain the formation of the complex T = 3 and T = 4 structures). These data corroborate the importance of the dimeric interaction between core protein subunits described above. Second, assembly requires a critical threshold concentration of core protein, which in this system is in the micromolar range (Seifer et al. 1993). Protease accessibility experiments are fully consistent with the notion that in the dimers the assembly domain is resistant to many proteases while the basic region is easily clipped off; in the assembled particle, by contrast, the C-terminal region is in its greater part shielded from protease attack (Seifer and Standring 1994). Co-expression of hepadnavirus core proteins from different species demonstrated that the core proteins from closely related mammalian hepadnaviruses form mixed particles (Chang et al. 1994). Hence their lateral interaction surfaces must be very similar in structure, while the DHBV core protein was not incorporated into such mosaic cores. Interestingly, the dimeric intermediates were preferentially formed from monomers of the same species; at least in part this *cis*-preference appears to be due to a very fast, perhaps cotranslational, dimerization process. Very likely, the pool of core protein from which capsids are formed in an HBV-infected cell consists also of dimers rather than monomers.

While the information to form a symmetrical capsid is encoded in the primary sequence of the core protein, cellular factors may be involved in the assembly process. Using in vitro translation of HBV core protein, Lingappa et al. (1994) have recently presented evidence for the association of a bona fide assembly intermediate with a cytoplasmic chaperone of the *Hsc*60/*Gro*EL family (Braig et al. 1994; for reviews: Georgopoulos and Welch 1993; Martin and Hartl 1994). This high molecular weight intermediate has sedimentation characteristics distinct from dimers and complete particles, and, importantly, it can be chased into complete cores with concomitant release of the chaperonin. That chaperones may be involved in the assembly process comes as no surprise given that *Gro*EL was originally characterized as a host factor in *E. coli* required for proper assembly of phage lambda particles. However, whether the observed association of *Hsc*60 with core protein in the reticulocyte system reflects a mechanistic necessity for core particle assembly in an infected hepatocyte remains to be determined. Given the ubiquitous existence of chaperones and their, by necessity, broad interaction specificities, they might also be involved in core particle assembly in the other systems used; hence it would be even more important to set up an assembly system comprising just the purified core protein monomers or dimers.

5.1.5 Replication-Competent Nucleocapsids

For the virus it is absolutely essential that the core particles produced are true replication-competent nucleocapsids. Thus specific incorporation of the correct RNA, and of the specialized enzyme able to transform this RNA into DNA, is of utmost importance for virus propagation. In retroviruses, protein-protein interactions between Gag, and the Gag-domain in the Gag-Pol fusion proteins, provide the fundamental mechanism for Pol incorporation into capsids, at the cost of a requirement for subsequent processing of the fusion proteins; RNA encapsidation is apparently mediated by Gag recognizing genomic viral RNA. That hepadnaviruses have evolved a completely different strategy became obvious with the discovery that, although the P ORF overlaps with the 3′ -part of the core ORF as in retroviruses, P protein is expressed as a separate entity (Schlicht et al. 1989b; Chang et al. 1989); the exact mechanism of P expression from the RNA pregenome is still not established (Chang et al. 1990; Lin and Lo 1992). P protein consists of three domains (Fig. 4A) in the order terminal protein (TP), which is covalently linked to (–)-strand DNA, polymerase and RNAseH (Radziwill et al. 1990). As outlined above, the nucleic acid binding capacity of core protein is rather broad and unlikely to account for the specific encapsidation of the hepadnaviral RNA pregenome. In retrospect, therefore, it came as no surprise that it is a specific interaction between P protein and the RNA pregenome that is responsible for the co-encapsidation of both components (Bartenschlager and Schaller 1992). The details of this interaction are the subject of ongoing studies in several laboratories; however, many of the basics of the process have been established (Nassal and Schaller 1993a). I will only summarize these fundamental aspects, and add new findings that corroborate the tight coupling of assembly and replication in hepadnaviruses.

Two key observations were that P protein and pregenome encapsidation are mutually dependent (Bartenschlager et al. 1990; Hirsch et al. 1990), and that a relatively short region close to the 5′-end of the RNA pregenome mediates specific RNA packaging, even if fused to foreign RNA molecules (Junker-Niepmann et al. 1990). This region, termed the encapsidation signal ε, was

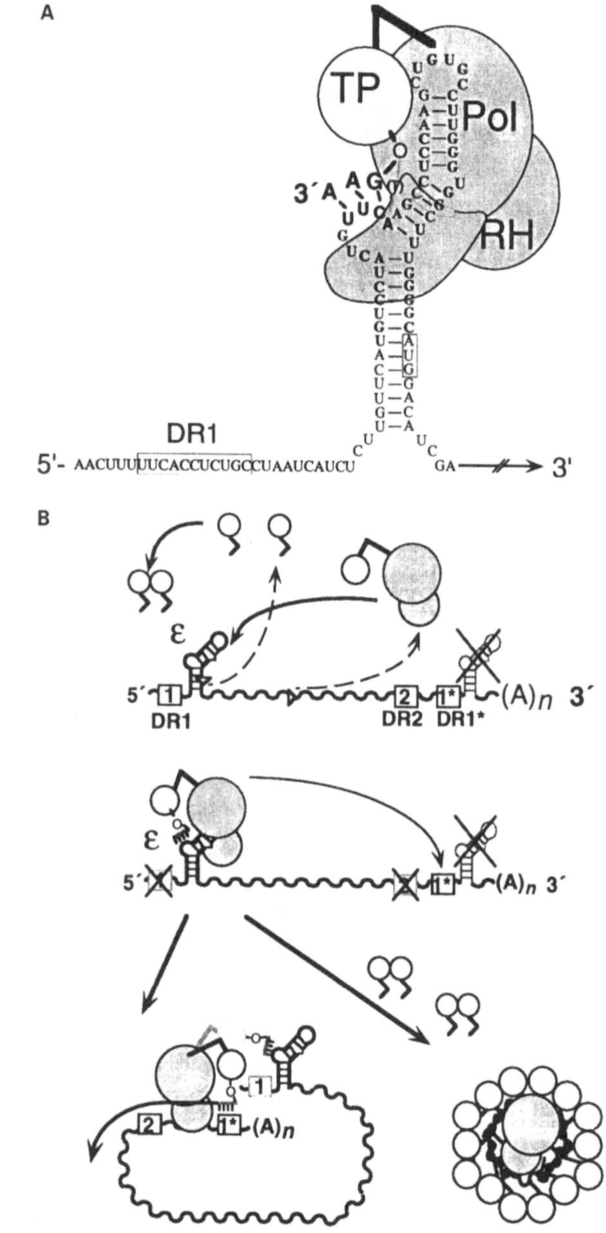

Fig. 4A,B. Central role for P protein/ε interaction in the hepadnaviral life-cycle. **A** Model for the interaction between P and ε. The secondary structure of ε and the three-domain structure of HBV P protein have been experimentally determined. The arrangement of P protein on the RNA is compatible with mutational studies analysing the competence of mutant ε-sequences for encapsidation and (–)-DNA initiation. Possibly, Hsp90 and other chaperones are associated with this complex (Hu and Seeger 1996) **B** Events mediated by P/ε interaction. Core and P protein are translated from the pregenome. The binding of P to ε in *cis* triggers formation of a short DNA primer for (–)-DNA synthesis from the ε-bulge, but also addition of core protein dimers, initiating capsid assembly. The temporal order of events is not yet known; however, P protein mutants incapable of primer synthesis also mediate RNA encapsidation

subsequently shown to act via a characteristic bipartite stem-loop structure (Fig. 4A), which is necessary and sufficient for the process (POLLACK and GANEM 1993; KNAUS and NASSAL 1993). The interaction between P and ε may already be initiated during translation, as in competition experiments a preferential encapsidation of the RNA used as template for P production was observed. In the preC RNAs serving as mRNAs for the precore protein, translating ribosomes compete with P protein binding, and hence prevent encapsidation of these transcripts which would not support viral replication (NASSAL et al. 1990). Further mutational analyses of ε using defined (POLLACK and GANEM 1993; KNAUS and NASSAL 1993), or pools of partially randomized, ε variants (NASSAL and RIEGER 1995a) revealed that in particular the upper part of the ε structure is important for encapsidation; the apical loop tolerates only a few nt exchanges, while in the bulge region separating the two base-paired stems especially positions 1 and 2, but not the following unpaired residues, are critical. Thus, as in other RNA-protein interactions, a specific protein-binding site is created by a combination of structural and sequence-specific features on the RNA (Fig. 4A). The biological relevance of the ε structure is strongly supported by sequence data from HBV variants isolated from infected patients in all of which the characteristic ε structure is preserved (LASKUS et al. 1994). Why the 3′-copy of the ε sequence, present on all viral transcripts, is functionally silent, is still an unresolved issue. However, the selective interaction of P with the 5′-copy of ε which contains the initiator codon for core protein, may provide an advantage for packaging the correct RNA: P binding should suppress translation of core (and P) protein, and in effect deprive the RNA from ribosomes which might interfere with encapsidation.

In summary, binding of P to ε appears to provide a nucleation centre for core particle assembly (Fig. 4B), most likely by providing a binding site for the first one or few core protein dimers; the high cooperativity of core protein association plus the increased substrate quality of the ribosome-deprived pregenome may then be sufficient to ensure the highly selective encapsidation of the correct viral transcript. This principle is known from other viral systems: for example, in vitro assembly of phage R17 capsids is relatively inefficient without any RNA; the critical concentration for self-assembly of coat protein is lowered by the presence of unspecific RNA, and further reduced by the presence of the specific operator fragment acting as encapsidation signal (BECKETT et al. 1988). The analogy should not be drawn too far, as in the phage it is the many subunits of coat protein that mediate RNA encapsidation (PEABODY 1993; OLSTHOORN et al. 1994), and copackaging of the viral replicase is not required. An alternative view with the same outcome would be that P protein binding to ε creates a high-affinity binding site on the RNA for core protein. Quantitative data on the role of RNA and/or chaperones as potential assembly mediators are, however, not yet available.

In DHBV, principally the same mechanism for specific co-encapsidation of P protein and pregenome is operating. However, the sequence homologous to HBV ε is not sufficient for encapsidation; a second region located inside the P

ORF is also required ("region II"; Calvert and Summers 1994), possibly due to the relatively low stability of the DHBV ε structure, which has not yet been experimentally analysed. According to computer predictions it is quite different from that in HBV while phylogenetic arguments (Junker-Niepmann et al. 1990) as well as functional homologies (Pollack and Ganem 1994) suggest that an HBV ε like structure can form. Interestingly, the lack of region II in the recently found major DHBV transcript that arises from splicing of the genomic RNA would prevent its encapsidation, similar as for subgenomic retroviral transcripts.

The general picture emerging from these data is that P protein interacts specifically with a structured RNA element on the RNA pregenome, forming a preassembly complex which lowers the critical concentration of core protein required to initiate self-assembly (Fig. 4B). This dependence favours the selective co-encapsidation of both the RNA substrate and the reverse transcriptase required to convert it into DNA, i.e. formation of replication-competent cores.

Intriguingly, the very same P/RNA interaction that triggers capsid assembly is also directly involved in viral replication. Until recently, it was believed that reverse transcription of the hepadnaviral pregenome (cf. Fig. 1C) would initiate, via protein priming, de novo at the 3′-proximal DR1*, and then proceed continuously, in contrast to the discontinuous mechanism employed by retroviruses and most other retro-elements. However, data obtained with an in vitro translation system capable of producing enzymatically active DHBV P protein (Wang and Seeger 1992, 1993; Tavis and Ganem 1993) suggested that part of region I of the DHBV encapsidation signal would represent the replication origin, by serving as template for a short DNA primer which is subsequently translocated to 3′-DR1*. We have recently obtained direct sequence evidence that indeed the 3′-part of the HBV ε bulge region is used as template for the synthesis of such a short DNA primer (Nassal and Rieger 1996). Exclusively the 5′-copy of ε and the 3′-copy of DR1 are involved (Rieger and Nassal 1996). Possibly, selective usage of the individual ε and DR copies is made possible by a close spatial proximity of the 5′- and the 3′-terminal regions of the RNA pregenome. In summary, the very same *cis*-acting RNA element is crucial for both encapsidation and hence nucleocapsid assembly, and for virus replication, as indicated in Fig. 4B.

The use of a *cis*-acting structured, 5′-proximal RNA element to correctly position the polymerase over the initiation site is a unique strategy which places the evolutionary origin of hepadnaviruses between that of primitive retro-elements like the Mauriceville plasmid from *Neurospora mitochondria*, which likewise uses a *cis*-acting, but 3′-proximal RNA structure to position its reverse transcriptase (Wang and Lambowitz 1993), and modern retroviruses in which the specificity of (–)-DNA initiation is determined by a *trans*-acting tRNA primer, its basepairing with the 5′-proximal primer-binding site and the affinity of the retroviral RT for tRNA, in particular the one used as primer (Mak et al. 1994). However, the discontinuity of (–)-DNA synthesis as such brings hepadnaviral replication into closer accord with the vast majority of retro-elements than previously thought.

While experiments in different heterologous systems have shown that the only viral gene products required for replication initiation are P protein and RNA (WANG and SEEGER 1992; SEIFER and STANDRING 1993; TAVIS and GANEM 1993), proper formation of the viral DNA genome has not been demonstrated in the absence of core protein. This suggests that the complex nucleic acid transformations during reverse transcription depend on the auxiliary functions of core protein.

Where in the cell do replication-competent core particles assemble? All available evidence suggests that this process takes place in the cytoplasm: P protein interacts probably cotranslationally with the pregenome, and encapsidation exclusion of the preC RNAs is translationally controlled; after transfection of an intron-containing ε reporter construct the spliced cytoplasmic but not the unprocessed nuclear transcript was encapsidated (RIEGER and NASSAL 1995). Hence, the major events during core particle assembly have to occur in the cytoplasm. In keeping with this, many immune fluorescence data show a cytoplasmic stain for core protein. Sometimes, however, staining is also observed in the nucleus, particularly with HBV (e.g. GUIDOTTI et al. 1994). Core protein does indeed contain nuclear localization signals (YEH et al. 1990; ECKHARDT et al. 1991) which might act in a cell cycle dependent manner (YEH et al. 1993). In transgenic mice expressing HBV core protein, the antigen is strictly nuclear in resting cells, becomes detectable in the cytoplasm during mitosis, and then remains cytoplasmic, indicating that intact core particles are unable to cross the nuclear membrane (GUIDOTTI et al. 1994). This suggests that if the net rate of synthesis is faster than that of nuclear transport assembly will occur in the cytoplasm; in the reverse case, unassembled core protein subunits will accumulate in the nucleus and assemble there. The exclusive cytoplasmic localization of the DHBV core protein (PUGH et al. 1989) may then be explained by a more efficient synthesis and/or lower critical concentration for assembly. The more general implication of these data is that after infection, or during the postulated intracellular cycling (TUTTLEMAN et al. 1986), the core particles delivering the viral genome will have to disassemble at the nuclear membrane.

5.1.6 Covalent Modifications of the Core Protein

The only known covalent modification of hepadnaviral core proteins is phosphorylation. Core particles isolated from infected livers contain, in addition to viral nucleic acid and P protein (see below), an endogenous kinase activity (ALBIN and ROBINSON 1980; GERLICH et al. 1982) that, upon addition of ATP, phosphorylates the core protein at several sites in the basic C-terminal region; similarly, a kinase activity is associated with cores produced from recombinant baculovirus (LANFORD and NOTVALL 1990). Heterologously expressed cores from *Xenopus* oocytes are also heavily phosphorylated; those from *E. coli* or yeast are not (MACHIDA et al. 1991; HATTON et al. 1992). Studies with specific kinase inhibitors, and an in vitro reconstitution system KANN and GERLICH (1994), have recently suggested that the major fraction of the associated kinase activity is attributable to protein kinase C, estimated to be present in about two copies per particle. A general conclusion from these studies is that core protein phosphorylation at the multiple SPRRR clusters decreases or abolishes nucleic acid binding, similarly to the known effect of phosphorylation at the SPKK motifs in histones (Hill et al. 1991); it may hence be involved in replication of the viral genome and/or uncoating after infection. However, only recently have functional data been obtained with the DHBV system, revealing a complex pattern of preferences for phosphorylated or non-phosphorylated Thr or Ser residues during the steps of core particle maturation; importantly, intracellular DHBV capsids are heterogeneously phosphorylated while capsids isolated from extracellular virions appear homogeneous with a mobility corresponding to unphosphorylated protein (Yu and Summers 1994). It therefore seems conceivable that different activities of the capsid in the viral life-cycle are regulated by sequential phosphorylation dephosphorylation.

5.2 Subviral Particles

Envelopment of hepadnaviruses differs in several respects fundamentally from that of retroviruses: (1) all hepadnaviruses generate more than one envelope protein; (2) there is no evidence for proteolytic processing; (3) the surface proteins by themselves, i.e. without requiring interactions with the capsid, form empty, non-infectious lipoprotein particles, usually in vast excess over complete virions (cf. Figs. 1B, 2A); (4) hepadnaviral S proteins do not accumulate to any appreciable extent at the plasma membrane of the host cell; (5) the S proteins are required for export of capsids, i.e. core particles cannot, by themselves, leave their host cell. This latter statement is based on the absence of naked capsids in serum, even of immune-compromised patients (Possehl et al. 1992); however, the supernatants of cells transfected with efficient expression plasmids for core protein often contain such naked cores (Bruss and Ganem 1991b; Nassal 1992; Lenhoff and Summers 1994). It is believed but not proven that these capsids are released from dead cells.

After summarizing earlier data on the composition of S particles, I will emphasize recent findings that in particular concern the influence on membrane topology of the additional N-terminal domains in the L protein. The reader interested in additional information on the biochemistry of S particles should consult the reviews by Heermann and Gerlich (1991) and Ganem (1991).

5.2.1 S Particle Composition

Earlier electron microscopic studies, using negative staining, of serum samples from HBV-infected individuals showed abundant coreless spherical particles about 22 nm in diameter (from preliminary cryo-EM data, however, it appears that for HBV the round objects may represent disks rather than spheres; S.D. Fuller, personal communication), fewer filamentous structures with the same diameter but of variable length, and even fewer virions with their characteristic inner core (Fig. 2A). Typical particle numbers in the serum of chronic carriers are in the range of 10^{13}/ml for the spheres, 10^{10}/ml for filaments and up to 10^{9}/ml for virions. These values, however, are extremely variable for different patients, and within one patient during the course of infection (HEERMANN and GERLICH 1991). The 22- nm particles of HBV are of historical interest as they represent the overwhelming majority of the classical "Australia antigen" (BLUMBERG et al. 1965) now known as HBsAg; also the first-generation HBV vaccine was derived from S particles purified from chronic carriers but has now been superseded by recombinant S protein preparations isolated from yeast (McALEER et al. 1984). For DHBV, the empty envelopes appear, at least by negative staining, as less-ordered "bags" of similar, or larger size as complete virions (diameter approximately 40 nm), and filamentous forms are absent. The biological function of S particles is not clear; it is generally believed that they represent decoys that trap neutralizing antibodies in unproductive interactions.

The three envelope proteins produced by mammalian hepadnaviruses (L, M and S) and the two avian proteins (L and S) are schematically represented in Fig. 5A. All HBV, but not the DHBV, surface proteins occur also in glycosylated form. For HBV, the apparent molecular masses of the proteins are p24/gp27 (S), p30/gp33/ggp36 (M) and p39/gp42 (L); the DHBV S protein appears as p17, the L protein as p35, sometimes as doublet p35/p36. While for HBV the overall ratio of HBV S, M and L proteins in serum is approximately 1000:10:1, the individual particles, enriched from serum by biochemical fractionation, differ substantially in their protein composition, in particular regarding L (HEERMANN and GERLICH 1991). In all of them, the small S protein is the predominant species, and M constitutes some 5%–10% of the total protein mass. However, spheres, estimated to consist of about 100 protein subunits (PETERSON 1987), contain very little L protein (up to maximally 5%); by contrast, L is enriched in filaments (approximately 10%–20% by weight) and even more abundant in virions (from 20% up to 50%); the exact composition is difficult to measure and may vary with the source of S particles (W.H. Gerlich, personal communication). In DHBV, the difference in composition between S particles and virions is much less pronounced (both contain some 10%–25% of L protein; SUMMERS et al. 1991) and, owing to the very similar properties of empty and core-containing particles, a complete separation of virions and S particles has not yet been achieved.

S particles from patient serum contain some 25% total lipid by weight, with about 67% phospholipids (PL), 15% free cholesterol, 14% cholesterol esters and 4% triglycerides (TG; GAVILANES et al. 1982). As the lipid composition of internal membranes is different from that of the plasma membrane (VAN MEER 1989; ALLAN and KALLEN 1994), the high proportion of phosphatidylcholine and relatively low proportions of phosphatidylserine and sphingomyelin suggest that the S particle lipid is derived from the ER; however, phosphatidylinositol (PI) is usually high in the ER and lower in the plasma membrane, yet PI could not be detected at all in HBsAg. Even more striking is its high content of total cholesterol. Typical ratios of cholesterol/phospholipid are about 0.08 for ER membrane,

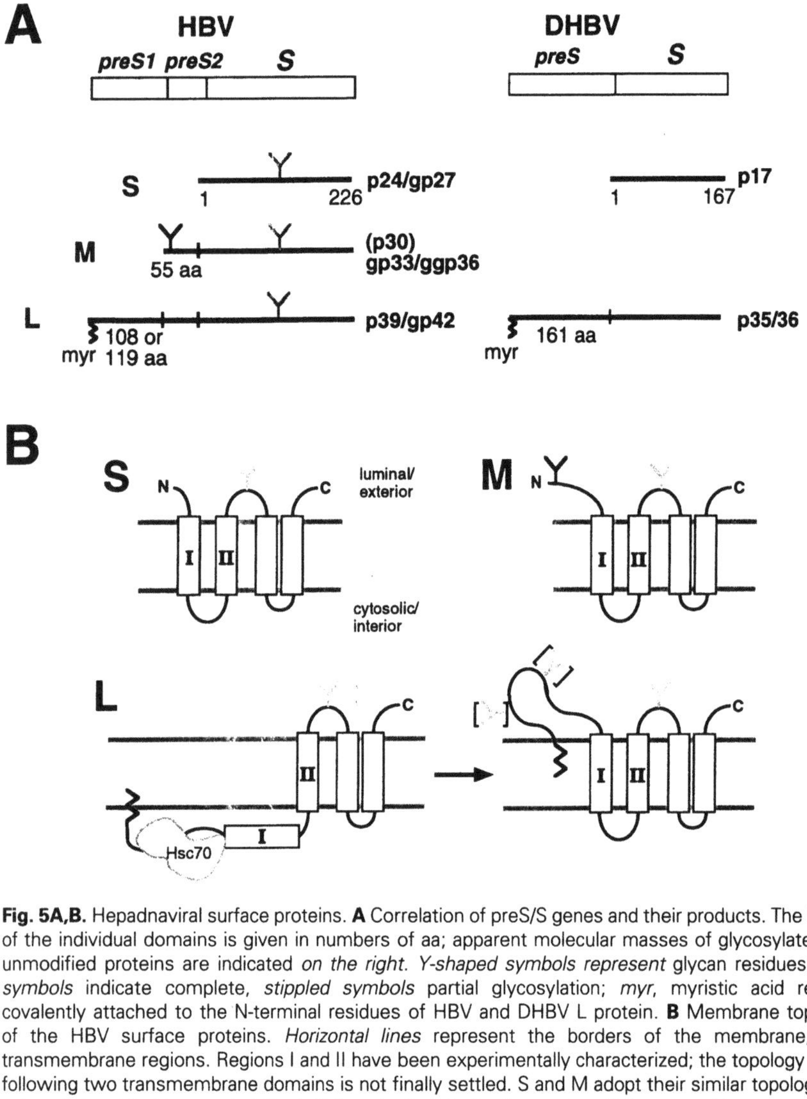

Fig. 5A,B. Hepadnaviral surface proteins. **A** Correlation of preS/S genes and their products. The length of the individual domains is given in numbers of aa; apparent molecular masses of glycosylated and unmodified proteins are indicated *on the right*. *Y-shaped symbols represent* glycan residues; *filled symbols* indicate complete, *stippled symbols* partial glycosylation; *myr*, myristic acid residue covalently attached to the N-terminal residues of HBV and DHBV L protein. **B** Membrane topology of the HBV surface proteins. *Horizontal lines* represent the borders of the membrane, *bars* transmembrane regions. Regions I and II have been experimentally characterized; the topology of the following two transmembrane domains is not finally settled. S and M adopt their similar topology in a cotranslational fashion; part of the initially cytoplasmically disposed preS domains of L are apparently post-translationally translocated across the membrane. Note that in L the glycosylation sites in the preS1 and preS2 domain are only used in artificial fusion proteins with a cleavable N-terminal signal sequence, while authentic M is completely glycosylated in its preS2 domain. The overall organization of the DHBV S and L proteins is probably very similar. The presence of *Hsc*70 at the indicated region is inferred from data obtained with the DHBV L protein. See text for details

and between 0.40 and 0.76 for plasma membrane; in HBsAg a ratio of 0.48 was found. The lipid composition of S particles from permanent cell lines (i.e. relatively freshly synthesized particles, in part obtained from cells grown in serum-free medium) with PL 91%, 4% TG and, strikingly, only about 5% cholesterol, is in much better agreement with ER derivation (Satoh et al. 1990). Altogether, however, the variability of the data suggests that lipids may be easily exchanged, and that lipid composition is of limited value for determining the site of S particle or virion biogenesis. It should be noted that the overall lipid content of S particles is very low, and that it has not been demonstrated that the lipids are arranged in a typical unit membrane bilayer.

5.2.2 Structure and Function of the Surface Proteins

As for the core protein, transfection of genetically altered preS/S genes, either alone or in the context of complete genomes as well as in in vitro experiments, has shed light on the roles of the individual S proteins in S particles, and virion assembly. In the DHBV system, infectivity is also a relatively easily addressable parameter. Before describing in more detail the newer data, focused on the surprising multifunctionality of the L protein, the major functions of the different surface proteins will be briefly summarized.

S alone is able to form 22- nm particles that are efficiently secreted, at least in cells from higher eukaryotes (not in insect cells or yeast; Lanford et al. 1989). The same is true for M protein (Sheu and Lo 1992) but not for L, which is retained in the endoplasmic reticulum (ER), or a later compartment, unless S protein is also present; in fact, excess L protein *trans*-inhibits S secretion (Persing et al. 1986). This suggests that normal L export is mediated by formation of mixed particles, and that the ratio of S to L protein is a critical factor determining the fate of L protein. Retention of excess L protein in the ER, or a subsequent compartment, can explain the extreme dilation of membranous structures observed in liver cells from transgenic mice overexpressing L, and may be one factor in hepatocellular carcinogenesis (Chisari et al. 1989). While, except for one report (Ueda et al. 1991), all other studies agree that M protein has no vital functions (e.g. Bruss and Ganem 1991b; Fernholz et al. 1993), L, and in particular its preS1 domain, is absolutely essential for two critical steps in the viral life-cycle: attachment to the target cell, and formation of complete virions, i.e. envelopment of the capsid (see below). Intriguingly, this dual function requires the preS1 domain to be located once on the exterior and once in the interior of the virion. Recent studies on the membrane topology of the hepadnaviral surface proteins may hold a clue to this structural puzzle.

5.2.3 Membrane Topology of the Hepadnaviral Surface Proteins

All hepadnaviral surface proteins are integral membrane proteins; as in their host cell counterparts, topology, i.e. the orientation of individual polypeptide chain segments relative to the membrane, is of utmost functional importance (Fig. 5B). In contrast to most viral glycoproteins (Doms et al. 1993), they contain more than one transmembrane region. N-Linked glycosylation provides clear biochemical evidence that all HBV surface proteins are initially targeted to the ER membrane, in accord with ultrastructural data (Patzer et al. 1986). The characteristic modifications of the glycan side chains demonstrate that surface protein export follows the normal secretory pathway. Initial ER targeting is mediated by an internal signal-anchor sequence ("signal I" in Fig. 5B) close to the N-terminus of S (aa residues 4–28), presumably in a signal recognition particle (SRP) dependent fashion. Such signal-anchor sequences usually consist of some 10–20 hydrophobic aa residues, mediating membrane targeting, and flanking hydrophilic regions. Combinations of signal anchor and so-called stop-transfer sequences allow multispanning membrane proteins to adopt a very specific topology (for review: HIGH and DOBBERSTEIN 1992). In the S protein, a second hydrophobic region ("signal II" in Fig. 5B; aa residues 80–100), separated from signal I by a cytoplasmic loop of about 50 aa, has also been shown to be embedded into the membrane, and is followed by a stretch of some 70 luminally exposed residues which carry the major S-specific epitope and a glycosylation site (EBLE et al. 1987), which for as yet unknown reasons is only used in about half of the S protein molecules. The following 50 residues are again hydrophobic and may be arranged in two more transmembrane domains (STIRK et al. 1992; but see also PRANGE and STREECK 1995). Signal II, in the absence of signal I, is able to mediate ER-targeting of corresponding N-terminal deletion variants and to correctly orient the remaining C-terminal part of S; such proteins are, however, secretion incompetent (EBLE et al. 1987; BRUSS and GANEM 1991a).

The topology of the M protein appears to be essentially the same as that of S; its larger N-terminal PreS2-containing domain is obviously efficiently translocated into the ER lumen by the downstream signals in the S-domain (EBLE et al. 1990), as demonstrated by its complete glycosylation at Asn-4 of the preS2 domain.

5.2.4 Special Role of the L Protein

Although L protein contains the complete M and S domains, its properties are distinctly different from those of the smaller surface proteins: it is not by itself secreted and instead retained in the ER, or a subsequent compartment, and it is required for both virion formation and for infectivity. These characteristics are obviously correlated with the preS1 domain, and possibly with the myristic acid modification present at the N-terminal Gly residue in mature L (PERSING et al. 1987). Deletion studies had indicated that the retention signal in L (subtype adw) is exclusively located in the very N-terminal region of the PreS1 domain (aa 2–19; KUROKI et al. 1989), and independent of myristoylation; other studies, however, concluded that both myristoylation (PRANGE et al. 1991) and the primary sequence of the carboxyl-proximal preS1 amino acids are important (YU 91; NEMECKOVA et al. 1994). Probably, these conflicting data are due to the different HBV subtypes used in these studies; e.g. subtype adw contains an insert of 11 aa at its N-terminus. While L protein retention may enhance virion production in vivo, it is not a prerequisite for virion formation in transfected cells (BRUSS and

THOMSSEN 1994); however, the fatty acid modification is certainly essential for infectivity of both DHBV (MACRAE et al. 1991) and HBV (GRIPON et al. 1995).

Previous data on the accessibility to antibodies and proteases of the preS1 domain in virions and filaments showed that, in accord with the proposed function of the domain as viral attachment protein, virions can be precipitated with preS1-specific antibodies and that the preS2 domain in the L protein is accessible to protease (HEERMANN et al. 1984), indicating an exterior location. Surprisingly, however, and in contrast to M, the L protein is only glycosylated in its S part (to about 50%, as also observed for S), i.e. neither the preS2-specific nor the preS1-specific (Asn-4 of preS1) sites are used. Also, the requirement for L protein in capsid envelopment (see below) is hard to understand if all of its extra sequence were located on the outside of the virion.

Recent data from three groups indicate now that HBV L protein is initially synthesized with a cytoplasmic disposition of the preS1 domain, and that a fraction of the preS1 domains is post-translationally translocated across the membrane (Fig. 5B). Thus it appears that hepadnaviruses exploit an unusual mechanism which allows them to use the same protein domain for functions requiring a luminal/exterior (host cell attachment) as well as a cytoplasmic/interior location (capsid envelopment).

In a coupled in vitro translation/translocation system, which is incompetent for particle formation, most if not all of the preS1 and preS2 domains of L remained accessible to protease, while the preS2 domain of M was protected, indicating that the N-terminal region of M, but not that of L, was efficiently translocated (OSTAPCHUK et al. 1994). Similar experiments with microsomes from in vivo labelled transfected cells showed that also in vivo the preS1/preS2 domains of L are initially cytoplasmically exposed whereas in secreted particles derived from transfected cells and patient serum about half of the preS domains of L are accessible on the particle exterior and hence must have traversed the membrane (BRUSS et al. 1994). Translocation is likely to occur in a post-translational fashion, since the protease-resistant fraction of L molecules increases slowly with time; however, it is not accompanied by glycosylation of the preS1 and/or preS2-specific glycan attachment sites (PRANGE and STREECK 1995). By contrast, these sites are efficiently used if cotranslational translocation is artificially brought about by fusing a cleavable N-terminal signal sequence to L (BRUSS et al. 1994). The region responsible for the unusual topology of L appears to be located in the very C-terminal part of the preS1 domain since an internal deletion of the last 38 preS1 aa gave an L protein whose preS domains are efficiently translocated and glycosylated (PRANGE and STREECK 1995). Complementary, cotranslational translocation was prevented in N-terminally truncated L variants as long as at least the last 17 preS1 encoded aa were present, while further deletions led to an M-like topology (BRUSS and THOMSSEN 1994). Interestingly, the same region preventing cotranslational translocation is also essential for interaction with the capsid.

What could be the mechanism for the delayed and apparently to some extent regulated translocation of about half of the preS1 domains? The answer is not yet clear; however, recent observations with the DHBV system may hold some clues. Although the topology of the DHBV L protein has not been studied in detail, the available data on accessibility to antibodies (YUASA et al. 1991) and proteases (KLINGMÜLLER 1992) suggest a similar organization to its HBV counterpart. Pulse-chase experiments indicate that the proportion of protease-protected and hence luminally disposed preS domains of intracellular L protein increases with time, and in secreted virions about one-half of the preS domains is accessible to protease (I. Swameye and H. Schaller, personal communication). By ortho-phosphate labelling of DHBV-infected primary duck hepatocytes it was shown that a fraction of the L protein is phosphorylated (GRGACIC and ANDERSON 1994). The exact phosphorylation sites are not yet known but may be located in the preS domain, which in HBV L protein can face either the interior or the exterior of the particle. Conceivably, in the phosphorylated molecules the extra charges would interfere with membrane translocation and could thus account for the initial cytoplasmic disposition of the preS domain. As the electrophoretic band pattern changes with time after infection one might envisage regulation of translocation by sequential phosphorylation/dephosphorylation events. A number of obvious mutational experiments should soon clarify this point.

However, post-translational translocation of an already folded protein domain is not a trivial task. An interesting observation is therefore that a cellular chaperone is associated with DHBV particles (I. SWAMEYE, C. KUHN, M. HILD, U. KLINGMÜLLER and H. SCHALLER, in preparation). The protein of about 72 kDa consistently copurifies with DHBV S particles and is also present in preparations enriched for virions; a similar protein is found in preparations of HBV filaments and virions. Direct sequencing followed by molecular cloning showed the duck protein to be the homologue of the constitutively expressed heat shock protein Hsc70, which is highly conserved between species. The major roles for the cytosolic heat shock proteins of the Hsp70/DnaK family, at least in bacteria usually performed in cooperation with DnaJ, are the prevention of premature folding of nascent polypeptides, inhibition of unproductive aggregation, and maintaining proteins in a translocation-competent state, likely mediated by the chaperones recognizing exposed hydrophobic protein regions; this interaction is released by ATP (for reviews: GEORGOPOULOS and WELCH 1993; MARTIN and HARTL 1994). Co-immunoprecipitations suggest that it is the preS domain that interacts with Hsc70 (Fig. 5B). Proteolytic cleavage and ATP-mediated release of the particle-associated Hsc70 require the presence of detergent, indicating that the protein is located in the interior of the particle.

Whether a causal relationship exists between Hsc70 association and the unusual topology of the L protein remains to be established; however, it is tempting to speculate that both are mechanisticly linked. That Hsc70 remains associated with the preS domain despite the relatively high ATP concentration in the cytoplasm is per se remarkable. One explanation could be that the primary sequence of PreS contains (a) region(s) of very high affinity to the chaperone (the C-terminal part of HBV preS1 preventing cotranslational translocation would be a candidate); the chaperone would then keep the preS domain competent for later translocation. Such an explanation is consistent with the finding that replacing preS by the tightly folding globin domain abrogates secretion competence, which is restored if an N-terminal signal sequence mediating cotranslational translocation is added to the artificial protein (BRUSS and GANEM 1991a). Alternatively, the preS domain might be unable to fold properly in the reducing environment of the cytoplasm and hence remain associated with the chaperone; those chains that manage to be translocated might be stabilized by disulphide bonds inside the ER lumen (or later compartment) with its much higher redox potential (DOMS et al. 1993). All S proteins contain abundant Cys residues, which are indeed extensively used for intra- and intermolecular disulphide bridges (PETERSON 1987). Correct disulphide bonding is not only important for further secretion but also for maintaining the antigenic, and hence presumably structural, integrity of the proteins (MANGOLD and STREECK 1993). Further experiments are required to elucidate the functional significance of Hsc70 in the hepadnaviral life-cycle; however, the principal parallel to the recent finding that another cytosolic chaperone, cyclophilin A (peptidyl-prolyl-*cis-trans*-isomerase), is associated with and required for the infectivity of HIV-1 virions (FRANKE et al. 1994; THALI et al. 1994) is certainly remarkable.

5.2.5 S Particle Assembly

The data presented above demonstrate that all hepadnaviral surface proteins are initially synthesized as transmembrane proteins at the ER (SIMON et al. 1988), with S and M acquiring their final topology in a cotranslational fashion, while in at least a fraction of L the preS domains are post-translationally reoriented. The absence of detectable surface proteins on the plasma membrane indicates that somewhere during their intracellular journey the surface proteins must bud, as S particles, into the lumen of one of the compartments of the secretory pathway, or, in the case of excess L protein, be retained in the membrane of such a compartment. Usually, the ER itself is considered the budding compartment, based on EM studies that show S particles in the lumen of smooth intracellular membranes (PATZER et al. 1986), and the observation that the glycans on all intracellular surface proteins are sensitive to endoglycosidase H while on the extracellular forms they are resistant. The enzyme digests only immature, high-mannose-type glycans (PATZER et al. 1984). This indicates that the rate-limiting step of export must lie prior to the medial Golgi.

Meanwhile, there is accumulating evidence for (an) additional "intermediate" compartment(s) between the ER and the Golgi complex (for review: HENDRIKS and FULLER 1993); thus, budding into this compartment would also be compatible with the glycosylation data. Double immunofluorescence of stably transfected, HBV S protein producing cells indeed showed a fraction of S in a compartment devoid of resident ER markers like BiP and protein disulphide isomerase (PDI) but containing Rab2, an established intermediate compartment marker (HUOVILA et al. 1992). When disulphide formation was followed by pulse chase experiments, initially only dimer cross-links were observed; these dimers were then slowly chased into heavily cross-linked multimers. At least in vitro, the oligomeric but not the dimer cross-links were unstable in the presence of PDI. Brefeldin A, a drug that via interaction with ARF (ROTHMAN 1994) blocks exit from the ER and leads to redistribution of *cis* and medial Golgi markers to the ER (LIPPINCOTT-SCHWARTZ et al. 1990), led to accumulation of dimers but prevented oligomer formation, suggesting that their generation requires exit from the ER. The PDI-excluding compartment, but not the ER contained the typical S particles. Thus, S particle formation appears to proceed by a two-step mechanism (Fig. 6A), in which S-S-linked S protein dimers are generated in the ER in the presence of PDI, but assemble into particles in a later, PDI-deficient compartment which is discontinuous with ER and Golgi (HUOVILA et al. manuscript submitted). Oligomerization, and stabilization of the multimers by oxidative cross-linking, could facilitate budding into the lumen of the compartment by excluding host components that mediate normal transport vesicle formation towards the cytoplasmic side (ROTHMAN 1994).

As the half-times of oligomer-cross-link formation, carbonate extractability (indicating transition from a membrane-bound into the soluble, particulate form) and particle secretion are very similar (SIMON et al. 1988), the budding process in this compartment appears to be the rate-limiting step in spherical S particle secretion. Whether the same holds true for the L-containing filaments (as implied in Fig. 6C) has not yet been established, and it is not clear where translocation of the preS domains occurs. The lack of glycosylation in the translocated preS domains as well as the inability of isolated microsomes to support translocation suggests that the process takes place in a post-ER compartment, possibly the intermediate compartment mentioned above. Determining the preS topology in cells kept at low temperature, or treated with brefeldin A, could help to address this question.

5.3 Virion Formation

Envelopment of the cytoplasmically produced nucleocapsids is the least understood of the morphogenetic events in the hepadnaviral life-cycle. Attempts to directly visualize budding virions by EM, i.e to show profiles in which the viral envelope is contiguous with the unit membrane of the corresponding compartment, have not led to convincing results, possibly because budding is very rapid. Based on analogies to S particle formation, it is generally assumed that budding occurs into the ER, as with flavi- and pestiviruses; in view of the newer data on empty envelope formation, the intermediate compartment may as well be the site of virion formation, as seen, e.g. for corona viruses (KRIJNSE-LOCKER et al. 1994; for reviews: PETTERSON 1991; GRIFFITHS and ROTTIER 1992). However, it is by no means proven that empty envelopes and virions are generated by exactly the same mechanism (as speculated in Fig. 6D). One might expect that capsid-triggered budding is in fact a more efficient process than S particle formation, otherwise only empty envelopes may be generated.

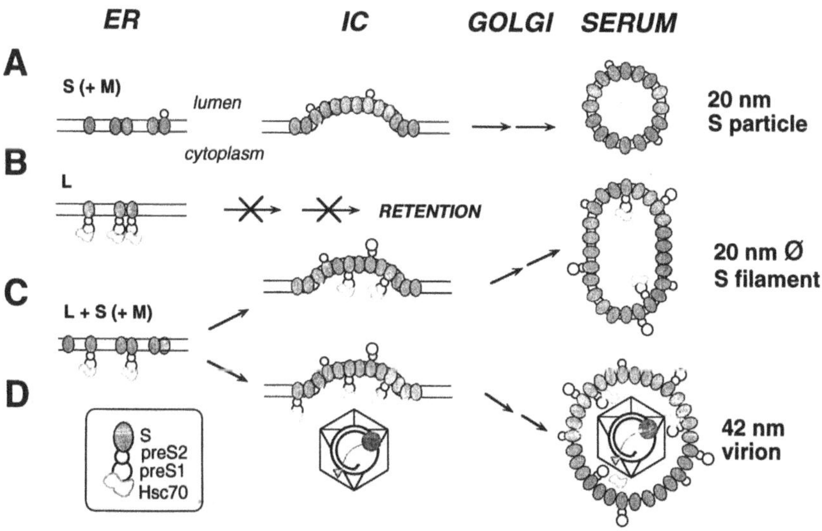

Fig. 6A-D. Model for the formation of S particles and virions. **A** Spherical S particles. S protein plus a small amount of M dimerize in the ER and oligomerize in the intermediate compartment (*IC*). Presumably triggered by interactions between subunits, spherical particles bud into the lumen of the IC and are exported through the subsequent compartments of the secretory pathway. **B** L protein retention. L protein is initially synthesized with cytoplasmically disposed preS domains which are probably associated with *Hsc*70. In the absence of S, L is not exported but retained in the ER or IC. **C** Filamentous S particles. If both L and S protein are present, mixed oligomers are formed which bud as filamentous subviral particles. About half of the preS domains are post-translationally translocated, the others probably remain associated with *Hsc*70. The different composition of spherical and filamentous particles suggests that L protein is responsible for their different morphologies. **D** Virions. Cytoplasmic nucleocapsids interact with the non-translocated preS domains of L, possibly mediated by *Hsc*70, while the translocated preS domains in the mature virion are available for later attachment to the host cell. There is good experimental evidence for the two-step assembly mechanism of spherical S particles which contain no or little L protein. The models shown in **C** and **D** assume similar pathways for the L-containing particles; inferred details of the sites of assembly and preS translocation, or the role of *Hsc*70 in HBV filament or virion formation, are compatible with the limited data available from the DHBV system but should be regarded as speculative at present

While it is not clear how capsids are targeted to the membrane containing the surface proteins, and what regions on the capsid are the determinants for interaction with the envelope proteins, recent data have revealed some details for the surface proteins. First, L protein is absolutely required for virion formation (Bruss and Ganem 1991b), implying a matrix protein-like function for the preS1 domain in this interaction. As outlined above, the paradox that this domain is found on the virion exterior has been resolved by demonstrating that about half of the preS domains remain on the interior side of the envelope. As shown using a series of deletions in the preS1 domain, the same relatively short C-terminal region of the preS1 domain that prevents cotranslational translocation is sufficient to produce replication-competent particles that can be immunoprecipitated with anti-S antibodies, as expected from enveloped virions (Bruss and Thomssen 1994). This suggests that an initial cytoplasmic disposition of preS is important for virion formation, and that the C-terminal one-sixth of preS1 is involved in capsid binding. Both myristoylation and L protein retention in an early secretory compartment are not per se essential for the process in transfected cells (Bruss and Ganem 1991b). However, the fatty acid modification is apparently critical for virion infectivity (see above), possibly reflecting a role in attachment to the host cell, or subsequent steps such as membrane fusion. Also, L protein retention might at least contribute to efficient virion formation in vivo.

In general, the functional roles of the DHBV S and L proteins appear to be similar to those of their HBV counterparts. Both proteins are required to form enveloped virus particles (Summers et al. 1991). Truncated L proteins lacking up to 52 N-terminal aa still support formation of enveloped virions which, however, are non-infectious, like a mutant lacking the myristoylation signal (Macrae et al. 1991). A linker-scan analysis of the preS region (Lenhoff and Summers 1994) showed that the primary sequence between aa residues 117 and 136 is important for capsid envelopment, while almost all of preS is essential for infectivity. A defect for capsid envelopment could also be induced by replacing the authentic aa residues at positions 128 and 131, suggesting that this region comprises a determinant for capsid/envelope interaction. Hence the preS domain appears to be functionally bipartite, with the N-terminal two-thirds being required for interaction with the host cell (i.e. infectivity), and the C-terminal one-third for interaction with the capsid (i.e. envelopment).

The regions directly implicated in these interactions are small compared to the overall size of the preS domains; for HBV, the viral attachment site has been proposed to reside in the preS1 sequence comprising aa 21–47 (Neurath et al. 1986, 1992), for DHBV the corresponding region has been mapped to the preS aa 81–121 (Klingmüller 1992). It should be stressed, however, that infectivity not only requires the physical presence of the viral attachment site but its being exposed on the virion exterior, while for envelopment the C-terminal part of preS must be cytoplasmically disposed. Thus, the information for preS translocation must also be encoded in the preS sequence. Moreover, the time course of the process must be coordinated with export, as neither topological state with all preS domains on one side is compatible with assembly of infectious virions. The as yet mysterious retention of L protein might contribute to this coordination.

Retroviruses are long known to form pseudotypes with unrelated viruses, e.g. vesicular stomatitis virus (cf. Hopkins 1993), indicating that the interaction between the capsid and Env proteins is relatively unspecific. The apparent dependence of core particle export on the presence of L protein implicates that hepatitis B virus envelopment, by contrast, relies on specific recognition events between the capsid and the surface proteins. Phenotypic mixing experiments support this view. Concerning envelope formation as such, only the closely related avian, or the mammalian envelope proteins can substitute for each other, since surface proteins from HBV and WHV, but not DHBV coassemble into mixed S particles. Regarding interaction with the capsid, a chimeric L protein with the preS region of WHV and the S domain of HBV supports virus envelopment, whereas a corresponding DHBV-preS/HBV-S chimera does not although it forms S particles (GERHARDT and BRUSS 1995). These data suggest that even if the lateral interactions in the envelope are undisturbed the preS domains can discriminate between different capsids. Complementary, HHBV cores with a DHBV envelope produced by cotransfection infect duck cells much more efficiently (ISHIKAWA and GANEM 1995) than authentic HHBV virions (SPRENGEL et al. 1988). Hence the preS domains discriminate between the different target cells but not the very similar capsids. Overall it appears that in its specificity the mechanism of hepadnaviral envelopment is more closely related to that of alphaviruses than that of retroviruses (SUOMALAINEN and GAROFF 1994; CHENG et al. 1995; for review: GAROFF et al. 1994).

The interaction between envelope proteins and capsids may also be responsible for the apparent regulation of cccDNA levels in DHBV-producing cells (SUMMERS et al. 1990). Nuclear cccDNA is the transcriptionally active template; the pool of cccDNA is believed to be replenished by an intracellular recycling of cytoplasmic cores to the nucleus (TUTTLEMAN et al. 1986; WU et al. 1990). If no envelope protein, or only S protein, is present, the copy number of cccDNA molecules per infected primary duck liver cell can increase from some 20 to several hundred (SUMMERS et al. 1990; LENHOFF and SUMMERS 1994). Downregulation depends on the L protein, and mutations in L that prevent capsid envelopment upregulate cccDNA levels. This suggests that the interaction of L with the capsid is involved in both processes. According to this model, core particles deliver the partially ds DNA genome to the nucleus by default; once sufficient amounts of L and S protein have accumulated they are instead sequestered into the secretory pathway. Recent in vitro evidence for direct binding between capsid and L protein supports this attractive model (DYSON and MURRAY 1995). Expectedly, higher cccDNA levels result in increased amounts of total viral RNA; however, for unknown reasons the amounts of replicative intermediates are not correspondingly amplified (SUMMERS et al. 1991).

A still unresolved question is whether all capsids, or capsids in all stages of maturation, are equally well enveloped and exported. At least for DHBV, exported virions contain much more mature DNA molecules (i.e. RC and full-length linear genomes) than intracellular DHBV cores. This has been interpreted to indicate the presence of some kind of maturation signal on the core particle surface which would trigger preferential envelopment of capsids with more mature genomes (SUMMERS and MASON 1982). The nature of that signal is not known to date, but it might be related to core protein phosphorylation (see above), as in virions core protein with fewer phosphoryl groups is enriched compared to the intracellular forms (YU and SUMMERS 1994). Genome maturation is, however, not an absolute requirement for HBV envelopment, as for instance HBV cores which contain immature genomes due to a defect in core protein can still be exported as enveloped particles, albeit with lower efficiency (NASSAL 1992).

5.4 Hepatitis Delta Virus

The interactions between hepadnaviral cores and envelope proteins appear to be specific; however, hepatitis Delta virus (HDV) provides a naturally occurring, and medically important, example of pseudotyping, using the surface proteins of HBV and, at least experimentally, WHV, but not DHBV (for review: LAZINSKI and TAYLOR 1994). HDV is a subviral agent with a small rod-shaped single-stranded RNA genome of about 1700 nt that bears structural features found in viroids and virusoids from plants, and encodes a nucleic acid binding protein. This delta antigen (δAg) occurs in two forms which are generated by RNA editing; the larger δAg-L differs from the smaller δAg-S by a 19-aa C-terminal extension. While δAg-S supports genome replication, δAg-L inhibits it and instead promotes genome packaging, a process correlated with isoprenylation of a Cys residue present in the unique C-terminal extension of δAg-L. The hepadnavirus helper is not required for HDV replication but rather for its efficient transmission, by supplying an envelope for the RNA/δAg complex. The resulting virions are spherical with a diameter similar to that of Dane particles (35–41 nm). The RNP complex, thought to consist of one molecule of genomic RNA and some 60–90 molecules of δAg (both forms), appears to be roughly spherical with a diameter of about 18 nm (RYU et al. 1993), and hence is substantially smaller than a hepadnaviral core particle. Other differences exist in morphogenesis: envelopment is independent of L protein since HDV core-containing, secreted particles can be produced in the presence of only S protein; δAg-L can also be packaged into S particles in the absence of viral RNA. Thus specific interactions between δAg-L and surface proteins appear to mediate HDV virion assembly, as in hepadnaviral morphogenesis; however, the contact regions on the surface proteins are apparently different, and the value of HDV as a model for HBV morphogenesis is disputable. By contrast, infectivity of HDV is apparently as dependent on the presence of L protein on the envelope as that of the hepadnaviruses, and independent of the M protein (SUREAU et al. 1993). Hence, the several 100 000 copies of δRNA/infected cell may provide a convenient marker for studies aimed at analysing the role of the hepadnaviral envelope in infection.

6 Model for Hepadnavirus Morphogenesis

In combination, the salient features of the individual assembly steps described above lead to a plausible though in some respects still speculative model for hepatitis B virus morphogenesis. The RNA pregenome is translated in the cytoplasm to yield core and P protein (Fig. 4B), while the surface proteins, made from the subgenomic preS/S and S mRNAs, are targeted to the ER. Core protein monomers rapidly dimerize and provide the pool of assembly intermediates. P protein binds to the encapsidation signal on the pregenome (Fig. 4A), forming a pre-assembly complex that triggers core protein association at a lower critical concentration than in its absence (Fig. 4B). The majority of S protein, plus M and very little L protein, dimerize in the ER and oligomerize and bud into the lumen of a post-ER/pre-Golgi compartment, forming spherical (Fig. 6A) and, in the presence of more L protein, filamentous subviral particles (Fig. 6C). In a relatively slow process, part of the preS1 domains of L are translocated across the membrane while the other part remains in its initial cytoplasmic location, presumably associated with the Hsc70 chaperon. Virion formation initiates, perhaps mediated by Hsc70, via specific interactions between core particles and the cytoplasmically located preS domains of L, and leads to budding into the ER or the intermediate compartment (Fig. 6D). L retention in an early compartment of the secretory pathway (Fig. 6B) may increase, at least in vivo, the efficiency of virus formation; the interaction with the core particle, on the other hand, could relax some sterical constraints that apparently prevent the formation of empty subviral particles with a high L protein content. After budding, the virions are secreted via vesicular transport through the remaining compartments of the secretory pathway, and are eventually released into the blood-stream. Overall, this mechanism ensures the formation of infectious particles: the triggering function of the pre-assembly complex of P protein and pregenome prevents formation of replication-incompetent cores, and the unusual topological properties of L warrant specific envelopment of these cores and the presence of a viral attachment domain for interaction with the target cell. HDV envelopment, by contrast, is L independent, which may be the reason that more than 90% of the HDV virions are non-infectious (Lazinski and Taylor 1994). This less elaborate strategy is compensated for by the 1000- to 10 000-fold higher number of HDV genomes per infected cells.

7 Summary and Perspectives

The studies summarized above have established that many steps of hepadnaviral morphogenesis rely on molecular mechanisms that are uniquely distinct from their retroviral counterparts. Prominent examples are the coupled encapsidation of the separately expressed P protein and the RNA pregenome, its concomitant reverse transcription, the lack of protein processing, the production of empty envelopes, the multifunctionality of the large S protein which depends on its most unusual topological properties, or the specificity of the interactions between capsid and surface proteins. Overall, a consistent picture of the apparently unique hepatitis B virus assembly pathway is emerging; however, many details as well some of the basic features are not yet fully understood.

The resolution of the structural data available for the hepadnaviral particles and their constitutent components is still low. Quick results from X-ray diffraction are not in sight, but advanced electron microscopic techniques combined with mutational studies might reveal the location of individual secondary structure elements in the core protein, the interior organization of the capsid including its nucleic acid contents as well as the architecture of the envelope, along with the arrangement of the lipids. Besides their importance for the further development of core and S particles into suitable carriers for foreign antigenic determinants, such studies might also provide information on some of the dynamic aspects of assembly, e.g. what structural changes accompany reverse transcription of the RNA pregenome into DNA, whether the T = 3 or the T = 4 capsids are the biologically relevant species, and whether the interaction between core and envelope is so tight and specific that the surface proteins are forced into a similar icosahedrally symmetrical arrangement as the underlying core protein subunits. Other biophysical techniques may help to elucidate the three-dimensional structure of the RNA encapsidation signal.

Some of the following open questions may yield to the continued application of reverse genetics and DNA transfection; others will require the development of in vitro systems allowing the reconstitution of the assembly of capsids, S particles and eventually complete virions from purified components. Of particular importance for a systematic study of the early steps of infection, which require a reversal of the assembly process, will be the establishment of a feasible in vitro infection system; in the meantime, microinjection techniques might be helpful. Some of the essential interactions in the assembly process still await experimental confirmation, e.g. between core protein and the preassembly complex of P protein and ε or between the capsid and the envelope; further studies might then reveal the specific regions involved. It is not clear how formation of the complex T = 4 or T = 3 structures of the capsid is determined or how the selective interaction between P protein and the 5′-copy, but not the 3′-copy of ε, is controlled. The entire process of RNA transformation into dsDNA within the restricted space of the capsid and with a single P protein molecule covalently linked to its substrate is a mechanistic puzzle. The unusual topological properties of L protein need to be further clarified. How and where does translocation of the preS domain occur and how does it influence the shape of S particles? Is the mechanism of virion formation identical to that of S particle production, and where in the cell does virus budding occur? The functional role of the chaperones implied in assembly will have to be elucidated; more than likely, further cellular proteins will be identified that play a crucial role in the virus life-cycle, not only for assembly but also for initiating a new round of infection. The cellular receptor for none of the hepadnaviruses is known. It is unclear where penetration of the capsid into the host cell cytoplasm occurs, what the mechanism of fusion is, where in the surface proteins the fusogenic activity resides and how it is triggered. Similarly, we do not know how capsid disassembly and delivery of the viral genome into the nucleus are regulated. Phosphorylation of the core and surface proteins may be involved, but there are as yet no experimental data to directly support this assumption.

This list of open questions is certainly still incomplete; however, the data collected so far are encouraging enough to initiate studies aimed at interfering with individual steps of the morphogenetic pathway. The frightening prospects of liver cirrhosis and hepatocellular carcinoma for long-term chronic HBV carriers certainly warrant that the potential of such novel antiviral concepts is explored.

Acknowledgements. I wish to thank many colleagues for providing data prior to publication, John M. Kenney for critically reading the manuscript, and Heinz Schaller for helpful discussions as well as financial support. Work in the author's laboratory has been supported by grants from the Bundesministerium für Forschung und Technologie, and the Deutsche Forschungsgemeinschaft (SFB229, Na-154/3).

References

Albin C, Robinson WS (1980) Protein kinase activity in hepatitis B virus. J Virol 34: 297–302

Allan D, Kallen KJ (1994) Is plasma membrane lipid composition defined in the exocytic or endocytic pathway? Trends Cell Biol 4: 350–353

Argos P, Fuller SD (1988) A model for the hepatitis B virus core protein: prediction of antigenic sites and relationship to RNA virus capsid proteins. EMBO J 7: 819–824

Bartenschlager R, Schaller H (1992) Hepadnaviral assembly is initiated by polymerase binding to the encapsidation signal in the viral RNA genome. EMBO J 11: 3413–3420

Bartenschlager R, Junker-Niepmann M, Schaller H (1990) The P gene product of hepatitis B virus is required as a structural component for genomic RNA encapsidation. J Virol 64: 5324–5332

Beckett D, Wu HN, Uhlenbeck, OC (1988) Roles of operator and nonoperator RNA sequences in bacteriophage R17 capsid assembly. J Mol Biol 204: 939–947

Bichko V, Schödel F, Nassal M, Gren E, Berzinsh I, Borisova G, Miska S, Peterson DL, Gren E, Pushko P, Will H (1993) Epitopes recognized by antibodies to denatured core protein of hepatitis B virus. Mol Immunol 30: 221–231

Birnbaum F, Nassal M (1990) Hepatitis B virus nucleocapsid assembly: primary structure requirements in the core protein. J Virol 64: 3319–3330

Blum HE (1993) Hepatitis B virus: significance of naturally occurring mutants. Intervirology 35: 40–50

Blumberg BS, Alter HJ, Visnich S (1965) A "new" antigen in leukemia sera. JAMA 191: 541–546

Borisova G, Arya B, Dislers A, Borschukova O, Tsibinogin V, Skrastina D, Eldarov MA, Pumpens P, Skryabin KG, Grens E (1993) Hybrid hepatitis B virus nucleocapsid bearing an immuno-dominant region from hepatitis B virus surface antigen. J Virol 67: 3696–3701

Braig K, Otwiniwski Z, Hegde R, Boisvert DC, Joachimiak A, Horwich AL, Sigler PB (1994) The crystal structure of the bacterial chaperonin GroEL at 2.8 Å resolution. Nature 371: 578–586

Bruss V, Ganem D (1991a) Mutational analysis of hepatitis B surface antigen particle assembly and secretion. J Virol 65: 3813–3820)

Bruss V, Ganem D (1991b) The role of envelope proteins in hepatitis B virus assembly. Proc Natl Acad Sci USA 88: 1059–1063

Bruss V, Thomssen R (1994) Mapping a region of the large envelope protein required for hepatitis B virion maturation. J Virol 68: 1643–1650

Bruss V, Lu X, Thomssen R, Gerlich WH (1994) Post-translational alterations in transmembrane topology of the hepatitis B virus large envelope protein. EMBO J 13: 2273–2279

Büscher M, Reiser W, Will H, Schaller H (1985) Transcripts and the putative RNA pregenome of duck hepatitis B virus: implications for reverse transcription. Cell 40: 717–724

Calvert J, Summers J (1994) Two regions of an avian hepadnavirus RNA pregenome are required in cis for encapsidation. J Virol 68: 2084–2090

Chang LJ, Pryciak P, Ganem D, Varmus HE (1989) Biosynthesis of the reverse transcriptase of hepatitis B viruses involves de novo translational initiation not ribosomal frameshifting. Nature 337: 364–368

Chang LJ, Ganem D, Varmus HE (1990) Mechanism of translation of the hepadnaviral polymerase (P) gene. Proc Natl Acad Sci USA 87: 5158–5162

Chang C, Zhou S, Ganem D, Standring DN (1994) Phenotypic mixing between different hepadnavirus nucleocapsid proteins reveals C protein dimerization to be cis preferential. J Virol 68: 5225–5231

Chen HS, Kaneko S, Girones R, Anderson RW, Hornbuckle WE, Tennant BC, Cote PJ, Gerin JL, Purcell RH, Miller RH (1993) The woodchuck hepatitis virus X gene is important for establishment of virus infection in woodchucks. J Virol 67: 1218–1226

Cheng RH, Kuhn RJ, Olson NH, Rossmann MG, Choi H-K, Smith TJ, Baker TS (1995) Nucleocapsid and glycoprotein organization in an enveloped virus. Cell 80: 621–630

Chisari FV, Klopchin K, Moriyama T, Pasquinelli C, Dunsford HA, Brinster RL, Palmiter RD (1989) Molecular pathogenesis of hepatocellular carcinoma in hepatitis B virus transgenic mice. Cell 59: 1145–1156

Cohen BJ, Richmond JE (1982) Electron microscopy of hepatitis B core antigen synthesized in E. coli. Nature 296: 677–678

Crowther RA, Kiselev NA, Böttcher B, Berriman JA, Borisova GP, Ose V, Pumpens P (1994) Three-dimensional structure of hepatitis B virus core particles determined by electron cryomicroscopy. Cell 77: 943–950

Dalseg R (1990) Expression of hepatitis core antigen in E. coli, its characterization and properties as a carrier for foreign antigenic determinants. PhD thesis, University of Heidelberg

Doms RW, Lamb RA, Rose JK, Helenius A (1993) Folding and assembly of viral membrane proteins. Virology 193: 545–562

Dubochét J, Adrian M, Chan JJ, Homo JC, Lepault J, McDowall AW, Schultz P (1988) Cryo-electron microscopy of vitrified specimens. Q Rev Biophys 21: 129–228

Dyson MR, Murray K (1995) Selection of peptide inhibitors of interactions involved in complex protein assemblies: association of the core and surface antigens of hepatitis B virus. Proc Natl Acad Sci USA 92: 2194–2198

Eble BE, MacRae DR, Lingappa VR, Ganem D (1987) Multiple topogenic sequences determine the transmembrane orientation of hepatitis B surface antigen. Mol Cell Biol 7: 2591–3601

Eble BE, Lingappa VR, Ganem D (1990) The N-terminal (preS2) domain of a hepatitis B virus surface glycoprotein is translocated across membranes by downstream signal sequences. J Virol 64: 1414–1419

Eckhardt SG, Milich DR, McLachlan A (1991) Hepatitis B virus core antigen has two nuclear localization sequences in the arginine-rich carboxy terminus. J Virol 65: 575–582

Fernholz D, Galle PR, Stemler M, Brunetto M, Bonino F, Will H (1993) Infectious hepatitis B virus variant defective in preS2-expression in a chronic carrier. Virology 194: 137–148

Franke EK, Yuan HEH, Luban J (1994) Specific incorporation of cyclophilin A into HIV-1 virions. Nature 372: 359–362

Galle PR, Hagelstein J, Kommerell B, Volkmann M, Schranz P, Zentgraf H (1994) In vitro experimental infection of primary human hepatocytes with hepatitis B virus. Gastroenterology 106: 664–673

Gallina A, Bonelli F, Zentilin L, Rindi G, Muttini M, Milanesi G (1989) A recombinant hepatitis B core antigen polypeptide with the protamine-like domain deleted self-assembles into capsid particles but fails to bind nucleic acids. J Virol 63: 4645–4652

Ganem D (1991) Assembly of hepadnaviral virions and subviral particles. In: Mason WS, Seeger C (eds) Hepadnaviruses. Molecular biology and pathogenesis. Springer, Berlin Heidelberg New York, pp 61–83 (Current topics in microbiology and immunology, vol 168)

Ganem D, Varmus HE (1987) The molecular biology of hepatitis B virus Annu Rev Biochem 56: 651–693

Garoff H, Wilschut J, Liljestrom P, Wahlberg JM, Bron R, Suomalainen M, Smyth J, Salminen A, Barth BU, Zhao H et al (1994) Assembly and entry mechanisms of Semliki Forest virus. Arch Virol [Suppl] 9: 329–338

Gavilanes F, Gonzalez-Ros JM, Peterson DL (1982) Structure of hepatitis B surface antigen. Characterization of the lipid components and their association with the viral proteins. J Biol Chem 257: 7770–7777

Georgopoulos C, Welch WJ (1993) Role of the major heat shock proteins as molecular chaperones. Annu Rev Cell Biol 9: 601–634

Gerhardt E, Bruss V (1995) Phenotypic mixing of rodent but not avian hepadnavirus surface proteins into human hepatitis B virus particles. J Virol 69: 1201–1208

Gerlich WH, Goldmann U, Müller R, Stibbe W, Wolff W (1982) Specificity and localization of the hepatitis B virus-associated kinase. J Virol 42: 761–766

Grgacic EVL, Anderson DA (1994) The large surface protein of duck hepatitis B virus is phosphorylated in the preS domain. J Virol 68: 7344–7350

Griffiths G, Rottier P (1992) Cell biology of viruses that assemble along the biosynthetic pathway. Semin Cell Biol 3: 367–381

Gripon P, Diot C, Guguen-Guillouzo C (1993) Reproducible high level infection of cultured adult human hepatocytes by hepatitis B virus: effect of polyethylene glycol on adsorption and penetration. Virology 192: 534–540

Gripon P, Le Seyec J, Rumin S, Guguen-Guillouzo C (1995) Myristylation of the hepatitis B virus large surface protein is essential for viral infectivity. Virology 213: 292–299

Guidotti LG, Martinez V, Loh YT, Rogler CE, Chisari FV (1994) Hepatitis B virus nucleocapsid particles do not cross the hepatocyte nuclear membrane in transgenic mice. J Virol 68: 5469–5475

Hatton T, Zhou S, Standring DN (1992) RNA- and DNA-binding activities in hepatitis B virus capsid protein: a model for their roles in viral replication. J Virol 66: 5232–5241

Heermann KH, Gerlich WH (1991) Surface proteins of hepatitis B viruses. In: McLachlan A (ed) Molecular biology of the hepatitis B virus. CRC Press, Boca Raton, pp 145–169

Heermann KH, Goldmann I, Schwartz W, Seyffarth T, Baumgarten H, Gerlich WH (1984) Large surface proteins of hepatitis B virus containing the preS sequence. J Virol 52: 396–402

Hendriks RJM, Fuller SD (1993) Compartments of the early secretory pathway. In: Maddy AH, Harris JR (eds) Subcellular biochemistry: membrane biogenesis. Plenum, New York, 22: 101–149

High S, Dobberstein B (1992) Mechanisms that determine the transmembrane disposition of proteins. Curr Opin Cell Biol 4: 581–586

Hill CS, Rimmer JM, Green BN, Finch JT, Thomas JO (1991) Histone-DNA interactions and their modulation by phosphorylation of Ser-Pro-X-Lys/Arg motifs. EMBO J 10: 1939–1948

Hirsch RC, Lavine JE, Chang LJ, Varmus HE, Ganem D (1990) Polymerase gene products of hepatitis B viruses are required for genomic RNA packaging as well as for reverse transcription. Nature 344: 552–555

Hopkins N (1993) High titers of retrovirus (vesicular stomatitis virus) pseudotypes, at last. Proc Natl Acad Sci USA 90: 8759–8760

Hu J, Seeger C (1996) Hsp90 is required for the activity of a hepatitis B virus reverse transcriptase. Proc Natl Acad Sci USA 93: 1060–1064

Huovila AP, Eder AM, Fuller SD (1992) Hepatitis B surface antigen assembles in a post-ER, pre-Golgi compartment. J Cell Biol 118: 1305–1320

Huovila AP, Gowen BE, Fuller SD (manuscript submitted) An ER-Golgi intermediate compartment is not continuous with the endoplasmic reticulum.

Ishikawa T, Ganem D (1995) The pre-S domain of the large viral envelope protein determines host range in avian hepatitis B viruses. Proc Natl Acad Sci USA 92: 6259–6263

Junker-Niepmann M, Bartenschlager R, Schaller H (1990) A short cis-acting sequence is required for hepatitis B virus pregenome encapsidation and sufficient for packaging of foreign RNA. EMBO J 9: 3389–3396

Kann M, Gerlich WH (1994) Effect of core protein phosphorylation by protein kinase C on encapsidation of RNA within core particles of hepatitis B virus. J Virol 68: 7993–8000

Kenney JM, von Bonsdorff C-H, Nassal M, Fuller SD (1995) Conformational flexibility and evolutionary conservation in the hepatitis B virus core structure. Structure 3: 1009–1019

Klingmüller U (1992) Interactions of DHBV with primary duck hepatocytes: characterization of the viral ligand and detection of the cellular receptor. PhD thesis, University of Heidelberg

Klingmüller U, Schaller H (1993) Hepadnavirus infection requires interaction between the viral pre-S domain and a specific hepatocellular receptor. J Virol 67: 7414–7422

Knaus T, Nassal M (1993) The encapsidation signal on the hepatitis B virus RNA pregenome forms a stem-loop structure that is critical for its function. Nucleic Acids Res 21: 3967–3975

Köck J, Schlicht HJ (1993) Analysis of the earliest steps of hepadnavirus replication: genome repair after infectious entry into hepatocytes does not depend on viral polymerase activity. J Virol 67: 4867–4874

Krijnse-Locker J, Ericsson M, Rottier PJ, Griffiths G (1994) Characterization of the budding compartment of mouse hepatitis virus: evidence that transport from the RER to the Golgi complex requires only one vesicular transport step. J Cell Biol 124: 55–70

Kuroki K, Russnack, Ganem D (1989) Novel N-terminal amino acid sequence required for retention of a hepatitis B virus glycoprotein in the endoplasmic reticulum. Mol Cell Biol 9: 4459–4466

Lanford RE, Notvall L (1990) Expression of hepatitis B virus core and precore antigens in insect cells and characterization of a core-associated kinase activity. Virology 176: 222–233

Lanford RE, Luckow V, Kennedy RC, Dreesman GR, Notvall L, Summers MD (1989) Expression and characterization of hepatitis B virus surface antigen polypeptides in insect cells with a baculovirus expression system. J Virol 63: 1549–1557

Laskus T, Rakela J, Persing DH (1994) The stem-loop structure of the cis-encapsidation signal is highly conserved in naturally occurring hepatitis B virus variants. Virology 200: 809–12
Lazinski DW, Taylor JM (1994) Recent developments in hepatitis delta research. Adv Virus Res 43: 187–231
Lenhoff RJ, Summers J (1994) Coordinate regulation of replication and virus assembly by the large envelope protein of an avian hepadnavirus. J Virol 68: 4565–4571
Lin CG, Lo SJ (1992) Evidence for involvement of a ribosomal leaky scanning mechanism in the translation of the hepatitis B virus pol gene from the viral pregenome RNA. Virology 188: 342–52
Lingappa JR, Martin RL, Wong ML, Ganem D, Welch WJ, Lingappa VR (1994) A eukaryotic cytosolic chaperonin is associated with a high molecular weight intermediate in the assembly of hepatitis B virus capsid, a multimeric particle. J Cell Biol 125: 99–111
Lippincott-Schwartz J, Donaldson JG, Schweizer A, Berger EG, Hauri HP, Yuan LC, Klausner RD (1990) Microtubule-dependent retrograde transport of proteins into the ER in the presence of Brefeldin A suggests an ER recycling pathway. Cell 60: 821–836
Lok ASF (1994) Treatment of chronic hepatitis B. J Viral Hepatitis 1: 105–124
Machida A, Ohnuma H, Tsuda F, Yoshikawa A, Hoshi Y, Tanaka T, Kishimoto S, Akahane Y, Miyakawa Y, Mayumi M (1991) Phosphorylation in the carboxyl-terminal domain of the capsid protein of hepatitis B virus: evaluation with a monoclonal antibody. J Virol 65: 6024–6030
Macrae DR, Bruss V, Ganem D (1991) Myristylation of a duck hepatitis B virus envelope protein is essential for infectivity but not for virus assembly. Virology 181: 359–363
Mak J, Jiang M, Wainberg MA, Hammarskjold ML, Rekosh D, Kleiman L (1994) Role of Pr160 gag-pol in mediating the selective incorporation of tRNA(Lys) into human immunodeficiency virus type 1 particles. J Virol 68: 2065–2072
Mangold CM, Streeck RE (1993) Mutational analysis of the cysteine residues in the hepatitis B virus small envelope protein. J Virol 67: 4588–4597
Martin J, Hartl FU (1994) Molecular chaperones in cellular protein folding. Bioessays 16: 689–692
Mason WS, Seeger C (eds) (1991) Hepadnaviruses. Molecular biology and pathogenesis. Springer, Berlin Heidelberg New York (Current topics in microbiology and immunology, vol 168)
McAleer WJ, Buynak EB, Maigetter RZ, Wampler DE, Miller WJ, Hilleman MR (1984) Human hepatitis B vaccine from recombinant yeast. Nature 307: 178–180
McLachlan A (ed) (1991) Molecular biology of the hepatitis B virus. CRC Press, Boca Raton
Miska S, Will H (1993) Hepatitis B virus C-gene variants. Arch Virol [Suppl] 8: 155–169
Nassal M (1992) The Arg-rich domain of the HBV core protein is required for pregenome encapsidation and productive positive-strand DNA synthesis but not for virus assembly. J Virol 66: 4107–4116
Nassal M, Schaller H (1993a) Hepatitis B virus replication. Trends Microbiol 1: 221–226
Nassal M, Schaller H (1993b) Hepatitis B virus nucleocapsid assembly. In: Dörfler W, Böhm P (eds) Virus strategies. VCH, Weinheim, pp 41–75
Nassal M, Rieger A (1993) An intramolecular disulfide bridge between Cys-7 and Cys61 determines the structure of the secretory core gene product (HBeAg) of hepatitis B virus. J Virol 67: 4307–4315
Nassal M, Rieger A (1996) A bulged region of the hepatitis B virus RNA encapsidation signal contains the replication origin for discontinuous first strand DNA synthesis. J Virol 70: 2764–2773
Nassal M, Junker-Niepmann M, Schaller H (1990) Translational inactivation of RNA function: discrimination against a subset of genomic transcripts during HBV nucleocapsid assembly. Cell 63: 1357–1363
Nassal M, Rieger A, Steinau O (1992) Topological analysis of the hepatitis B virus core particle by cysteine-cysteine crosslinking. J Mol Biol 225: 1013–1025
Nemeckova S, Kunke D, Press M, Nemecek V, Kutinova L (1994) A carboxy-terminal portion of the PreS1 domain of hepatitis B virus (HBV) occasioned retention in endoplasmic reticulum of HBV envelope proteins expressed by recombinant vaccinia viruses. Virology 202: 1024–1027
Neurath AR, Kent SBH, Strick N, Parker K (1986) Identification and chemical synthesis of a host cell receptor binding site on hepatitis B virus. Cell 446: 429–436
Neurath AR, Strick N, Sproul P (1992) Search for hepatitis B virus cell receptors reveals binding sites for interleukin 6 on the virus envelope protein. J Exp Med 175: 461–469
Obert S, Zachmann-Brand B, Deindl E, Tucker W, Bartenschlager R, Schaller H (1996) A spliced hepadnavirus RNA that is essential for virus replication. EMBO J (in press)
Olsthoorn RC, Licis N, van Duin J (1994) Leeway and constraints in the forced evolution of a regulatory RNA helix. EMBO J 13: 2660–2668

Onodera S, Ohori H, Yamaki M, Ishida N (1982) Electron microscopy of human hepatitis B virus cores by negative staining-carbon film technique. J Med Virol 10: 147–155
Ostapchuk P, Hearing P, Ganem D (1994) A dramatic shift in the transmembrane topology of a viral envelope protein accompanies hepatitis B viral morphogenesis. EMBO J 13: 1048–1057
Pasek M, Goto T, Gilbert W, Zink B, Schaller H, MacKay P, Leadbetter G, Murray K (1979) Hepatitis B virus genes and their expression in E. coli. Nature 282: 575–579
Patzer EJ, Nakamura GR, Yaffe A (1984) Intracellular transport and secretion of hepatitis B surface antigen in mammalian cells. J Virol 51: 346–353
Patzer EJ, Nakamura GR, Simonsen CC, Levinson AD, Brands R (1986) Intracellular assembly and packaging of hepatitis B surface antigen particles occur in the endoplasmic reticulum. J Virol 58: 884–892
Peabody DS (1993) The RNA binding site of bacteriophage MS2 coat protein. EMBO J 12: 595–600
Persing DH, Varmus HE, Ganem D (1986) Inhibition of secretion of hepatitis B surface antigen by a related presurface polypeptide. Science 234: 1388–1391
Persing DH, Varmus HE, Ganem D (1987) The preS1 protein of hepatitis B virus is acylated at its N-terminus with myristic acid. J Virol 61: 1672–1677
Peterson DL (1987) The structure of hepatitis B surface antigen and its antigenic sites. Bioessays 6: 258–262
Petterson RF (1991) Protein localization and virus assembly at intracellular membranes. In: Compans RW (ed) Protein traffic in eukaryotic cells. Selected reviews. Springer, Berlin Heidelberg New York, pp 67–107 (Current topics in microbiology and immunology, vol 170)
Pollack JR, Ganem D (1993) An RNA stem-loop structure directs hepatitis B virus genomic RNA encapsidation. J Virol 67: 3254–3263
Pollack JR, Ganem D (1994) Site-specific RNA binding by a hepatitis B virus reverse transcriptase initiates two distinct reactions: RNA packaging and DNA synthesis. J Virol 68: 5579–5587
Possehl C, Repp R, Heermann KH, Korec E, Uy A, Gerlich WH (1992) Absence of free core antigen in anti-HBc negative viremic hepatitis B carriers. Arch Virol [Suppl] 4: 39–41
Prange R, Streeck RE (1995) Novel transmembrane topology of the hepatitis B virus envelope proteins. EMBO J 14: 247–256
Prange R, Clemen A, Streeck RE (1991) Myristylation is involved in intracellular retention of hepatitis B virus envelope proteins. J Virol 65: 3919–3923
Pugh J, Zweidler A, Summers J (1989) Characterization of the major duck hepatitis B virus core particle protein. J Virol 63: 1371–1376
Pushko P, Sällberg M, Borisova G, Ruden U, Bichko V, Wahren B, Pumpens P, Magnius L (1994) Identification of hepatitis B virus core protein regions exposed or internalized at the surface of HBcAg particles by scanning with monoclonal antibodies. Virology 202: 912–20
Radziwill G, Tucker W, Schaller H (1990) Mutational analysis of the hepatitis B virus P gene product: domain structure and RNase H activity. J Virol 64: 613–620
Rieger A, Nassal M (1995) Distinct requirements for primary sequence in the 5′- and 3′- part of a bulge in the hepatitis B virus RNA encapsidation signal revealed by a combined in vivo selection/in vitro amplification system. Nucl Acids Res 23: 3309–3315
Rieger A, Nassal M (1996) Specific hepatitis B virus minus-strand DNA synthesis requires only the 5′-encapsidation signal and the 3′- proximal direct repeat DR1*. J Virol 70: 585–589
Rigg RJ, Schaller H (1992) Duck hepatitis B virus infection of hepatocytes is not dependent on low pH. J Virol 66: 2829–2836
Rossmann MG, Johnson JE (1989) Icosahedral RNA virus structure. Annu Rev Biochem 58: 533–573
Rothman JE (1994) Mechanisms of intracellular protein transport. Nature 372: 55–63
Rothnie HM, Chapdelaine Y, Hohn T (1994) Pararetroviruses and retroviruses: a comparative review of viral structure and gene expression strategies. Adv Virus Res 44: 1–67
Ryu WS, Netter HJ, Bayer M, Taylor J (1993) Ribonucleoprotein complexes of hepatitis delta virus. J Virol 67: 3281–3287
Sällberg M, Ruden U, Magnius LO, Harthus HP, Noah M, Wahren B (1991) Characterisation of a linear binding site for a monoclonal antibody to hepatitis B core antigen. J Med Virol 33: 248–252
Salfeld J, Pfaff E, Noah M, Schaller H (1989) Antigenic determinants and functional domains in core antigen and e antigen from hepatitis B virus. J Virol 63: 798–808
Satoh O, Umeda M, Imai H, Tunoo H, Inoue K (1990) Lipid composition of hepatitis B virus surface antigen particles and the particle-producing human hepatoma cell lines. J Lipid Res 31: 1293–1300

Schlicht HJ, Bartenschlager R, Schaller H (1989a) The duck hepatitis B virus core protein contains a highly phosphorylated C terminus that is essential for replication but not for RNA packaging. J Virol 63: 2995–3000
Schlicht HJ, Radziwill G, Schaller H (1989b) Synthesis and encapsidation of duck hepatitis B virus reverse transcriptase do not require formation of core-polymerase fusion proteins. Cell 56: 85–92
Schödel F, Moriarty AM, Peterson DL, Zheng JA, Hughes JL, Will H, Leturcq DJ, McGee JS, Milich DR (1992) The position of heterologous epitopes inserted in hepatitis B virus core particles determines their immunogenicity. J Virol 66: 106–114
Schödel F, Peterson D, Zheng J, Jones JE, Hughes JL, Milich DR (1993) Structure of hepatitis B virus core and e-antigen. A single precore amino acid prevents nucleocapsid assembly. J Biol Chem 268: 1332–1337
Seifer M, Standring DN (1993) Recombinant human hepatitis B virus reverse transcriptase is active in the absence of the nucleocapsid or the viral replication origin, DR1. J Virol 67: 4513–4520
Seifer M, Standring DN (1994) A protease-sensitve hinge linking the two domains of the hepatitis B virus core protein is exposed on the viral capsid surface. J Virol 68: 5548–5555
Seifer M, Standring DN (1995) Assembly and antigenicity of hepatitis B virus core particles. Intervirology 38: 47–62
Seifer M, Zhou S, Standring DN (1993) A micromolar pool of antigenically distinct precursors is required to initiate cooperative assembly of hepatitis B virus capsids in Xenopus oocytes. J Virol 67: 249–257
Sheu SY, Lo SJ (1992) Preferential ribosomal scanning is involved in the differential synthesis of the hepatitis B viral surface antigens from subgenomic transcripts. Virology 188: 353–357
Simon K, Lingappa VR, Ganem D (1988) Secreted hepatitis B surface antigen polypeptides are derived from a transmembrane precursor. J Cell Biol 107: 2163–2168
Sprengel R, Kuhn C, Will H, Schaller H (1985) Comparative sequence analysis of duck and human hepatitis B virus genomes. J Med Virol 15: 323–333
Sprengel R, Kaleta EF, Will H (1988) Isolation and characterization of a hepatitis B virus endemic in herons. J Virol 62: 3832–3839
Standring DN (1991) The molecular biology of the hepatitis B virus core protein. In: McLachlan A (ed) Molecular biology of the hepatitis B virus. CRC Press, Boca Raton, pp 145–169
Stirk HJ, Thornton JM, Howard CR (1992) A topological model for hepatitis B surface antigen. Intervirology 33: 148–158
Summers J, Mason WS (1982) Replication of the genome of hepatitis B-like virus by reverse transcription of an RNA intermediate. Cell 29: 403–415
Summers J, Smith PM, Horwich AL (1990) Hepadnavirus envelope proteins regulate covalently closed circular DNA amplification. J Virol 64: 2819–2824
Summers J, Smith PM, Huang MJ, Yu MS (1991) Morphogenetic and regulatory effects of mutations in the envelope proteins of an avian hepadnavirus. J Virol 65: 1310–1317
Suomalainen M, Garoff H (1994) Incorporation of homologous and heterologous proteins into the envelope of Moloney murine leukemia virus. J Virol 68: 4879–4889
Sureau C, Guerra B, Lanford R (1993) Role of the large hepatitis B virus envelope protein in infectivity of the hepatitis delta virion. J Virol 67:366–372
Tavis JE, Ganem D (1993) Expression of functional hepatitis B virus polymerase in yeast reveals it to be the sole viral protein required for correct initiation of reverse transcription. Proc Natl Acad Sci USA 90: 4107–4111
Thali M, Bukovsky A, Kondo E, Rosenwirth B, Walsh CT, Sodroski J, Göttlinger HG (1994) Functional association of cyclophilin A with HIV-1 virions. Nature 372: 363–365
Tuttleman J, Pourcel C, Summers J (1986) Formation of the pool of covalently closed circular viral DNA in hepadnavirus-infected cells. Cell 47: 451–460
Ueda K, Tsurimoto T, Matsubara K (1991) Three envelope proteins of hepatitis B virus: large S, middle S, and major S proteins needed for the formation of Dane particles. J Virol 65: 3521–3529
van Meer G (1989) Lipid traffic in animal cells. Annu Rev Cell Biol 5: 247–275
Wang GH, Seeger C (1992) The reverse transcriptase of hepatitis B virus acts as a protein primer for viral DNA synthesis. Cell 71: 663–670
Wang GH, Seeger C (1993) Novel mechanism for reverse transcription in hepatitis B viruses. J Virol 67: 6507–6512
Wang H, Lambowitz AM (1993) The Mauriceville plasmid reverse transcriptase can initiate cDNA synthesis de novo and may be related to reverse transcriptase and DNA polymerase progenitor. Cell 75: 1071–1081

Wasenauer G, Köck J, Schlicht HJ (1993) Relevance of cysteine residues for biosynthesis and antigenicity of human hepatitis B virus e protein. J Virol 67: 1315–1321

Wu TT, Coates L, Aldrich C, Summers J, Mason W (1990) In hepatocytes infected with duck hepatitis B virus, the template for viral RNA synthesis is amplified by an intracellular pathway. Virology 175: 255–261

Yang W, Guo J, Ying Z, Hua S, Dong W, Chen H (1994) Capsid assembly and involved function of twelve core protein mutants of duck hepatitis B virus. J Virol 68: 338–345

Yeh CT, Liaw YF, Ou JH (1990) The arginine-rich domain of hepatitits B virus core and precore proteins contains a signal for nuclear transport. J Virol 64: 6141–6147

Yeh CT, Wong SW, Fung YK, Ou JH (1993) Cell cycle regulation of nuclear localization of hepatitis B virus core protein. Proc Natl Acad Sci USA 90: 6459–6463

Yu M, Summers J (1991) A domain of the hepadnavirus capsid protein is specifically required for DNA maturation and virus assembly. J Virol 65: 2511–2517

Yu M, Summers J (1994) Multiple functions of capsid protein phosphorylation in duck hepatitis B virus replication. J Virol 68: 4341–4348

Yu XM (1991) The C-terminal half of the preS1 region is essential for the secretion of human hepatitis B virus large S protein devoid of the N-terminal retention sequence. Virology 181: 386–389

Yuasa S, Cheung RC, Pham Q, Robinson WS, Marion PL (1991) Peptide mapping of neutralizing and nonneutralizing epitopes of duck hepatitis B virus pre-S polypeptide. Virology 181: 14–21

Zheng J, Schödel F, Peterson DL (1992) The structure of hepadnaviral core antigens. Identification of free thiols and determination of the disulfide bonding pattern. J Biol Chem 267: 9422–942

Zhou S, Standring DN (1992) Hepatitis B virus capsids are assembled from core protein dimers. Proc Natl Acad Sci USA 89: 10046–10050

Zhou S, Standring DN (submitted) Mutational analysis of the disulfide bonds between hepatitis B virus core protein subunits in core protein dimers and core particles

Subject Index

Current Topics in Microbiology and Immunology

Volumes published since 1989 (and still available)

Vol. 174: **Fleischer, Bernhard; Sjögren, Hans Olov (Eds.):** Superantigens. 1991. 13 figs. IX, 137 pp. ISBN 3-540-54205-1

Vol. 175: **Aktories, Klaus (Ed.):** ADP-Ribosylating Toxins. 1992. 23 figs. IX, 148 pp. ISBN 3-540-54598-0

Vol. 176: **Holland, John J. (Ed.):** Genetic Diversity of RNA Viruses. 1992. 34 figs. IX, 226 pp. ISBN 3-540-54652-9

Vol. 177: **Müller-Sieburg, Christa; Torok-Storb, Beverly; Visser, Jan; Storb, Rainer (Eds.):** Hematopoietic Stem Cells. 1992. 18 figs. XIII, 143 pp. ISBN 3-540-54531-X

Vol. 178: **Parker, Charles J. (Ed.):** Membrane Defenses Against Attack by Complement and Perforins. 1992. 26 figs. VIII, 188 pp. ISBN 3-540-54653-7

Vol. 179: **Rouse, Barry T. (Ed.):** Herpes Simplex Virus. 1992. 9 figs. X, 180 pp. ISBN 3-540-55066-6

Vol. 180: **Sansonetti, P. J. (Ed.):** Pathogenesis of Shigellosis. 1992. 15 figs. X, 143 pp. ISBN 3-540-55058-5

Vol. 181: **Russell, Stephen W.; Gordon, Siamon (Eds.):** Macrophage Biology and Activation. 1992. 42 figs. IX, 299 pp. ISBN 3-540-55293-6

Vol. 182: **Potter, Michael; Melchers, Fritz (Eds.):** Mechanisms in B-Cell Neoplasia. 1992. 188 figs. XX, 499 pp. ISBN 3-540-55658-3

Vol. 183: **Dimmock, Nigel J.:** Neutralization of Animal Viruses. 1993. 10 figs. VII, 149 pp. ISBN 3-540-56030-0

Vol. 184: **Dunon, Dominique; Mackay, Charles R.; Imhof, Beat A. (Eds.):** Adhesion in Leukocyte Homing and Differentiation. 1993. 37 figs. IX, 260 pp. ISBN 3-540-56756-9

Vol. 185: **Ramig, Robert F. (Ed.):** Rotaviruses. 1994. 37 figs. X, 380 pp. ISBN 3-540-56761-5

Vol. 186: **zur Hausen, Harald (Ed.):** Human Pathogenic Papillomaviruses. 1994. 37 figs. XIII, 274 pp. ISBN 3-540-57193-0

Vol. 187: **Rupprecht, Charles E.; Dietzschold, Bernhard; Koprowski, Hilary (Eds.):** Lyssaviruses. 1994. 50 figs. IX, 352 pp. ISBN 3-540-57194-9

Vol. 188: **Letvin, Norman L.; Desrosiers, Ronald C. (Eds.):** Simian Immunodeficiency Virus. 1994. 37 figs. X, 240 pp. ISBN 3-540-57274-0

Vol. 189: **Oldstone, Michael B. A. (Ed.):** Cytotoxic T-Lymphocytes in Human Viral and Malaria Infections. 1994. 37 figs. IX, 210 pp. ISBN 3-540-57259-7

Vol. 190: **Koprowski, Hilary; Lipkin, W. Ian (Eds.):** Borna Disease. 1995. 33 figs. IX, 134 pp. ISBN 3-540-57388-7

Vol. 191: **ter Meulen, Volker; Billeter, Martin A. (Eds.):** Measles Virus. 1995. 23 figs. IX, 196 pp. ISBN 3-540-57389-5

Vol. 192: **Dangl, Jeffrey L. (Ed.):** Bacterial Pathogenesis of Plants and Animals. 1994. 41 figs. IX, 343 pp. ISBN 3-540-57391-7

Vol. 193: **Chen, Irvin S. Y.; Koprowski, Hilary; Srinivasan, Alagarsamy; Vogt, Peter K. (Eds.):** Transacting Functions of Human Retroviruses. 1995. 49 figs. IX, 240 pp. ISBN 3-540-57901-X

Vol. 194: **Potter, Michael; Melchers, Fritz (Eds.):** Mechanisms in B-cell Neoplasia. 1995. 152 figs. XXV, 458 pp. ISBN 3-540-58447-1

Vol. 195: **Montecucco, Cesare (Ed.):** Clostridial Neurotoxins. 1995. 28 figs. XI., 278 pp. ISBN 3-540-58452-8

Vol. 196: **Koprowski, Hilary; Maeda, Hiroshi (Eds.):** The Role of Nitric Oxide in Physiology and Pathophysiology. 1995. 21 figs. IX, 90 pp. ISBN 3-540-58214-2

Vol. 197: **Meyer, Peter (Ed.):** Gene Silencing in Higher Plants and Related Phenomena in Other Eukaryotes. 1995. 17 figs. IX, 232 pp. ISBN 3-540-58236-3

Vol. 198: **Griffiths, Gillian M.; Tschopp, Jürg (Eds.):** Pathways for Cytolysis. 1995. 45 figs. IX, 224 pp. ISBN 3-540-58725-X

Vol. 199/I: **Doerfler, Walter; Böhm, Petra (Eds.):** The Molecular Repertoire of Adenoviruses I. 1995. 51 figs. XIII, 280 pp. ISBN 3-540-58828-0

Vol. 199/II: **Doerfler, Walter; Böhm, Petra (Eds.):** The Molecular Repertoire of Adenoviruses II. 1995. 36 figs. XIII, 278 pp. ISBN 3-540-58829-9

Vol. 199/III: **Doerfler, Walter; Böhm, Petra (Eds.):** The Molecular Repertoire of Adenoviruses III. 1995. 51 figs. XIII, 310 pp. ISBN 3-540-58987-2

Vol. 200: **Kroemer, Guido; Martinez-A., Carlos (Eds.):** Apoptosis in Immunology. 1995. 14 figs. XI, 242 pp. ISBN 3-540-58756-X

Vol. 201: **Kosco-Vilbois, Marie H. (Ed.):** An Antigen Depository of the Immune System: Follicular Dendritic Cells. 1995. 39 figs. IX, 209 pp. ISBN 3-540-59013-7

Vol. 202: **Oldstone, Michael B. A.; Vitković, Ljubiša (Eds.):** HIV and Dementia. 1995. 40 figs. XIII, 279 pp. ISBN 3-540-59117-6

Vol. 203: **Sarnow, Peter (Ed.):** Cap-Independent Translation. 1995. 31 figs. XI, 183 pp. ISBN 3-540-59121-4

Vol. 204: **Saedler, Heinz; Gierl, Alfons (Eds.):** Transposable Elements. 1995. 42 figs. IX, 234 pp. ISBN 3-540-59342-X

Vol. 205: **Littman, Dan R. (Ed.):** The CD4 Molecule. 1995. 29 figs. XIII, 182 pp. ISBN 3-540-59344-6

Vol. 206: **Chisari, Francis V.; Oldstone, Michael B. A. (Eds.):** Transgenic Models of Human Viral and Immunological Disease. 1995. 53 figs. XI, 345 pp. ISBN 3-540-59341-1

Vol. 207: **Prusiner, Stanley B. (Ed.):** Prions Prions Prions. 1995. 42 figs. VII, 163 pp. ISBN 3-540-59343-8

Vol. 208: **Farnham, Peggy J. (Ed.):** Transcriptional Control of Cell Growth. 1995. 17 figs. IX, 141 pp. ISBN 3-540-60113-9

Vol. 209: **Miller, Virginia L. (Ed.):** Bacterial Invasiveness. 1996. 16 figs. IX, 115 pp. ISBN 3-540-60065-5

Vol. 210: **Potter, Michael; Rose, Noel R. (Eds.):** Immunology of Silicones. 1996. 136 figs. XX, 430 pp. ISBN 3-540-60272-0

Vol. 211: **Wolff, Linda; Perkins, Archibald S. (Eds.):** Molecular Aspects of Myeloid Stem Cell Development. 1996. 98 figs. XIV, 298 pp. ISBN 3-540-60414-6

Vol. 212: **Vainio, Olli; Imhof, Beat A. (Eds.):** Immunology and Developmental Biology of the Chicken. 1996. 43 figs. IX, 281 pp. ISBN 3-540-60585-1

Vol. 213/I: **Günthert, Ursula; Birchmeier, Walter (Eds.):** Attempts to Understand Metastasis Formation I. 1996. 35 figs. XV, 293 pp. ISBN 3-540-60680-7

Vol. 213/II: **Günthert, Ursula; Birchmeier, Walter (Eds.):** Attempts to Understand Metastasis Formation II. 1996. 33 figs. XV, 288 pp. ISBN 3-540-60681-5

Vol. 213/III: **Günthert, Ursula; Schlag, Peter M.; Birchmeier, Walter (Eds.):** Attempts to Understand Metastasis Formation III. 1996. 14 figs. XV, 262 pp. ISBN 3-540-60682-3

Springer-Verlag and the Environment

We at Springer-Verlag firmly believe that an international science publisher has a special obligation to the environment, and our corporate policies consistently reflect this conviction.

We also expect our business partners – paper mills, printers, packaging manufacturers, etc. – to commit themselves to using environmentally friendly materials and production processes.

The paper in this book is made from low- or no-chlorine pulp and is acid free, in conformance with international standards for paper permanency.

MIX
Papier aus verantwortungsvollen Quellen
Paper from responsible sources
FSC® C105338

If you have any concerns about our products,
you can contact us on
ProductSafety@springernature.com

In case Publisher is established outside the EU,
the EU authorized representative is:
Springer Nature Customer Service Center GmbH
Europaplatz 3, 69115 Heidelberg, Germany

Printed by Libri Plureos GmbH
in Hamburg, Germany